FORENSIC FIRE SCENE RECONSTRUCTION

3rd Edition

David J. Icove, Ph.D., P.E.
The University of Tennessee

John D. DeHaan, Ph.D.
Fire-Ex Forensics, Inc.

Gerald A. Haynes, M.S., P.E.
Forensic Fire Analysis, LLC

Boston   Columbus   Indianapolis   New York   San Francisco   Upper Saddle River   Amsterdam
Cape Town   Dubai   London   Madrid   Milan   Munich   Paris   Montreal   Toronto   Delhi
Mexico City   São Paulo   Sydney   Hong Kong   Seoul   Singapore   Taipei   Tokyo

Publisher: Julie Levin Alexander
Publisher's Assistant: Regina Bruno
Editor-in-Chief: Marlene McHugh Pratt
Senior Acquisitions Editor: Stephen Smith
Associate Editor: Monica Moosang
Assistant Editor: Samantha Sheehan
Development Editor: Carol Lazerick
Director of Marketing: David Gesell
Marketing Manager: Brian Hoehl
Marketing Specialist: Michael Sirinides
Marketing Assistant: Crystal Gonzalez
Managing Production Editor: Patrick Walsh

Production Liaison: Julie Boddorf
Production Editor: Lisa S. Garboski, bookworks
Senior Media Editor: Matt Norris
Media Project Manager: Lorena Cerisano
Manufacturing Manager: Lisa McDowell
Creative Director: Jayne Conte
Cover Designer: Bruce Kenselaar
Cover Photo: Courtesy of Inv. Shawn Short, Knox County Fire Marshal's Office, Knoxville, TN
Composition: S4Carlisle Publishing Services
Printing and Binding: LSC Communications
Cover Printer: LSC Communications

Dedicated to Dr. John L. Bryan, for his inspiration, leadership, and support.

Credits and acknowledgments borrowed from other sources and reproduced, with permission, in this textbook appear on appropriate pages within text.

Many of the designations by manufacturers and sellers to distinguish their products are claimed as trademarks. Where those designations appear in this book, and the publisher was aware of a trademark claim, the designations have been printed in initial caps or all caps.

Note: A portion of this book was prepared by David J. Icove, a former employee of the U.S. Tennessee Valley Authority (TVA). The views expressed in this book are the views of Dr. Icove and his coauthors. They are not necessarily the views of TVA or the United States Government. Reference in this book to any product, process, or service by trade name, trademark, manufacturer, or otherwise does not necessarily constitute its endorsement or recommendation by TVA or the United States Government.

Disclaimer: The authors and Pearson Education make no warranty, expressed or implied, to the readers of *Forensic Fire Scene Reconstruction*, and accept no responsibility for its use. The readers assume sole responsibility for determining the appropriateness of its suggested use in any particular methodology, calculation, or fire modeling application; for any conclusions drawn from the results of its use; and for any actions taken or not taken as a result of analyses performed using these tools.

Readers are warned that this textbook is intended for use only by those competent in the fields including, but not limited to, fire investigation, fluid dynamics, thermodynamics, combustion, and heat transfer; and the textbook is intended only to supplement the informed judgment of the qualified user. This textbook provides methodologies that may or may not have predictive capability when applied to a specific set of factual circumstances. Lack of accurate predictions by any particular methodology, calculation, or fire modeling application could lead to erroneous conclusions with regard to fire safety. All results should be evaluated by an informed user.

Throughout this document, the mention of products or commercial software does not constitute endorsement by the authors or Pearson Education, nor does it indicate that the products are necessarily those best suited for the intended purpose.

Cataloging-in-Publication data on file with the Library of Congress.

www.pearsonhighered.com

ISBN 10:   0-13-260577-5
ISBN 13: 978-0-13-260577-9

CONTENTS

More has changed in the last year in the field of fire investigation than in all the years since the 2004 publication of the First Edition of *Forensic Fire Scene Reconstruction*. No longer is the investigation of fires just limited to inspecting the ruins, asking questions of the witnesses, and applying basic common sense and observations to conclude the fire's origin and cause. Fire investigators must now keep in step with the rapid changes in the forensic sciences, the innovations in fire scene documentation, and challenges in the court stressing precise defensible expert testimonies.

This book was designed from the outset to meet these forensic challenges as well as to be a companion text to *Kirk's Fire Investigation*, which is presently in its seventh edition. *Kirk's* is still considered to be the foremost authoritative text for training as well as an expert treatise for fire investigation professionals. Both *Kirk's* and *Forensic Fire Scene Reconstruction* also serve as bridge textbooks providing interoperation of the concepts presented in the latest editions of the National Fire Protection Association (NFPA) *Guide for Fire and Explosion Investigation (NFPA 921)* and related standards of care.

Because *forensic fire scene documentation* is emerging as an indispensable cornerstone in nearly all judicially contested investigations, modest reconstructive efforts such as repositioning furniture and other postfire artifacts into prefire positions no longer suffice. Therefore, during scene documentation, today's fire investigator must possess skill in scientifically based fire pattern analysis and must employ analytical methodologies based on expert interpretation of discernible patterns and fire dynamics principles.

Delving deep into forensic fire engineering, *Forensic Fire Scene Reconstruction* covers engineering calculations and fire modeling, and features several exhaustive case studies that leverage the current technology explained in depth throughout the text. Several specialized topic areas are also covered, including use of scanning and panoramic photography, preparation of comprehensive reports, the ignition matrix, expert testimony, and computer modeling and laser imaging. Also included is an advanced discussion of tenability, which is briefly introduced in *Kirk's Fire Investigation*.

Paving the way for those investigators who seek to take their forensic investigation skills to the next level, *Forensic Fire Scene Reconstruction* explores the use of the scientific method beyond the scope of *Kirk's* and details the need-to-know information behind developing a working hypothesis. Readers are also thoroughly prepared for the courtroom with this guide, which teaches investigators how to form and authoritatively present their expert testimony. Additionally, *Forensic Fire Scene Reconstruction* serves as the most in-depth textbook in the field of fire investigation on the subject of arson crime scene analysis and examining motive as it relates to fire setting. For investigators seeking to gain a deeper and more advanced understanding of forensic fire engineering, technology, and analysis in the field of fire investigation beyond the broad overview given in *Kirk's,* its companion textbook, *Forensic Fire Scene Reconstruction* targets several specialized areas and takes readers to the next level.

Dedication: This third edition of *Forensic Fire Scene Reconstruction* is dedicated to Dr. John L. Bryan, who in 1956 established the Department of Fire Protection Engineering within the School of Engineering at the University of Maryland. Dr. Bryan, known affectionately as the "Prof." by his students, served as chair and professor of the department from 1956 until his retirement in 1993.

Under Dr. Bryan's leadership, the department evolved from a modest, one-person operation to a mature and vital program serving the fire protection needs of the nation. Through all his innovative work at the University of Maryland, College Park, the National Fire Protection Association, and the Society of Fire Protection Engineers, he has advanced the science through his development and establishment of the first undergraduate ABET-accredited Fire Protection Engineering curriculum in the United States.

From that department have come innovative ideas; a cadre of professors who have advanced the areas of forensic fire investigation so that fire dynamics has become commonplace in investigation classes; and graduates who chose forensic fire investigation as their profession, including those who went on to leadership positions at federal and state agencies, including NIST, ATF and the FBI. Professor Bryan is the author of two textbooks and numerous technical papers and has served as editor of a number of publications. He remains active in the development of national and international fire safety engineering standards. Dr. Bryan has participated as a peer reviewer of *Forensic Fire Scene Reconstruction* since its inception.

About This Book

Forensic fire scene reconstruction goes well beyond establishing what furnishings were present and where they were placed. Reconstruction involves identification and documentation of all relevant features of the fire scene—materials, dimensions, location, and physical evidence—that help identify fuels and establish human activities and contacts. This information is then placed in context with principles of fire engineering and human behavior and is used to evaluate various scenarios of the origin, cause, and development of the fire and the interaction of people with it.

This textbook is intended for use by a wide range of both public and private sector individuals in the investigative, forensic, engineering, and judicial sectors. In particular, these intended users include

- public safety officials charged with the responsibility of investigating fires;
- prosecutors of arson and fire-related crimes who seek to add to their capabilities of evaluating evidence and presenting technical details to a nontechnical judge and jury;
- judicial officials seeking to better comprehend the technical details of cases over which they preside;
- private-sector investigators, adjusters, and attorneys representing the insurance industry who have the responsibility of processing claims or otherwise have a vested interest in determining responsibility for the start or cause of a fire;
- citizens and civic community service organizations committed to conducting public awareness programs designed to reduce the threat of fire and its devastating effect on the economy; and
- scientists, engineers, academicians, and students engaged in the education process pertaining to forensic fire scene reconstruction.

A thorough understanding of fire dynamics is valuable in applying these forensic engineering techniques. This book describes and illustrates the latest interpretations of systematic approaches for reconstructing fire scenes. These approaches apply the principles of fire protection engineering along with those of forensic science and behavioral science.

Using historical fire cases, both in the textbook and in the online *Resource Central*, the authors provide new lessons and insight into the ignition, growth, development, and outcome of those fires. All documentation in the case examples follows or exceeds the methodology set forth by the NFPA in *NFPA 921: Guide for Fire and Explosion Investigations*, 2011 edition, and its companion standard, *NFPA 1033: Standard for Professional Qualifications for Fire Investigator*, 2009 edition. The best features of this text include the use of real-world case examples to illustrate the concepts discussed herein. These examples shed new light on the forensic science, fire engineering, and human factor issues. Each example is illustrated using the guidelines from *NFPA 921* and *Kirk's Fire Investigation*. In cases where fire engineering analysis or fire modeling is applicable, these techniques are explored.

The authors acknowledge the numerous essential references from the Society of Fire Protection Engineers (SFPE) that form the basis for much of the material included in this textbook. These essential references, ranging from core principles of fire science to human behavior, include the *SFPE Handbook of Fire Protection Engineering* and many of SFPE's engineering practice guides.

New to This Edition

Fully updated and streamlined from the previous edition, the third edition offers the latest information on investigative technologies and innovative documentation techniques.

- Meets the FESHE guidelines for Fire Investigation and Analysis, with correlations to *NFPA 921 and NFPA 1033*.
- The textbook is required reading for several professional certification study courses for the International Association of Arson Investigators (IAAI).
- Includes the latest information on applying the scientific method to fire investigations.
- Nearly all photos are now in color, making interpretation by the reader easier.
- Updated, in-depth case examples demonstrate the use of forensic fire engineering analysis approaches.
- Comprehensive glossary provides the latest definitions of common forensic fire investigation terminologies.
- Provides the essential information required in preparing for the International Association of Arson Investigator's (IAAI) *Fire Investigation Technician (IAAI-FIT)* and *Evidence Collection Technician (IAAI-ECT)* Examinations.
- New Resource Central solution provides online resources for both instructors and students, including problem sets from NIST, NFPA, U.S. Department of Justice, and the U.S. Chemical Safety and Hazard Assessment Board.

Hallmark Features

- Describes a systematic approach to reconstructing fire scenes in which investigators rely on the combined principles of fire protection engineering along with forensic and behavioral science.
- Describes the underpinnings of how fire patterns are produced and how they can be used by investigators in assessing fire damage and determining a fire's origin.
- Details the systematic approach needed to support forensic analysis and reports.
- Reviews the techniques used in the analysis of arsonists' motives and intents.
- Discusses the use of various mathematical, physical, and computer-assisted techniques for modeling fires, explosions, and the movement of people.
- Provides an in-depth examination of the impact and tenability of fires on humans.
- Reviews the applicable standard fire testing methods that are important to forensic fire scene analysis and reconstruction.
- Contains an epilogue written by an experienced career firefighter and attorney.

Scope of the Book

Forensic Fire Scene Reconstruction is divided into the following chapters, which flow logically and build on one another:

Chapter 1, "Principles of Reconstruction," describes a systematic approach to reconstructing fire scenes in which investigators rely on the combined principles of fire protection engineering along with forensic and behavioral science. Using this approach, the investigator can more accurately document a structural fire's origin, intensity, growth, direction of travel, and duration as well as the behavior of the occupants.

Chapter 2, "Basic Fire Dynamics," provides the investigator with a firm understanding of the phenomenon of fire, heat release rates of common materials, heat transfer, growth and development, fire plumes, and enclosure fires.

Chapter 3, "Fire Pattern Analysis," describes the underpinnings of how fire patterns are used by investigators in assessing fire damage and determining a fire's origin. Fire patterns are often the only remaining visible evidence after a fire is extinguised. The ability to document and interpret fire pattern damage accurately is a skill of paramount importance to investigators when they are reconstructing fire scenes.

Chapter 4, "Fire Scene Documentation," details a systematic approach needed to support forensic analysis and reports. The purpose of forensic fire scene documentation includes recording visual observations, emphasizing fire development characteristics, and authenticating and protecting physical evidence. The underlying theme is that thorough documentation produces sound investigations and courtroom presentations. The chapter includes commentaries on new technologies useful in improving the accuracy and comprehensiveness of documentation.

Chapter 5, "Arson Crime Scene Analysis," reviews the techniques used in the analysis of arsonists' motives and intents. It presents nationally accepted motive-based classification guidelines along with case examples of the crimes of vandalism, excitement, revenge, crime concealment, and arson-for-profit. The geography of serial arson is also examined, along with techniques for profiling the targets selected by arsonists.

Chapter 6, "Fire Modeling," discusses the use of various mathematical, physical, and computer-assisted techniques for modeling fires, explosions, and the movement of people. Numerous models are explored, along with their strengths and weaknesses. Several case examples are also presented.

Chapter 7, "Fire Death and Injuries," provides an in-depth examination of the impact and tenability of fires on humans. The chapter examines what kills people in fires, particularly their exposure to by-products of combustion, toxic gases, and heat. It also examines the predictable fire burn pattern damage inflicted on human bodies and summarizes postmortem tests and forensic examinations desirable in comprehensive death investigations.

Chapter 8, "Fire Testing," reviews the applicable standard fire testing methods that are important to forensic fire scene analysis and reconstruction. These can range from benchtop "lab" tests to full-scale fire reconstructions of varying complexity.

The appendix is a short refresher lesson on scientific notations and calculations. At the end of this book you will find a comprehensive glossary of fire science terms. On the Resource Central website is a directory of current websites of potential value to the reader, and it includes vendor contacts for many of the products mentioned in the text. Through Resource Central, this text also offers instructors a full complement of online supplemental teaching materials such as test banks, PowerPoint lectures and instructor outlines to aid in the classroom.

Peer Reviewers

Peer review is important for ensuring that a textbook is well balanced, useful, authoritative, and accurate. The following individuals, agencies, institutions, and companies provided invaluable support during the peer-review process in the first, second, and third editions.

Dr. Vytenis (Vyto) Babrauskas, Fire Science and Technology Inc., Issaquah, Washington

Richard L. Bennett, Associate Professor of Fire Protection & Emergency Services, The University of Akron, Akron, Ohio

Dr. John L. Bryan, Professor Emeritus, University of Maryland, Department of Fire Protection Engineering, College Park, Maryland

Guy E. "Sandy" Burnette, Jr., Attorney, Tallahassee, Florida

Steven W. Carman, BS, Physical Science, IAAI, Owner, Carman & Associates Fire Investigation, Grass Valley, California

Jody Cooper, IAAI-CFI, CVFI, Owner/investigator, JJMA Investigations LLC, and Instructor, Oklahoma State University, Poteau, Oklahoma

Carl E. Chasteen, BS, CPM, FABC, Chief of Forensic Services, Florida Division of State Fire Marshal, Havana, Florida

Community College of South Nevada, Nevada

Coosa Valley Technical College, Georgia

Robert F. Duval, National Fire Protection Association, Quincy, Massachusetts

Fishers Fire Department, New York

Christopher Gauss, IAAI-CFI, Captain, Baltimore County Fire Investigation, Towson, Maryland

Gregory E. Gorbett, MScFPE, CFEI, IAAI-CFI, Assistant Professor/FPSET Program Coordinator, Eastern Kentucky University, Richmond, Kentucky

Brian P. Henry, Esq., CFEI, Attorney, Smith, Rolfes & Skavdahl, Sarasota, Florida

Gary S. Hodson, Utah Valley University, Utah Fire and Rescue Academy, Utah

Patrick M. Kennedy, National Association of Fire Investigators, Sarasota, Florida

Frederick J. Knipper, Fayetteville State University, Durham, NC

Daniel Madrzykowski, National Institute of Standards and Technology, Gaithersburg, Maryland

John E. "Jack" Malooly, Bureau of Alcohol, Tobacco, Firearms and Explosives, Chicago, Illinois (retired)

Michael Marquardt, Bureau of Alcohol, Tobacco, Firearms and Explosives, Grand Rapids, Michigan

J. Ron McCardle, Bureau of Fire and Arson Investigations, Florida Division of State Fire Marshal (retired)

Lamont "Monty" McGill, McGill Investigations, Gardnerville, Nevada (deceased)

C. W. Munson, Captain, Long Beach Fire Department, California (retired)

Bradley E. Olsen, BS, Fire Science Management, Southern Illinois University; Fire Captain, City of Madison Fire Department, Madison, Wisconson

Robert R. Rielage, former State Fire Marshal, Ohio Division of State Fire Marshal, Reynoldsburg, Ohio

James Ryan, Investigator, New York State Office of Fire Prevention & Control, New York

Michael Schlatman, Fire Consulting International Inc., Shawnee Mission, Kansas

Carina Tejada, BA Mathematics QAS, Marietta, Georgia

Robert K. Toth, Iris Fire, LLC, Parker, Colorado

Luis Velazco, Bureau of Alcohol, Tobacco, Firearms and Explosives, Brunswick, Georgia, (retired)

ACKNOWLEDGMENTS

We have many people to thank for both their help on and their inspiration for one or more of the three editions of this book, including the many individuals and present and past employees at the following institutions and agencies.

David M. Banwarth Associates, LLC, Dayton, Maryland: David M Banwarth, P. E.

Bureau of Alcohol, Tobacco, Firearms and Explosives (ATF): Steve Carman (retired), Steve Avato, Dennis C. Kennamer, Dr. David Sheppard, Jack Malooly (retired), Wayne Miller (retired), Michael Marquardt, Luis Velaszco (retired), John Mirocha (retired), Ken Steckler (retired), Brian Grove, and John Allen.

J. H. Burgoyne & Partners (UK): Robin Holleyhead, and Roy Cooke

California State Fire Marshal's Office: Joe Konefal (retired) and Jim Allen (retired).

Eastern Kentucky University: Gregory E. Gorbett, James L. Pharr, Andrew Tinsley, and Ronald L. Hopkins (retired)

Federal Bureau of Investigation (FBI): Richard L. Ault (retired), Steve Band, S. Annette Bartlett, John Henry Campbell (retired), R. Joe Clark (retired), Roger L. Depue (retired), Joseph A. Harpold (retired), Timothy G. Huff (retired), Sharon A. Kelly (retired), John L. Larsen (retired), James A. O'Connor (retired), John E. Otto (retired), William L. Tafoya (retired), and Arthur E. Westveer

Mark A. Campbell, Fire Forensic Research, Colorado.

Fire Safety Institute: Dr. John M. "Jack" Watts, Jr.

Fire Science and Technology: Dr. Vyto Babrauskas

Forensic Fire Analysis: Lester Rich

Gardiner Associates: Mick Gardiner, Jim Munday, Jack Deans

Rodger H. Ide

Iris Fire, LLC, Parker, Colorado: Robert K. Toth

Kent Archaeological Field School, UK: Dr. Paul Wilkinson

Knox County (Tennessee) Sheriff's Office and Knox County Fire Investigation Unit: Det. Michael W. Dalton (retired); Inv. Shawn Short; Inv. Greg Lampkin; Inv. Aaron Allen; Inv. Michael Patrick; Inv. Daniel Johnson; Kathy Saunders, Knox County Fire Marshal

Leica Geosystems: Rick Bukowski, Tony Grissim

Mesa County Sheriff's Office, Colorado, Benjamin J. Miller

Metropolitan Police Forensic Lab, Fire Investigation Unit, London (UK): Roger Berrett (retired)

McGill Consulting: Monty McGill (deceased)

McKinney (Texas) Fire Department: Chief Mark Wallace

National Institute of Standards and Technology (NIST): Richard W. Bukowski, Dr. William Grosshandler, Dan Madrzykowski, and Dr. Kevin McGrattan

New South Wales Fire Brigades: Ross Brogan (retired)

Novato Fire Protection District: Assistant Chief Forrest Craig

Ohio State Fire Marshal's Office, Reynoldsburg: Eugene Jewell (deceased), Charles G. McGrath (retired), Mohamed M. Gohar (retired), J. David Schroeder (retired), Jack Pyle (deceased), Harry Barber, Lee Bethune, Joseph Boban, Kenneth Crawford, Dennis Cummings, Dennis Cupp, Robert Davis, Robert Dunn, Donald Eifler, Ralph Ford, James Harting (retired), Robert Lawless, Keith Loreno, Mike McCarroll, Matthew J. Hartnett, Brian Peterman, Mike Simmons, Rick Smith, Stephen W. Southard, and David Whitaker (retired)

Panoscan: Ted Chavalas

Precision Simulators, Inc.: Kirk McKenzie

Richland (Washington) Fire Department: Glenn Johnson and Grant Baynes
Sacramento County (California) Fire Department: Jeff Campbell (retired)
   Saint Paul (Minnesota) Fire Department: Jamie Novak
Santa Ana (California) Fire Department: Jim Albers (retired) and Bob Eggleston
   (retired)
Seneca College School of Fire Protection Engineering Technology: David McGill
Smith, Rolfes & Skavdahl: Brian P. Henry
Tennessee State Fire Marshal's Office: Richard L. Garner (retired), Robert Pollard,
   Eugene Hartsook (deceased), and Jesse L. Hodge (retired)
Tennessee Valley Authority (TVA): Carolyn M. Blocher, James E. Carver (retired),
   R. Douglas Norman (retired), Larry W. Ridinger (retired), Sidney G. Whitehurst
   (retired), and Norman Zigrossi (retired)
Underwriters Laboratories, Inc. (UL): Dr. J. Thomas Chapin
University of Arkansas, Department of Anthropology: Dr. Elayne J. Pope, now at
   Office of the Chief Medical Examiner Tidewater District, Norfolk, Virginia
University of Edinburgh, Department of Civil Engineering: Professor Emeritus
   Dr. Dougal Drysdale
University of Maryland, Department of Fire Protection Engineering: Dr. John
   L. Bryan (Emeritus), Dr. James A. Milke, Dr. Frederick W. Mowrer (Emeritus),
   Dr. James G. Quintiere, Dr. Marino di Marzo, and Dr. Steven M. Spivak
   (Emeritus)
University of Tennessee, College of Engineering: Dr. A. J. Baker, J. Douglas
   Birdwell, Samir M. El-Ghazaly, Dr. Rafael C. Gonzalez, Dr. M. Osama Soliman,
   Dr. Jerry Stoneking (deceased), Dr. Tse-Wei Wang, and the many students in
   the ME 495 and ECE 599 courses, Enclosure Fire Dynamics and Computer Fire
   Modeling
U.S. Consumer Products Safety Commission: Gerard Naylis (retired) and Carol Cave
U.S. Fire Administration, Federal Emergency Management Agency (FEMA):
   Edward J. Kaplan, Kenneth J. Kuntz (retired), Robert A. Neale, and
   Dr. Denis Onieal.

Our sincere, heartfelt thanks go to Carolyn M. Blocher, Dr. Angi M. Christensen, Shirley Runyan, Vivian Woodall, Edith De Lay, Wendy Druck, and Larry Harding, who reviewed the early manuscripts, addressed technical questions, and made many beneficial suggestions as to the format and content of this textbook.

A very special mention goes to our development editor Carol Lazerick; our production editor, Lisa Garboski; our copyeditor and proofreader, Barbara Liguori; our ever-patient associate editor at Pearson/Brady, Monica Moosang; and Samantha Sheehan, Editorial Assistant.

Also, special thanks go to our families and close friends over the years for their patient support.

Past and Present Special Acknowledgments

In addition to our ongoing admiration of Bud Nelson, to whom the first edition was dedicated, and to Professor Emeritus Dougal Drysdale to whom the second edition was dedicated, the authors specifically acknowledge in this third edition the contributions and encouragement of Professor Emeritus John L. Bryan, one of the most influential fire protection engineering educators of the twentieth century.

ABOUT THE AUTHORS

This textbook is coauthored by three of the most experienced fire scientists in the United States. Their combined talents total more than 100 years of experience in the fields of fire service, behavioral science, fire protection engineering, fire behavior, investigation, criminalistics, and crime scene reconstruction.

An internationally recognized forensic fire engineering expert with more than 40 years of experience, Dr. Icove is coauthor of *Kirk's Fire Investigation,* 7th ed., a leading treatise in the field, and *Combating Arson-for-Profit,* the leading textbook on the crime of economic arson. Since 1992 he has served as a principal member of the NFPA 921 Technical Committee on Fire Investigations. As a retired career federal law enforcement agent, Dr. Icove served as a criminal investigator on the federal, state, and local levels. He is a Registered Professional Engineer, a Certified Fire and Explosion Investigator (CFEI), and a Fellow in the Society of Fire Protection Engineers.

David J. Icove, PhD, PE, CFEI, FSFPE

Photo courtesy of the University of Tennessee, College of Engineering.

He retired in 2005 as an Inspector in the Criminal Investigations Division of the U.S. Tennessee Valley Authority (TVA) Police, Knoxville, Tennessee, where he was assigned for his last 2 years to the Federal Bureau of Investigation (FBI) Joint Terrorism Task Force (JTTF). In addition to conducting major case investigations, Dr. Icove oversaw the development of advanced fire investigation training and technology programs in cooperation with various agencies, including the Federal Emergency Management Agency's (FEMA's) U.S. Fire Administration.

Before transferring to the U.S. TVA Police in 1993, he served 9 years as a program manager in the elite Behavioral Science and Criminal Profiling Units at the FBI, Quantico, Virginia. At the FBI, he implemented and became the first supervisor of the Arson and Bombing Investigative Support (ABIS) Program, staffed by FBI and ATF criminal profilers. Prior to his work at the FBI, Dr. Icove served as a criminal investigator at arson bureaus of the Knoxville Police Department, the Ohio State Fire Marshal's Office, and the Tennessee State Fire Marshal's Office.

His expertise in forensic fire scene reconstruction is based on a blend of on-scene experience, conduction of fire tests and experiments, and participation in prison interviews of convicted arsonists and bombers. He has testified as an expert witness in civil and criminal trials, as well as before U.S. congressional committees seeking guidance on key arson investigation and legislative initiatives.

Dr. Icove holds BS and MS degrees in Electrical Engineering and a PhD in Engineering Science and Mechanics from the University of Tennessee. He also holds a BS degree in Fire Protection Engineering from the University of Maryland–College Park. He is presently a Research Professor in the Department of Electrical Engineering and Computer Science at the University of Tennessee, Knoxville; serves on the faculty of the University of Maryland's Professional Master of Engineering in Fire Protection program; and serves as an Investigator in the Knox County Fire Investigation Unit as a Reserve Deputy Sheriff for the Knox County, Sheriff's Office, Knoxville, Tennessee.

John D. DeHaan, PhD, FABC, CFI–IAAI, CFEI, FFSS, FSSDip

Photo courtesy of Fire-Ex Forensics, Inc.

An internationally recognized forensic science expert, Dr. DeHaan is the coauthor of *Kirk's Fire Investigation,* the leading textbook in the field of fire and arson investigation. He is also a former principal member of the NFPA 921 Technical Committee on Fire Investigations.

Dr. DeHaan has been a criminalist for more than 42 years and has gained considerable expertise in fire and explosion evidence as well as human hair, shoe print, and instrumental analysis, and crime scene reconstruction. He has been employed as a criminalist by the Alameda County Sheriff's Office; U.S. Department of Justice, Bureau of Alcohol, Tobacco, Firearms and Explosives; and the California Department of Justice.

His research into forensic fire scene reconstruction is based on firsthand fire experiments on fire behavior involving more than 500 observed full-scale structure, and 100 vehicle fires under controlled conditions, as well as laboratory-scale studies. Dr. DeHaan has testified as an expert witness in civil and criminal trials across the United States and overseas. He is currently the president of Fire-Ex Forensics Inc. and consults on civil and criminal fire and explosion cases across the United States and Canada and overseas.

Dr. DeHaan graduated from the University of Illinois—Chicago Circle in 1969 with a BS degree in Physics and a minor in Criminalistics. He was awarded a PhD in Pure and Applied Chemistry (Forensic Science) by Strathclyde University in Glasgow, Scotland, in 1995. Dr. DeHaan is a Fellow of the American Board of Criminalistics (Fire Debris), a Fellow of the Forensic Science Society and the Institution of Fire Engineers. He is a Certified Fire Investigator from the International Association of Arson Investigators and is also a Certified Fire and Explosion Investigator through the National Association of Fire Investigators.

Gerald A. Haynes is a Registered Professional Engineer with more than 35 years of experience in the field of fire protection. He holds both BS and MS degrees in Fire Protection Engineering from the University of Maryland—College Park. He has worked in fire suppression operations, fire safety inspections, fire investigations, and fire protection engineering. He is currently the Vice President of Forensic Fire Analysis, LLC, and serves as an adjunct lecturer for the Office of Advanced Engineering Education at the University of Maryland.

Gerald A. Haynes, MSFPE, PE, CFEI

Photo courtesy of Gerald A. Haynes, PE.

Mr. Haynes's extensive experience includes work at the National Institute of Standards and Technology (NIST) Building and Fire Research Laboratory and the U.S. Department of Justice, Bureau of Alcohol, Tobacco, Firearms and Explosives (ATF). His experience is multidimensional, including municipal fire service, fire and explosion investigations, complex case analysis and engineering support, expert testimony and military service. He has provided testimony as an expert witness in state and federal courts in cases related to fire cause and fire testing protocols. He has conducted numerous presentations in the United States and internationally in the areas of fire dynamics, fire modeling, and forensic fire engineering analysis.

He currently is a member of numerous professional organizations, including the Society of Fire Protection Engineers (SFPE), the National Fire Protection Association (NFPA), the International Association of Arson Investigators (IAAI), and the Institute of Forensic Investigation. He is a former member of the technical committees of NFPA 921 Fire and Explosion Investigations, and NFPA 1033 Standard for Professional Qualifications for Fire Investigator.

COURSE DESCRIPTION

This course examines the technical, investigative, legal, and social aspects of arson, including principles of incendiary fire analysis and detection, environmental and psychological factors of arson, legal considerations, intervention, and mitigation strategies.

The National Fire Academy–developed Fire and Emergency Services Higher Education (FESHE) curriculum serves as a national training program guideline that is a requirement for many fire service organizations and training programs. The following grid outlines Fire Investigation and Analysis course requirements and where specific content can be located within this text.

Course Requirements	1	2	3	4	5	6	7	8
Demonstrate a technical understanding of the characteristics and impacts of fire loss and the crime of arson necessary to conduct competent fire investigation and analysis.	X	X	X	X	X	X	X	X
Document the fire scene, in accordance with best practice and legal requirements.	X	X	X	X		X	X	X
Analyze the fire scenario utilizing the scientific method, fire science, and relevant technology.	X	X	X	X	X	X	X	X
Analyze the legal foundation for conducting a systematic incendiary fire investigation and case preparation.	X	X	X	X	X		X	X
Design and integrate a variety of arson related intervention and mitigation strategies.	X				X		X	

	Professional Levels of Job Performance for Fire Investigators as Cited in *NFPA 1033*, 2009 Edition
General Requirements for a Fire Investigator	**4.1.2** Employ all elements of the scientific method as the operating analytical process **4.1.3** Complete site safety assessments on all scenes **4.1.4** Maintain necessary liaison with other interested professionals and entities. **4.1.5** Adhere to all applicable legal and regulatory requirements **4.1.6** Understand the organization and operation of the investigative team and incident management system
Scene Examination	**4.2.1** Secure the fire ground **4.2.3** Conduct an interior survey **4.2.4** Interpret fire patterns **4.2.5** Interpret and analyze fire patterns **4.2.6** Examine and remove fire debris **4.2.7** Reconstruct the area of origin **4.2.8** Inspect the performance of building systems **4.2.9** Discriminate the effects of explosions from other types of damage

(continued)

Professional Levels of Job Performance for Fire Investigators as Cited in *NFPA 1033*, 2009 Edition (*continued*)	
Documenting the Scene	**4.3.1** Diagram the scene **4.3.2** Photographically document the scene **4.3.3** Construct investigative notes
Evidence Collection and Preservation	**4.4.1** Utilize proper procedures for managing victims and fatalities **4.4.2** Locate, collect, and package evidence **4.4.3** Select evidence for analysis **4.4.4** Maintain a chain of custody **4.4.5** Dispose of evidence
Interview	**4.5.1** Develop an interview plan **4.5.2** Conduct interviews **4.5.3** Evaluate interview information
Post-Incident Investigation	**4.6.1** Gather reports and records **4.6.2** Evaluate the investigative file **4.6.3** Coordinate expert resources **4.6.4** Establish evidence as to motive and/or opportunity **4.6.5** Formulate an opinion concerning origin, cause, or responsibility for the fire
Presentations	**4.7.1** Prepare a written report **4.7.2** Express investigative findings verbally **4.7.3** Testify during legal proceedings **4.7.4** Conduct public informational presentations

Source: NFPA 2009. *NFPA 1033: Standard for Professional Qualifications for Fire Investigator.* Quincy, MA: National Fire Protection Association.

1

Principles of Reconstruction

❝ It is a capital mistake to theorize before one has data. Insensibly one begins to twist facts to suit theories, instead of theories to suit facts. ❞

—Sir Arthur Conan Doyle
A Scandal in Bohemia

KEY TERMS

abductive reasoning, *p. 13*

arson, *p. 36*

cause, *p. 34*

corpus delicti, *p. 33*

data, *p. 6*

deductive reasoning, *p. 6*

explosion, *p. 2*

fire, *p. 2*

hypothesis, *p. 6*

incendiary fire, *p. 43*

inductive logic or reasoning, *p. 3*

opinion, *p. 6*

peer review, *p. 2*

scientific method, *p. 4*

suspicious, *p. 44*

technical expert, *p. 4*

OBJECTIVES

After reading this chapter, the student should be able to:

■ Explain the need for science in fire investigation and scene reconstruction.

■ Define and explain the scientific method.

■ List the seven basic steps used in the systematic process of the scientific method.

■ Explain the concept of developing alternative working hypotheses.

■ Describe the foundations of expert testimony as applied to fire scene investigation.

■ Explain the impact of *NFPA 921* on scientific fire investigations and expert testimony.

■ List and explain the steps used to perform a thorough forensic fire scene reconstruction.

The goal of this chapter is to introduce a new viewpoint on the principles and concepts of science-based fire scene reconstruction. The use of the scientific method in forming an expert opinion as to a fire's origin and cause, development, and impact on human lives is discussed from several aspects.

The process of fire scene reconstruction is used to determine the most likely development of a fire using a science-based methodology. Reconstruction begins with an understanding of the conditions and actions just prior to the fire and follows the fire from ignition to extinguishment. It explains aspects of the fire and smoke development, the role of fuels, effects of ventilation, the impact of manual and automatic extinguishment, the performance of the building, life safety features, and the manner of injuries or death.

Underlying principles of forensic fire scene reconstruction rely firmly on a comprehensive review of the fire pattern damage, sound fire protection engineering principles, human factors, physical evidence, and an appropriate application of the scientific method. These factors often form the basis for an expert opinion as to the most probable origin and/or the cause of the **fire** or **explosion**. The expert opinion may be part of a written report or the basis for oral testimony in depositions or courtrooms.

To be effective, these expert opinions must be able to pass the eventual scrutiny of cross-examination during **peer review**, sworn depositions, *Daubert* or similar types of hearings, and courtroom testimony. Recent court decisions have placed more weight on expert forensic testimony based on scientific knowledge, rather than knowledge that is merely experience-based.

In the determination of the origin and cause of a fire, a comprehensive reconstruction should involve a fire engineering analysis that tests various scenarios. This analysis may use fire modeling to compare actual events with predicted outcomes by varying causes and growth scenarios. This engineering analysis adds value, understanding, and clarity to the investigation of complex fire scenes.

Need for Science in Fire Scene Reconstruction

An international conference in November 1997 assessed the current state of the art and identified technical gaps in fire investigation (Nelson and Tontarski 1998). The International Conference on Fire Research for Fire Investigation concluded that many scientific gaps existed in the methodologies and principles used to reconstruct fire scenes.

This conference cited research, development, training, and education needs, including the following:

- *Fire incident reconstruction:* laboratory methods for testing ignition source hypotheses
- *Fire and burn pattern analysis:* methods for validation of and training in the evaluation of patterns on walls and ceilings, and patterns resulting from liquids on floors, and comparison with patterns generated by radiant heat from solid fuels

fire ▪ Rapid oxidation with the evolution of heat and light; uncontrolled combustion. A rapid oxidation process, which is a chemical reaction resulting in the evolution of light and heat in varying intensities (*NFPA 921,* 2011 ed., pt. 3.3.58).

explosion ▪ The sudden conversion of potential energy (chemical or mechanical) into kinetic energy with the extremely rapid production or release of heat, gases, or mechanical pressure.

- *Burning rates:* determination of burning rates for different items, and development of a burning rate database
- *Electrical ignition:* validation of means of identifying electrical faults as an ignition source
- *Flashover:* impacts of flashover on fire patterns and other indicators
- *Ventilation:* effects of ventilation on fire growth and origin determination
- *Fire models:* methods for validation of, training in, and education on the use of fire models
- *Health and safety:* methods for evaluating fire investigator occupational health and safety
- *Certification:* development of training and certification programs for investigators and laboratory personnel
- *Train-the-trainers:* methods for teaching training objectives and expertise to groups of qualified trainers

The primary goal of the conference was to encourage the development and use of scientific principles by fire investigators and to determine what methodologies could improve investigations by public sector investigators. Postconference work resulted in the issuance of a white paper under the auspices of the Fire Protection Research Foundation (FPRF 2002). That study concluded that many methods now in use lack a basic scientific foundation for identifying the *area of origin* and the cause of fires. Complex types and forms of burning materials, building geometry, ventilation, and firefighting actions were listed as contributing effects that complicate these determinations.

In 2009, the Committee on Identifying the Needs of the Forensic Sciences Community of the National Research Council issued a report published through the National Academies Press (NAP 2009) titled "Strengthening Forensic Science in the United States: A Path Forward." The study addressed challenges facing the forensic science community, including those of fire investigation.

The study specifically stated:

> By contrast, much more research is needed on the natural variability of burn patterns and damage characteristics and how they are affected by the presence of various accelerants. Despite the paucity of research, some arson investigators continue to make determinations about whether or not a particular fire was set. However, according to testimony presented to the committee, many of the rules of thumb that are typically assumed to indicate that an accelerant was used (e.g., "alligatoring" of wood, specific char patterns) have been shown not to be true. Experiments should be designed to put arson investigations on a more solid scientific footing.

The National Research Council report addressed the need for integrated governance of crime scene investigation, the forensic science system, laboratories, research funding, and professional associations. Finally, the report recommended improving methods, practice, and performance in forensic science, accreditation, proficiency testing, training, education, medical examiner/coroner systems, automated fingerprint identification systems, and homeland security.

The Scientific Method

Although he began his professional career as a barrister, Francis Bacon (1561–1626) is best known for establishing and popularizing the method of **inductive logic or reasoning** for conducting critical scientific inquiries. Using this method, Bacon constructed and evaluated propositions or principles that were abstractions of observations of a series of specific events. The method of inductive logic extracts knowledge through observation, experimentation, and testing of hypotheses.

peer review ■ Peer review is a formal procedure generally employed in prepublication review of scientific or technical documents and screening of grant applications by research-sponsoring agencies. Peer review carries with it connotations of both independence and objectivity. Peer reviewers should not have any interest in the outcome of the review. The author does not select the reviewers, and reviews are often conducted anonymously. As such, the term *peer review* should not be applied to reviews of an investigator's work by coworkers, supervisors, or investigators from agencies conducting investigations of the same incident. Such reviews are more appropriately characterized as "technical reviews" (*NFPA 921,* 2011 ed., pt. 4.6.3).

inductive logic or reasoning ■ The process by which a person starts from a particular experience and proceeds to generalizations. The process by which hypotheses are developed based on observable or known facts and the training, experience, knowledge, and expertise of the observer (*NFPA 921,* 2011 ed., pt. 3.3.104).

The underpinning of forensic fire scene reconstruction is the use and application of relevant scientific principles and research in conjunction with a systematic examination of the scene. This is particularly important for cases that later require expert witness opinions.

ASTM International, formerly the American Society for Testing and Materials, succinctly defines a **technical expert** as "a person educated, skilled, or experienced in the mechanical arts, applied sciences, or related crafts" (ASTM 1989). A properly trained and qualified fire investigator who applies and practices the principles of fire science can clearly be considered a technical expert under the ASTM definition. These and related definitions unfortunately do not appear in the 10th edition of the *ASTM Dictionary of Engineering Science and Technology* (ASTM 2005a).

The basic concepts of the **scientific method** are simply to *observe, document, hypothesize, test, reassess,* and *conclude.* The scientific method, which embraces sound scientific and fire protection engineering principles combined with peer-reviewed research and testing, is considered the best approach for conducting fire scene analysis and reconstruction. Recently, the National Academy of Sciences, National Academies, published their third edition of the *Reference Manual on Scientific Evidence,* which cites the major impact of the use of the scientific method, particularly in the forensic and engineering sciences (NAP 2011).

Furthermore, present job performance requirements under *NFPA 1033: Standard for Professional Qualifications for Fire Investigator,* pt. 4.1.2 (NFPA 2009), specify that fire investigators "shall employ all elements of the scientific method as the operating analytical process throughout the investigation and for drawing of conclusions."

NFPA 921: Guide for Fire and Explosion Investigations, 2011 ed., pt. 3.3.139 (NFPA 2011), defines the application of the scientific method as

> the systematic pursuit of knowledge involving the recognition and formulation of a problem, the collection of data through observation and experiment, and the formulation and testing of a hypothesis.

When utilizing the scientific method, a fire investigator continuously refines and explores a working hypothesis until he or she reaches a final expert conclusion or opinion, as illustrated in Figure 1-1. An important concept in the application of the scientific method to fire scene investigations is that the investigator should approach all fire scenes without a presumption of the cause. Until sufficient data have been collected in the examination of the scene, no specific hypothesis should be formed (*NFPA 921,* 2011 ed., pt. 4.3.6.7) (NFPA 2011).

The following list is an annotated seven-step systematic process, based on *NFPA 921,* 2011 ed., pt. 4.3 (NFPA 2011), for applying the scientific method to fire investigation and reconstruction. Properly applied, this systematic process addresses how to correctly formulate, test, and validate a hypothesis.

1. *Recognize the need.* A paramount responsibility of first responders is to protect a scene until a full investigation can be started. After initial notification, the investigator should proceed to the scene as soon as possible and determine the resources needed to conduct a thorough investigation. It is necessary to determine not only the origin and cause of this event but also whether future fires, explosions, or loss of life can be prevented through new designs, codes, or enforcement strategies. In public safety sectors, determination of the origin and cause of every fire is often a statutory requirement. The identification of unsafe products related to the cause of the fire can also result in recalls and prevention of future incidents.

2. *Define the problem.* Devise a tentative investigative plan to preserve and protect the scene, determine the cause and nature of the loss, conduct a needs assessment, formulate and implement a strategic plan, and prepare a report. Also determine who has the primary responsibility and authority to interview witnesses, protect evidence, and review preliminary findings and documents describing the loss.

technical expert ■ A person educated, skilled, or experienced in the mechanical arts, applied sciences, or related crafts (*ASTM E1138-89,* 1989).

scientific method ■ The systematic pursuit of knowledge involving the recognition and formulation of a problem, the collection of data through observation and experiment, and the formulation and testing of a hypothesis (*NFPA 921,* 2011 ed., pt. 3.3.139).

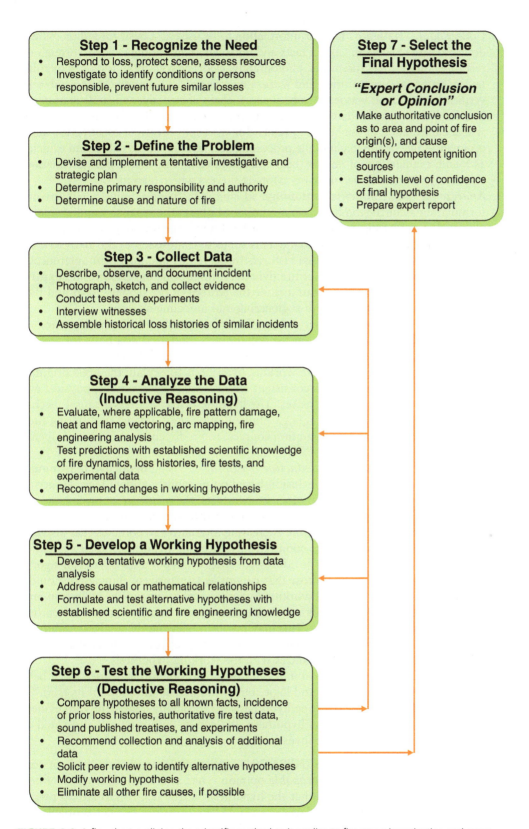

FIGURE 1-1 A flowchart outlining the scientific method as it applies to fire scene investigation and reconstruction. *Derived from* NFPA 921, *2011 ed., pt. 4.3* (*NFPA 2011*).

3. *Collect data.* **Data** are "facts or information to be used as a basis for a discussion or decision" (ASTM 1989). Collect facts and information about the incident through direct observations, measurements, photography, evidence collection, testing, experimentation, historical case histories, and witness interviews. All collected data are subject to verification of how it was legally obtained, its chain of custody, and confirmation of its reliability and authoritative nature. Data collection should include the comprehensive documentation of the building, vehicle, or wildlands involved; construction and occupancy; fuel or fire loading; the processing and layering of the debris and evidence as found; examination of the fire (heat and smoke) patterns; char depths, calcination, and the application of electrical arc mapping surveys where relevant.

4. *Analyze the data (inductive reasoning).* Using inductive reasoning which involves formulating a conclusion based on a number of observations, analyze all the data collected. The investigator relies on his or her knowledge, training, and experience in evaluating the totality of the data. This subjective approach to the analysis may include knowledge of similar loss histories (observed or obtained from references), training in and understanding of fire dynamics, fire testing experience, and experimental data. Data evaluation can include the pattern of fire damage, heat and flame vectoring, arc mapping, and fire engineering and modeling analysis.

5. *Develop a working hypothesis.* A **hypothesis** is defined as "a supposition or conjecture put forward to account for certain facts, and used as a basis for further investigation by which it may be proved or disproved" (ASTM 1989). Based on the data analysis, develop a tentative working hypothesis to explain the fire's origin, cause, and development that is consistent with on-scene observations, physical evidence, and testimony from witnesses. The hypothesis may address a causal mechanism or a mathematical relation (e.g., plume flame height, impact of differing fuel loads, locations of competent ignition sources, room dimensions, impact of open or closed doors and windows).

6. *Test the working hypothesis (deductive reasoning).* Using **deductive reasoning**, which involves coming to a conclusion based on previously known facts, compare the working hypothesis with all other known facts, the incidence of prior loss histories, relevant fire test data, authoritative published treatises, and experiments. A critical feature of hypothesis testing is creating alternative hypotheses that also can be tested. If the alternatives are counter to the working hypothesis, their evaluation may reveal issues that need to be addressed. After all hypotheses have been rigorously tested against the data, those that cannot be conclusively eliminated should be considered viable. Evaluate all working hypotheses, recommend the collection and analysis of additional data, seek new information from witnesses, and develop or modify the working hypothesis. This may involve reviewing the analysis with other investigators with relevant experience and training (peer review). Repeat steps 4, 5, and 6 until there are no discrepancies with the working hypothesis.

7. *Select final hypothesis (conclusion or opinion).* An **opinion** is defined as "a belief or judgment based upon facts and logic, but without absolute proof of its truth" (ASTM 1989). When the working hypothesis is thoroughly consistent with evidence and research, it becomes a final hypothesis and can be authoritatively presented as a conclusion or opinion of the investigation.

NFPA 921, 2011 ed., pt. 18.6 (NFPA 2011), currently suggests two levels of confidence with respect to opinions A *probable* opinion ranks at a level of certainty of being more likely true, or the likelihood that the hypothesis is true is greater than 50 percent. A *possible* opinion ranks at the level of certainty of being feasible, but not probable, particularly if two or more hypotheses are equally likely.

If the level of certainty of the opinion is only "possible" or "suspected," the fire cause is unresolved and a single final hypothesis cannot be determined, then the cause should be classified as "undetermined." *NFPA 921,* 2011 ed., pt.18.7.4 (NFPA 2011), notes also that the decision as to the level of confidence in data collected in the investigation or of any hypothesis drawn from an analysis of the data rests solely with the investigator.

Levels of confidence may also vary with the profession of the person preparing the report. For example, engineers commonly state in their written reports that their *expert opinion* is "to a reasonable degree of engineering certainty." Forensic opinions should be offered only at a very high level of certainty; that is, no other logical solutions offer the same level of agreement with the available data. Curiously, *NFPA 921* currently does not include such a characterization in its hierarchy. In criminal cases, such opinions are scrutinized under the same "beyond a reasonable doubt" standard as are the ultimate decisions of the triers of fact. In civil cases, opinions may rest on an evaluation as to whether the event was more likely than not.

It is important to emphasize that when applying the scientific method in stating an expert opinion on the origin and cause of a fire or explosion, the investigator sets a level for the degree of confidence in forming the opinion at a prescribed threshold. Specifically, when this degree of confidence in the data or hypothesis is at the possible or suspected level, "undetermined" should be cited as the cause of the fire or explosion. An opinion is expected to withstand the challenges of a reasonable examination by peer review or a thorough cross-examination in the courtroom (*NFPA 921,* 2011 ed., pt. 18.7) (NFPA 2011).

THE WORKING HYPOTHESIS

The concept of the *working hypothesis* is central within the framework of the scientific method. In the case of fire scene reconstruction, the working hypothesis is based on how an investigator describes or explains the fire's origin, cause, and subsequent development. It is subject to continuous review and modification and is not the final conclusion.

As shown in Figure 1-2, the investigator molds a working hypothesis then modifies and refines it by drawing on and examining many sources of information, including past investigative experiences. The working hypothesis may include technical reviews by other qualified investigators, who bring institutional knowledge and experiences. Alternative hypotheses should be created and tested. See *NFPA 921,* pt. 4.6.2 (NFPA 2011) for a detailed discussion of technical reviews.

New hypotheses are often built on their predecessors, particularly when the knowledge base of a discipline broadens. When an accepted working hypothesis cannot explain new data, investigators strive to construct a new hypothesis and in some cases generate completely new insight into the problem. A historical example of the application of this procedure dealt with proving whether the geocentric or the heliocentric concept of the solar system was correct, or simply, Does the earth revolve around the sun—or vice versa? The early hypothesis that the sun revolved around the earth was replaced with a new model proposing that the earth revolved around the sun that better fit new data.

One of the underpinnings of the scientific method is the concept of reproducibility, whereby a hypothesis can be independently tested and verified by others. Even when a working hypothesis agrees with all known experimental evidence, it may, nevertheless, be disproved through subsequent experimentation or an emergent discovery. For example, recent experimental evidence

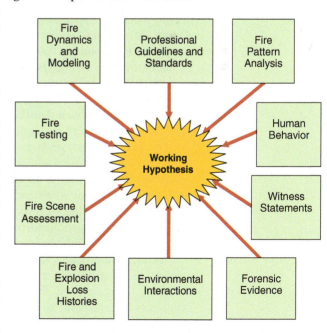

FIGURE 1-2 Sources of information that may contribute to a working hypothesis. *Courtesy of D. J. Icove.*

brought new insights to the interpretation of fire patterns (FEMA 1997a; Icove and De-Haan 2006; Hopkins 2008; Hopkins, Gorbett, and Kennedy 2009). Certain classes of fire pattern damage once thought to be produced solely by accelerants at fire scenes have been disproved by new principles, practice, and testing (DeHaan and Icove 2012; NFPA 2011). These include, for example, spalling of concrete, collapsed furniture springs, and crazed glass (DeHaan and Icove 2012; Tobin and Monson 1989; Lentini 1992).

Knowledge of the application of *fire dynamics and modeling* is a significant advantage that can clearly set expert fire investigators apart from novices. Such application is not necessarily new or novel, since the basic tenets of fire spread are based on the established principles of fire dynamics. The understanding of fire dynamics is improving steadily and is often improved by modeling fire situations. Because investigators apply fire science in their profession, the use of consistent terminology, descriptions of the scene, and interpretations are important in developing and communicating the working hypothesis and in conducting peer review.

Professional guidelines and standards, particularly those that are authoritative, are often relied on in developing hypotheses. Engineering handbooks in wide use include the *Fire Protection Handbook* (NFPA 2008); the *SFPE Handbook of Fire Protection Engineering* (SFPE 2008); and the *Ignition Handbook: Principles and Applications to Fire Safety Engineering, Fire Investigation, Risk Management and Forensic Science* (Babrauskas 2003). Widely recognized guides include *NFPA 921: Guide for Fire and Explosion Investigations* (NFPA 2011); the National Institute of Justice's *Fire and Arson Scene Evidence: A Guide for Public Safety Personnel* (NIJ 2000); *Kirk's Fire Investigation,* 7th ed. (DeHaan and Icove 2012); and *Combating Arson-for-Profit: Advanced Techniques for Investigators,* 2nd ed. (Icove, Wherry, and Schroeder 1998). Other NIJ national guidelines cover such topics as general crime scenes, death investigations, eyewitness identification, bombings, digital images, and electronic evidence. ASTM has several standard practices applicable to fire investigations, which are discussed later in this chapter.

The Society of Fire Protection Engineers (SFPE) publishes a series of engineering guides, many of which are directly applicable to better understanding fire and human behavior and to making realistic assessments. These publications include the *SFPE Engineering Guide for Assessing Flame Radiation to External Targets from Pool Fires, Predicting 1st and 2nd Degree Skin Burns, Piloted Ignition of Solid Materials under Radiant Exposure, Performance-Based Fire Protection Analysis and Design of Buildings, Human Behavior in Fire, Fire Exposures to Structural Elements, Evaluation of the Computer Fire Model DETACT-QS, Fire Risk Assessment in Fire Protection Design, Peer Review in the Fire Protection Design Process,* and *Substantiating Whether a Computer Model Is Appropriate for a Given Application.* The SFPE offers these references and other publications for a nominal fee through their website, www.sfpe.org/.

Fire pattern analysis is an important aspect of interpreting the impact of the fire plume on the structure and contents of the building, vehicle, wildland, victim, or other property. Because liquid accelerants are frequently used by arsonists, authoritative fire testing continues to extend the body of knowledge into fire pattern analysis, particularly in comparing burn patterns produced by flammable liquids and combustible liquids with those produced by ordinary combustibles (DeHaan 1995; Shanley 1997; Putorti, McElroy, and Madrzykowski 2001; DeHaan and Icove 2012).

An understanding of *human behavior* is paramount when evaluating the movement and tenability of persons involved in fires. Not all people respond to perceptions of fire danger in the same way. The investigator must be able to understand and describe how occupants may respond to smoke and fire conditions that change in time and location, resulting in either their successful escape or their death from exposure to various by-products of combustion. The evaluation should include not only behavior patterns of individuals but also interactions with their primary group (family) or with strangers. This understanding should include gender-based differences of persons in their response to the interpretation of the fire cues, decision making, decisions, and their resulting coping actions. For additional information on this topic, see the *Engineering Guide: Human Behavior in Fire,* published by the SFPE (2003).

Witness statements giving accounts of actions taken prior to, during, and after the fire are important to include when developing and testing the working hypothesis. Often, witnesses may need to be reinterviewed to elicit details left out of the initial statements given to authorities. This information may include observations seen by a witness prior to the fire and as the fire was discovered and progressed.

Forensic science can add additional information to a working hypothesis. Examples include the use of traditional forensic examinations of fire debris, the analysis of impression and trace evidence, and, lately, the increased use of DNA in cases of death investigations. Knowledge of *environmental interactions* of the initial temperature, humidity, wind speed, and direction on the developing fire may play an important role in developing the working hypothesis. For example, in a high-rise building fire, the environmental factors of wind and temperature along with information about the location of the fire, whether it occurred above or below the neutral plane, may affect the initial discovery of odors, smoke, and fire development or the spread or intensity of the fire (SFPE 2008; Madrzykowski and Kerber 2009). Lightning strikes are an environmental interaction and potential cause for ignition that should be considered. U.S. meteorological studies identify Louisiana and Florida as the leading states for the highest density of lightning strikes, followed by the adjacent states of Tennessee, Mississippi, and Kentucky (Orvill and Huffines 1999).

Fire and explosion loss histories are significant to the development of any working hypothesis. The knowledge of similar historical case histories may shed new knowledge in developing the hypotheses that should be tested before selecting the final hypothesis. Such statistical data should not be relied on to prove the cause of a particular fire under investigation; rather, they serve a better purpose in helping formulate, examine, or refine the working hypotheses.

Many public and private organizations collect fire data for statistically profiling occupancies to develop risk assessment and fire prevention strategies. Table 1-1 is a profile of average annual fire losses in eating and drinking establishments by cause for the years 2004–2008 (Evarts 2010).

NFPA also profiles fires in residences and publishes statistics on the forms of material first ignited, data that are essential when developing hypotheses (Rohr 2001, 2005). Other industry sources of fire loss experience data are also available, such as the periodically issued FM Global "Property Loss Prevention Data Sheets" (FM Global 2012).

TABLE 1-1	Profile by Cause for Average Annual Fire Losses in Eating and Drinking Establishments, 2004–2008	
FIRE CAUSE	**NO. OF LOSSES**	**PERCENTAGE**
Cooking equipment (stove, fryer, grill)	4410	54
Electrical distribution (lights, lamps, wiring)	620	8
Arson	410	5
Heating equipment	830	10
Smoking materials	610	7
Clothes dryer or washer	130	2
Other	1150	14
Total losses	8160	100.0

Source: Data calculated from Evarts 2010.

These loss histories may also be included in the investigator's experience of *fire scene assessments* while he or she is working on, reviewing, or reading about similar cases. The investigator may draw on heuristic knowledge, training, and experience, combined with professional investigative guidelines and standards, during the development of a working hypothesis. Such statistical histories are useful for developing alternative hypotheses to be tested but not for eliminating a cause for a specific fire because it does not appear on the list.

The combination of generally accepted scientific principles and research, when supported by empirical *fire testing*, can yield valuable insight into fire scene reconstruction. Scientifically conducted test results can normally be reproduced and their error rates validated. Such results are considered to be *objective*—neither prejudiced nor biased by the researcher. Accepted fire testing procedures are developed and maintained by the Committee on Fire Standards (E05) of ASTM International. Fire testing will be discussed later in greater detail in Chapter 8.

APPLYING THE WORKING HYPOTHESIS

It is vital for every experienced investigator to become proficient in the development of the working hypothesis. Proficiency can be accomplished by consistently documenting and evaluating evidence from the fire scene, recording observations and behaviors of witnesses to the fire, comparing similar case histories, reading the fire literature, actively participating in authoritative fire testing, and writing and publishing peer-reviewed research.

The scientific method as it applies to forensic fire scene reconstruction provides fundamental principles for developing a modifiable (working) hypothesis about a fire, testing the hypothesis, and deriving a sound theory as to its origin, development, damage, and cause. Through an iterative process, the working hypothesis is revisited, revised, and reformulated until it is reconciled with all the available data. At that point, the hypothesis becomes a final theory and can be presented as a conclusive opinion of the investigation.

Struggling with the concept of alternative hypotheses is not new in fire investigations. Example 1-1 illustrates the many hypotheses that arose during the investigation of the 1942 Cocoanut Grove Night Club fire. Some of the theories appeared far-fetched, but investigators examined each one during the case. Clearly, the authorities pursued all theories that surfaced in this case. This is an example of how the scientific method can be used to discount those theories that cannot be supported by analytical data, fire protection principles, and witness observations.

EXAMPLE 1-1 ■ Hypothesis Testing

The Cocoanut Grove Night Club fire of November 28, 1942, in Boston cost the lives of 492 people, with many others seriously injured, including 131 treated at nearby hospitals. The fire was first discovered in the basement area called the Melody Lounge. It spread quickly through the lounge and up the stairs to the entrance lobby at street level.

Investigating authorities probing the fire were flooded with publicly expressed opinions explaining how the fire started and progressed. Even though the exact source of ignition is still disputed today, many of the publicly expressed opinions had to be addressed in the report. The major issues were the use of combustible decorations, the rapid rate of fire spread, the lack of adequate exits, and the cause of so many postfire deaths. What is known is that witnesses first reported that the fire originated in the lounge, ignited adjacent combustible decorations and cloth-covered ceilings and walls, and progressed up the stairs, thereby blocking the only visible means of exiting the building.

Many of the theories on the fire's origin, cause, and rapid spread were reported in the official NFPA (1943) report issued January 11, 1943. Even though the published report does not go into sufficient technical detail on these theories, it still serves as an example of a working hypothesis in fire scene reconstruction and analysis. The initial NFPA report listed the following theories and eliminated some based on their improbability and others on the lack of supporting physical evidence.

Ordinary Fire Theory. The ordinary fire theory stated that the cloth and paper decorations on the ceiling and walls, and other available combustible materials, generated an abundance of smoke and carbon monoxide and a means for rapid fire spread. The report noted the possibility of another unexplained element that could have accelerated the growth of the fire. The report did not further discuss the unexplained element. In retrospect, the high number of deaths—hours after the fire—of some who had initially survived led to suspicion of unusual factors in the toxicity of the smoke.

Alcohol Theory. The alcohol theory suggested that vapors from the mouths of heavy drinkers contributed to the rapid fire spread. This included the proposal that warm alcohol evaporating from drinks on tables could also have contributed to the fire. This theory was discounted as a possibility, as the concentration of such vapors would be far too low to enhance flame spread.

Pyroxylin Theory. The pyroxylin (nitrocellulose) theory suggested that the extensive use of artificial leather used on the walls of the building caused the rapid fire spread and production of nitrous by-products of combustion, accounting for the numerous fatalities. A scientific assessment led to the elimination of this theory. These materials contained only a small percentage of pyroxylin, which was insufficient to contribute to the fire spread or to the number of fatalities.

Motion Picture Film Theory. The motion picture film theory was raised when it was determined that the building had previously been used as a motion picture film exchange. The theory suggested that a quantity of decomposing scrap nitrocellulose film might have remained in a hidden storage area and was ignited by a source of ignition. The observations of witnesses did not support this theory.

Refrigerant Gas Theory. The air conditioning supplying the Melody Lounge was located in a false corner wall. Historically, air-conditioning systems used ammonia or sulfur dioxide as refrigerant. These refrigerant gases are combustible and toxic. A theory suggesting that these refrigerant gases ignited was discounted. Ammonia is a flammable gas but at concentrations far above the tolerable level for human exposure. Also, survivors did not report ammonia odors. Furthermore, the refrigerant gas ran in copper tubing, which would have needed to be thermally or mechanically breached to release the gas during the initial stages of the fire.

Flameproofing Theory. The flameproofing chemicals theory was based on the suggestion that additives to the combustible decorations may have given off ammonia and other gases when heated. The investigation failed to determine whether any flame retardants were ever applied to the decorations, and the surviving victims did not report any symptoms associated with this type of exposure. Fire inspectors at the time routinely exposed the edge of a portion of the decorative fabric to a match flame as a field test. Decorations tested prior to the fire reportedly passed this limited test.

Fire Extinguisher Theory. The fire extinguisher theory came about when investigators learned that fire extinguishers were used on an incipient fire in an artificial palm tree and that the extinguishing liquid may have emitted toxic gases. The investigation found no supporting evidence for this theory.

Insecticide Theory. The insecticide theory emerged when investigators learned that insecticides were used in the basement kitchen. They suggested that if flammable vapors had collected in concealed wall spaces, the vapors may have caused the reported initial flash fire. There was no information as to whether the insecticide used contained an ignitable liquid, and there were no indications of excess quantities of any insecticide.

Gasoline Theory. The gasoline theory was suggested when investigators determined that the building had previously been used as a garage and that gasoline tanks still existed under the basement floor. The theory that gasoline vapor emerged from these tanks was discounted, since such fumes are heavier than air and would have remained at floor level. Furthermore, the physical scene examination proved that there was no source to create an adequate concentration of vapors to support the flame spread seen in this fire. Occupants would have detected vapors by their odor at levels far below ignitable concentrations.

Electric Wiring Theory. The electric wiring theory came about when investigators learned that an unlicensed electrician had installed part of the wiring in the building. The theory suggested that the heated insulation on overloaded wiring was responsible for forming flammable and toxic vapors. No evidence was found that could support the theory of overloading or that an electrical failure was responsible for ignition.

Smoldering Fire Theory. The smoldering fire theory was suggested when investigators learned that witnesses reported that several walls were hot to the touch and that a smoldering fire went undetected. Investigators physically examining the fire scene found no physical evidence to support this theory.

As a postscript to these theories, more than half a century later the cause of the fire is still listed as "undetermined." Some witnesses reported that the initial fire was in an artificial tree, where an errant flame from a match, lighter, or candle could have started it. Hypotheses are still being tested as to the cause of the Cocoanut Grove fire. Reviews of the investigation by the NFPA begun in 1982 along with the latest reviews suggest that fire modeling combined with the use of the scientific method may shed new insights into this case (Reinhardt 1982; Grant 1991; Beller and Sapochetti 2000). For further reading, see Schorow (2005) and Esposito (2005).

BENEFITS OF USING THE SCIENTIFIC METHOD

There are numerous benefits to using the scientific method to examine fire and explosion cases, including the following:

- Acceptance of the methodology in the scientific community
- Use of a uniform, peer-reviewed protocol of practice, such as *NFPA 921*
- Improved reliability of testimony from opinions formed using the scientific method

First, the use of the scientific method is well received in both the technical and the research communities. An investigation conducted using this approach is more likely to be embraced by those who would tend to doubt a less thoroughly conducted probe.

Second, the scientific method is a generally accepted protocol of practice in the literature. Those persons ignoring or deviating from recommended practices will be subject to closer scrutiny of their reports and opinions. Modern fire investigation texts endorse the application (DeHaan and Icove 2012).

The NIJ's (2000) *Fire and Arson Scene Evidence: A Guide for Public Safety Personnel* repeatedly cites the scientific method. The guide notes that actions taken at the scene of a fire or arson investigation can play a pivotal role in the resolution of a case. A thorough investigation is key to ensuring that potential physical evidence is neither tainted nor destroyed, nor potential witnesses overlooked.

Third, and most important, expert testimony in fire and explosion cases has come to rely more heavily on opinions formed using the scientific method. Recent U.S. Supreme Court decisions underscore these principles, with many state courts following the trend. The following section discusses at length many issues surrounding expert testimony.

ALTERNATIVES TO THE SCIENTIFIC METHOD

There are alternative approaches to the scientific method that use other forms of logic apart from inductive and deductive reasoning. For example, **abductive reasoning** involves the process of reasoning to the best explanations for a phenomenon. An example of using abductive reasoning to explain an observation might be, "If it rains, the grass is wet." This abductive rule explains why the grass is wet. An alternative explanation might answer the question: Can the grass be wet if it has not rained? Yes, the lawn sprinklers were used, or the dew was heavy. In this approach, the reasoning starts with a set of given facts and derives their most likely explanations or hypotheses. Such reasoning leads to testable alternatives and is therefore key to proper application of the scientific method to origin and cause determination.

> **abductive reasoning** ■ The process of reasoning to the best explanations for a phenomenon.

Modern-day applications of abductive reasoning include artificial intelligence, fault tree diagnosis, and automated planning. For a thorough discussion on the comparison of inductive, deductive, and abductive reasoning, see the National Defense Intelligence College paper titled "Critical Thinking and Intelligence Analysis" (Moore 2007). The admissibility of abductive reasoning and inference has been discussed but has not yet been included in *NFPA 921*, but it has been cited in fire symposiums (Brannigan and Buc 2010).

Foundations of Expert Testimony

For the purposes of this textbook, emphasis is placed on testimony given primarily in the federal court system. The *Federal Rules of Evidence* (FRE 2012) clearly state who may offer testimony in federal court. Many state courts model their guidelines on the federal rules.

The following amendments to the FRE, which were effective December 1, 2011, affect the admissibility of evidence and opinion testimony under *Rules 701, 702, 703, 704, 705, and 706:*

Rule 701. Opinion Testimony by Lay Witnesses

If the witness is not testifying as an expert, the witness' testimony in the form of opinions or inferences is limited to those opinions or inferences which are (a) rationally based on the perception of the witness, (b) helpful to a clear understanding of the witness' testimony or the determination of a fact in issue, and (c) not based on scientific, technical, or other specialized knowledge within the scope of Rule 702.

Rule 702. Testimony by Experts

If scientific, technical, or other specialized knowledge will assist the trier of fact to understand the evidence or to determine a fact in issue, a witness qualified as an expert by knowledge, skill, experience, training, or education may testify thereto in the form of an opinion or otherwise, if (1) the testimony is sufficiently based upon reliable facts or data, (2) the testimony is the product of reliable principles and methods, and (3) the witness has applied the principles and methods reliably to the facts of the case.

Rule 703. Bases of Opinion Testimony by Experts

The facts or data in the particular case upon which an expert bases an opinion or inference may be those perceived by or made known to the expert at or before the hearing. If of a type reasonably relied upon by experts in the particular field in forming opinions or inferences upon the subject, the facts or data need not be admissible in evidence in order for the opinion or inference to be admitted.

Facts or data that are otherwise inadmissible shall not be disclosed to the jury by the proponent of the opinion or inference unless the court determines that their probative value in assisting the jury to evaluate the expert's opinion substantially outweighs their prejudicial effect.

Rule 704. Opinion on Ultimate Issue

(a) Except as provided in subdivision (b), testimony in the form of an opinion or inference otherwise admissible is not objectionable because it embraces an ultimate issue to be decided by the trier of fact.

(b) No expert witness testifying with respect to the mental state or condition of a defendant in a criminal case may state an opinion or inference as to whether the defendant did or did not have the mental state or condition constituting an element of the crime charged or of a defense thereto. Such ultimate issues are matters for the trier of fact alone.

Rule 705. Disclosure of Facts or Data Underlying Expert Opinion

The expert may testify in terms of opinion or inference and give reasons therefor without first testifying to the underlying facts or data, unless the court requires otherwise. The expert may in any event be required to disclose the underlying facts or data on cross-examination.

Rule 706. Court Appointed Experts

(a) Appointment.

The court may on its own motion or on the motion of any party enter an order to show cause why expert witnesses should not be appointed, and may request the parties to submit nominations. The court may appoint any expert witnesses agreed upon by the parties, and may appoint expert witnesses of its own selection. An expert witness shall not be appointed by the court unless the witness consents to act. A witness so appointed shall be informed of the witness' duties by the court in writing, a copy of which shall be filed with the clerk, or at a conference in which the parties shall have opportunity to participate. A witness so appointed shall advise the parties of the witness' findings, if any; the witness' deposition may be taken by any party; and the witness may be called to testify by the court or any party. The witness shall be subject to cross-examination by each party, including a party calling the witness.

(b) Compensation.

Expert witnesses so appointed are entitled to reasonable compensation in whatever sum the court may allow. The compensation thus fixed is payable from funds which may be provided by law in criminal cases and civil actions and proceedings involving just compensation under the Fifth Amendment. In other civil actions and proceedings the compensation shall be paid by the parties in such proportion and at such time as the court directs, and thereafter charged in like manner as other costs.

(c) Disclosure of appointment.

In the exercise of its discretion, the court may authorize disclosure to the jury of the fact that the court appointed the expert witness.

(d) Parties' experts of own selection.

Nothing in this rule limits the parties in calling expert witnesses of their own selection.

Note that these federal guidelines do not apply to all state courts or international situations, and the reader is encouraged to seek guidance from an appropriate legal advisor.

SOURCES OF INFORMATION FOR EXPERT TESTIMONY

In general, experts normally use four major sources of information on which to base their opinions (Kolczynski 2008):

- Firsthand observations
- Facts presented to experts prior to trial
- Facts supplied in court
- Testing of alternative hypotheses

Examples of *firsthand observations* include those observed at a fire scene, in a laboratory examination or fire test, or from evaluation of evidence under *FRE 703*. It is important that experts try to visit the fire scene and not merely rely on sketches and photographic evidence taken by another party. Although valuable, it is not always required or possible to visit the scene, particularly if the scene has been considerably altered or destroyed. If photographic documentation is sufficiently comprehensive and can be documented or validated by other means, the expert can rely on it (*United States of America v. Ruby Gardner* 2000).

Experts usually critically review the case and gain insight and knowledge of *facts prior to trial*. This knowledge may be based on facts gleaned from reports of other experts, scientific manuals, learned treatises, or results of historical testing. Examples include information learned through expert treatises such as *NFPA 921, Kirk's Fire Investigation,* and the *SFPE Handbook.*

Rule 701 allows for lay witnesses to offer nonexpert opinion testimony as evidence. These opinions might include the state of intoxication, a vehicle's speed, and the memory of an identifiable odor of gasoline. For example, firefighters who are not experts in fire investigation may be allowed to testify that they detected a strong odor of gasoline when they entered a certain room of a burning structure.

Rule 702 requires that the expert's testimony must assist the trier of fact, and its admissibility depends in part on the connection between the scientific research or test to be presented and the particular disputed factual issues in this case (*In re Paoli Railroad Yard PCB Litigation* (1994) citing *U.S. v. Downing* 1985). Furthermore, each step in the expert's analysis must reliably connect the work of the expert to that particular case.

A lay witness may also offer expert testimony under *Rule 702* when the opinion is based on an accepted evaluation method, as long as the witness is qualified to make this evaluation. Examples of accepted evaluation methods include the guidelines in *NFPA 921* and various ASTM standards.

Since its enactment in 1975, *Rule 703* allowed experts to form an opinion based on facts, whether or not in evidence, as long as those facts were necessary to form professional judgments from a nonlitigation point of view. Just because an expert witness uses information to form an opinion, its use does not automatically make that information admissible in court. Under *FRE 703*, experts may also rely on the hearsay statements, such as those of other witnesses. For example, the expert may use information from the persons who witnessed the fire, testimony of other witnesses, documents and reports, videotapes of the fire in progress, and related written reports submitted in the case, to help form an opinion (*United States of America v. Ruby Gardner* 2000).

Facts supplied in court may also be a basis for expert witness testimony. These facts may include information from testimony of other witnesses and evidence presented. The expert witness may also be asked a hypothetical question. For example, an expert may be asked, "Assume that three separate gasoline-filled plastic jugs with partially burned wicks were found in separate rooms of the house. What would be your conclusions based on these facts and your 25 years of experience as a fire investigator?"

The *testing of alternative hypotheses* is an important concept in expert witness testimony, whether it be physical laboratory tests or critical reviews of other reasonable hypotheses raised in a case. In *ASTM E620-11,* pt. 4.3.2 (ASTM 2011a), the expert is required to document in his or her report the logic and reasoning used to reach each of the opinions and conclusions. In a recent case, an expert did not undertake a process to eliminate another possible cause of a fire that might otherwise seem on par with his hypothesis. The court found that although the expert was qualified in fire research, his testimony lacked objective, evidence-based support for his conclusions (*Wal-Mart Stores, Petitioner, v. Charles T. Merrell, Sr., et al.* 2010).

DISCLOSURE OF EXPERT TESTIMONY

The following is an excerpt regarding guidelines on the requirement to disclose expert testimony in civil cases from *Rule 26(a)(2)(B)* of the *Federal Rules of Civil Procedure* (FRCP 2010) effective December 1, 2010, as amended.

2. *Disclosure of Expert Testimony*
    A. *In General.* In addition to the disclosures required by Rule 26(a)(1), a party must disclose to other parties the identity of any witness it may use at trial to present evidence under Federal Rule of Evidence *702, 703,* or *705.*
    B. *Witnesses Who Must Provide a Written Report.* Unless otherwise stipulated or ordered by the court, this disclosure must be accompanied by a written report—prepared and signed by the witness—if the witness is one retained or specially employed to provide expert testimony in the case or one whose duties as the party's employee regularly involve giving expert testimony. The report shall contain:
        (i) a complete statement of all opinions the witness will express and the basis and reasons for them;
        (ii) the facts or data considered by the witness in forming them;
        (iii) any exhibits that will be used to summarize or support them;
        (iv) the witness's qualifications, including a list of all publications authored in the previous ten years;
        (v) and a list of all other cases in which, during the previous 4 years, the witness has testified as an expert at trial or by deposition; and
        (vi) a statement of the compensation to be paid for the study and testimony in the case.
    C. *Witnesses Who Do Not Provide a Written Report.* Unless otherwise stipulated or ordered by the court, if the witness is not required to provide a written report, this disclosure must state:
        (i) the subject matter on which the witness is expected to present evidence under Federal Rule of Evidence 702, 703, or 705; and
        (ii) a summary of the facts and opinions to which the witness is expected to testify.
    D. *Time to Disclose Expert Testimony.* A party must make these disclosures at the times and in the sequence that the court orders. Absent a stipulation of a court order, the disclosures must be made:
        (i) at least 90 days before the date set for trial or for the case to be ready for trial; or
        (ii) if the evidence is intended solely to contradict or rebut evidence on the same subject matter identified by another party under Rule 26(a)(2)(B) or (C), within 30 days after the other party's disclosure.
    E. *Supplementing the Disclosure.* The parties must supplement these disclosures when required under Rule 26(e).

FRE Rule 705 addresses the disclosure of facts or underlying data on which an expert forms his/her expert opinion. The expert may testify as to an opinion without first introducing the underlying facts or data used to develop that opinion, unless otherwise directed by the court. However, the expert may be required to later disclose the underlying facts or data used during a cross-examination.

Under *FRCP Rule 26(a)(2)(B)*, experts are required to provide a written expert witness report. This report, which must be prepared and signed by the expert witness, must contain the following.

- Complete statement of all opinions to be expressed along with the basis for these opinions
- Data or information considered by the expert used in forming the opinions
- Exhibits planned to be used in summarizing or supporting the opinions
- Qualifications of the expert witness, including a list of all publications authored within the preceding 10 years
- Compensation to be paid for the report and testimony
- Listing of all cases in which the expert has testified in trial or deposition within the preceding 4 years

The outlined format of written expert disclosure reports varies. Table 1-2 is an example of an outline for a summary disclosure letter or report sent by an expert to an attorney. Table 1-3 lists areas to cover when qualifying as a fire expert (Beering 1996).

TABLE 1-2	Suggested Headings in an Expert Disclosure Report

1. Disclosure and scope of engagement
2. Personal background and qualifications
3. Summary of professional opinion(s)
4. Description of the property destroyed
5. The standard of care and methodology used in fire investigations
6. The fire incident (include names, table of events)
7. An analysis of other investigations
   ☐ Training and experience in fire investigations
   ☐ Postsecondary education
   ☐ Knowledge and certification
   ☐ Expertise in applying a reliable methodology
   ☐ Qualifications
8. Impact of spoliation (if present)
   ☐ Definition, responsibility, and documentation
   ☐ Realization timeframe of potential litigation
   ☐ Notification of interested parties
   ☐ Loss and destruction of critical evidence
   ☐ Timeframe of spoliation of evidence
9. Impact of conducting an independent fire investigation
10. Executive summary of opinions with signature/seal

Appendix A – List of information reviewed

Appendix B – Professional qualifications, past testimony

Appendix C – Compensation received

Appendix D – Supplementary data

Appendix E – Program listings (computer models)

TABLE 1-3	Areas to Address When Qualifying as a Fire Expert
CATEGORY	**EXAMPLE**
Identification—Who is the expert and how long has he or she worked in the field?	Name
	Title, department
	Employment—current and previous
Education and training—What formal specialized training has the expert received?	Formal education
	Fire training academies
	National Fire Academy
	FBI Academy
	Federal Law Enforcement Training Center
	Annual international and state seminars
	Specialized training
Certifications—Is the expert licensed, certified, or otherwise credentialed in his or her field?	Fire investigator
	Police officer
	Fire protection specialist
	Professional engineer
	Instructor
Experience—How has the expert acquired firsthand knowledge and experience about fire?	Fire suppression
	Fire investigations
	Fire testing
	Laboratory
Prior testimony—Has the expert given prior testimony?	Criminal, civil, and administrative courts
	Lay and expert witness
	Jurisdictions (state, federal, international)
Professional associations—Of which professional associations is the expert a member, and what standing (e.g., fellow, life member)?	Professional fire investigation
	Fire and forensic engineering
Teaching experience—Has the expert delivered certified and credible training to others in the field?	Local, state, federal, international schools and organizations
	National academies
	Colleges and universities
Professional publications—Has the expert written and published peer-reviewed quality articles?	Peer-reviewed articles, technical papers
	Books
	National standards
Awards and honors—What significant honors and awards has the expert received?	Local, state, federal, and international professional organizations and societies

Source: Summarized and updated from Beering 1996.

Recognition of Fire Investigation as a Science

DAUBERT CRITERIA

In its recent decisions the U.S. Supreme Court has continued to define and refine the admissibility of expert scientific and technical opinions, particularly those concerning scene investigations. These decisions affect how expert testimony is interpreted and accepted.

Although much controversy exists over these Court decisions, the result is that fire and explosion investigation is being considered more as a science and less as an art. This is particularly true when the scientific method is combined with relevant engineering principles and research in providing expert witness testimony (Kolar 2007; Ogle 2000).

A judge has the discretion to exclude testimony that is speculative or based on unreliable information. In the case *Daubert v. Merrell Dow Pharmaceuticals* (1993), the Court placed the responsibility on a trial judge to ensure that expert testimony was not only relevant but also reliable. The judge's role is to serve as a "gatekeeper" to determine the reliability of a particular scientific theory or technology. The Court defined four criteria to be used by the gatekeeper to determine whether the expert's theory or underlying technology should be admitted. *Daubert* allows the judge to gauge whether the expert testimony aligns with the facts of the case as presented. The central issue is often whether the expert has followed an identified, accepted, and peer-reviewed method as the basis for his or her conclusion.

In a more recent decision, *Kumho Tire Co. Ltd. v. Carmichael* (1999), the Court applied the four *Daubert* criteria to expert testimony to determine whether the testimony was based on science or experience. The *Daubert* criteria examine testability, peer review and publication, error rates and professional standards, and general acceptability, as explained in Table 1-4. In short, expert testimony must rely on a balance of valid peer-reviewed literature, testing, and acceptable practices and a demonstration that the courts followed those practices if they considered the testimony credible.

A recent federal case citing *Daubert* and *Rules 702* and *703* allowed the expert testimony of Dr. John DeHaan pertaining to the origin and cause of an arson fire based on his review of case materials relied on by other experts (*United States of America v. Ruby Gardner* 2000). Dr. DeHaan relied on a review of reports, photographs, and third-party observations, some of which may not have been directly admissible as evidence.

The court determined that this procedure was reliable and appropriate, confirming the conviction. Furthermore, the court drew a parallel to other testimony, including hearsay and third-party observations used by an arson investigation expert in forming an opinion on the ability of a psychiatrist to testify as an expert by relying only on staff reports, interviews with other doctors, and background information.

In *United States of America v. John W. Downing* (1985), a Third Circuit case, both *Daubert* and *Downing* factors were used in determining whether an expert's methodology was reliable. *Downing* factors consider (1) the relationship of the technique to methods that have been established to be reliable; (2) the qualifications of the expert witness testifying based on methodology; and (3) the nonjudicial issues to which the method has been put.

TABLE 1-4	Criteria and Pertinent Issues Examined by *Daubert*
CRITERIA	**PERTINENT ISSUES EXAMINED**
Testing	Has the methodology, theory, or technique been tested?
Peer review and publication	Has the methodology, theory, or technique been subjected to peer review and publication?
Error rates and professional standards	What are the known or potential error rates of this methodology, theory, or technique?
	Does the methodology, theory, or technique comply with controlling standards, and how are those standards maintained?
General acceptance	Is the methodology generally accepted in the scientific community?

FRYE STANDARD

At present 16 states and the District of Columbia (Gianelli and Imwinkelried 2007, 1–15) rely on the *Frye* test (or a local variant of it such as the *Kelly-Frye* test in California, *Reed-Frye* in Maryland, and *Davis-Frye* in Michigan) as the standard for admissibility of expert testimony instead of the *Daubert* standard. The original *Frye* decision was reached in 1923 over the issue of admissibility of the (then) new scientific discipline of polygraph examinations, for which a limited number of adherents maintained its reliability in the face of general skepticism. The U.S. Court of Appeals for the District of Columbia stated:

> Just when a scientific principle or discovery crossed the line between experimental and demonstrable stages is difficult to define. Somewhere in this twilight zone, the evidential force of the principle must be recognized and, while courts will go a long way in admitting expert testimony deducted from a well-recognized scientific principle or discovery, the thing from which the deduction is made must be sufficiently established to have gained general acceptance in the particular field in which it belongs. (*Frye v. United States* 1923)

The *Frye* decision established that the judge is the gatekeeper responsible for excluding junk science. It has been suggested (Graham 2009) that the reasoning behind the "general acceptance" standard is to assist the trier of fact in six ways:

1. Promote a degree of uniformity of decision across the judiciary
2. Avoid the interjection of a time-consuming and often misleading determination of the reliability of a scientific technique during litigation
3. Assure that the scientific evidence introduced will be reliable and thus relevant
4. Ensure that a pool of experts exists who can critically examine the validity of a scientific determination
5. Provide a preliminary screening to protect against the natural inclination of the jury to assign significant weight to scientific techniques presented as evidence where the trier of fact is in a poor position to evaluate the reliability of that evidence accurately
6. Impose a threshold standard of reliability for scientific expert testimony in new areas when cross-examination by opposing counsel may be unable or unlikely to bring inaccuracies to light

This standard is a conservative one, when strictly applied, requiring new technologies or techniques to have been in use long enough for them to have been tested by a number of practitioners. It is the offering party's burden to prove the concept of general acceptance. There are no methods suggested in *Frye* by which a judge should determine general acceptance. The testimony of a single expert, particularly one who might have a bias or vested interest, is rarely acceptable. Presentations of multiple peer-reviewed, published papers; testimony of an independent expert retained by the court; and testimony of multiple experts have all been used.

"The trial judge determines whether the *Frye* test has been satisfied under a rule comparable to *Federal Rules of Evidence 104(a)*. The 'preponderance' standard applies" (Gianelli and Imwinkelried 2007, 1–5). Some courts have applied *Daubert*-type considerations to establishing "hallmarks" such as independent peer review, thorough testing, establishing error rates, or promulgation of objective standards applicable to the technique. One general principle is that general acceptance is met if the use of the technique is supported by a clear majority of the scientific community.

Another major issue with *Frye* is the difficulty in establishing what is the "relevant" scientific field or field in which the technique falls. Fire investigation and reconstruction both rely on multiple fields—fire dynamics, fire engineering, chemistry, and, in some cases, human behavior; they do not fall within a single academic discipline or professional discipline. The judge may turn to the published fire investigation literature or to resources in fire engineering, fire safety, or chemistry (including local college instructors who may or may not have the necessary familiarity with the concepts). Prior judicial decisions have also been used as a basis for such decisions (even when they did not necessarily apply to the technique at hand).

Some courts have extended the *Frye* standard beyond the reliability of the technique to considerations of the underlying theory or theories on which the technique is based. That extension can create real problems for fire investigators if called on to prove the validity of underlying theories for events such as V patterns, spalling, failure of glass, flashover, or halo or ring patterns whose underlying theories, if known, are couched in the fire dynamics and materials science disciplines.

Other courts have avoided the *Frye* issue by deciding that some evidence does not fall within the "scientific" realm but rather is simply demonstrative evidence (physical models, photographs, X-rays, and the like). In those cases, the jury do not have to place blind faith in the assurances of a testifying expert but can verify the results by their own observations. Such reliance raises the specter of the multitude of courtroom tricks of past decades in which unscrupulous counsel used substitutions, mischaracterizations, and even sleight of hand to sway a jury.

IMPACT OF *NFPA 921* ON SCIENCE-BASED EXPERT TESTIMONY

The importance of *NFPA 921* has also been cited along with *Daubert* as an interlinking element of expert testimony, since it establishes guidelines for reliable and systematic investigations and the analysis of fire and explosion incidents (Campagnolo 1999; DeHaan and Icove 2012). Several clusters of recent federal court opinions and rules fall into the following areas:

- Use of investigative protocols, guidelines, and peer-reviewed citations
- Methodological explanations for burn patterns
- Qualifications to testify

Ongoing professional education is critical for any profession, particularly fire investigators. It is incumbent on all professional fire investigators to continuously read and keep abreast of all relevant fire, engineering, and legal publications and to critically evaluate their conclusions with this ever-changing knowledge. As with science, today's knowledge may change with new developments and impact on the reviews of an established hypothesis.

References to NFPA 921 In 1997, a federal judge upheld a motion by a defendant insurance company in a civil case barring the plaintiff's expert fire investigator from mentioning *NFPA 921* during his testimony. *NFPA 921* was first published in 1992, well after the November 16, 1988, fire in question. The judge requested that the plaintiffs provide him a copy of *NFPA 921*. The judge found that under *Rule 703* of the *Federal Rules of Evidence,* any reference to *NFPA 921* would have little probative value and would confuse the jury (*LaSalle National Bank et al. v. Massachusetts Bay Insurance Company et al.* 1997). This ultraconservative finding has not been widely accepted, and in fact, most court decisions since then have noted the value of *NFPA 921*.

For example, following a November 18, 1999, motor vehicle accident, an individual was trapped for approximately 45 minutes in the front passenger seat after the car in which he was riding collided with a utility pole. Eyewitnesses reported a bluish flickering and fire that engulfed the vehicle's interior and severely burned the trapped passenger. A product liability lawsuit against the vehicle's manufacturer alleged a defective design, since a proposed safety device would have disconnected the battery after the collision (*John Witten Tunnell v. Ford Motor Company* 2004). The plaintiffs alleged that the energized wiring harness suffered a high-resistance fault, which could have caused a fire without blowing a fuse. The judge reviewed and overruled motions to exclude the testimony of fire origin experts and repeatedly cited the applicability of the *NFPA 921* methodology to the case.

Investigative Protocols   In a 1999 federal criminal case, the defendant moved that the court exclude a state fire marshal's testimony on the incendiary nature of a fire, arguing that the testimony did not meet the standards for admissibility under *Daubert* (*U.S. v. Lawrence R. Black Wolf, Jr.* 1999). The defendant argued that the testimony should be excluded because the investigator did not arrive at the scene until 10 days after the fire occurred and did not take samples for laboratory analysis, the theory of the fire's origin and cause had not been subject to accurate and reliable peer review, and no scientific methods or procedures were used to reach the conclusions. The defendant's Exhibit A was the 1998 edition of *NFPA 921.*

The federal magistrate reviewing this dispute determined that the investigator's proffered testimony was reliable, since he indeed used an investigative protocol consistent with the basic methodologies and procedures recommended by *NFPA 921,* and he passed the requisite scrutiny demanded of fire origin and cause experts (Figure 1-3).

Regarding the relevancy test, the magistrate determined that the investigator's testimony would help a jury understand the circumstances surrounding the fire's origin and cause. The magistrate further found that *Rule 703* had been met, since the investigator's proffered testimony relied on facts and data normally collected by experts when forming opinions as to the fire's origin and cause. The magistrate concluded that the defendant would have ample opportunity to address the expert's opinion during cross-examination, through presentation of contrary evidence, and in instructions to the jury.

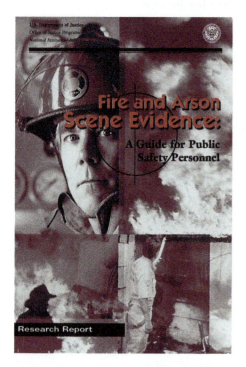

FIGURE 1-3 An investigator's proffered testimony may be deemed reliable if it is in accordance with an investigative protocol, such as the one published by the U.S. Department of Justice that cites methodologies and procedures recommended by *NFPA 921. Courtesy of U.S. Department of Justice, www.usdoj.gov.*

Use of Guidelines and Peer-Reviewed Citations   In a case involving a November 16, 1994, residential fire, a number of independent investigators attempted to determine the exact origin and cause of the fire. One investigator, an electrical engineer, offered the opinion that a television set located in the basement family room caused the fire. The plaintiffs sued the manufacturer of the television, alleging product liability, negligence, and breach of warranties (*Andronic Pappas et al. v. Sony Electronics, Inc. et al.* 2000).

A federal judge adjudicating the case held a two-day *Daubert* hearing and concluded that the one investigator's causation testimony was inadmissible. The judge cited that the primary concern with the testimony stemmed from the second *Rule 702* factor, that an expert's opinion should be based on a reliable methodology. The judge noted that the investigator did not use a fixed set of guidelines in determining the cause of the fire. Notably, the investigator did confirm that even though he was aware of the existence of established guidelines in *NFPA 921* and *Kirk's Fire Investigation,* he relied on his own experience and knowledge. During the hearing, the plaintiffs did not submit any books, articles, or witnesses on proper fire causation techniques other than the testimony of one investigator. The judge specifically commented on the fact that the one investigator referred to his reliance on a burn pattern inside the television set. No citations were made to peer-reviewed sources to support this opinion.

In a product liability case arising from a fire on July 16, 1998, which destroyed a business, the plaintiff's only expert placed the origin of the fire at a product manufactured by the defendant (*Chester Valley Coach Works et al. v. Fisher-Price Inc.* 2001). Based on testimony and arguments received at a *Daubert* hearing held June 14, 2001, the court excluded the expert's testimony and his opinion regarding the fire's origin and cause. At the hearing, the expert indicated his conclusions were based primarily on his experience and education and not based on any testing, experimentation, or generally accepted texts or treatises. Several of the steps he undertook in his investigation were contrary to the *NFPA 921* guidelines.

In a federal court decision on April 16, 2003, a plaintiff's *Daubert* motion to exclude the testimony of a defendant's expert witness was overruled (*James B. McCoy et al. v. Whirlpool Corp. et al.* 2003). The plaintiffs alleged that a dishwasher caused a fire at their residence on February 16, 2000, and argued that the expert's report on the origin and cause of the fire did not explicitly cite the methodology set forth in *NFPA 921*. In its decision, the court noted that the expert did not need to cite *NFPA 921*, since he had adequately provided a complete statement of all opinions and a substantive rationale for the basis of these opinions.

Peer Review of Theory In a case involving a residence fire that occurred around December 22, 1996, an insurance company hired an investigator to determine the cause of the fire, which all parties agreed had originated in an area to the right of a fireplace firebox. The defendants disputed the cause of the fire and filed a motion to exclude the testimony of the plaintiff's expert (*Allstate Insurance Company v. Hugh Cole et al.* 2001).

Even though the plaintiff filed his complaint on December 21, 1998 prior to the December 1, 2000, effective date of the amended *Rule 702*, the judge applied the amended *Rule 702* to the admissibility of the testimony (FRCP 2010). The plaintiff asserted that in his trial testimony the investigator would rely on data from *NFPA 921* to support his opinion.

The opinion of the plaintiff's investigator was that heat from a metal pipe ignited adjacent combustibles. The judge noted that the theory, based on principles embodied in *NFPA 921,* had already been subjected to peer review, was generally accepted in the scientific community, and met the sufficient reliability guidelines to satisfy the admissibility requirements of *Rule 702*. Peer review can also include consultation with other qualified fire experts, both scientific and investigative.

Methodology Needed On September 13, 1996, a residential fire started in the corner of a kitchen containing a dishwasher, toaster oven, and microwave oven. The municipal fire marshal concluded that the fire was caused by the microwave oven, whereas in a federal civil case, the plaintiff's expert asserted that a defective toaster oven caused the fire (*Jacob J. Booth and Kathleen Booth v. Black and Decker, Inc.* 2001).

A senior federal judge held a *Daubert* hearing in which evidentiary and testimonial records were reviewed, and he ruled that the opinion of the plaintiff's investigator was not admissible under *Rule 702*. The judge concluded that the plaintiff's investigator did not provide sufficient reliable evidence to support the methodology for investigating the cause of the fire. The judge also noted that the comprehensive nature of *NFPA 921* contained a methodology that could have supported the opinion and could have been utilized as a basis for testing the hypothesis for the fire's cause.

On December 8, 1999, a circuit court judge in the State of Michigan granted a plaintiff's motion *in limine* to exclude the testimony of the defendant's expert witness (*Ronald Taepke v. Lake States Insurance Company* 1999). In testimony and written reports, the defendant's expert admitted the authoritative nature of *NFPA 921* yet admitted that he did not follow its accepted methodology. Furthermore, the methodology used to conclude that the fire was arson is not recognized in any reliable source of literature in the fire investigation field. In other areas, the expert admitted that he speculated that high-temperature accelerants were used without a reliable evidentiary basis for the opinion.

In a federal civil case opinion issued on March 28, 2002, involving automobile ignition switch fires, a federal judge examined the proposed use of a statistical vehicle fire loss database in litigation. The judge also examined in detail the importance of reliance on *NFPA 921* in fire investigations when forming a hypothesis using the scientific method (*Snodgrass et al. v. Ford Motor Company and United Technologies Automotive, Inc.* 2002). Note that since this court case was within the Third Circuit, a determination of whether an expert's methodology was reliable used the combined *Daubert* and *Downing* factors.

In *United States of America v. Downing* (1985), the court considered the admissibility of expert testimony concerning the reliability of eyewitness identification. The court held that in appropriate cases, an eyewitness expert can indeed assist the trier of fact, and thus the expert's testimony is admissible under the *Federal Rules of Evidence*. An expert's opinion testimony is one of "helpfulness" to the jury, because many of the factors the expert will testify to go beyond common knowledge and at times may directly contradict common sense and may be able to refute jurors' erroneous assumptions about eyewitness reliability.

In a state court of appeals opinion decided on June 16, 2005, the court ruled that the testimony of the appellee's independent fire investigator was correctly permitted by the trial court under *Daubert* as to the origin and cause of the fire (*Abon Ltd. et al. v. Transcontinental Insurance Company* 2005). The court noted that the independent investigator followed the *NFPA 921* methods and principles to reach his opinion as to the incendiary nature of the fire's origin and cause.

In a fatal apartment fire on October 23, 2000, personal representatives and relatives of the decedent brought an action against the manufacturer of an electric blanket, alleging it had a defective safety circuit that caused the deadly fire (*David Bryte v. American Household Inc.* 2005). The fire marshal who determined the origin and cause of the fire was not certified as a fire investigator, yet he had attended training in the fire and explosion investigation field. He traced the fire's path to the deceased's left side, took photographs, made a fire scene sketch, and took oral statements from witnesses. While he did not inspect the electrical wiring or take any evidence, he noted an electrical cord draped across the victim's body and concluded the cause of the fire to be the improper use of an electric blanket. During the trial on December 8, 2003, a *Daubert* analysis was conducted, and the court disallowed the fire marshal and another expert from providing expert testimony because they did not have a sufficiently reliable basis for their testimony.

In a federal court of appeals ruling in a product liability case concerning a fire-related death, both the magistrate judge and district court concluded that pursuant to the *Daubert* and *Federal Rules of Evidence* standards, the methodologies employed by the plaintiff's expert witnesses were too unreliable to serve as the basis for admissible testimony (*Bethie Pride v. BIC Corporation, Société BIC, S.A.* 2000). The case alleged that the death of a 60-year-old man who sustained burns on 95 percent of his body was caused by an explosion of a disposable lighter he was carrying while inspecting a pipe behind his house. The court noted that the plaintiff's experts failed to conduct timely, replicable laboratory experiments demonstrating how the explosion was consistent with a manufacturing defect in the lighter. One of the plaintiff's experts testified he began a laboratory experiment to test his hypothesis but "chickened out and shut the experiment down."

In a federal case in which an insurance company attempted to recover in subrogation for fire damage to the property of one of its insureds, the defendant construction company filed a motion to preclude the trial testimony of the plaintiff's fire expert on origin and cause (*Royal Insurance Company of America as Subrogee of Patrick and Linda Magee v. Joseph Daniel Construction, Inc.* 2002). The plaintiff's insureds had hired a construction company to work on a garage on their property on December 14, 1998. The work involved the use of an acetylene torch to install new beams above the second floor. During the operation of the torch, two fires occurred and were extinguished. Approximately 14 hours after the construction company left, a fire was discovered in the garage, which sustained damage.

Approximately one year after the fire, the plaintiffs hired an expert to investigate the origin and cause of the fire. Based on his investigation, the expert concluded that the fire was caused by the careless use of the welding and cutting equipment by the construction company employees. The court noted that the expert conducted the investigation in accordance with methodology set forth by *NFPA 921*. To develop his hypothesis that the garage fire occurred as a result of molten slag dropped by careless construction workers, the investigator used data consisting of information gleaned from copies of photographs, insurance investigative and engineering reports, reviews of depositions, and his own witness interviews. He also eliminated other common causes for the fire. Citing the *Daubert* standard, the court ruled that the expert's testimony was relevant in establishing a direct relationship between the proffered theory concerning the carelessness of the defendant with regard to molten slag and the cause of the fire.

An unpublished case from the U.S. Court of Appeals for the Third Circuit concluded that a district court did not err in granting and excluding the testimony of a plaintiff's expert witness (*State Farm and Casualty Company as subrogee of Rocky Mountain and Suzanne Mountain v. Holmes Products; J.C. Penney Company, Inc.* 2006). On April 5, 1999, a fire destroyed a residence. The insurance company's investigator placed the fire's area of origin at a halogen lamp that lacked a safety guard to prevent contact between the high-temperature portions of the lamp and nearby curtains, clothing, or other flammable materials. After eliminating all other potential causes, the investigator concluded that the fire was caused by nearby draperies' coming into contact with the defective energized lamp. The investigator further hypothesized that the dog belonging to the residence's owners might have accidentally pulled the window draperies over the energized lamp or knocked the lamp into contact with the draperies. The judge concluded that the fire investigator's testimony did not satisfy *Daubert* because the conclusion on causation was based on assumptions and was not supported by any methodology.

In a subrogation lawsuit, an insurance company brought an action against a clothes dryer manufacturer to recover damages in connection with 23 fires allegedly caused by a design defect (*Travelers Property and Casualty Corporation v. General Electric Company* 2001). A *Daubert* hearing was necessary when the defendant manufacturer moved to prevent the testimony of the plaintiff's expert, who had written and issued a three-page report alleging that accumulated lint in an undetectable location could be ignited by the dryer's heating elements. The judge overruled the motion, finding that the expert's proffered opinion testimony met the requirements of both *Rule 702* and *Daubert,* that the opinion was consistent with the principles of *NFPA 921* and the scientific method, and the expert's qualifications were sufficient. However, the judge found that the plaintiff's three-page expert disclosure report amounted to bad faith, and he ordered the plaintiff to reimburse the defendant for one-third of its costs and expenses for taking the first 12 days of the expert's deposition and permitted up to 2 more days of depositions.

Methodological Explanations for Burn Patterns   In the case of an April 23, 1997, building fire, a federal magistrate denied the plaintiff's motion to bar opinion testimony at the time of trial as to the origin and cause of the fire (*Eid Abu-Hashish and Sheam Abu-Hashish v. Scottsdale Insurance Company* 2000). In this case, both a municipal fire department investigator and a private insurance investigator concluded from their examinations of the scene that the fire was incendiary in origin. No physical evidence was taken for laboratory examination.

The plaintiff argued that these investigators were not reliable and their testimonies were inadmissible under *Rule 702* because they did not use the scientific method as outlined in *NFPA 921* and relied only on physical evidence observed at the scene. The case also contained parallels with the *Benfield* case (*Michigan Miller's Mutual Insurance Company v. Benfield* 1998). In denying the plaintiff's motion, the magistrate noted that the investigators were able to provide an adequate methodological explanation for the analysis of burn patterns that led to how they reached their conclusion as to the fire's incendiary origin (Figure 1-4).

FIGURE 1-4 Investigators should be able to provide adequate and methodical explanations of the fire pattern analysis methods they use to reach conclusions as to the origin of a fire. *Courtesy of D. J. Icove.*

In a product liability case arising from a fire on October 16, 2000, which destroyed a video rental store, two of the plaintiff's fire causation experts determined the origin and cause of the fire to be a copier machine (*Fireman's Fund Insurance Company v. Canon U.S.A., Inc.* 2005). The two experts relied on burn patterns inside the copier and testing of an exemplar heater assembly for the copier. A federal appeals court upheld the district court's ruling that documentation of the experimentation by the plaintiff's experts did not meet the standards of *NFPA 921* and was unreliable and potentially confusing to a jury. The court found that tests must be carefully designed to replicate actual conditions, conducted properly, and documented thoroughly.

A state civil court opinion and order entered February 1, 2006, precluded *in limine* the testimony of the defendant's expert, who proposed to testify regarding the incendiary nature of a fire he investigated (*Marilyn McCarver v. Farm Bureau General Insurance Company* 2006). On August 16, 2003, a fire damaged the plaintiff's residence. A report authored August 23, 2003, by the defendant insurance company's expert concluded that the fire was incendiary in origin after visiting and viewing the remains of the residence, taking photographs, taking samples for laboratory examination, and performing other techniques. The state court noted that although the defendant's expert stated his investigation complied with *NFPA 921*, he did not follow procedure in several areas. The expert failed to document any depth-of-char physical measurements; failed to document his vector analysis, which formed the basis for flame patterns; and failed to retrieve samples submitted for chemical testing, which resulted in their destruction. After a *Daubert* hearing, the court concluded that a fatal flaw in the expert's failure to collect and record depth-of-char data undermined his methodology for determining the origin and cause of the fire.

Methodology and Qualifications In the case of a July 1, 1998, residential fire, which resulted in a federal lawsuit for product liability, the judge granted the defendant's motion to exclude the testimony of the electrical engineer citing *Daubert*, *Rule 702*, and

Rule 704 (*American Family Insurance Group v. JVC America Corp.* 2001). In this case, an insurance company investigator called on an electrical engineer to remove and examine the charred remains of a bathroom exhaust fan, clock, lamp, timer, compact disc player, computer with printer and monitor, ceiling fan, power receptacle, and power strip. The engineer came to an opinion that a defect in the compact disc player caused the fire. His observations were based on burn patterns in the room, on appliance remains, and on his experience, education, and training. In his ruling, the judge noted that the training and experience of the engineer did not qualify him to offer an analysis of burn patterns and theory of fire origin. Furthermore, the judge noted that the engineer did not use the scientific method recommended by *NFPA 921* to form a hypothesis from the analysis of the data, nor did he satisfy the requirements for expert testimony under *Daubert*.

A March 1, 2001, fire originating in a building's janitorial closet was alleged to have been caused by a short circuit in a contaminated electrical bus duct (*103 Investors I, L.P. v. Square D Company* 2005). The plaintiff retained a licensed mechanical engineer who was also a certified fire investigator to investigate the origin and cause of the fire. The engineer testified that he investigated the fire in accordance with the scientific method protocol set forth in *NFPA 921*. In an opinion dated May 10, 2005, a federal judge ruled that the engineer's testimony could include that the short circuit resulted from the presence of contaminants. However, since the engineer failed to follow established methods of inquiry for hypothesis testing of how the contamination actually occurred, he could not testify that the fire specifically resulted from short circuiting in the bus duct.

In a case before a federal magistrate, a third-party defendant's motion to exclude the testimony of the plaintiff's expert was denied (*TNT Road Company et al. v. Sterling Truck Corporation, Sterling Truck Corporation, Third-Party Plaintiff v. Lear Corporation, Third-Party Defendant* 2004). The third-party defendant alleged that the witness was not qualified to render an expert opinion, lacked a college degree, was not a certified fire investigator, was not a licensed private investigator, did not interview every conceivable witness, quickly came to his conclusion, relied heavily on theory and circumstantial evidence, had testified only once as an expert witness, and that his investigation was too faulty to be admissible under *FRE Rule 702*. After significant review, the court ruled that the plaintiff's witness was indeed qualified to present expert testimony on the origin and cause of the vehicle fire.

The magistrate noted that the witness had demonstrated sufficient knowledge, skill, training, and education to qualify as an expert and that his methodology appeared reliable. The judge decided that the witness's investigation substantially complied with *NFPA 921* in that with the vehicle's owners he had systematically examined and photographed the scene along with all the components that may have been potential causes of the fire, obtained and reviewed the maintenance records, systematically removed and photographed the fire debris, formed no immediate opinions as to the cause of the fire, and narrowed his focus onto a suspected ignition switch only after several hours. Once he suspected the switch may have been the cause of the fire, the expert suspended his investigation to preserve the evidence until other interested parties could be available to also inspect the vehicle.

On February 15, 2001, a fire broke out in the kitchen of a residence in the vicinity of a coffeemaker, causing extensive fire, smoke, and water damage. The insurance company paid the claim and pursued subrogation against the manufacturer of the coffeemaker (*Vigilant Insurance v. Sunbeam Corporation* 2005). The testimonies of several witnesses were limited based on a *Daubert* hearing on March 1, 2004. The testimony of the fire investigator was found inadmissible under *Rule 702* because his opinion was not based on data that he might have gleaned from testing the coffeemaker and toaster cords, safety devices, or the fluorescent light. He was able to testify under *Rule 701* as a lay witness as to his personal observations that the coffeemaker was at the base of the V burn pattern. During the hearing, the court found that another witness's burn test was simply a reenactment of the kitchen fire and was not substantially similar owing to differences in the heat applied to the coffeemaker, the amount of water in the coffeepot, and the thickness of the countertop. A jury trial held from March 2, 2005, to March 17, 2005, resulted in a verdict in favor of the plaintiff's insurance company.

In a federal civil lawsuit, the plaintiffs sought to hold the defendant, the manufacturer of a freezer, liable for damages stemming from a fire that allegedly originated in the freezer (*Mark and Dian Workman v. AB Electrolux Corporation et al.* 2005). Despite using *NFPA 921* to guide his investigative process, the investigator for the plaintiff's insurance company failed to determine what caused the fire. The investigator then arranged for a mechanical engineer, who conducted a further destructive examination and determined that the fire originated inside the freezer as the result of a short circuit in the evaporator compartment. The court's review of the engineer's testimony based on the criteria set forth in *Daubert* and *Rule 702* found the engineer's testimony reliable. However, the engineer did not mention or describe a defect in the freezer.

In a federal civil lawsuit, a June 1, 2001, fire damaged the home, garage, and property of the plaintiffs (*Theresa M. Zeigler, individually; and Theresa M. Zeigler, as mother and next friend of Madisen Zeigler v. Fisher-Price, Inc.* 2003). The fire allegedly originated in a toy vehicle parked in the garage and plugged into a charger. Two expert witnesses looked into the causation of the fire. The court determined that the insurance company investigator followed *NFPA 921* and an appropriate and generally accepted methodology, and found that his opinions were reliable. The second investigator was an engineer who did not perform an origin and cause analysis but had an opinion that the fire's ignition source was located on a connector to the toy vehicle. The court ruled that the engineer could not give an opinion as to the manufacturer's record keeping, or the origin or cause of the fire.

AUTHORITATIVE SCIENTIFIC TESTING

The *Daubert* criteria list "testability" as one of the primary considerations in evaluating and demonstrating the reliability of a scientific theory or technique. In fire investigations, the scientific theory relies heavily on established and proven aspects of nature, such as those demonstrated in the science of fire dynamics. The combination of generally accepted scientific principles and research, when supported by empirical fire testing, can yield valuable insight into fire scene reconstruction. Much of the accumulated knowledge of fire scene reconstruction is impossible for individual investigators to gain from personal experience alone, regardless of the skill and diligence with which their analyses are conducted.

There are cases demonstrating that reliance on tests conducted prior to the event is less likely to be biased. Some experts develop opinions based solely on the results of tests conducted specifically to support expert testimony. The *Daubert II* (*Daubert v. Merrell Dow Pharmaceuticals Inc.* 1995) court placed greater weight on testimony based on preexisting research that used the scientific method, as it is considered more reliable (Clifford 2008).

Scientifically conducted test results can normally be reproduced and their error rates validated or estimated. Such results are considered to be *objective*, neither prejudiced nor biased by the researcher. An example of peer-reviewed published guidelines is incorporated in the accepted fire testing procedures developed and maintained by the Committees on Fire Standards (E05) and Forensic Sciences (E30) of ASTM International.

It is important to understand the hierarchy of the interrelationships of these ASTM standards and guidelines, particularly with *NFPA 921*. Table 1-5 lists the main ASTM testing and forensic standards applicable to fire investigations, particularly when civil or criminal litigation is anticipated. The ASTM standards form a hierarchy, as shown in Figure 1-5. The capstone standard is *ASTM E1188-11* (ASTM 2011b). Shown under that overarching standard are the physical standards on the left-hand side and the

TABLE 1-5	Testing Standards Applicable to Fire Investigations	
TECHNIQUE	**INSTRUCTION**	**STANDARD**
Reporting	Standard Practice for Reporting Opinions of Scientific or Technical Experts	ASTM E620
	Standard Practice for Evaluation of Scientific or Technical Data	ASTM E678
	Standard Practice for Examining and Preparing Items That Are or May Become Involved in Criminal or Civil Litigation	ASTM E860
	Standard Practice for Reporting Incidents That May Involve Criminal or Civil Litigation	ASTM E1020
	Terminology of Technical Aspects of Products Liability Litigation	ASTM E1138
	Standard Practice for Collection and Preservation of Information and Physical Items by a Technical Investigator	ASTM E1188
	Standard Guide for Physical Evidence Labeling and Related Documentation	ASTM E1459
	Standard Practice for Receiving, Documenting, Storing, and Retrieving Evidence in a Forensic Science Laboratory	ASTM E1492
Laboratory Testing	Debris Samples by Steam Distillation	ASTM E1385
	Debris Samples by Solvent Extraction	ASTM E1389
	Debris Samples by Gas Chromatography	ASTM E1387
	Debris Samples by Headspace Vapors	ASTM E1388
	Debris Samples by Passive Headspace	ASTM E1412
	Debris Samples by GC-MS	ASTM E1618
Flash and Fire Point	Tag Closed Tester	ASTM D56
	Cleveland Open Cup	ASTM D92
	Pensky-Martens Closed Tester	ASTM D93
	Tag Open Cup Apparatus	ASTM D1310
	Setaflash Closed Tester	ASTM D3278
	Ignition Temperature of Plastic	ASTM D1929
Autoignition Temperature	Liquid Chemicals	ASTM E659
Heat of Combustion	Hydrocarbon Fuels by Bomb Calorimeter	ASTM D2382
Flammability	Apparel Textiles	ASTM D1230
	Apparel Fabrics by Semi-restraint Method	ASTM D3659
	Finished Textile Floor Covering	ASTM D2859
	Aerosol Products	ASTM D3065
	Chemical Concentration Limits	ASTM E681
	Standard Methods of Fire Tests for Flame Propagation of Textiles and Films	NFPA 701
	Recommended Practice for a Field Flame Test for Textiles and Films	NFPA 705
Cigarette Ignition	Mockup Upholstered Furniture Assemblies	ASTM E1352
	Upholstered Furniture	ASTM E1353
Surface Burning	Building Materials	ASTM E84

(continued)

TECHNIQUE	INSTRUCTION	STANDARD
Fire Tests and Experiments	Roof Coverings	ASTM E108
	Floor/Ceiling, Floor/Roof; Walls, Columns	ASTM E119
	Room Fire Experiments	ASTM E603
	Measurement of Gases Present or Generated	ASTM E800
	Windows	ASTM 2010
	Doors	ASTM 2074
Critical Radiant Flux	Floor Covering Systems	ASTM E648
Release Rates	Heat and Visible Smoke	ASTM E906
	Heat and Visible Smoke Using an Oxygen Consumption Calorimeter	ASTM E1354
Rate of Pressure Rise	Combustible Dusts	ASTM E1226
Electrical	Dielectric Withstand Voltage	ASTM D495
	Insulation Resistance	MILSTD202F
	Investigation of Electrical Incidents	ASTM E2345
Self-Heating	Spontaneous Heating Values of Liquids and Solids (Differential Mackey Test)	ASTM D3523 UN N.1, N.4

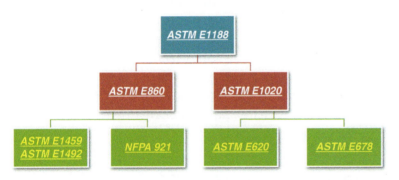

FIGURE 1-5 ASTM standards cited when civil or criminal litigation is anticipated. *D. J. Icove and B. P. Henry. Expert Report Writing: Best Practices for Producing Quality Reports. ISFI 2010, International Symposium on Fire Investigation Science and Technology, University of Maryland, College Park, Maryland, September 27–29, 2010.*

informational standards on the right-hand side. The number appended to the right of each standard indicates its year of last issuance. For example, *ASTM E620-11* was last issued in 2011.

ASTM E1188-11 (Standard Practice for Collection and Preservation of Information and Physical Items by a Technical Investigator) presents an outline of evidence documentation and preservation that is reflected in good crime scene practices that establish a defensible chain of custody for any physical evidence recovered.

The first-tier physical evidence standard on the left is *ASTM E860-07* (ASTM 2007b). It governs the second-tier standards *ASTM E1459-92(2005)* (ASTM 2005b), *ASTM E1492-11* (ASTM 2011c), and *NFPA 921: Guide for Fire and Explosion Investigations* (NFPA 2011).

ASTM E860-07 (Standard Practice for Examining and Preparing Items That Are or May Become Involved in Criminal or Civil Litigation) is the leading standard practice for examining and testing items that are or may become involved in litigation. The standard addresses actions that need to be taken if the testing methods will alter or destroy the evidence in any way. Such destructive testing may limit how additional testing may occur. It recommends documenting the condition of the evidence prior to and after any examination or testing and establishing the chain of custody. If proposed tests are expected to alter the item, procedures are outlined for notifying the client and other interested parties, giving them the opportunity to respond or attend the proposed tests.

ASTM E1459-92(2005) (*Standard Guide for Physical Evidence Labeling and Related Documentation*) (ASTM 2005b) sets procedures for producing a traceable audit trail so that any item of physical evidence will have documentation of its origin, past history, treatment, and analysis. The guide also describes methods for labeling physical evidence collected during field investigations; as it is received in a forensic laboratory; or derived from other items submitted for laboratory examination. The guide points out that some physical evidence may be hazardous; therefore, personnel handling this evidence must be properly trained and outfitted to handle this task.

ASTM E1492-11 (*Standard Practice for Receiving, Documenting, Storing, and Retrieving Evidence in a Forensic Laboratory*) (ASTM 2011c) describes the appropriate procedures and techniques that must be in place in a forensic laboratory to protect and document the integrity of physical evidence.

The information standard *ASTM E1020-06* (ASTM 2006) governs *ASTM E620-11* (ASTM 2011a) and *ASTM E678-07* (ASTM 2007a). Thorough documentation of all phases of testing will help the courts evaluate the reliability and admissibility of the test results.

ASTM E1020-96(2006) (*Standard Practice for Reporting Incidents That May Involve Criminal or Civil Litigation*) is a guide to the information that should be included in reports of accidents or events that may become the subject of an investigation or litigation. Its provisions are very similar to the content of fire incident report forms such as those reprinted in Appendix H of *Kirk's Fire Investigation* (DeHaan and Icove 2012).

ASTM E620-11 (*Standard Practice for Reporting Opinions of Scientific or Technical Experts*) outlines the content that should be included in a typical report of a technical expert: names and addresses of authors, descriptions of items examined, date and location of exam, scope of activities performed, pertinent facts relied on and sources for them, all opinions and conclusions rendered, logic and reasoning of the expert by which each of the conclusions was reached, and signature of the expert with affiliations. These requirements follow *FRCP Rule 26(a)*.

ASTM E678-07 (*Standard Practice for Evaluation of Scientific or Technical Data*) is essentially a description of the scientific method. It recommends including the definition of the problem or issue addressed, identification and explanation of hypotheses addressed (including alternatives), a description of the data collected and analyzed (and sources for that data), an estimation of the reliability of data used, and the opinion reached. Conclusions expressed must be consistent with known facts and accepted scientific principles.

PEER REVIEW AND PUBLICATIONS

A credible, reliable theory must take into account the body of research that has been compiled, verified, and published by experts in the field. The need for credibility underscores the advantages of using the scientific method. Recently, there has been a tendency for courts to hold experts to the same standards that scientists use in evaluating each other's work, sometimes referred to as peer review (Ayala and Black 1993).

Peer review is defined in *NFPA 921*, 2011 ed., pt. 4.6.3 (NFPA 2011), as

> a formal procedure generally employed in prepublication review of scientific or technical documents and screening of grant applications by research-sponsoring agencies. Peer review carries with it connotations of both independence and objectivity. Peer reviewers should not have any interest in the outcome of the review. The author does not select the reviewers, and reviews are often conducted anonymously. As such, the term "peer review" should not be applied to reviews of an investigator's work by coworkers, supervisors, or investigators from agencies conducting investigations of the same incident. Such reviews are more appropriately characterized as "technical reviews, . . .

Peer review is common among respected journals that publish the results of theories, testing, and methodology. The common approach to peer review in technical journals is to use reviewers who are quite familiar with the published scientific data and methodologies to determine whether the author's views are of the quality of these publications and whether the data presented support the conclusions of the author. Peer-reviewed journals useful in the field of forensic fire investigation include *Fire Technology, Journal of Fire Protection Engineering, Fire and Materials, Fire Safety Journal, Combustion and Flame,* and *Journal of Forensic Sciences.* Another valuable resource is the *Fire and Arson Investigator,* a journal produced by the International Association of Arson Investigators.

The SFPE's *Guidelines for Peer Review in the Fire Protection Design Process* (2002) cover the scope and standard of care related to a peer review, the confidentiality of the results of a peer review, and the manner in which the results should be reported. Although these guidelines concentrate on the design of fire protection systems, they include concepts with significant parallels to the fire investigation field.

Investigators can participate in a technical peer review when their cases are submitted to supervisors for review (*NFPA 921,* 2011 ed., pt. 4.6.2). *NFPA 921* cautions that a technical review can have multiple facets and that the reviewer should be qualified and familiar with all aspects of proper fire investigation and have access to all the documentation available to the investigator whose work is being reviewed (NFPA 2011).

In cases in which a technical reviewer is asked to critique only specific aspects of the investigator's work product, then the technical reviewer should be qualified and familiar with those specific aspects. For example, an engineer may be asked to conduct a technical peer review of another engineer's report of investigation. In some states this type of technical peer review is considered the practice of engineering and may require registration as a licensed professional engineer.

NFPA 921 cautions that although technical review may add value to an investigation, there are limitations concerning reviewers who might have a bias or an interest in the outcome of the case. These interests, conscious or subconscious, may introduce confirmation bias (*NFPA 921,* 2011 ed., pt. 4.3.9) or expectation bias (pt. 4.3.8) (NFPA 2011). Any one of these biases may result in the failure of the both the investigator and the reviewer to consider alternative hypotheses, thus undermining the case under review. In law enforcement, the primary function of supervisory review is to assure that all questions, logical investigative leads, laboratory examinations, and plausible theories are addressed. Private investigation companies working in civil cases often have their supervisors peer-review fire reports. Some corporations use qualified experts to peer-review fire investigation reports when their companies are too small or do not have an established peer-review process. Other equally qualified employees in the same agency or company often carry out peer review.

In these instances, it must be stressed that the initial investigator still has the primary responsibility to collect and report relevant information in a manner consistent with the scientific method. The investigator, not the peer reviewer, will still be the primary witness in any litigation.

Peer review also applies to establishing the credibility of published information (Icove and Haynes 2007). Several peer-reviewed textbooks and references are recommended for use in both conducting and assessing complex fire investigations. These texts are written by prominent credentialed fire scientists and engineers, are widely cited in scholarly publications, are used and distributed in the fire investigation community, are included in association publication lists, are cited in court cases, and have often been reviewed by forensic journals. Although not totally inclusive, Table 1-6 lists several peer-reviewed reference documents that are considered expert treatises in the fire investigation field.

These texts have become or have the potential to become expert or learned treatises in the field of fire investigation. Learned treatises are those texts that rise to the level of authoritative acceptance, so that they can be admitted as evidence in court to support or rebut the

TABLE 1-6	**Expert Peer-Reviewed Treatises in the Fire Investigation Field**

- Babrauskas, V. 2003. *Ignition Handbook: Principles and Applications to Fire Safety Engineering, Fire Investigation, Risk Management and Forensic Science*. Issaquah, WA: Fire Science Publishers, Society of Fire Protection Engineers.

- Babrauskas, V., & Grayson, S. J. 1992. *Heat Release in Fires*. Basingstoke, UK: Taylor and Francis.

- Beveridge, A. D. 2012. *Forensic Investigation of Explosions*, 2nd ed. CRC Press/Taylor and Francis, Boca Raton, FL.

- Brannigan, F. L., & Corbett, G. P. 2007. *Brannigan's Building Construction for the Fire Service*, 4th ed. Sudbury, MA: National Fire Protection Association; Jones and Bartlett.

- Cole, L. S. 2001. *The Investigation of Motor Vehicle Fires*, 4th ed. San Anselmo, CA: Lee Books.

- Cooke, R. A., & Ide, R. H. 1985. *Principles of Fire Investigation*. Leicester, UK: Institution of Fire Engineers.

- DeHaan, J. D., & Icove, D. J. 2012. *Kirk's Fire Investigation*, 7th ed. Upper Saddle River, NJ: Pearson-Prentice Hall.

- Drysdale, D. 2011. *An Introduction to Fire Dynamics*, 3rd ed. Chichester, West Sussex, UK: Wiley.

- Icove, D. J., & DeHaan, J. D. 2009. *Forensic Fire Scene Reconstruction*, 2nd ed. Upper Saddle River, N.J.: Pearson/Prentice Hall.

- Icove, D. J., Wherry, V. B., & Schroeder, J. D. 1998. *Combating Arson-for-Profit: Advanced Techniques for Investigators*, 2nd ed. Columbus, OH: Battelle Press.

- Iqbal, N., & Salley, M. H. 2004. *Fire Dynamics Tools (FDTs): Quantitative Fire Hazard Analysis Methods for the U.S. Nuclear Regulatory Commission Fire Protection Inspection Program*. Washington, DC: U.S. Nuclear Regulatory Commission.

- Janssens, M. L., & Birk, D. M. 2000. *An Introduction to Mathematical Fire Modeling*, 2nd ed. Lancaster, PA: Technomic.

- Karlsson, B., & Quintiere, J. G. 2000. *Enclosure Fire Dynamics*. Boca Raton, FL: CRC Press.

- Quintiere, J. G. 1998. *Principles of Fire Behavior*. Albany, NY: Delmar.

- Quintiere, J. G. 2006. *Fundamentals of Fire Phenomena*. Chichester, West Sussex, UK: Wiley.

testimony given by an expert witness. Furthermore, authoritative references are those that the fire investigator frequently uses as a reference or guide because they have been found to be reliable. When asked in court what they consider to be an authoritative text, some fire investigators will list only those that have a legal or administrative impact on their work.

NEGATIVE CORPUS

The use of the *negative corpus* or *arson by default* approach (the process of ruling out all accidental causes for a fire without sufficient scientific and factual basis to determine what did cause the fire) is rarely an acceptable methodology for determining that a fire was intentionally set. The term *negative corpus* has no recognized definition but its use is obvious. It *implies* that the **corpus delicti** of the event (not necessarily of a crime) has not been proven.

NFPA 921 first mentions negative corpus in its 2011 edition as follows:

Rendering an opinion in the absence of any evidence of fire cause, or if there is only ambiguous evidence to support it, is known as "negative corpus" and does not follow the Scientific Method. The process of eliminating all other fire causes and thereby determining that the fire cause classification must be incendiary, accidental, or natural does not conform to the Scientific Method, is inappropriate and should not be done. In such a case, the only appropriate fire cause classification is undetermined. (*NFPA 921*, 2011 ed., pt. 18.6.5.2) (NFPA 2011)

corpus delicti ■ Literally, the body of the crime. The fundamental facts necessary to prove the commission of a crime.

The logic of negative corpus cause determination was examined in great detail by Smith (2006). He pointed out that if it has any application at all, it is only where the origin is "clearly defined" (*NFPA 921*) or is identified as the "exact point of origin" (DeHaan and Icove 2012).

The U.S. Court of Appeals for the Eleventh Circuit applied *Daubert* and excluded the testimony of a fire investigator in the *Benfield* case (*Michigan Miller's Mutual Insurance Company v. Benfield* 1998). In that case, the investigator was not able to articulate the methodology he used to eliminate possible ignition sources and had no scientific basis for his opinion. The court held that the investigation of fires is science-based and that the *Daubert* criterion applies.

It is argued that if the fire damage is severe, the surrounding materials will be too degraded, and a clearly defined point of origin cannot be determined. This may be true when postfire burn patterns are the entirety of the evidence of origin location. However, when a reliable witness or several witnesses observe an established fire in only one area and can view other nearby areas and reliably conclude there was no observable fire there, then that should be taken as strong evidence that an ignition occurred in that location. Sometimes, color and density of smoke or height of flames will offer a clue as to the nature of the first fuel package involved, further narrowing the area of interest. In many fires that have caused great destruction, surveillance cameras have recorded the area, or even the point of first ignition. Once a narrow area of origin can be established, then possible sources of heat can be identified. An origin determination is a hypothesis that requires testing, like any other inductive conclusion. This testing involves careful physical examination of the area for remains of either fuel packages or processes known to have been in that vicinity.

Before possible ignition sources are blindly pursued only within a defined area, the scientific method demands testing by posing and testing alternatives. Such efforts could take the form of asking, This looks like the origin, but I have a heat source over *there*. Could the damage I'm seeing (or the events a witness saw) have been the result of something (wall covering, drapery, cabinet) igniting from *that* and falling into *this* area? Smith (2006) points out that the better the knowledge base of the investigator, the more likely he or she will be to see (and test) these alternatives. The poorly prepared investigator is far more likely to seize on an apparent point of origin based on burn pattern and then look for heat sources only in that area, never considering that his or her hypothesis concerning the area of origin is wrong.

All aspects of origin and cause determination should be subjected to comprehensive and repetitive analysis by hypothesis testing by all means available, including the formation and testing of alternatives—both consistent and contrary ones. The **cause** of a fire is a determination of the first fuel ignited (not necessarily an identification of the cause), the source of heat, and the circumstances that brought them together. The determination of first fuel ignited can rely on laboratory tests (ignitable liquid or chemical incendiary residues), physical remains, or witness observations. In a criminal case, where the standard of proof is very high, circumstantial evidence of first fuel or even heat source may not be adequate. In a civil case, where the standard of proof may be "preponderance of evidence" or "more likely than not" (*NFPA 921*, 2011 ed., pt. 11.5.5), circumstantial exclusion of all but one competent ignition source and one susceptible fuel may be adequate (NFPA 2011). The tests for "competent" and "susceptible" are part of the fire engineering analysis conducted as part of the scientific method of fire investigation.

Presently, the only reference to negative corpus appears in *NFPA 921*, 2011 ed., pt. 18.6.5.2 (NFPA 2011). If pursued vigorously and within bounded known experimental limits, the scientific method can be used to demonstrate successfully that the only mechanism for ignition had to be deliberate by demonstrating that all relevant accidental mechanisms were specifically evaluated, tested, and eliminated, and that deliberate ignition (and its expected aftermath) fits all the available data. If new data are presented, the

cause ■ The circumstances, conditions, or agencies that brought about or resulted in the fire or explosion incident, damage to property resulting from the fire or explosion incident, or bodily injury or loss of life resulting from the fire or explosion incident (*NFPA 921,* 2011 ed., pt. 3.3.22).

conclusion must be reevaluated, possibly resulting in a different conclusion. For a more detailed discussion on negative corpus, see Smith (2006).

ERROR RATES, PROFESSIONAL STANDARDS, AND ACCEPTABILITY

Under *Daubert,* a court may consider the known or potential rate of error and the existence of acceptable professional standards for the techniques used by the expert. Error rates from repeated tests are available for many equations, relationships, and models used to describe fire and explosion dynamics. These error rates are evaluated during fire test development, such as those listed in Table 1-5. Error rates have not yet been established for the entire process of origin and cause determination.

ASTM E860-07 (Standard Practice for Examining and Preparing Items That Are or May Become Involved in Criminal or Civil Litigation) has a broad impact on investigators. This practice covers evidence (actual items or systems) that may have future potential for testing or disassembly and are involved in litigation, including:

■ Documentation of evidence prior to removal and/or disassembly, testing, or alteration
■ Notification of all parties involved
■ Proper preservation of evidence after testing

This practice also stresses the importance of safety concerns associated with testing and disassembly of evidence, which is particularly important when dealing with energized equipment or evidence containing potentially hazardous chemicals. The ASTM E30 committee has jurisdiction over these standards and is presently revising them to make them suitable for all litigation, both civil and criminal.

Forensic Fire Scene Reconstruction

Forensic fire scene reconstruction extends beyond the systematic evaluation of the scene by using the scientific method in conducting a comprehensive review of the fire pattern damage, witness accounts, and evidence supporting human activities. According to *NFPA 921,* 2011 ed., pt. 3.3.60 (NFPA 2011), *fire scene reconstruction* is defined as

> The process of recreating the physical scene during fire scene analysis through the removal of debris and the replacement of contents of structural elements in their pre-fire positions.

Specifically, the *forensic* fire scene reconstruction approach is much more comprehensive than that described in *NFPA 921,* as it goes well beyond the concept of merely physically placing the items back into the scene. The scientific method mandates the testing of hypotheses by applying appropriate tools for their evaluation. This includes not only the testing of the physical processes that create visible fire patterns but also the many methods of applying fire engineering and fire dynamics principles.

These principles are well documented in the peer-reviewed literature, often much more so than are the mechanisms that produce the postfire physical effects on which investigators regularly rely. If sufficient, accurate, reliable detail is available to create reasonable hypotheses to test, there is no reason that analyses to predict layer temperatures, plume heights, flame spread, toxic gas concentrations, smoke movement, and other similar factors should not be carried out. In fact, the scientific method demands that such conceptual tests be carried out where appropriate. Fire investigators have long applied some of the "rules" expressed by the engineering relationships to origin and cause determinations but have lacked the specialist knowledge to fully apply them accurately with regard to important variables and constraints. These steps may be taken in different sequences when conditions require. Some may be conducted in parallel with others; yet

TABLE 1-7	Steps in Forensic Fire Scene Reconstruction

- **Step 1**—Document the fire growth, burn pattern indicators, and processing of the scene.
- **Step 2**—Evaluate the heat transfer damage by fire plumes and gases.
- **Step 3**—Establish the starting conditions.
- **Step 4**—Correlate the human observations and factors.
- **Step 5**—Conduct a fire engineering analysis.
- **Step 6**—Formulate and evaluate conclusions.

others form an iterative process in which the results of one step may require repetition of earlier ones.

The applicability of using the science of modern fire dynamics to fire scene investigations has been advanced over the years. The impact is documented in key textbooks and articles by Babrauskas (1997), Drysdale (2011), Grosselin (1998), Lilley (1995), and Nelson (1987).

The steps in Table 1-7 form the basis for the remaining portion of this textbook and are illustrated by example and references. In cases where suitable fire patterns or damage exist, investigators can often accurately determine the area and point of origin of the fire after extinguishment. In this process, the investigator traces the impact of fire plumes on the fire's area, growth, development, and direction of travel, taking into account the identifiable fire patterns of smoke deposits and damage due to heat transfer. Other areas of concern taken into consideration include human factors, forensic physical evidence, effects of fire suppression, and a broad knowledge of case histories of similar scenarios.

Reconstruction may point to obvious areas to collect fire debris evidence, particularly when investigators suspect that flammable or combustible liquids have been used to start the fire (DeHaan and Icove 2012). In evidence collection, reconstruction should be combined with knowledge of fire dynamics. Armed with this knowledge, an investigator who suspects that an ignitable liquid was used to start a fire will look for an area at floor level close to the edge of a fire plume's fuel source that may have remained cooler than the surrounding area during the fire. Sometimes, residues of liquids survive in these cooler areas, especially when absorbed and protected by a floor covering such as carpet. These areas are usually evidenced by demarcation patterns on the material (Mealy, Benfer, and Gottuk 2011, 142–51). The analysis of fire patterns left by fire plumes can also provide valuable information, especially in complicated fires where multiple points of origin may exist, as is often the case in **arson** fires.

arson ■ The intentional setting of a fire with intent to damage or defraud.

Forensic fire scene reconstruction involves establishing what the scene looked like at the start of the fire, not only physically (what was present and where it was located) but also considering environmental conditions (temperatures, weather, ventilation, etc.). A very useful concept for testing accidental scenarios is for the investigator to consider what was different the day of or just before the fire. A change in environmental conditions, equipment operation, time of use, or source of material can sometimes be the factor that shifts a marginally stable system into an unstable state, triggering a fire by accidental means. Reconstruction may also indicate specific evidence of accidental or natural fires.

STEP 1: DOCUMENT THE FIRE PATTERNS OF THE FIRE SCENE AND ITS PROCESSING

A combination of photographs, sketches, and fire scene analysis is necessary to record accurately the information obtained during the fire scene reconstruction. This step is explored in depth in Chapter 4.

NFPA 921 and *Kirk's Fire Investigation* provide numerous examples of how to graphically record and photographically document, witness, and interpret fire scene analysis. This documentation should also identify the operation of fire protection equipment, such as heat and smoke detectors, sprinklers, and audible alarms.

The systematic documentation of a fire scene from its initial stages can play an essential role in a professional effort to record the events for all parties involved. This approach captures all available information that will be needed for later use in criminal, civil, or administrative matters.

Fire scenes often contain complex information that must be documented, a task of paramount importance for the investigator. A single photograph or scene diagram is usually not sufficient to capture information on fire dynamics, building construction, evidence collection, and avenues of escape for its occupants, as illustrated in Figure 1-6.

Systematic documentation, when correctly conducted, will be better suited to pass the *Daubert* standard for courtroom admissibility. Forensic fire scene documentation is properly accomplished through a comprehensive effort of forensic photography, sketches, drawings, and analysis. In summary, this task includes documenting visual observations, emphasizing development characteristics, and authenticating physical evidence found at the scene. On fire scene sketches, it is important that the investigator record his or her observations as to the significant fire pattern damage, location of the fire plumes, and the area and point of suspected fire origin. The evidence and explanations contained in documentation should be sufficiently clear and precise so that an independent and qualified investigator who is unfamiliar with the fire scene can independently arrive at the same conclusions as to the fire's origin and cause.

FIGURE 1-6 A single photograph may not be sufficient to document the fire scene processing. *Courtesy of D. J. Icove.*

STEP 2: EVALUATE THE HEAT TRANSFER DAMAGE BY FIRE PLUMES

Heat is transferred by three fundamental methods: conduction, convection, and radiation. The relative importance of these three methods that dominate a fire depends on not only the intensity and size of the fire but also the physical environment. The heat transferred increases the surface temperature of the target, which can cause observable and measurable effects. Heat transfer is covered in more detail in Chapter 2.

Rapid detection and suppression normally account for fires' being contained within their rooms or compartments of origin. In fires of limited extension, the heat transfer relationships of conduction, convection, and radiation can be used to explain further the fire plume burn pattern damage (Quintiere 1994). Understanding plume damage is a key to reconstruction, and a simplified diagram showing a typical plume-damaged room is shown in Figure 1-7. These processes are examined in more detail in Chapter 3.

STEP 3: ESTABLISH STARTING CONDITIONS

Nearly all rooms contain some sort of fuel. The location, distribution, and type of fuel are critical in evaluating the ignition and spread of a fire. Fuels may be in the form of furniture (live load) carpet and pad, draperies, and wall and ceiling coverings, as well as the building materials themselves (e.g., wood floors, walls and ceilings, fiberboard walls). It is vital to know what these materials were, where they were located in the room, and what are their ignition and combustion properties.

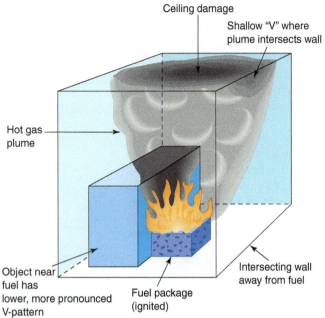

Ceiling damage

Shallow "V" where plume intersects wall

Hot gas plume

Object near fuel has lower, more pronounced V-pattern

Fuel package (ignited)

Intersecting wall away from fuel

FIGURE 1-7 Typical fire plume damage may produce a circular burn pattern on the ceiling and V or inverted-V patterns on intersecting vertical surfaces. The area of most intense damage on the ceiling is centered over the highest-temperature region of the plume's flame. *From J. D. DeHaan, Kirk's Fire Investigation, 7th ed., 2012. Reprinted by permission of Pearson Education, Inc., Upper Saddle River, NJ.*

It is also important to know the conditions before the fire. For example, the investigation may need to document the prefire conditions of environment (temperature, wind, humidity), ventilation (doors, windows, furnace, HVAC), and operability of potential ignition sources. The first fuel ignited is a vital piece of information in most reconstructions. This fuel should be assessed to determine what it was, how it could have been ignited, how long ignition took, and how large a fire would be produced.

STEP 4: CORRELATE THE HUMAN OBSERVATIONS AND FACTORS

It is important to document and correlate the observation of witnesses and victims to the fire scene reconstruction. Investigators' fire analyses should include eyewitness testimony, news videos, and firefighters' descriptions as to the fire's growth during their presence. This documentation can be used to develop and test hypotheses correlating the actions of persons prior to, during, and after the fire, as well as those related to origin and cause.

Early research work evaluated the decision-making behaviors of the occupants and the various factors that affected their decisions in large fires involving high-rise buildings (Paulsen 1994). An investigator involved with forensic fire scene reconstructions should become knowledgeable about the existing research into the wide range of human behavior in fire situations. Reactions during violent crimes are generally fight, hide, or run. The range of reactions in fire environments is more extensive. People can ignore, investigate, approach and observe, run away and call for help, run away and return for property, fight the fire, and so on.

Human factors may include acts or omissions with respect to fires ignited either deliberately or by acts of carelessness, improper extinguishment of cigarettes, impaired judgment while under the influence of drugs or alcohol, or reactions to the environment. Other factors include the reactions and decisions (either proper or improper) to trigger alarms and escape to areas of refuge or to remain to fight the fire.

Sometimes, in criminal cases, the confession of an arsonist comes into question, particularly when investigators attempt to determine the veracity of the statement. For example, the type, quantity, and distribution of an accelerant used to set the fire or its identification can arise from the admissions of someone involved in the ignition of the fire and can be compared with the physical evidence.

The location and intensity of the fire origin can sometimes be established by combining witness observations with an understanding of fire dynamics and physical evidence. This strategy is particularly important when the fire has caused extensive damage to the building or compartment or when the damage extends beyond the compartment of origin, making the determination of the location of the initial fire plume difficult.

Thermal burns on a fire victim's or arsonist's body or carboxyhemoglobin (COHb) levels in his or her blood may be indicative of his or her location at the time of the initial incipient fire, direction of travel, activities, and escape route. For example, an arsonist running away from a premature and rapidly developing flash fire (often characteristic of a flammable liquid) may suffer radiant heat burns on exposed skin of the neck and upper arms or to the backsides of legs and clothing, whereas someone facing an oncoming fire will tend to have burns to the face, arms, and fronts of the lower legs.

STEP 5: CONDUCT A FIRE ENGINEERING ANALYSIS

The behavior of many fires can be analyzed by applying fundamental fire protection engineering calculations to estimate a fire's size, growth, and damage effects from heat and smoke. Even though these calculations are never 100 percent accurate, they can assist in the interpretation and explanation of fire behavior (Quintiere 1998, 2006). Common factors used in fire engineering calculations useful to investigators include heat release rates, flame height of a fire plume, location of the virtual source, heat flux, and conditions for flashover.

Fire modeling can also be an important tool in fire engineering analysis. A fire model is generally a computer program used to predict the environment in single- and multi-compartment structures subjected to a fire. Models usually calculate temperatures associated with the distribution of smoke and fire gases throughout a building. Fire calculations and models are discussed in Chapter 6.

STEP 6: FORMULATE AND EVALUATE THE CONCLUSION

The final step, the comprehensive conclusion, should include the most complete documentation of how each of the hypotheses, tests, and witness observations points to the final conclusion for the case. The information that helped exclude other hypotheses should also be described.

Benefits of Fire Engineering Analysis

The primary goal of this text is to extend and explain scientific approaches to fire analysis for use by the fire investigator. However, investigators need a reasonable degree of training in the principles of fire protection engineering and fire dynamics to construct and correctly interpret the results in their cases. Analysis by applying these principles greatly increases the hypothesis testing ability of a capable investigator, as it may provide answers to queries about a fire's ignition, spread, or effects that would otherwise go unresolved.

FIRE ENGINEERING ANALYSIS

Fire engineering analysis, which can range from a basic back-of-the-envelope first-level calculation to a sophisticated fire model, provides numerous benefits to fire scene reconstruction. Historically, investigations using fire engineering analysis have accurately assessed fire development, measured the performance of fire protection features and systems, and predicted the survivability and behavior of people during the incident.

Over the years, NIST has undertaken several significant fire engineering analysis studies. These detailed reports include the Dupont Plaza fire (Nelson 1987), First Interstate Bank Building fire (Nelson 1989), Pulaski Building fire (Nelson 1994), Hillhaven Nursing Home fire (Nelson and Tu 1991), Happyland Social Club fire (Bukowski and Spetzler 1992), 62 Watts Street fire (Bukowski 1995, 1996), Cherry Road fire (Madrzykowski and Vettori 2000), Cook County Administration Building Fire (Madrzykowski and Walton 2004), and the Station Nightclub Fire (Grosshandler et al. 2005). Some of these cases will be described in Chapter 8.

Some form of fire engineering analysis is now considered a prudent step in comprehensive fire scene analysis and reconstruction (Schroeder 2004). Combined with modeling, which is explored in detail later in this text, fire engineering analysis offers the following benefits:

- Establishes the basis for the collection of data needed to construct event timelines, ignition sequences, and failure modes and effects analysis (FMEA)
- Invites application of the scientific method when testing hypotheses and validating fire scene reconstruction
- Provides a viable alternative to full-scale fire testing and can extend full-scale test results to differing ranges of conditions (sensitivity analysis)
- Provides answers to many important questions raised about human factors, ignition sequences, equipment failure, and fire protection systems (detectors, alarms, and sprinklers)
- Identifies important future research areas in fire investigation

Historically, the use of engineering analysis and modeling in fire scene reconstruction was conducted on a case-by-case basis, owing mainly to the complexity of the process. A vast amount of knowledge and time were required to collect and analyze the information on the development of the fire. Engineering analysis does not necessarily involve mathematical modeling, but modeling whether by computer or hand calculation offers important support to most analyses. All engineering analyses are based on the fundamental physics of mass and energy transfer and convection.

Engineering analysis and modeling are becoming more commonplace. Despite the complexity of the processes, fire engineering analysis and modeling should be applied to cases involving multiple deaths, cases in which code deficiencies contributed to the fire, or cases that will probably involve extensive civil or criminal litigation.

FIRE MODELING

A fire engineering analysis often uses fire modeling to compare actual events with predicted outcomes using varying fire causes and growth scenarios, as illustrated in Figure 1-8, involving a chair fire in an apartment. Surface temperatures on the walls are calculated by the fire model and displayed using a color-coded gradient scale. Results from the surface temperatures can be compared with burn patterns to confirm the area of origin and sequence of the fire. This analysis often adds value, understanding, and clarity to complex fire scene investigation. Fire modeling, a less expensive and more environmentally sensitive approach, can also serve as an alternative to full-scale fire testing to explain burn patterns and the fire dynamics.

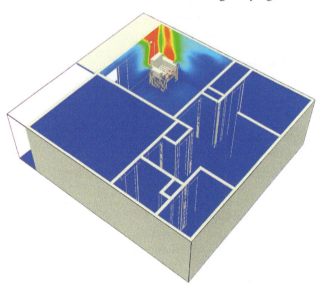

FIGURE 1-8 A fire engineering analysis often uses fire modeling to compare actual events with predicted outcomes. Heat transfer contact areas from a fire initiated in a chair in the upper-right corner of the living room are shown at 2.3 minutes of the computer simulation. This pattern can be compared with witness accounts and post-fire patterns. *Courtesy of D. J. Icove.*

Through a process called *sensitivity analysis,* results can be expanded to predict the effects of different conditions on the fire. For example, a model based on heat release rates, plume height, fuel configuration, and placement of the origin within the room (middle, wall, or corner) could potentially resolve whether the fire was the result of an accidental trash can fire or a fire caused by a pool of intentionally spread flammable liquid.

The model can also predict the impact of an open door or window on the fire. In addition, the comparison of a timeline is a beneficial product of the fire model. Example 1-2 shows the successful application of fire modeling as a tool in litigation. However, it should be understood that modeling is less realistic than full-scale fire testing. The exact fuel arrangement that was present can be accurately tested in full scale, whereas fire modeling is often dependent on fuel data such as heat release rates, which are available in the published literature and which may be quite different from that of the supposedly similar item.

EXAMPLE 1-2 ▪ **Fire Model Helps Defend U.S. Government against Lawsuit**

At 6:45 a.m. in a three-bedroom town house on a military reservation, just before the winter holidays, a five-year-old was involved in a storage closet fire. The smoke alarm notified the parents, and the father tried unsuccessfully to extinguish the fire. Because their telephone was out of order, the husband told his wife to awaken their children while he went next door to call the fire department.

The wife and two youngest children perished in the fire. Three older children survived. The responding fire department was slightly delayed, by 2.5 minutes, when they had to yield to troops on the road. This delay formed the basis for a $36 million wrongful death lawsuit against the U.S. Government. The U.S. Department of Justice asked the National Institute of Standards and Technology (NIST) to model the fire to determine what, if any, role the reported delay played in the deaths. A NIST fire protection engineer used Fire Hazard Assessment Method (HAZARD I) which at the time contained the earlier version 2.0 of Consolidated Model of Fire and Smoke Transport (CFAST) to model the fire. He relied on construction plans, fire incident reports, radio logs, and witness statements to input the variables into the model.

The fire model accurately predicted areas of fire and smoke damage, the safe egress routes of the survivors, and the impact that attempts by neighbors to gain access by opening a bedroom window had on the resulting carbon monoxide and heat intensity levels. These predictions agreed with the autopsy findings, blood toxicity tests, and thermal injuries to the victims.

A further timeline analysis showed that the delayed response played no significant role in the fatalities. After depositions were taken from the NIST engineer, the lawsuit was settled out of court for less than $200,000 (Bukowski 1991).

Fire engineering analysis and modeling may also provide answers to important questions raised during fire investigations. These questions might include the following:

- What was the most probable cause of the fire (i.e., can several possible causes be eliminated)?
- How long did it take to activate fire or smoke detectors and fire sprinklers?
- What were the smoke and carbon monoxide levels in each room after incremental times in minutes?
- Why didn't the occupants of the building survive the fire?
- Was an accelerant used in the fire and, if so, what type?
- How much time elapsed from when the occupant of the structure left the building until the fire reached flashover?
- Did a negligent building design or failed fire detection and suppression system contribute to the growth of the fire?
- What changes to policy, building, or fire codes are necessary to ensure that similar fires will not occur in the future?

Discussion and cooperative research between fire investigators and fire protection engineers underscore the need to include fire modeling as an integral and required step in fire scene reconstruction. Fire investigators and engineering professionals need to collaborate further to expand their knowledge and validate the methods shared in common by both groups.

Daubert and the Fire Investigator

The debate over the application of *Daubert* to fire scene investigations has intensified and focused on whether origin and cause determination is to be considered scientific evidence or nonscientific technical evidence. The advocates of the strict scientific approach bristle at the suggestion that fire scene investigation is in any way nonscientific, pointing to the many misconceptions previously used by fire investigators (spalling, floor penetrations, etc.) that were corrected by fire scientists only in recent years. They advocate the use of *Daubert* in fire scene analysis as the only means of preventing a return to the improper fire scene methodologies employed by unqualified investigators lacking proper scientific training.

In contrast, the technicians argue that fire scene investigation has never been a pure science like chemistry or physics, as it employs elements of both disciplines. The term *nonscientific* in the context of *Daubert* is a legal distinction, rather than a scientific one. That is not to say that fire investigation is *unscientific* or devoid of any application of scientific principles. Instead, it is a recognition of the objective and subjective components that form a part of every fire scene investigation, more precisely, the human component in examining, analyzing, and, ultimately, interpreting fire scene evidence to reach a conclusion about the fire's origin and cause.

Early on in *Daubert*'s history, the U.S. Court of Appeals for the Tenth Circuit directly addressed the testimony of a fire investigator in an arson case. In *United States v. Markum* (1993), the court found the admission of a fire chief's testimony that a fire was the result of arson to be proper based primarily on his extensive experience in fire investigations. The court said:

> Experience alone can qualify a witness to give expert testimony. See *Farner v. Paccar, Inc.*, 562 F. 2d 518, 528–29 (8th Cir. 1977); *Cunningham v. Gans*, 501 F. 2d 496, 500 (2nd Cir. 1974).
>
> Chief Pearson worked as a firefighter and Fire Chief for 29 years. In addition to observing and extinguishing fires throughout that period, he attended arson schools and received arson investigation training. The trial court found that Chief Pearson possessed the experience and training necessary to testify as an expert on the issue whether the second fire was a natural rekindling of the first fire or was deliberately set. That finding was not clearly erroneous. (*Markum* at 896)

Another case that directly addressed the application of *Daubert* to fire investigation was *Polizzi Meats, Inc.* (PMI) *v. Aetna Life and Casualty* (1996). In that case, the Federal District Court of New Jersey ruled:

> PMI's counsel argues that because of a lack of "scientific proof" of the fire's causation, none of Aetna's witnesses may testify at trial. This astounding contention is based on a seriously flawed reading of the United States Supreme Court's decision in *Daubert v. Merrell Dow Pharmaceuticals, Inc. Daubert* addresses the standards to be applied by a trial judge when faced with a proffer of expert scientific testimony based upon a novel theory or methodology. Nothing in *Daubert* suggests that trial judges should exclude otherwise relevant testimony of police and fire investigators on the issues of the origins and causes of fires. (*Polizzi* at 336–37; citations omitted)

These two decisions were the only reported cases considering *Daubert* in the specific context of fire investigation until the Eleventh Circuit announced its decision in the Joiner case, which took an entirely different view on the process of fire scene investigation.

The *Markum* case analysis was provided courtesy of Guy E. Burnette, Jr., Tallahassee, Florida (2003).

The *Joiner* Case; A Clarification of *Daubert*

As courts from various jurisdictions continued trying to shed light on the full meaning of *Daubert,* the U.S. Supreme Court took up the issue again and provided some guidance and insights. In *General Electric Company v. Joiner* (1997), the Court reviewed a case in which the trial judge had entered summary judgment in favor of the defendant in a lawsuit alleging that the plaintiff had contracted cancer as the result of exposure to PCB chemicals. The scientific evidence in support of the plaintiff's claim was derived from laboratory studies of mice that had been injected with massive doses of PCB chemicals and certain limited epidemiological studies suggesting a causal connection between PCB chemicals and cancer in humans. The trial judge ruled that the evidence offered by the plaintiff failed to satisfy the requirements of *Daubert,* describing the evidence offered by the plaintiff's experts as "subjective belief or unsupported speculation." It was noted that *Joiner* failed to present any credible scientific evidence of a direct causal connection between exposure to PCB chemicals and cancer.

On appeal, the ruling was reversed by the U.S. Court of Appeals for the Eleventh Circuit, which held that the evidence should have been presented to the jury for a decision. The appellate court observed that the *Federal Rules of Evidence* favor the admissibility of expert testimony as a general rule. Furthermore, the appellate court applied a more stringent standard of review of the trial court's ruling, since the ruling was "outcome determinative" (i.e., resolved the entire case).

The U.S. Supreme Court overturned the decision of the Eleventh Circuit and reinstated the ruling of the trial court. In doing so, the Court reiterated and clarified some of the points made in the *Daubert* decision. First, the role of the trial judge as "gatekeeper" was reaffirmed. In particular, the trial judge not only was allowed to draw his own conclusions about the weight of evidence offered by an expert witness but was expected to do so. It was noted that this had been a function of the trial judge long before the *Daubert* decision itself. Since this was a proper role of the trial judge, the decision to accept or reject expert testimony would not be subject to a more stringent standard of review on appeal. The decision of the trial judge would be given deference on appeal and it would require showing an "abuse of discretion" for the decision of the trial judge to be overturned.

The Supreme Court held that the application of *Daubert* to expert testimony is not merely a review and approval of the methodology employed but also includes scrutiny of the ultimate conclusions reached by the expert witness based on the methodologies and data employed to reach those conclusions. Notably, the Court did not clarify the controversy over scientific evidence versus technical evidence. The Court did not address the issue in *Joiner* because it was clearly a "scientific evidence" case. That remains a major part of the controversy in construing *Daubert* and the admissibility of expert testimony. The *Benfield* decision did directly address this issue and demonstrated a new perspective on this critical aspect of fire investigation.

This case history was provided courtesy of Guy E. Burnette, Jr., Tallahassee, Florida (2003).

The *Benfield* Case

In *Michigan Miller's Mutual Insurance Company v. Janelle R. Benfield* (1998), the U.S. Court of Appeals for the Eleventh Circuit applied the *Daubert* analysis to a fire scene investigation. This case has attracted great attention within the fire investigation community and has become a focal point of the *Daubert* controversy.

In January 1996 the *Benfield* case was tried in federal district court in Tampa, Florida. The case involved a house fire in which the insurance company, Michigan Miller's Mutual, refused to pay on the policy based, in part, on the fire's being **incendiary** and on the apparent involvement of an insured party in setting the fire. As a part of the insurer's case, a fire investigator with over 30 years of experience in fire investigations was called as an expert witness to present his opinion as to the origin and cause of the fire. He testified that the fire was started on top of the dining room table where some clothing, papers, and ordinary combustibles had been piled together. He examined the fire scene primarily by visual observation and concluded that the fire was incendiary based on the absence of any evidence of an accidental cause, along with other evidence and factors noted at the scene. After cross-examining the investigator, the plaintiff moved to exclude the testimony under *Daubert*. The trial court agreed. In the trial court's ruling striking the expert's testimony, the judge specifically found that the witness

incendiary fire ■
A deliberately set fire.

> cites no scientific theory, applies no scientific method. He relies on his experience. He makes no scientific tests or analyses. He does not list the possible causes, including arson, and then using scientific methods exclude all except arson. He says no source or origin can be found on his personal visual examination and, therefore, the source and origin must be arson. There is no question but that the conclusion is one to which *Daubert* applies, a conclusion based on the absence of accepted scientific method.
>
> And finally, it must be noted that [his] conclusion was not based on a scientific examination of the remains, but only on his failure to be able to determine a cause and origin from his unscientific examination. This testimony is woefully inadequate under *Daubert* principles and pre-*Daubert* principles, and his testimony will be stricken and the jury instructed to disregard the same. (*Daubert* motion hearing transcript at 124–26)

Interestingly, the court in *Benfield* initially found the expert to be qualified to render opinions in the area of origin and cause of fires and allowed him to testify, based on his qualifications and credentials as a fire investigator. However, the judge struck the expert's testimony after it was presented based on his methodologies in conducting the particular fire scene investigation in that case. Having stricken the expert's testimony, the judge then found that, as a matter of law, arson had not been proved by Michigan Miller's and directed a verdict against them on the arson issue.

The evidence was undisputed that the area of origin was on top of the dining room table. Therefore, the only issue was the cause of the fire. The expert testified that while he was conducting his investigation, he spoke with Ms. Benfield, who told him that when she was last in the house before the fire there was a hurricane lamp and a half-full bottle of lamp oil on the top of the table. He further testified that he examined photographs taken by the fire department before the scene was disturbed and observed an empty, undamaged bottle of lamp oil lying on the floor with the cap removed (also undamaged), indicating that it had been opened and moved from the table prior to the setting of the fire. He also explained his observations at the fire scene that enabled him to rule out all possible accidental causes. He concluded that the fire was incendiary, using the "elimination method" long recognized as a valid method of determining fire origin and cause. He could not, however, determine the source of ignition for the fire. More importantly, he did not "scientifically document" his findings on various points and relied primarily on his 30 years of experience as a fire investigator, even as he held himself out as an expert in fire science adhering to the scientific method in conducting his investigation.

On cross-examination by Ms. Benfield's attorney, the expert was asked to define the scientific method and was asked the "scientific basis" for the taking of certain photographs apparently unrelated to the fire itself. The cross-examination continued by attacking each piece of evidence used by the expert that could not be said to be scientifically objective and scientifically verified. The investigator's determination of the smoldering nature of the fire and the time he estimated it burned before being discovered were discredited as not being based on scientific calculations of heat release rate and fire spread but merely the investigator's observations of the smoke damage and other physical evidence. The court noted those points from the cross-examination in finding that the expert's methodology was not in conformity with the scientific method, relying instead almost exclusively on the expert's own training and experience, which it held to be inadmissible under *Daubert*.

On May 4, 1998, the Eleventh Circuit issued its ruling in the *Benfield* case. Contrary to the Tenth Circuit decision in *Markum* and the federal district case in *Polizzi Meats,* the court found the investigator's fire scene analysis to be subject to the *Daubert* test of reliability. In reaching this conclusion, the court noted that the investigator in *Benfield* held himself out as an expert in the area of "fire science" and claimed that he had complied with the "scientific method" under *NFPA 921*. Thus, by his own admission he was engaged in a "scientific process," which the court held to be subject to *Daubert*.

Under the *Daubert* test of reliability, the appeals court upheld the decision of the trial judge striking the expert's testimony. Noting that it is within the trial court's discretion to admit expert testimony, such a decision will be affirmed on appeal absent a showing of an "abuse of discretion" or that the decision was "manifestly erroneous." Under such a daunting standard, the trial judge is effectively given the "final word" on whether both the qualifications and findings of an expert witness will be considered sufficient and reliable enough to be presented to the jury. It is not simply a matter of having the power to decide whether a witness is qualified to testify as an expert; the substance of the expert's testimony and his or her professional conclusions first have to meet the approval of the trial judge before they can be presented to the jury. The trial judge acting as gatekeeper can summarily reject the findings and conclusions of an expert witness, preempting the jury from making that decision.

In the *Benfield* decision, various scientifically unsupported and scientifically undocumented conclusions of the investigator were cited as grounds for the determination that his observations and findings failed the *Daubert* reliability test. A chandelier hanging over the dining room table where the fire started showed no signs of having caused the fire, but the investigator had not conducted any tests or examinations to eliminate it scientifically as a potential cause of the fire. His observations alone were held inadequate. Similarly, his opinion that the fire had likely been accelerated with the lamp oil contained in the bottle was rejected under *Daubert,* since he could not scientifically prove that there had been oil in the bottle before the fire and he had not taken any samples from the fire debris to prove scientifically its presence or absence at the time the fire was ignited. These and other observations of the investigator were held to demonstrate that there was no scientific basis for his conclusions, only his personal opinion from experience in investigating other fires.

It was not that the investigator was found to be "wrong" in the *Benfield* case. Indeed, there was never any evidence of an accidental cause of the fire. Ironically, although the appeals court upheld the decision of the trial judge to strike the testimony of the insurance investigator, it granted a new trial for Michigan Miller's on the arson defense. The appeals court felt that a prima facie case of arson had been shown at trial through the fire department investigator, who initially classified the fire only as **suspicious** (with virtually no challenge to the scientific documentation of his opinion), and the many incriminating circumstances surrounding the fire itself.

Circumstances included the fact that Ms. Benfield claimed that she had not locked the deadbolts when she left the house, yet the deadbolts were locked when she returned to discover the fire. Ms. Benfield and her daughter (who had been out of town) had the only keys to those locks. Her assertion that her boyfriend extinguished the fire with a garden hose was refuted by the observations of the responding firefighters. Her insurance claim appeared to be significantly inflated. She had tried to sell the house and could not do so. She was trying to convince her estranged husband to transfer the house to her but could not do so. Ms. Benfield had given conflicting and contradictory accounts of her activities immediately before the fire.

In listing all these reasons, the appeals court found that there was compelling evidence of arson even as it discredited the findings of the insurance investigator that the fire was incendiary. The critical point in the case appears to be the distinction between the insurance investigator's testifying as an expert in fire science and the fire department representative's testifying as an expert in fire investigation.

This case history was provided courtesy of Guy E. Burnette, Jr., Tallahassee, Florida (2003).

suspicious ■ Fire cause has not been determined, but there are indications that the fire was deliberately set and all accidental fire causes have been eliminated.

MagneTek: A Legal Case Analysis

A decision of the U.S. Court of Appeals for the Tenth Circuit underscored the importance of properly evaluating and presenting expert testimony in fire litigation cases. *Truck Insurance Exchange v. MagneTek, Inc.* (2004) not only demonstrated the application of the *Daubert* standard for the admissibility of expert testimony but effectively contradicted a long-standing principle of fire science that had been used in a significant number of cases to prove the cause of a fire.

The *MagneTek* case involved a subrogation action filed by Truck Insurance Exchange against the manufacturer of a fluorescent light ballast that was alleged to have caused a fire that destroyed a restaurant in Lakewood, Colorado. Responding to the alarm, the fire department first found heavy smoke in the restaurant but no open flames. The fire subsequently broke through the kitchen floor in the restaurant from the ceiling of a storage room in the basement. Before the fire could be controlled and extinguished, it destroyed the restaurant and caused damages in excess of $1.5 million.

Investigations by both the local fire protection district and a private fire investigation firm hired by the insurer determined that the fire had started in a void space between the basement storage room ceiling and the kitchen floor. In the basement, the investigators found the remains of a fluorescent light fixture that had been mounted to the ceiling of the storage room. They determined the light fixture had been located in the area of origin of the fire and concluded that the fire had been caused by an apparent failure of the ballast in the light fixture.

The investigators and a physicist examined the remains of the fluorescent light fixture. They determined that MagneTek manufactured the ballast. They observed oxidation patterns on the light fixture indicating an internal failure, along with discoloration of the heating coils of the ballast suggesting it had shorted to cause overheating, which resulted in the fire. The ballast contained a thermal protector designed to shut off power to the fixture when the internal temperature exceeded 111°C (232°F). The thermal protector in the ballast appeared to function properly, even after the fire. However, the investigators remained convinced that the ballast had somehow failed, overheated, and started the fire.

Tests were conducted with similar ballasts manufactured by MagneTek, which showed that at least one of the exemplar ballasts when shorted would not cut off power to the fixture until the internal temperatures had reached at least 171°C (340°F) and would continue to provide power to the fixture even when the ballast maintained constant temperatures of 148°C (300°F) or more.

The investigators theorized that the heat from the ballast had caused pyrolysis to occur in the adjacent wood structure of the ceiling, causing pyrophoric carbon to form over a prolonged time, which would be capable of ignition at temperatures substantially below the normal ignition range of 204°C (400°F) or more for the fresh whole wood. The phenomenon of pyrolysis in the formation of pyrophoric carbon has been the subject of a number of studies, reviews, and articles by fire investigators and fire scientists. It has been cited as the cause of a number of fires having no other apparent explanation, often linked to heated pipes in structures within walls, ceilings, and floor areas. As the MagneTek court case would note, however, the validity of this phenomenon has been discussed and debated by fire investigators and fire scientists for a number of years.

The investigators in this case acknowledged that electrical wiring ran through the ceiling area of the storage room near the fluorescent light fixture but discounted the possibility of a failure in the electrical wiring. They reported finding no evidence of arcing or shorting in the electrical wiring, although the fire at the restaurant resulted in the destruction of much of the electrical wiring and other evidence in the area. Because the investigations concluded the fire had originated in the immediate area of the light fixture, they concluded the only source of ignition for the fire was the light fixture and its ballast.

Pyrolysis and the formation of pyrophoric carbon was the foundation of the plaintiff's case against MagneTek. The ballast in the light fixture showed no signs of failure in the thermal protector, which would have limited the heat generated by the ballast to about 111°C (232°F). Even with the exemplar ballast whose thermal protector failed to perform as it had been designed, the temperatures generated did not exceed 171°C (340°F). The investigators admitted the ignition temperature of fresh *whole wood* is typically at least 204°C (400°F), and the temperatures from the ballast alone would not have been sufficient to cause ignition. Their theory that the ballast had caused the fire depended on the concept of pyrolysis to allow ignition to occur at a lower temperature within the range of the temperatures generated by the ballast.

Following discovery in the case, MagneTek filed a *Daubert* motion to exclude the testimony of the experts that the ballast had caused the fire. MagneTek asserted that the theory of pyrolysis was not sufficiently reliable and scientifically verifiable to be offered by the experts in support of their conclusion for the cause of the fire. It was a challenge to the reliability of the experts' theory, which required a consideration of whether the reasoning or methodology underlying the testimony was scientifically valid, as mandated by *Daubert* and *FRE Rule 702.*

The Supreme Court in *Daubert* had outlined a number of factors that, although not an exclusive list of considerations for a trial court, should be examined in making the determination of reliability (see Table 1-4).

In proving the scientific validity of an expert's reasoning or methodology, the Court noted that "the plaintiff need not prove that the expert is indisputably correct or that the expert's theory is 'generally accepted' in the scientific community. Instead, the plaintiff must show that the method employed by the expert in reaching the conclusion is scientifically sound and that the opinion is based on facts which sufficiently satisfy *Rule 702*'s reliability requirements."

The Tenth Circuit applied the standard of appellate review for a trial court ruling on a *Daubert* issue: the showing of an "abuse of discretion" demonstrating that the appellate court has "a definite and firm conviction that the lower court made a clear error of judgment or exceeded the bounds of permissible choice in the circumstances" *United States v. Ortiz* (1986). The court then looked to the ruling of the trial judge finding that the testimony of the experts failed to satisfy the reliability standard under *Rule 702* and the *Daubert* decision. The physicist testifying on behalf of Truck Insurance Exchange was a highly credentialed expert with an advanced degree in physics from Oxford University and over 20 years of experience in the study of fire and explosion incidents. Both the trial court and the appellate court observed that this expert was unquestionably qualified to testify as an expert witness under *Rule 702*. However, his hypothesis of pyrolysis and pyrophoric carbon as the cause of ignition of the wood in the area surrounding the light fixture could not be considered a reliable basis for the admission of his expert testimony on the cause of the fire.

The reliability standard under *Daubert* applies to both the theory offered and its application to the facts of the case. The court focused on the first component of these reliability criteria. In support of the theory of its experts, the insurer had introduced into evidence three publications on the theory of pyrolysis and pyrophoric carbon. Those articles were written by some of the most respected fire scientists in the world, but those articles and case studies acknowledged that the process of pyrolysis occurs over an undefined period of time described as "a period of years" or "a very long time" with no specific parameters for the timing and sequence of events involved in pyrolysis. One of those articles acknowledged that there are "a number of things not known about the process" and that "it may be many decades before it will be solved by sufficiently improving theory." The article concluded by stating "the phenomenon of long-term, low-temperature ignition of wood has neither been proven nor successfully disproven at this time."

The plaintiff's engineer had stated in his deposition that the process of pyrolysis "depends on a lot of factors, as yet quantitatively unidentified." He went on to testify "you would have to have a good theory of pyrophoric carbon and formation and the chemical kinetics of that; and there isn't one. . . . " The other experts testifying for the plaintiff in this case based their theory of pyrolysis and "pyrophoric carbon" on their experience in the investigation of fires without any reference to a specific scientific basis for that theory.

The court noted that under *NFPA 921* an investigator offering the hypothesis of an appliance fire must first determine the ignition temperature of the available fuel in the area and then must determine the ability of the appliance or *device* to generate temperatures at or above the ignition temperature of the fuel. In that regard, the experts failed on both counts. The ignition temperature of the fuel (wood) exposed to the fixture in this case could not be scientifically proved to be below the 204°C (400°F) threshold for the ignition of most types of wood, and the tests of the ballasts had shown that even with a failed thermal protector, a ballast could not generate temperatures anywhere near that range. Their hypothesis would have to be based on either an unreliable theory (pyrolysis) or unsubstantiated assumptions and speculation about the temperature of the ballast that contradicted their own test results. As such, their testimony could not be admitted.

The appellate court affirmed the ruling of the trial court that the testimony of all the experts had been shown to be not sufficiently reliable to be admitted under the standards of the *Daubert* decision. Without the testimony of the experts, Truck Insurance Exchange could not make a prima facie case for establishing the cause of the fire. Accordingly, the trial court entered a summary judgment in favor of MagneTek, and the appellate court affirmed.

This decision has significant implications for the litigation of fire cases everywhere. It demonstrates the importance of developing sound and scientifically verifiable theories for proving the cause of a fire as required by the standards of the *Daubert* decision. Moreover, it provides a compelling example of how a theory that has not been validated and generally recognized by others in the scientific community may not withstand a *Daubert* challenge in court.

The lessons from this decision are many. First and foremost, experts must be prepared to prove the reliability of their investigative methodologies and theories to the satisfaction of the trial court. Experience alone is not sufficient. Even a theory that appears on the surface to be a reasonable and logical theory for the cause of a fire must be shown to be scientifically verifiable. Without a foundation in science, even the most experienced investigator will never be allowed to testify in court. Parties hiring investigators in their cases must be aware of the requirements for proving a case using investigative methodologies and theories that will meet the reliability standards of the *Daubert* decision, to guide them in both the selection of the expert used to investigate the fire and the decision to litigate the case. Attorneys handling cases must be aware of these issues to successfully litigate that case at trial. The investigator, the client, and the attorney all have a responsibility to ensure that cases are properly investigated and properly litigated. The MagneTek case is a striking example of the consequences of not doing so.

This case history was provided courtesy of Guy E. Burnette, Jr., Tallahassee, Florida (2003).

MagneTek: **A Scientific Case Analysis**

A scientific analysis and assessment of the MagneTek case decision indicates that the court may not have understood several important aspects of fire science, including pyrolysis, which is not newly discovered nor scientifically disputed. In the MagneTek case, the court rejected under *Daubert* the pyrolysis theory of long-term, low-temperature ignition as being unreliable. Fire investigators who wish to explain this phenomenon in court must be prepared to make clear and convincing arguments using authoritative treatises and clearly defined definitions.

In the *Ignition Handbook* (Babrauskas 2003), pyrolysis is clearly defined as "the chemical degradation of a substance by the action of heat." This definition includes both oxidative and nonoxidative pyrolysis, taking into account the cases where oxygen plays a role. Pyrolysis is a process that has been studied for many decades. Without pyrolysis to break down their molecular structures, most solid fuels will not burn. Unfortunately, the published court decision declares pyrolysis to be inadequately studied and documented (presumably intending to refer to the pyrophoric process). It is unclear what effect this misstatement of the court will have on future deliberations.

The scientific method states that a description of the conditions under which ignition can take place (1) may use a scientific theory, test, or calculation or (2) refer to authoritative data on the substance. For example, in the second case, the peer-reviewed literature (Babrauskas 2004, 2005,) could have been used to show that the ignition temperatures of wood exposed to long-term, low-temperatures are documented as low as 77°C (170°F). Excluding other competent sources of energy, it could have been substantiated in the MagneTek case that a heat-producing product in excess of 77°C (170°F) in contact with wood for an extended period of time created a potential risk. Some substances have spontaneous combustion models. Unfortunately for the fire investigator, there does not exist at this time a model that predicts ignition of wood by pyrolysis with time/temperature curves that would allow the accurate prediction of necessary conditions, primarily owing to the complex chemical nature of wood as a fuel and the wide variability in the physical nature of the types and forms of wood.

Even more problematically, the court effectively denied the existence of any ignition due to self-heating, because it declared that ignitable substances have a "handbook" ignition temperature, and a hot object below that temperature will not be a competent source of ignition. In actual fires, self-heating substances do not have a unique ignition temperature. Furthermore, if a substance (e.g., wood) can ignite owing to external, short-term heating or to a prolonged process involving self-heating, the minimum temperature for the latter has no relation to the published ignition temperatures for the former.

Finally, in the future, fire investigators who testify as to pyrophoric (i.e., low-temperature) ignition of wood should be prepared to discuss the phenomenon by referring to this collected set of observations on the topic and the hard scientific data. In addition to the *Ignition Handbook*, there exist ample peer-reviewed publications and case studies on the ignition process of wood. Also, additional information is available on the charring rate and ignition of wood (Babrauskas 2005; Babrauskas, Gray, and Janssens 2007).

This scientific case analysis is primarily based on information courtesy of Dr. Vytenis Babrauskas, Fire Science and Technology, Inc., Issaquah, Washington. For additional information, Zicherman and Lynch (2006) and Babrauskas (2004) also address the technical merits of this case.

Summary

This chapter introduced the application of the scientific method formed by validated principles and research (fire testing, dynamics, suppression, modeling, pattern analysis, and historical case data) to fire investigation and reconstruction. This approach relies on combined principles of fire protection engineering, forensic science, and behavioral science. In using the scientific method, the investigator can more accurately estimate a fire's origin, intensity, growth, direction of travel, and duration, as well as understand and interpret the behavior of the occupants.

Fire scene reconstruction can also bring in additional fire testing and similar cases to support the final theory. In addition to *NFPA 921,* the fire investigator should become familiar with other authoritative references and sources of data and information such as the job performance requirements of *NFPA 1033* (Table 1-8). It should be remembered that *NFPA 921* is a summary of practices, extracted and developed from thousands of research efforts, some extending back over a century. There is a universe of reliable information available to fire investigators that is not included in *NFPA 921*.

The strengths of this approach make a thorough examination of many hypotheses more effective. The reliability of the result of a specific method of investigation is a function of the number of alternative hypotheses that have been tested and eliminated. Fire engineering analysis also offers many advantages. In particular, it allows the investigator to evaluate the potential of several scenarios without repeated full-scale fire testing. A sensitivity analysis can also extend the analysis to different ranges of conditions.

TABLE 1-8	Professional Levels of Job Performance for Fire Investigators as Cited in *NFPA 1033*, 2009 Edition
General Requirements for a Fire Investigator	4.1.1 Employ all elements of the scientific method as the operating analytical process
	4.1.2 Complete site safety assessments on all scenes
	4.1.3 Maintain necessary liaison with other interested professionals and entities
	4.1.4 Adhere to all applicable legal and regulatory requirements
	4.1.5 Understand the organization and operation of the investigative team and incident management system
Scene Examination	4.2.1 Secure the fire ground
	4.2.2 Conduct an interior survey
	4.2.3 Interpret fire patterns
	4.2.4 Interpret and analyze fire patterns
	4.2.5 Examine and remove fire debris
	4.2.6 Reconstruct the area of origin
	4.2.7 Inspect the performance of building systems
	4.2.8 Discriminate the effects of explosions from other types of damage
Documenting the Scene	4.3.1 Diagram the scene
	4.3.2 Photographically document the scene
	4.3.3 Construct investigative notes

TABLE 1-8	Professional Levels of Job Performance for Fire Investigators as Cited in *NFPA 1033*, 2009 Edition (*continued*)	
Evidence Collection/ Preservation	4.4.1 Utilize proper procedures for managing victims and fatalities	
	4.4.2 Locate, collect, and package evidence	
	4.4.3 Select evidence for analysis	
	4.4.4 Maintain a chain of custody	
	4.4.5 Dispose of evidence	
Interview	4.5.1 Develop an interview plan	
	4.5.2 Conduct interviews	
	4.5.3 Evaluate interview information	
Post-Incident Investigation	4.6.1 Gather reports and records	
	4.6.2 Evaluate the investigative file	
	4.6.3 Coordinate expert resources	
	4.6.4 Establish evidence as to motive and/or opportunity	
	4.6.5 Formulate an opinion concerning origin, cause, or responsibility for the fire	
Presentations	4.7.1 Prepare a written report	
	4.7.2 Express investigative findings verbally	
	4.7.3 Testify during legal proceedings	
	4.7.4 Conduct public informational presentations	

Sources: Derived from NFPA 2009; DeHaan and Icove 2012; S. Sklar, personal communications, 2008.

Problems

1.1. Obtain an adjudicated case from your local fire marshal's office involving the conviction of an arsonist who pleaded guilty to setting a fire. Analyze the fire investigator's opinion as to the origin and cause of the fire. Demonstrate how it did or did not meet the guidelines for the scientific method.

1.2. Obtain a photograph of a fire plume against a wall. Trace the lines of demarcation for char, scorching, and smoke deposition onto a plastic overlay. What can you learn from this exercise?

1.3. Search the Internet or current published periodical fire investigation literature and list three examples of peer-reviewed fire testing research.

1.4. Search the Internet or current published periodical fire investigation literature and list three federal and state court cases mentioning *NFPA 921* and *Daubert*.

1.5. Identify scientific treatises that describe pyrolysis. What is the earliest date?

1.6. Review recent fire investigation publications and find five references to the scientific method. What is the described impact on the investigations?

Suggested Readings

Cooke, R. A., and R. H. Ide. 1985. *Principles of Fire Investigation,* chaps. 8 and 9. Leicester, UK: Institution of Fire Engineers.

DeHaan, J. D., and D. J. Icove. 2012. *Kirk's Fire Investigation,* 7th ed., chaps. 4 and 5. Upper Saddle River, NJ: Pearson-Prentice Hall.

References

ASTM. 1989. *ASTM E1138-89: Terminology of technical aspects of products liability litigation.* (Withdrawn 1995). West Conshohocken, PA: ASTM International.

———.2005a. *ASTM dictionary of engineering science and technology,* 10th ed. West Conshohocken, PA: ASTM International.

———. 2005b. *ASTM E1459-92(2005): Standard guide for physical evidence labeling and related documentation.* West Conshohocken, PA: ASTM International.

———. 2006. *ASTM E1020-96(2006): Standard practice for reporting incidents that may involve criminal or civil litigation.* West Conshohocken, PA: ASTM International.

———. 2007a. *ASTM E678-07: Standard practice for evaluation of scientific or technical data.* West Conshohocken, PA: ASTM International.

———. 2007b. *ASTM E860-07: Standard practice for examining and preparing items that are or may become involved in criminal or civil litigation.* West Conshohocken, PA: ASTM International.

———. 2011a. *ASTM E620-11: Standard practice for reporting opinions of scientific or technical experts.* West Conshohocken, PA: ASTM International.

———. 2011b. *ASTM E1188-11: Standard practice for collection and preservation of information and physical items by a technical investigator.* West Conshohocken, PA: ASTM International.

———. 2011c. *ASTM E1492-11: Standard practice for receiving, documenting, storing, and retrieving evidence in a forensic science laboratory.* West Conshohocken, PA: ASTM International.

Ayala, F. J., & Black, B. 1993. Science and the courts. *American Scientist* 81:230–39.

Babrauskas, V. 1997. The role of heat release rate in describing fires. *Fire & Arson Investigator* 47 (June): 54–57.

———. 2003. *Ignition handbook: Principles and applications to fire safety engineering, fire investigation, risk management and forensic science.* Issaquah, WA: Fire Science Publishers, Society of Fire Protection Engineers.

———. 2004. *Truck Insurance v. MagneTek:* Lessons to be learned concerning presentation of scientific information. *Fire & Arson Investigator* 55 (October): 2.

———. 2005. Charring rate of wood as a tool for fire investigations. *Fire Safety Journal* 40 (6): 528–54, doi: 10.1016/j.firesaf.2005.05.006.

Babrauskas, V., Gray, B. F., & Janssens, M. L. 2007. Prudent practices for the design and installation of heat-producing devices near wood materials. *Fire and Materials* 31:125–35.

Babrauskas, V., & Grayson, S. J. 1992. *Heat release in fires.* Basingstoke, UK: Taylor & Francis.

Beering, P. S. 1996. Verdict: Guilty of burning—What prosecutors should know about arson. *Insurance Committee for Arson Control.* Indianapolis, IN.

Beller, D., & Sapochetti, J. 2000. Searching for answers to the Cocoanut Grove Fire of 1942. *Fire Journal* 94 (3): 84–86.

Beveridge, A. D. 2012. *Forensic investigation of explosions,* 2nd ed. Boca Raton, FL: CRC Press/Taylor & Francis.

Brannigan, F. L., & Corbett, G. P. 2007. *Brannigan's building construction for the fire service,* 4th ed. Sudbury, MA: National Fire Protection Association; Jones and Bartlett.

Brannigan, V. M., & Buc, E. C. 2010. The admissibility of forensic fire investigation testimony: Justifying a methodology based on abductive inference under *NFPA 921.* Paper presented at the ISFI 2010 International Symposium on Fire Investigation Science and Technology, September 27–29, College Park, MD.

Bukowski, R. W. 1991. Fire models: The future is now! *Fire Journal* 85(2): 60–69.

———. 1995. Modelling a backdraft incident: The 62 Watts Street (New York) fire. *NFPA Journal* (November/December): 85–89.

———. 1996. Modelling a backdraft incident: The 62 Watts Street (New York) fire. *Fire Engineers Journal* 56 (185): 14–17.

Bukowski, R. W., & Spetzler, R. C. 1992. Analysis of the Happyland Social Club fire with HAZARD I. *Fire and Arson Investigator* 42 (3): 37–47.

Burnette, G. E. 2003. Fire scene investigation: The *Daubert* challenge. Personal communication.

Campagnolo, T. 1999. The Fourth Amendment at fire scenes. *Arizona Law Review* 41 (Fall): 601–50.

Clifford, R. C. 2008. *Qualifying and attacking expert witnesses.* Costa Mesa, CA: James Publishing.

Cole, L. S. 2001. *The investigation of motor vehicle fires,* 4th ed. San Anselmo, CA: Lee Books.

Cooke, R. A., & Ide, R. H. 1985. *Principles of fire investigation.* Leicester, UK: Institution of Fire Engineers.

DeHaan, J. D. 1995. The reconstruction of fires involving highly flammable hydrocarbon liquids. PhD diss., University of Strathclyde, Glasgow, Scotland, UK.

DeHaan, J. D., & Icove, D. J. 2012. *Kirk's fire investigation,* 7th ed. Upper Saddle River, NJ: Pearson-Prentice Hall.

Drysdale, D. 2011. *An introduction to fire dynamics,* 3rd ed. Chichester, West Sussex, UK: Wiley.

Esposito, J. C. 2005. *Fire in the Grove: The Cocoanut Grove tragedy and its aftermath.* Cambridge, MA: Da Capo Press.

Evarts, B. 2010. U.S. structure fires in eating and drinking establishments. Quincy, MA: National Fire Protection Association.

FM Global. 2012. FM global property loss prevention data sheets. Retrieved January 16, 2012, from www.fmglobaldatasheets.com/.

FPRF. 2002. The recommendations of the research advisory council on post-fire analysis: A white paper, iv. Quincy, MA: Fire Protection Research Foundation.

FRCP. 2010. *Federal rules of civil procedure*, from www
.law.cornell.edu/rules/frcp/.

FRE. 2012. *Federal rules of evidence. Federal Evidence
Review.* Arlington, VA: http://FederalEvidence.com/.

Giannelli, P. C., & Imwinkelried, E. J. 2007. *Scientific
evidence*, 4th ed. Newark, NJ: LexisNexis.

Graham, M. H. 2009. *Cleary and Graham's handbook of
Illinois evidence,* 9th ed. Austin TX: Aspen Publishers.

Grant, C. C. 1991. Last dance at the Cocoanut Grove. *Fire
Journal* 85 (3): 74–80.

Grosselin, S. D. 1998. The application of fire dynamics to
fire forensics. Master's thesis, Worcester Polytechnic
Institute, Worcester, MA

Grosshandler, W. L., Bryner, N. P., Madrzykowski, D., &
Kuntz, K. J. 2005. Report of the technical investigation
of the Station Nightclub fire, *NIST NCSTAR 2*, vol. 1.
Gaithersburg, MD: National Institute of Standards and
Technology.

Hopkins, R. L. 2008. Fire pattern research in the US: Current status and impact. Richmond, KY: TRACE Fire
and Safety.

Hopkins, R. L., Gorbett, G. E., & Kennedy, P. M. 2009.
Fire pattern persistence and predictability during full
scale comparison fire tests and the use for comparison
of post fire analysis. Paper presented at Fire and
Materials 2009, 11th International Conference,
January 26–28, San Francisco, CA.

Icove, D. J., & DeHaan, J. D. 2006. Hourglass burn pat-
terns: A scientific explanation for their formation.
Paper presented at the International Symposium on
Fire Investigation Science and Technology, June 26–28,
Cincinnati, OH.

Icove, D. J., & Haynes, G. A. 2007. Guidelines for conducting peer reviews of complex fire investigations. Paper
presented at Fire and Materials 2007, 10th International Conference, January 29–31, San Francisco, CA.

Icove, D. J., Wherry, V. B., & Schroeder, J. D. 1998.
*Combating arson-for-profit: Advanced techniques for
investigators,* 2nd ed. Columbus, OH: Battelle Press.

Iqbal, N., & Salley, M. H. 2004. *Fire dynamics tools
(FDTs): Quantitative fire hazard analysis methods for
the U.S. Nuclear Regulatory Commission Fire Protection Inspection Program.* Washington, DC: Nuclear
Regulatory Commission.

Janssens, M. L., & Birk, D. M. 2000. *An introduction to mathematical fire modeling,* 2nd ed. Lancaster, PA: Technomic.

Karlsson, B., & Quintiere, J. G. 2000. *Enclosure fire
dynamics.* Boca Raton, FL: CRC Press.

Kolar, R. D. 2007. Scientific and other expert testimony:
Understand it; keep it out; get it in. *FDCC Quarterly*
(Spring): 207–35. Tampa, FL: The Federation of
Defense & Corporate Counsel, Inc.

Kolczynski, P. J. 2008. *Preparing for trial in federal court.*
Costa Mesa, CA: James Publishing.

Lentini, J. J. 1992. Behavior of glass at elevated temperatures. *Journal of Forensic Sciences* 37 (5): 1358–62.

Lilley, D. G. 1995. Fire dynamics. Paper AIAA-95-0894,
33rd Aerospace Sciences Meeting and Exhibits,
January 9–12, Reno, NV.

Madrzykowski, D., & Kerber, S. 2009. Fire fighting tactics
under wind driven conditions: Laboratory experiments.
Gaithersburg, MD: National Institute of Standards and
Technology.

Madrzykowski, D., & Vettori, R. L. 2000. Simulation of
the dynamics of the fire at 3146 Cherry Road NE,
Washington, DC, May 30, 1999. *NISTIR 6510.*
Gaithersburg, MD: National Institute of Standards
and Technology, Center for Fire Research.

Madrzykowski, D., & Walton, W. D. 2004. Cook County
Administration Building fire, October 17, 2003. *NIST
SP 1021.* Gaithersburg, MD: National Institute of
Standards and Technology.

Mealy, C. L., Benfer, M. E., & Gottuk, D. T. 2011. Fire
dynamics and forensic analysis of liquid fuel fires.
Baltimore, MD: Hughes Associates, Inc.

Moore, D. T. 2007. Critical thinking and intelligence analy-
sis. Occasional paper no. 14, 2nd printing with rev.
Washington, DC: Center for Strategic Intelligence
Research, National Defense Intelligence College.

NAP. 2009. *Strengthening forensic science in the United
States: A path forward.* Washington, DC: National
Academy of Sciences, National Academies Press.

———. 2011. *Reference manual on scientific evidence,*
3rd ed. Washington, DC: National Academy of
Sciences, National Academies Press.

Nelson, H. E. 1987. An engineering analysis of the early stages
of fire development: The fire at the Dupont Plaza Hotel
and Casino on December 31, 1986, 119. Gaithersburg,
MD: National Institute of Standards and Technology.

———. 1989. An engineering view of the fire of May 4,
1988, in the First Interstate Bank Building, Los
Angeles, California. *NISTIR 89-4061.* Gaithersburg,
MD: National Institute of Standards and Technology,
Center for Fire Research.

———. 1994. Fire growth analysis of the fire of March
20, 1990, Pulaski Building, 20 Massachusetts Avenue,
NW, Washington, DC. *NISTIR 4489,* 51. Gaithersburg,
MD: National Institute of Standards and Technology,
Center for Fire Research.

Nelson, H. E., & Tontarski, R. E. 1998. *Proceedings of the
international conference on fire research for fire investigation.* HAI Report 98-5157-001, 353. Washington,
DC: Department of Treasury, Bureau of Alcohol,
Tobacco & Firearms.

Nelson, H. E., & Tu, K. M. 1991. Engineering analysis of
the fire development in the Hillhaven Nursing Home
fire, October 5, 1989. *NISTIR 4665.* Gaithersburg,
MD: National Institute of Standards and Technology,
Center for Fire Research.

NFPA. 1943. The Cocoanut Grove Night Club fire, Boston,
November 28, 1942. Quincy, MA: National Fire
Protection Association.

———. 2008. *Fire protection handbook,* 20th ed. Quincy, MA: National Fire Protection Association.

———. 2009. *NFPA 1033: Standard for professional qualifications for fire investigator.* Quincy, MA: National Fire Protection Association.

———. 2011. *NFPA 921: Guide for fire and explosion investigations.* Quincy, MA: National Fire Protection association.

NIJ. 2000. Fire and arson scene evidence: A guide for public safety personnel, 48. Washington, DC: Technical Working Group on Fire/Arson Scene Investigation (TWGFASI), Office of Justice Programs, National Institute of Justice.

Ogle, R. A. 2000. The need for scientific fire investigations. *Fire Protection Engineering,* 8 (Fall): 4–8.

Orville, R. E., & Huffines, G. R. 1999. Annual summary: Lightning ground flash measurements over the contiguous United States: 1995–97. *Monthly Weather Review* 127:2693–2703.

Putorti Jr., A. D., McElroy, J. A., & Madrzykowski, D. 2001. Flammable and combustible liquid spill/burn patterns, 52. Rockville, MD: National Institute of Standards and Technology

Quintiere, J. G. 1994. A perspective on compartment fire growth. *Combustion Science and Technology* 39:11–54.

———. 1998. *Principles of fire behavior.* Albany, NY: Delmar.

———. 2006. *Fundamentals of fire phenomena.* Chichester, West Sussex, UK: Wiley.

Reinhardt, W. 1982. Looking back at the Cocoanut Grove. *Fire Journal* 76 (11): 60–63.

Rohr, K. D. 2001. An update to what's burning in home fires. *Fire and Materials* 25 (2): 43–48, doi: 10.1002/fam.757.

———. 2005. Products first ignited in U.S. home fires. Quincy, MA: National Fire Protection Association.

Schorow, S. 2005. *The Cocoanut Grove Fire.* Beverly, MA: Commonwealth Editions.

Schroeder, R. A. 2004. Fire investigation and the fire engineer. *Fire Protection Engineering* 21(Winter): 4–9.

SFPE. 2002. Guidelines for peer review in the fire protection design process. Bethesda, MD: Society of Fire Protection Engineers.

———. 2003. *Engineering guide: Human behavior in fire.* Bethesda, MD: Society of Fire Protection Engineers.

———. 2008. *SFPE handbook of fire protection engineering,* 4th ed. Quincy, MA: National Fire Protection Association, Society of Fire Protection Engineers.

Shanley, J. H. 1997. Report of the United States Fire Administration Program for the study of fire patterns. Washington, DC: Federal Emergency Management Agency, USFA Fire Pattern Research Committee.

Smith, D. W. 2006. The pitfalls, perils, and reasoning fallacies of determining the fire cause in the absence of proof: The negative corpus methodology. Paper presented at the International Symposium on Fire Investigation Science and Technology, June 26–28, Cincinnati, OH.

Tobin, W. A., & Monson, K. L. 1989. Collapsed spring observations in arson investigations: A critical metallurgical evaluation. *Fire Technology* 25 (4): 317–35.

Zicherman, J., & Lynch, P. A. 2006. Is pyrolysis dead?—Scientific processes vs. court testimony: The recent 10th Circuit Court and associated appeals court decisions. *Fire and Arson Investigator* 56 (3): 46–52.

Legal References

103 Investors I, L.P. v. Square D Company, Case No. 01-2504-KHV, 372 F.3d 1213 (U.S. App. 2004). LEXIS 12439, U.S. Dist. LEXIS 8796, 10th Cir. Kan., 2004. Decided May 10, 2005.

Abon Ltd. v. Transcontinental Insurance Company, Docket No. 2004-CA-0029, 2005 Ohio 302 (Ohio App. 2005). LEXIS 2847. Decided June 16, 2005.

Eid Abu-Hashish and Sheam Abu-Hashish v. Scottsdale Insurance Company, Case No. 98 C4019, 88 F. Supp. 2d 906 (U.S. Dist. 2000). LEXIS 3663. Decided March 16, 2000.

Allstate Insurance Company, as Subrogee of Russell Davis v. Hugh Cole Builder Inc. Hugh Cole individually and dba Hugh Cole Builder Inc., Civil Action No. 98-A-1432-N, 137 F. Supp. 2d 1283 (U.S. Dist. 2001). LEXIS 5016. Decided April 12, 2001.

American Family Insurance Group v. JVC American Corp., Civil Action No. 00-27(DSD-JMM) (U.S. Dist. 2001). LEXIS 8001. Decided April 30, 2001.

Booth, Jacob J., and Kathleen Booth v. Black and Decker Inc., Civil Action No. 98-6352, 166 F. Supp. 2d 215 (U.S. Dist. 2001). LEXIS 4495, CCH Prod. Liab. Rep. P16, 184. Decided April 12, 2001.

Chester Valley Coach Works et al. v. Fisher-Price Inc., Civil Action No. 99-CV-4197 (U.S. Dist. 2001). LEXIS 15902, CCH Prod. Liab. Rep. P16,18. Decided August 29, 2001.

Commonwealth of Pennsylvania v. Paul S. Camiolo, Montgomery County, No. 1233 of 1999.

Cunningham v. Gans, 501 F.2d 496, 500 (2nd Cir. 1974).

David Bryte v. American Household Inc., Case No. 04-1051, CA-00-93-2, 429 F.3d 469 (4th Cir. 2005; 2005 U.S. App.). LEXIS 25052. Decided November 21, 2005.

Daubert v. Merrell Dow Pharmaceuticals Inc. 509 U.S. 579 (1993); 113 S. Ct. 2756, 215 L. Ed. 2d 469.

Daubert v. Merrell Dow Pharmaceuticals Inc. (Daubert II), 43 F.3d 1311, 1317 (9th Cir. 1995).

Farner v. Paccar Inc., 562 F.2d 518, 528–29 (8th Cir. 1977).

Fireman's Fund Insurance Company v. Canon U.S.A. Inc., Case No. 03-3836, 394 F.3d 1054 (U.S. App. 2005). LEXIS 471; 66 Fed. R. Evid. Serv. (Callaghan) 258; CCH Prod Liab. Rep. P17,274. Filed January 12, 2005.

Frye v. United States, 293 F.1013 (DC Cir. 1923).

General Electric Company v. Joiner, 66 U.S.L.W. 4036 (1997).

Kumho Tire Co. Ltd. v. Carmichael, 119 S. Ct. 1167 (1999). U.S. LEXIS 2199 (March 23, 1999).

LaSalle National Bank et al. v. Massachusetts Bay Insurance Company et al., Case No. 90 C 2005. (U.S. Dist. 1997). LEXIS 5253. Decided April 11, 1997. Docketed April 18, 1997.

Michigan Miller's Mutual Insurance Company v. Benfield, 140 F.3d 915 (11th Cir. 1998).

Marilyn McCarver v. Farm Bureau General Insurance Company, Case Number 2004-3315-CKM, State of Michigan, County of Berrien. Decided February 1, 2006.

Mark and Dian Workman v. AB Electrolux Corporation et al., Case No. 03-4195-JAR (U.S. Dist. 2005). LEXIS 16306. Decided August 8, 2005.

James B. McCoy et al. v. Whirlpool Corp. et al., Civil Action No. 02-2064-KHV; 214 F.R.D. 646 (U.S. Dist. 2003). LEXIS 6901; 55 Fed. R. Serv. 3d (Callaghan) 740.

In re Paoli Railroad Yard PCB Litigation, 35F.3d 717 (U.S. App. 1994). LEXIS 23722; 30 Fed. R. Serv.3d (Callaghan) 644; 25 ELR 20989; 35 F.3d at 743. Filed August 31, 1994.

Pappas,Andronic, et al. v. Sony Electronics Inc. et al., Civil Action No. 96-339J, 136 F. Supp. 2d 413 (U.S. Dist. 2000). LEXIS 19531, CCH Prod. Liab. Rep. P15,993. Decided December 27, 2000.

Polizzi Meats Inc. v. Aetna Life and Casualty, 931 F. Supp. 328 (D.N.J. 1996).

Bethie Pride v. BIC Corporation, Société BIC, S.A., Civil Case No. 98-6422; 218 F.3d 566 (U.S. App. 2000). LEXIS 15652; 2000 FED App. 0222P (6th Cir.); 54 Fed. R. Serv. 3d (Callaghan) 1428; CCH Prod. Liab. Rep. P15,844. Decided July 7, 2000.

Royal Insurance Company of America as Subrogee of Patrick and Linda Magee v. Joseph Daniel Construction Inc., Civil Action No. 00-Civ.-8706 (CM); 208 F. Supp. 2d 423 (U.S. Dist. 2002). LEXIS 12397. Decided July 10, 2002.

Snodgrass, Teri, Robert L. Baker, Kendall Ellis, Jill P. Fletcher, Judith Shemnitz, Frank Sherron, and Tamaz

Tal v. Ford Motor Company and United Technologies Automotive Inc., Civil Action No. 96-1814 (JBS) (U.S. Dist. 2002). LEXIS 13421. Decided March 28, 2002.

State Farm and Casualty Company, as subrogee of Rocky Mountain and Suzanne Mountain, v. Holmes Products; J.C. Penney Company, Case No. 04-4532 (3rd Cir. 2006). LEXIS 2370. Argued January 17, 2006. Filed January 31, 2006. (Unpublished).

Ronald Taepke v. Lake States Insurance Company, Fire No. 98-1946-18-CK, Circuit Court for the County of Charlevoix, State of Michigan. Entered December 8, 1999

TNT Road Company et al. v. Sterling Truck Corporation, Sterling Truck Corporation, Third-Party Plaintiff v. Lear Corporation, Third-Party Defendant. Civil No. 03-37-B-K (U.S. Dist. 2004). LEXIS 13463; CCH Prod. Liab. Rep. P17,063 (U.S. Dist. 2004). LEXIS 13462 (D. Me., July 19, 2004). Decided July 19, 2004.

Truck Insurance Exchange v. MagneTek Inc. (U.S. App. 2004). LEXIS 3557 (February 25, 2004).

John Witten Tunnell v. Ford Motor Company. Civil Action No. 4:03CV74; 330 F. Supp. 2d 731 (U.S. Dist. 2004). LEXIS 24598 (W.D. Va., July 3, 2004). Decided July 2, 2004.

Theresa M. Zeigler, individually; and Theresa M. Zeigler, as mother and next friend of Madisen Zeigler v. Fisher-Price Inc. No. C01-3089-PAZ (U.S. Dist. 2003). LEXIS 11184; Northern District of Iowa. Decided July 1, 2003. (Unpublished).

Travelers Property and Casualty Corporation. v. General Electric Company, Civil Action No. 3; 98-CV-50(SRU); 150 F. Supp. 2d 360 (U.S. Dist. 2001). LEXIS 14395; 57 Fed. R. Evid. Serv. (Callaghan) 695; CCH Prod. Liab. Rep. P16,181. Decided July 26, 2001.

United States of America v. John W. Downing, Crim. No. 82-00223-01 (U.S. Dist. 1985). LEXIS 18723; 753 F.2d 1224, 1237 (3d Cir. 1985). Decided June 20, 1985.

United States of America v. Ruby Gardner, No. 99-2193, 211 F.3d 1049 (U.S. App. 2000). LEXIS 8649, 54 Fed. R. Evid. Serv. (Callaghan) 788.

United States v. Markum, 4 F.3d 891 (10th Cir. 1993).

United States v. Ortiz, 804 F.2d 1161 (10th Cir. 1986).

United States of America v. Lawrence R. Black Wolf Jr., CR 99-30095 (U.S. Dist. 1999). LEXIS 20736. Decided December 6, 1999.

Vigilant Insurance v. Sunbeam Corporation. No. CIV-02-0452-PHX-MHM; 231 F.R.D. 582 (U.S. Dist. 2005). LEXIS 29198. Decided November 17, 2005.

Wal-Mart Stores, Petitioner, v. Charles T. Merrell,Sr., et al., Supreme Court of Texas, No. 09-0224, June 2010.

Courtesy of Capt. Sandra K. Wesson, Crystal Fire Department, Little Rock, Arkansas.

KEY TERMS

ambient, *p. 62*

autoignition, *p. 72*

British thermal unit (BTU), *p. 64*

conduction, *p. 58*

convection, *p. 59*

detonation, *p. 75*

endothermic, *p. 57*

exothermic, *p. 56*

flameover, *p. 80*

flame spread, *p. 66*

flashover, *p. 64*

flash point, *p. 75*

fully involved, *p. 69*

heat release rate, *p. 64*

piloted ignition, *p. 73*

radiation, *p. 61*

rollover, *p. 80*

self-heating, *p. 58*

smoldering combustion, *p. 58*

soot, *p. 60*

superposition, *p. 61*

thermal inertia, *p. 71*

thermoplastics, *p. 70*

vapor, *p. 56*

volatile, *p. 77*

watt, *p. 64*

OBJECTIVES

After reading this chapter, the student should be able to:

- Describe the basic dynamics of fire.
- Identify fire dynamics as it relates to forensic fire scene reconstruction.
- Illustrate the use of fire dynamics in an investigation.
- Determine dynamics in the fire's growth.

The body of fire dynamics knowledge as it applies to fire scene reconstruction and analysis is derived from the combined disciplines of physics, thermodynamics, chemistry, heat transfer, and fluid mechanics. Accurate determination of a fire's origin, intensity, growth, direction of travel, duration, and extinguishment requires that investigators rely on and correctly apply all principles of fire dynamics. Investigators must realize that variability in fire growth and development is influenced by a number of factors such as available fuel load, ventilation, and physical configuration of the room.

Fire investigators can benefit from a working knowledge of published research involving the application of the expertise that exists in the area of fire dynamics. The published research includes several recent textbooks and expert treatises (DeHaan and Icove 2012; Drysdale 2011; Karlsson and Quintiere 2000 Quintiere 1998, 2006), traditional fire protection engineering references (SFPE 2008), U.S. Government research done primarily by the National Institute of Standards and Technology (NIST) and the U.S. Nuclear Regulatory Commission (NRC) (Iqbal and Salley 2004), and fire research journals. These journals include peer-reviewed fire-related ones such as *Fire Technology*, the *Journal of Fire Protection Engineering*, and *Fire Safety Journal*. Fire-related research also appears in peer-reviewed forensic journals such as the *Journal of Forensic Science* and *Science and Justice*.

Many fire investigators do not routinely have access to all these resources. Bridging this knowledge gap is the primary purpose of this chapter. Many basic fire dynamics concepts are described in *NFPA 921,* 2011 ed., pt. 5 (NFPA 2011). These basic concepts and other, more advanced ones will be discussed later in this textbook. Existing real-world examples and training exercises illustrate their application and describe how basic fire dynamics principles apply to fuels commonly found at fire scenes. Case examples are offered in this chapter along with discussions on the application of accepted fire protection engineering calculations that can assist investigators in interpreting fire behavior and conducting forensic scene reconstruction.

Basic Units of Measurement

The common units of measurement and dimensions used in fire scene reconstruction come from two systems: United States (U.S.) and metric. The United States is in the gradual process of converting to the international standard, which is known as the International System of Units (SI, from the French, Système International). We show all measurements and calculations using the SI system, and in most cases, we give the equivalent English units in parentheses for reference. Table 2-1 lists many of these commonly used fundamental dimensions from fire dynamics along with their typical symbol, unit, and conversion factor.

The most fundamental property of fire dynamics is *heat*. All objects above absolute zero (−273°C, −460°F, 0 K) contain heat owing to the motion of their molecules.

TABLE 2-1	Fundamental Dimensions Commonly Used in Fire Dynamics		
DIMENSION	**SYMBOL**	**SI UNIT**	**CONVERSION**
Length	L	m	1 m = 3.2808 ft
Area	A	m^2	1 m^2 = 10.7639 ft^2
Volume	V	m^3	1 m^3 = 35.314 ft^3
Mass	M	kg	1 kg = 2.2046 lb
Mass density	ρ	kg/m^3	1 kg/m^3 = 0.06243 lb/ft^3
Mass flow rate	$\dot{m}$	kg/s	1 kg/s = 2.2 lb/s
Mass flow rate per unit area	$\dot{m}''$	$kg/(s \cdot m^2)$	1 $kg/(s \cdot m^2)$ = 0.205 $lb/(s \cdot ft^2)$
Temperature	T	°C	$T(°C) = [T(°F) - 32]/1.8$
			$T(°C) = T(K) - 273.15$
			$T(°F) = [T(°C) \times 1.8] + 32$
Temperature	T	K	$T(K) = T(°C) + 273.15$
			$T(K) = [T(F) + 459.7]0.56$
Energy, heat	Q, q	J	1 kJ = 0.94738 Btu
Heat release rate	$\dot{Q}, \dot{q}$	W	1 W = 3.4121 Btu/hr = 0.95 Btu/s, 1 kW = 1 kJ/s
Heat flux	$\dot{q}''$	W/m^2	1 W/cm^2 = 10 kW/m^2 = 0.317 $Btu/(hr \cdot ft^2)$

Sources: Derived from SFPE 2008; Quintiere 1998; Karlsson and Quintiere 2000; and Drysdale 2011.

vapor ■ The gaseous phase of a material that is a liquid or solid at normal temperatures and pressures.

exothermic ■ Generating or giving off heat during a chemical reaction.

Sufficient heat transfer (by conduction, convection, radiation) to an object increases its *temperature* and may ignite it, which may spread fire to neighboring objects.

The Science of Fire

Fire is simply a rapid oxidation reaction or combustion process involving the release of energy. Although a fire's energy release may be visible in the form of flames or *glowing combustion*, nonvisible forms exist. Nonvisible forms of energy transfer may include heat transfer in the form of radiation or conduction (Quintiere 1998, 49). Oxidation processes like rusting of iron or yellowing of newsprint generate heat but so slowly that the temperatures of the objects do not increase perceptibly.

FIRE TETRAHEDRON

For fires to exist, four conditions must coexist, which may be represented by what is commonly known as the *fire tetrahedron,* shown in Figure 2-1. The basic components of the tetrahedron are (1) the presence of *fuel* or combustible material, (2) sufficient *heat* to raise the material to its ignition temperature and release fuel **vapors**, (3) enough of an *oxidizing agent* to sustain combustion, and (4) conditions that allow uninhibited **exothermic** chemical *chain reactions* to occur. The science of fire

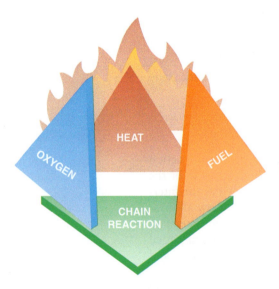

FIGURE 2-1 The fire tetrahedron.

prevention and extinguishment is based on the isolation or removal of one or more of these components.

The action of heat on a flammable liquid fuel causes simple evaporation (i.e., the liquid is transformed into a vapor). The action of heat on most solid fuels causes the chemical breakdown of the molecular structure, called *pyrolysis*, into vapors, gases, and residual solid (char). The pyrolysis also absorbs some heat from the reaction (i.e., it is an **endothermic** process).

The flaming combustion of the gases from a liquid or solid generally takes place within an area above the fuel's surface. Heat and mass transfer processes create a situation in which the fuel generates a region of vapors that can be ignited and sustained with the proper conditions (Figure 2-2).

endothermic ■
Absorbing heat during a chemical reaction.

TYPES OF FIRES

There are four recognized categories of combustion or fire types: (1) diffusion flames, (2) premixed flames, (3) smoldering combustion, and (4) spontaneous combustion (Quintiere 1998, 25). Diffusion flames make up the category of most natural fuel-controlled flaming fires such as candles, campfires, and fireplaces. These occur as fuel gases or vapors diffuse from the fuel surface into the surrounding air. In the zone where appropriate concentrations of fuel and oxygen are present, flaming combustion will occur. Oxygen in the air will move (diffuse) to the flame reaction zone as it is consumed in the combustion process.

If fuel and oxygen are combined prior to ignition, the resulting combustion process is known as *premixed flames*. These fuels generally consist of either vaporized liquid fuels

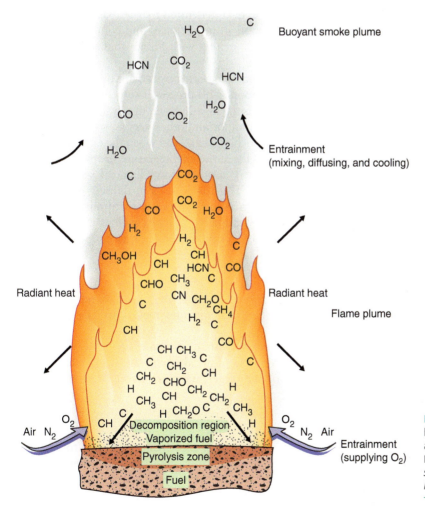

FIGURE 2-2 Schematic representation of the heat and mass transfer processes associated with a burning surface. *From J. D. DeHaan,* Kirk's Fire Investigation, *5th ed., 2002. Reprinted by permission of Pearson Education, Inc., Upper Saddle River, NJ.*

or gases. Ignition of these gas or vapor mixtures is possible within regions bounded by upper and lower fuel/oxygen concentrations. In a jet engine, a special type of flame can occur when the fuel vapor is released under pressure into the air and mixes by mechanical action to produce an ignitable mixture.

Smoldering is typically a slow-propagating, self-sustained exothermic process in which oxygen combines directly with the fuel on its surface or within it, if it is porous. If enough heat is being produced, the surface may glow with incandescent reaction zones. Smoldering fires are characterized by visible charring and absence of flame. A discarded or dropped cigarette on the surface of a couch, mattress, or other upholstered furniture can cause a smoldering fire. Although smoldering combustion does not emit a visible glow, it does generate heat detectable by touch.

The terms glowing combustion and **smoldering combustion** are often used interchangeably, but there is a distinction between the two. Both terms refer to flameless oxidative exothermic combustion where oxygen is combining directly with the solid fuel surface and creating heat. Glowing combustion occurs where direct surface reactions become dominant on charring materials (Babrauskas 2003, 315). If ventilation is aided by draft or extra oxygen, the glowing can be maintained (as observed when bellows are used on a smoldering wood fire). This type of combustion is dependent on external heating and may not be a self-sustaining reaction.

A torch flame played against a wood surface can cause it to char and glow where the flame is applied, but when the flame is removed, the glow stops and the combustion diminishes, often to a complete stop. Smoldering combustion can best be described as a self-sustaining flameless combustion where the heat generated by the reaction heats surrounding fuel to the point where the combustion front advances and continues to grow. The layperson often uses smoldering to mean "burning with flames that are not large," but this is not the correct scientific definition.

A similar misperception may occur when a surface heated by radiant or convective heat transfer is observed to darken and emit white vapors. This process is sometimes erroneously described as smoldering, but it is not. Rather, it is the heat-induced pyrolysis of the fuel surface that causes the formation of wisps of water vapor and other vapors, sometimes called *off-gassing*. This is an endothermic process (absorbing heat), and when the external heating is halted, the vapor or smoke emission stops as the surface temperature drops rapidly.

Like smoldering, *spontaneous combustion* is another slow chemical reaction. **Self-heating** can produce enough heat in a mass of fuel that results in flaming or sustained smoldering combustion at a critical point known as *thermal runaway*. Spontaneous combustion usually involves naturally occurring vegetable-based fuels, such as those found in peanut and linseed oils (DeHaan and Icove 2012, chap. 6; Babrauskas 2003).

Heat Transfer

During a fire, heat is usually transferred from the fire plume to surrounding objects, compartment surfaces, and occupants by three fundamental methods: (1) conduction, (2) convection, and (3) radiation.

The relative importance of these three methods that dominate a fire depends on not only the intensity and size of the fire but also the physical environment. The heat transferred increases the surface temperature of the target, which can cause observable and measurable effects.

CONDUCTION

Through the process of **conduction**, thermal energy passes from a warmer to a cooler area of a solid material. Heat conducted through walls, ceilings, and other adjacent objects often leaves signs of fire damage based on the varying temperature gradients (Figure 2-3).

smoldering combustion ■ The direct combination of a solid fuel with atmospheric oxygen to generate heat in the absence of gaseous flames; see glowing combustion.

self-heating ■ An exothermic chemical or biological process that can generate enough heat to become an ignition source; spontaneous ignition.

conduction ■ Process of transferring heat through a material or between materials by direct physical contact. The transfer of energy in the form of heat by direct contact through the excitation of molecules and/or particles driven by a temperature difference (*NFPA 921*, 2011 ed.).

The general equations used in conductive heat transfer through a solid whose faces are at different temperatures, T_1 (cooler) and T_2 (warmer), are

$$\dot{q} = \frac{kA(T_2 - T_1)}{l} = \frac{kA(\Delta T)}{l}, \qquad (2.1)$$

$$\dot{q}'' = \frac{\dot{q}}{A} = \frac{kA(\Delta T)}{Al} = \frac{k(\Delta T)}{l}, \qquad (2.2)$$

where

$\dot{q}$ = conduction heat transfer rate (J/s or W),

$\dot{q}''$ = conductive heat flux (W/m^2),

k = thermal conductivity of material [W/(m · K)],

A = area through which the heat is being conducted (m^2),

$\Delta T = (T_2 - T_1)$ wall face temperature difference between the warmer, T_2, and cooler, T_1, (K or °C) surfaces, and

l = length of material through which heat is being conducted (m).

FIGURE 2-3 Example of internal conductive heat transfer through the rear doors from a fire inside the vehicle. *Courtesy of Capt. Sandra K. Wesson, Crystal Fire Department, Little Rock, Arkansas.*

Conduction requires direct physical contact between the source and the target receiving the heat energy. For example, heat from a fire in a living room may be conducted through a poorly insulated common wall to an adjoining room, producing surface demarcation patterns. In an investigation, examine and record damage to both sides of interior and exterior walls and surface materials. Interior wall finishing materials such as wood paneling may also have to be removed to evaluate the effects of heat transfer. See Table 2-2 for the thermal characteristics of typical materials found at fire scenes.

The graphs in Figure 2-4 illustrate the temperature during short- and medium-duration conductions and steady-state conduction through a solid material. During short-duration conduction, heat flux events produce high surface temperatures, with the inner core staying much cooler. Increasing the time duration of heat transfer increases the internal temperature of the solid and eventually produces a nearly linear temperature gradient from front (hot) to rear (cool).

CONVECTION

The second process, **convection**, is the transfer of heat energy via movement of liquids or gases from a warmer to a cooler area, also known as *Newton's law of cooling*. An indicator of convective heat transfer damage may be found in the region where a fire plume contacts the ceiling directly above the plume, as shown in Figure 1-7.

The general equations used in convective heat transfer are

$$\dot{q} = h A (T_\infty - T_S) = h A (\Delta T), \qquad (2.3)$$

$$\dot{q}'' = \frac{\dot{q}}{A} = \frac{hA(\Delta T)}{A} = h \, \Delta T, \qquad (2.4)$$

where

$\dot{q}$ = convective heat transfer rate (J/s or W),

$\dot{q}''$ = convective heat flux (W/m^2),

h = convective heat transfer coefficient [W/(m^2 · K)],

A = area through which the heat is being convected (m^2), and

$\Delta T = (T_\infty - T_S)$ temperature difference between the fluid, T_∞, and the surface, T_S (K or °C).

convection ■ Process of transferring heat by movement of a fluid (typically gas). In convective flow, a warm fluid is less dense than the surrounding fluid and rises, inducing a circulation.

TABLE 2-2	Thermal Characteristics of Common Materials Found at Fire Scenes			
MATERIAL	THERMAL CONDUCTIVITY k [W/(m · K)]	DENSITY ρ (kg/m³)	SPECIFIC HEAT c_P [J/(kg · K)]	THERMAL INERTIA $k\rho c_P$ [W² · s/(m⁴ · K²)]
Copper	387	8940	380	1.30×10^9
Steel	45.8	7850	460	1.65×10^8
Brick	0.69	1600	840	9.27×10^5
Concrete	0.8–1.4	1900–2300	880	2×10^6
Glass	0.76	2700	840	1.72×10^6
Gypsum plaster	0.48	1440	840	5.81×10^5
PPMA	0.19	1190	1420	3.21×10^5
Oak	0.17	800	2380	3.24×10^5
Yellow pine	0.14	640	2850	2.55×10^5
Asbestos	0.15	577	1050	9.09×10^4
Fiberboard	0.041	229	2090	1.96×10^4
Polyurethane foam	0.034	20	1400	9.52×10^2
Air	0.026	1.1	1040	2.97×10^1

Sources: Data derived from Drysdale 2011, 37.

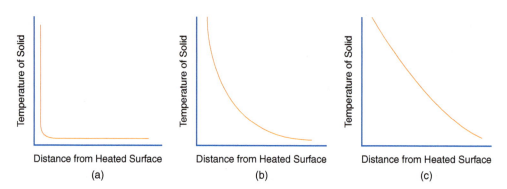

FIGURE 2-4 Temperature distribution through conduction for (a) short duration, (b) medium duration, and (c) steady state.

The term h, the convective heat transfer coefficient [W/(m² · K)], is dependent on the nature of the surface and the velocity of the gas. Typical values for air are 5–25 W/(m² · K) for free convection and 10–500 W/(m² · K) for forced convection (Drysdale 2011).

Closer examination of surface features will usually indicate the direction and intensity of the fire plume and ceiling jet, which contain heated products of combustion such as hot gases, **soot**, ash, pyrolysates, and burning embers. During the early stages of fire development the most extensive damage normally exists at the plume's impingement on the ceiling directly above the fire source. Generally, materials farther from the fire plume will exhibit less damage. In the later stages of fire development the more extensive convection-related damage may be located remote from the fire origin. This can be associated with the flow of heated combustion products through vents and adjacent compartments and may

soot ■ The carbon-based solid residue created by incomplete combustion of carbon-based fuels.

include the intense burning of pyrolysates (fuel) where adequate mixing with oxygen occurs. Figure 2-5 shows the external convective heating of a car door from a gasoline fire of short duration.

RADIATION

Radiation is the emission of heat energy via electromagnetic waves from a surface at a temperature above absolute zero (0 K). Heat transfer by radiation often damages surfaces facing the fire plume, including immobile objects such as furniture. Convection and radiation are responsible for most victim burn injuries. Radiant heat travels in straight lines. It can be reflected by some materials and transmitted by others. The total thermal effect on a material's target surface may be a combination of conductive, convective, and radiative heat transfer rates.

The equation for *radiative heat flux* received by a target surface is represented in general terms by

$$\dot{q}'' = \varepsilon\sigma T_2^4 F_{12}, \tag{2.5}$$

where

$\dot{q}''$ = radiative heat transfer flux (kW/m²),
ε = emissivity of the hot surface (dimensionless),
σ = Stefan-Boltzmann constant = 5.67×10^{-11} (kW/m²)/K⁴ *or* kW/(m² · K⁴),
T_2 = temperature of the emitting body (K), and
F_{12} = configuration factor (depends on the surface characteristics, orientation, and distance).

On heated solid and liquid surfaces the emissivity is commonly 0.8 ± 0.2. The emissivities for gaseous flames depend on their fuel and thickness, and may be very low if the layer is thin. Most are approximately 0.5–0.7 (Quintiere 1998).

Guidelines for interpreting the damage due to radiation are simple—the investigator should generally start with the least damaged surfaces farthest from the apparent fire plume, systematically advancing toward the apparent area of most intense fire. Areas of more intense damage to these surfaces often are closer to the most severe radiant heat source, as shown in Figure 1-5. By comparing these damaged surfaces, an investigator can often determine the direction of the fire movement and intensity as it developed. After this evaluation, the investigator assesses the fire development and spread to assist in determining an area and potential point of fire origin.

SUPERPOSITION

Superposition is the combined effect of two or more fires or heat transfer effects, a phenomenon that may generate confusing fire damage indicators. The predominant heat transfer mode through direct flame impingement (flame contact, for example) involves both radiant and convective heat transfer from the flaming gases to the surface as well as some conduction through the surface. Direct flame impingement is a case of superposition in which the high radiant heat flux and high convective transfer of the flame quickly affect the exposed fuels. Such combined effects can quickly ignite fuels and raise others to very high surface temperatures.

Because the goal of an arsonist is often the rapid destruction of a building, multiple-set fires are often encountered. These fires, which initially consist of separate plumes,

FIGURE 2-5 Example of external heating of a car door from a very short duration gasoline fire that only scorched the paint on the door. The gasoline was poured outside the door with its window closed. The glass shattered by convected and radiated heat. *Courtesy of J. D. DeHaan.*

radiation ■ Transfer of heat by electromagnetic waves.

superposition ■ The combined effects of two or more fires or heat transfer effects, a phenomenon that may generate confusing fire damage indicators can confuse even the most experienced investigator, particularly when their sequence of ignition is unknown, and the effects of separate fires overlap. Note also that separate multiple fires can occur when there is a collapse or "falldown" of burning materials, such as hanging draperies or window shades involved as fire spreads from a single origin. If multiple points of fire origin are suspected, the investigator should evaluate the combined heat transfer effects or superposition of these plumes.

EXAMPLE 2-1 ■ Heat Transfer Calculations

Problem. A fire in a storage room in an automobile repair garage raises the temperature of the interior surface of the brick walls and ceiling over a period of time to 500°C (932°F or 773 K). The **ambient** air temperature on the outside of the brick wall of the garage is 20°C (68°F or 293 K). The thickness of the brick wall is 200 mm (0.2 m), and its convective heat transfer coefficient is 12 W/(m² · K). Calculate the temperature to which the exterior of the wall will be raised.

ambient ■ Surrounding conditions.

Suggested Solution: Conductive-Convective Problem. In this solution, the configuration is considered a plane wall of uniform homogenous material (brick) having a constant thermal conductivity, a given interior surface temperature of $T_S = 500°C$ (932°F or 773 K and exposed to an external garage ambient air temperature of $T_a = 20°C$ (68°F or 293 K). The external surface temperature, T_E, of the wall is unknown. This temperature might be important later if combustible materials come into contact with this wall surface.

Since the steady-state heat transfer rate of conduction through the brick wall is equal to the heat transfer rate of convection passing into the ambient air outside the garage,

$$\dot{q}'' = \frac{\dot{q}}{A} = \frac{T_S - T_E}{L / k_a} = \frac{T_E - T_a}{1 / h_G},$$

where

$\dot{q} =$ conduction heat transfer rate (J/s or W),
$\dot{q}'' =$ conductive heat flux (W/m²),
$h =$ convective heat transfer coefficient [W/(m² · K)](brick to air),
$k =$ thermal conductivity of material [W/(m · K)] (through brick),
$L =$ path length through which heat is being conducted (m),
$A =$ area through which the heat is being conducted (m²), and
$\Delta T = (T_S - T_E)$ wall face temperature difference between the warmer T_S and cooler T_E (K).

$$\frac{773 \text{ K} - T_E}{0.2 \text{ m} / \left[0.69 \text{ W} / (\text{m} \cdot \text{K}) \right]} = \frac{T_E - 293 \text{ K}}{1 / \left[12 \text{ W} / \left(\text{m}^2 \cdot \text{K} \right) \right]}$$

$$2665 - 2.448 \, T_E = 12.05 \, T_E - 3530$$

$$T_E = \frac{6195}{15.49} = 399 \text{ K} = 126°C \left(258°F \right)$$

Suggested Alternative Solution: Electrical Analog. Assuming that the brick maintains its integrity during the fire, determine the steady-state temperature of the wall's surface in the garage (example modified from Drysdale 2011, 39). A one-dimensional electrical circuit analog, shown in Figure 2-6, is used in this solution, because we can look at the way heat transfers through a material in the same way that current flows through resistors.

(a)

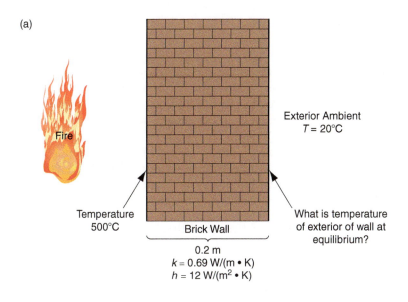

Exterior Ambient
$T = 20°C$

Fire

Temperature
500°C

Brick Wall

What is temperature
of exterior of wall at
equilibrium?

0.2 m
$k = 0.69$ W/(m • K)
$h = 12$ W/(m² • K)

(b)

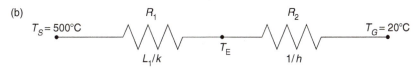

R_1 R_2

$T_S = 500°C$ $T_G = 20°C$

T_E

L_1/k $1/h$

FIGURE 2-6 (a) Heat transfer by conduction: surface temperature calculation. (b) Heat transfer by conduction: electrical analog.

Solutions by electrical analogs are often used. The steady-state heat transfer solution is best illustrated by a direct-current electrical circuit analogy, where the temperatures are voltages ($V = T$), the thermal resistances of heat convection and conduction are represented as electrical resistances ($R_1 = L_b/k_n$, and $R_2 = 1/h_b$, respectively). The heat flux, $\dot{q}''$, whose units are W/m², is analogous to the current, that is, $I = \dot{q}''$.

The problem can be represented as shown in Figure 2-6b, as a steady-state electrical circuit analog, where R_1 and R_2 are the thermal resistances of the brick wall and inner air of the garage, respectively.

Temperature in storage room $\quad T_S = 500°C$ (773 K)
Temperature in garage $\quad T_G = 20°C$ (293 K)
Thermal conductivity $\quad k = 0.69$ W/(m · K)
Thickness of the brick $\quad L = 200$ mm (0.2 m)
Conductive thermal resistance $\quad R_1 = \Delta L/k = (0.2$ m)/(0.69 W/ (m · K) = 0.290 (m² · K)/W

Convective heat transfer coefficient $\quad h = 12$ W/(m² · K)
Convective resistance $\quad R_2 = 1/h = (1)/[12$ W/(m² · K)] = 0.083 (m² ·K)/W

Total heat flux $\quad \dot{q}'' = (T_S - T_G)/(R_1 + R_2) = 1287$ W/m²

$\quad T_E = 773$ K $- (\dot{q}'')(R_1)$

Exterior wall temperature $\quad T_E = 773$ K $- 374$ K
$\quad = 399$ K
$\quad T_E = 399$ K $- 273 = 126°C$ (258°F)

Heat Release Rate

RELEASE RATES

heat release rate ■ The rate at which heat is generated by a source, usually measured in watts, joules/second, or Btu/second.

The **heat release rate** (HRR) of a fire is the amount of heat released per unit time by a heat source. The term is often expressed in **watts** (W), kilowatts (kW), megawatts (MW), kilojoules per second (kJ/s), or **BTUs** per second (BTU/s) and is denoted in formulas as $\dot{Q}$ (the dot over the Q means per unit time). For general conversion purposes, 3412 Btu/hr = 0.95 BTU/s = 1 kJ/s = 1000 W = 1 kW.

watt ■ Unit of power, or rate of work, equal to one joule per second, or the rate of work represented by a current of one ampere under the potential of one volt (*NFPA 921*, 2008 ed.).

The HRR is essentially the size or power of the fire. Three important effects of HRR make it the single most important variable in describing a fire and the hazards associated with its interaction with buildings and their occupants Bukowski 1995b; (Babrauskas and Peacock 1992):

- Heat release rate increase (increase the size of the developing fire)
- Correlation with other effects (smoke production, temperature)
- Occupant tenability

British thermal unit (BTU) ■ The quantity of heat required to raise the temperature of one pound of water 1°F at a pressure of 1 atmosphere; a British thermal unit is equal to 1055 joules, 1.055 kilojoules, or 252.15 calories (*NFPA 921*, 2011 ed., pt. 3.3.19).

First, and most important, heat released by a fire is the driving force that subsequently produces more fuel by evaporation or pyrolysis. This increased fuel production can lead to a further increase in heat release rate (a growing fire). As long as adequate quantities of fuel and oxygen are available, the transfer of heat from the burning fuel causes pyrolysis or an exothermic condition, and more heat is released from it and surrounding fuels. It is not guaranteed that all the fuel will burn. As with burning characteristics of fuels, there are physical limits to theoretical and actual heat release rates.

Second, another important role of the heat release rate in forensic fire scene reconstruction is that it directly correlates with many other variables. Examples include the production of smoke and toxic by-products of combustion, room temperatures, heat flux, mass loss rates, and flame height.

Third, the direct correlation of high heat release rates and lethality of a fire is significant. High HRRs produce high mass loss rates of burning materials, some of which may be very toxic. High heat fluxes, large volumes of high-temperature smoke, and toxic gases may overwhelm occupants, preventing their safe escape during fires (Babrauskas and Peacock 1992).

flashover ■ The final stage of the process of fire growth. When all combustible fuels within a compartment are ignited, the room is said to have undergone flashover.

The concept of heat release rates was introduced to fire investigators in 1985, when Drysdale introduced formulas for the calculation of estimated flame heights and onset time for **flashover**. Flashover is discussed in detail later in the chapter.

These equations attempted to answer the following important questions:

- How intense was the fire, and could it ignite nearby combustibles or produce thermal injuries to humans?
- Was an ignitable liquid of sufficient quantity used in the fire, and how high did the flames reach?
- Were conditions right for flashover to occur?
- When did the smoke detectors and/or sprinklers activate?

An investigator's most basic question centers on the peak heat release rate, which can range from several watts to thousands of megawatts. Table 2-3 lists peak HRRs for common burning items of potential interest to fire investigators.

The peak heat release rate is dependent on the fuel present, its surface area, and its physical configuration. The HRRs of many typical fuels are available on the NIST website, fire.nist.gov/. Other fire dynamics factors that relate to the HRR include mass loss, mass flux, heat flux, and combustion efficiency.

TABLE 2-3	Peak Heat Release Rates for Common Objects Found at Fire Scenes			
MATERIAL	TOTAL MASS (kg)	PEAK HEAT RELEASE RATE (kW)	TIME (s)	TOTAL HEAT RELEASED (MJ)
Cigarette	—	0.005 (5 W)	—	—
Wooden kitchen match or cigarette lighter	—	0.050 (50 W)	—	—
Candle	—	0.05–0.08 (50–80 W)	—	—
Wastepaper basket	0.94	50	350	5.8
Office wastebasket with paper	—	50–150	—	—
Pillow, latex foam	1.24	117	350	27.5
Small chair (some padding)	—	150–250	—	—
Television set (T1)	39.8	290	670	150
Armchair (modern)	—	350–750 (typical)–1.2 MW	—	—
Recliner (synthetic padding and covering)	—	500–1000 (1 MW)	—	—
Christmas tree (T17)	7.0	650	350	41
Pool of gasoline (2 qt, on concrete)	—	1 MW	—	—
Christmas tree (dry, 6–7 ft)	—	1–2 MW (typical)–5 MW	—	—
Sofa (synthetic padding and covering)	—	1–3 MW	—	—
Living room or bedroom (fully involved)	—	3–10 MW	—	—

Sources: Data derived from Babrauskas (SFPE 2008, chap. 3-1); Karlsson and Quintiere 2000, 26; Gorbett and Pharr 2010, 133; and DeHaan and Icove 2012, 35.

HEAT OF COMBUSTION

A material's heat of combustion is the amount of heat released per unit of mass during combustion, expressed in (MJ/kg) or kilojoules per gram (kJ/g). The term Δh_c denotes the *complete* heat of combustion and is used when all the material combusts, leaving no fuel behind. The effective heat of combustion, denoted by Δh_{eff}, is reserved for more realistic cases where combustion is not complete, and the heat lost to heating the fuel is considered.

MASS LOSS

The mass loss rate or burning rate of a fuel typically depends on three factors: type of fuel, configuration, and area involved. For example, a campfire constructed with small sticks of dry wood arranged in a tent configuration will burn more quickly and yield a higher heat release rate over a shorter period of time than one using wet, large-dimensioned wood placed in a haphazard arrangement. This burning rate is expressed in kilograms per second (kg/s) or grams per second (g/s).

The heat release rate can be calculated from the mass loss rate and heat of combustion of an object. The heat release rate can be calculated using the general equation (2.6). Mass loss rates can be experimentally determined by simply weighing a fuel package while it burns.

The heat release rate, $\dot{Q}$, can be calculated from

$$\dot{Q} = \dot{m}\,\Delta h_c,$$ (2.6)

where

$\dot{Q}$ = heat release rate (kJ/s or kW),
$\dot{m}$ = burning rate or mass loss rate (g/s), and
Δh_c = heat of combustion (kJ/g).

MASS FLUX

Another concept related to the heat release rate is a material's *mass flux* or *mass burning rate* per unit area, expressed in kg/(m^2 · s) and denoted by $\dot{m}''$ in equations. It is measured in laboratory tests where the fuel surface area (pool diameter) and orientation are controlled variables, as they can affect $\dot{m}''$ dramatically. When the horizontal burning area of a material is known along with the burning rate per unit area, the heat release rate equation is written

$$\dot{Q} = \dot{m}''\,\Delta h_c A,$$ (2.7)

where

$\dot{Q}$ = total heat release rate (kW),
$\dot{m}''$ = mass burning rate per unit area [g/(m^2 · s)],
Δh_c = heat of combustion (kJ/g), and
A = burning area (m^2).

The amount of heat needed to convert a solid or liquid fuel to a combustible form (gas or vapor) is called the *latent heat of vaporization* and represents heat that has to be created to initiate and sustain combustion. The energy needed to evaporate a mass unit of fuel, expressed in kilojoules per kilogram (kJ/kg), is denoted by H_L in equations. Solids, unlike liquids, have values of H_L that are grossly time dependent. It is very hard to obtain accurate H_L values for solids. Values for Δh_c, $\dot{m}''$, and H_L are listed in Table 2-4 and are available in the reference literature for various fuels.

HEAT FLUX

flame spread ■ The rate at which flames extend across the surface of a material (usually under specific conditions).

Another fire dynamics concept vital to interpreting ignition, **flame spread**, and burn injuries is *heat flux*. Heat flux is the rate at which heat is falling on a surface or passing through an area, measured in kilowatts per square meter (kW/m^2), and is represented by the symbol $\dot{q}''$.

TABLE 2-4	Typical Mass Flux, Heat of Vaporization, and Effective Heat of Combustion for Common Materials		
COMMON MATERIAL	MASS FLUX $\dot{m}''$ [g/(m^2 · s)]	HEAT OF VAPORIZATION H_L (kJ/g)	HEAT OF COMBUSTION (EFFECTIVE) ΔH_{eff} (kJ/g)
Gasoline	44–53	0.33	43.7
Kerosene	49	0.67	43.2
Paper	6.7	2.21–3.55	16.3–19.7
Wood	70–80	0.95–1.82	13–15

Sources: Derived from SFPE 2008 and Quintiere 1998.

TABLE 2-5	Impact of a Range of Radiant Heat Flux Values

RADIANT HEAT FLUX (kW/m²)	OBSERVED EFFECT ON HUMANS AND WOODEN SURFACES
170	Maximum heat flux measured in postflashover fires
80	Thermal protective performance (TPP) test
52	Fiberboard ignites after 5 seconds
20	Floor of residential family room at flashover
16	Pain, blisters, second-degree burns to skin at 5 seconds
7.5	Wood ignites after prolonged exposure (piloted or not)
6.4	Pain, blisters, second-degree burns to skin at 18 seconds
4.5	Blisters, second-degree burns to skin at 30 seconds
2.5	Exposure while firefighting, pain and burns after prolonged exposure
<1.0	Exposure to sun

Sources: Derived from *NFPA 921*, 2011 ed., table 5.5.4.2 (NFPA 2011) and Gratkowski, Dembsey, and Beyler 2006

The radiant heat output (emissive power) of a surface at temperature T is expressed as

$$\dot{q}'' = \varepsilon \sigma T^4, \qquad (2.8)$$

where

ε = emissivity of the source fire,
σ = Stefan-Boltzmann constant, and
T = Kelvin temperature of the surface,

so the higher the temperature of a source, the greater the heat flux from that source. Note that heat flux increases as a function of temperature, T (K), to the fourth power.

Two major pieces of information can be obtained if an investigator can estimate the heat flux and its duration: (1) whether ignition of a secondary or target surface can be achieved and (2) whether thermal injury to a victim exposed to a fire is possible. The minimum heat flux needed to produce thermal injuries and to ignite some common fuels is presented in Table 2-5.

Heat fluxes from various flames and fires are listed in detail in the *Ignition Handbook* (Babrauskas 2003, chap. 1). Because in most cases these values are difficult to measure and even more difficult to accurately estimate, the investigator should carefully study these published experimental data.

In the simplest of cases, radiant heat flux from a fire plume can be approximated as emerging from a single point source known as the *virtual origin*, a concept discussed in more detail in Chapter 3. A schematic illustration of the concept of radiant heat transfer from a point source of a flame to a target is shown in Figure 2-7 (NFPA 2008, chaps. 10–11).

The equation used to predict the incident heat flux falling on an object or victim from a point-source fire, as in Figure 2-7, is

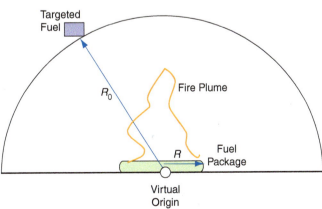

FIGURE 2-7 Schematic illustration of radiant heat transfer from a point source (virtual origin) of a fire plume to a target.

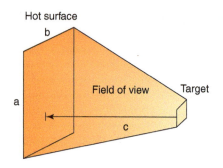

FIGURE 2-8 A small target facing a large radiant heat source (*a* by *b* in size) a distance *c* away is subject to heating from a wide angle of view.

$$\dot{q}_0'' = \frac{P}{4\pi R_0^2} = \frac{X_r \dot{Q}}{4\pi R_0^2}, \qquad (2.9)$$

where

$\dot{q}_0''$ = incident radiation flux on target (kW/m^2),
P = total radiative power of the flame (kW),
R_0 = distance to the target surface (m),
X_r = radiative fraction (typically 0.20–0.60), representing the percentage of heat released from the source as radiant heat, and
$\dot{Q}$ = total heat release rate of source (kW),

or, by rearrangement,

$$R_0 = \left(\frac{X_r \dot{Q}}{4\pi \dot{q}_0''} \right)^{1/2}. \qquad (2.10)$$

This equation is valid when the ratio of the distance to the targeted object to the radius (R) of the burning fuel package is

$$\frac{R_0}{R} > 4, \qquad (2.11)$$

that is, when the radius of the burning surface of the fire, R, is small compared with the distance between the target and the fire, R_0, as shown in Figure 2-7.

For cases where

$$\frac{R_0}{R} \leq 4, \qquad (2.12)$$

the source cannot be treated as a point, and the geometry of the target surface relative to the source becomes critical. For each geometry, a configuration factor will have to be calculated. The configuration factor (F_{12}) appears in the general radiant heat relation

$$\dot{q}'' = \varepsilon \sigma F_{12} T^4 \quad \text{or} \quad \dot{Q} F_{12}.$$

Because the calculations can be very complex, F_{12} for a geometry such as that in Figure 2-8 is usually determined from a nomograph such as that in Figure 2-9. Nomographs for various geometries are found in the *SFPE Handbook* (SFPE 2008, App. D, A44-47).

EXAMPLE 2-2 ■ Heat Flux Calculation

Problem. A 500 kW fire represented as a point source as shown in Figure 2-7 has a surface radius of 0.2 m. Assume the radiative fraction to be 0.60. What is the heat flux falling on the target fuel at distances of R_0 equal to 1, 2, and 4 m?

Suggested Solution. The expected heat release rate can be calculated from equation (2.6).

Heat release rate	$\dot{Q}$ = 500 kW
Radiative fraction	X_r = 0.60
Radius of the burning fuel surface	R = 0.20 m
Equation ratio test	$R_0/R > 4$
Incident radiation flux on target	$\dot{q}_0'' = X_r \dot{Q}/4\pi R_0^2$
At R_0 equal to 1, 2, and 4 m	$\dot{q}_0''$ = 23.87, 5.96, and 1.49 kW/m^2, respectively

Note that as the distance is doubled, the radiant heat flux decreases by a factor of one-fourth. This is known as the *inverse square rule*.

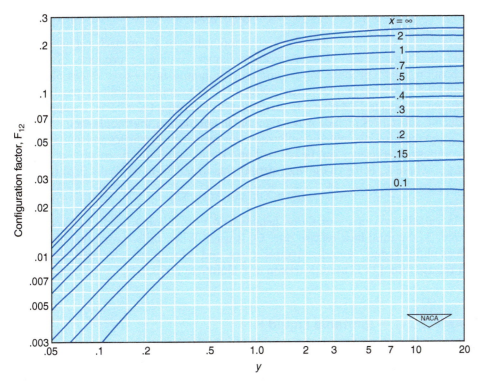

FIGURE 2-9 The configuration factor F_{12} for a geometry like that shown in Figure 2-8 is determined from this monograph, where $x = a/c$, and $y = b/c$. The curve with the closest value for x is selected. The intersection of that curve with the vertical line equal to the calculated value of y gives F_{12}. *From Hamilton and Morgan, NACA Tech. Note 2836, December 1952.*

STEADY BURNING

Once a material is exposed to sufficient incident heat on the fuel and an ample air supply, the fire may spread evenly, become **fully involved**, and burn at a steady state. The formulas for steady-state burning are

$$\dot{m}'' = \dot{q}'' / H_L \tag{2.13}$$

and

$$\dot{Q}'' = \frac{\dot{q}''}{L_V} \Delta h_c, \tag{2.14}$$

where

$\dot{Q}''$ = heat release rate per unit surface area of burning fuel (kW/m^2),
$\dot{m}''$ = mass burning rate per unit area [g/(m$^2 \cdot$ s)],
$\dot{q}''$ = net incident heat flux from the flame (kW/m^2),
L_V = latent heat of vaporization, the heat required to produce fuel vapors (kJ/g), and
Δh_c = heat of combustion (kJ/g).

These equations are useful because they allow investigators to make estimates of the steady-state burning rate. For many solid fuels, L_V is not constant, owing to char formation and complex heat and mass transfer processes. It can vary greatly with time and is not easily measured.

COMBUSTION EFFICIENCY

The *combustion efficiency* of a substance is the ratio of its effective heat of combustion to the complete heat of combustion and is expressed by the term X, where

fully involved ■ The entire area of a fire building so involved with heat, smoke, and flame that entry is not possible until some measure of control has been obtained with hose stream attacks.

$$X = \Delta h_{\text{eff}} / \Delta h_c. \tag{2.15}$$

Thus, for fuels with more complete combustion, values of X approach 1. Fuels with lower combustion efficiencies—characterized by flames containing products of incomplete combustion, which produce sootier and more luminous fires—have values of X between 0.6 and 0.8 (Karlsson and Quintiere 2000, 64). For example, gasoline typically has a Δh_c of 46, but its Δh_{eff} is 43.8 owing to incomplete combustion ($X = 0.95$). Examples of soot-producing substances include many flammable liquids and **thermoplastics**.

To account for incomplete combustion (for fuels such as plastics and ignitable liquids), the combustion efficiency can be inserted into the equation for the convective heat release rate, $\dot{Q}_c$, where X_r is the energy loss attributable to radiation (typically 0.20 to 0.40):

$$\dot{Q}_c = X\dot{m}'' \, \Delta h_c A (1 - X_r). \tag{2.16}$$

In calculations of mean flame heights and virtual origins, which are explored in Chapter 3, the total heat release rate, $\dot{Q}$, is used. When other properties of fire plumes such as plume radius, centerline temperatures, and velocity are calculated, the convective heat release rate, $\dot{Q}_c$, is used (Karlsson and Quintiere 2000, 64).

thermoplastics ■
Organic materials that can be melted and resolidified without chemical degradation.

EXAMPLE 2-3 ■ Ignitable Liquid Fire

Problem. An arsonist pours a quantity of gasoline onto a concrete floor, creating a 0.46-m (1.5-ft)-diameter pool. The arsonist ignites the gasoline, creating a fire very similar to a classic burning pool. Estimate the heat release rate from the burning pool using historical fire testing results.

Suggested Solution. Assuming an unconstrained pool fire, the expected heat release rate can be calculated from equation (2.8).

Heat release rate	$\dot{Q} = \dot{m}'' \, \Delta h_{\text{eff}} A$
Mass flux rate for gasoline	$\dot{m}'' = 0.036 \text{ kg}/(\text{m}^2 \cdot \text{s})$ (reduced by incomplete combustion)
Heat of combustion for gasoline	$\Delta h_{\text{eff}} = 43.7 \text{ MJ/kg}$
Area of the burning pool	$A = (\pi/4)(D^2) = (3.1415/4)(0.46^2) = 0.166 \text{ m}^2$
Heat release rate	$\dot{Q} = (0.036)(43,700)(0.166) = 261 \text{ kW}$

The heat release rate of a real fire under these conditions would be expected to be in the range 200–300 kW; but in actuality, a thin layer of gasoline will show a heat release rate of only approximately 25 percent of the value computed by the preceding equations, which were developed for pools of substantial depth. For a detailed discussion, see the work by Putorti, McElroy, and Madrzykowski (2001, 52) of the NIST.

Fire Development

Fires, particularly those within enclosed structures, burn in a predictable manner over time. For the purpose of fire scene reconstruction, development can be divided into four separate *fire phases*, each containing unique characteristics and its own time frame. These phases are applicable to both single fuel packages and well-ventilated compartment fires. These fire development curves are often referred to as the *fire signature*. The phases are as follows:

- ■ *Phase 1:* Incipient ignition
- ■ *Phase 2:* Growth
- ■ *Phase 3:* Fully developed fire
- ■ *Phase 4:* Decay

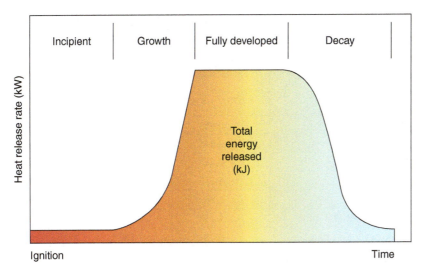

FIGURE 2-10 Fire development phases that often form a fire signature, consisting of (1) ignition, (2) growth, (3) fully developed fire, and (4) decay. *From the* SFPE Handbook of Fire Protection Engineering, *4th ed., 2008, by permission of Society of Fire Protection Engineers*

As shown in Figure 2-10, these phases form a characteristic fire signature or design curve fire, which is often studied for items identified as the source of the first fuel burned in typical structural fires. These phases are important to understanding a fire's growth and its physical effects on the compartment and its contents; calculating the activation of smoke, heat, and automated sprinkler systems; estimating evacuation times; and determining the risk of heat exposure to occupants.

The following physical features are associated with each fire development phase:

- *Phase 1:* Low heat, some smoke, no detectable flame
- *Phase 2:* More heat, lots of smoke, flame spread
- *Phase 3:* Massive heat and smoke output, extensive flame
- *Phase 4:* Decreased flame and heat, considerable smoke, with increased smoldering

At the transition from phase 2 to phase 3, flashover may occur, depending on the total amount of heat released, the amount of ventilation, the amount of fuel available, and the size and configuration of the compartment. In the absence of flashover, the fire may not become fully developed and instead may reach a maximum size because of limited available fuels, flame spread, or a ventilation limit.

PHASE 1: INCIPIENT IGNITION

The fundamental properties that influence an object's *ignitability* are its density, ρ, heat capacity, c_p, and thermal conductivity, k. Table 2-2 listed these characteristic values for some common materials found at fire scenes. The calculated product of these three characteristics $\left(k\rho c_p\right)$ is often referred to as an object's **thermal inertia** and the calculated value $k/\rho c_p$ is called the *thermal diffusivity*. As with mechanical inertia—where the higher the inertia of an object, the harder it is to move—the higher the thermal inertia, $k\rho c_p$, of a material, the harder it is to ignite—that is, more heat has to be applied or longer exposure is required for ignition to occur.

Table 2-6 lists the ignition properties of some common fuels, where T_{ig} is the approximate surface temperature observed at ignition, and $\dot{q}''_{critical}$ is the minimum heat flux needed to bring about ignition (Drysdale 2011, 33). These results are applicable to short-term exposures (10 to 30 minutes) of fresh, whole products. If heating is prolonged (hours), much lower $\dot{q}''_{critical}$ values may be found. For example, wood products may ignite at approximately 7.5 kW/m² under such conditions. These values do not include

thermal inertia ■ The properties of a material that characterize its rate of surface temperature rise when exposed to heat. Thermal inertia is related to the product of the material's thermal conductivity (k), its density (ρ), and its heat capacity (c_p) (*NFPA 921*, 2011 ed., pt. 3.3.170).

TABLE 2-6	Ignition Properties		
MATERIAL	**THERMAL INERTIA** $k\rho c_p$ KW/(m² · K)s²	**IGNITION TEMPERATURE** T_{ig} (°C)	**CRITICAL HEAT FLUX** $\dot{q}''_{critical}$ (kW/m²)
Plywood, plain (0.635 cm)	0.46	390	16
Plywood, plain (1.27 cm)	0.54	390	16
Plywood, FR (1.27 cm)	0.76	620	44
Hardboard (6.35 mm)	1.87	298	10
Hardboard (3.175 mm)	0.88	365	14
Hardboard, gloss paint (3.4 mm)	1.22	400	17
Hardboard, nitrocellulose paint	0.79	400	17
Particleboard (1.27-cm stock)	0.93	412	18
Douglas fir particleboard (1.27 cm)	0.94	382	16
Fiber insulation board	0.46	355	14
Polyisocyanurate (5.08 cm)	0.020	445	21
Foam, rigid (2.54 cm)	0.030	435	20
Foam, flexible (2.54 cm)	0.32	390	16
Polystyrene (5.08 cm)	0.38	630	46
Polycarbonate (1.52 mm)	1.16	528	30
PMMA, type G (1.27 cm)	1.02	378	15
PMMA, polycast (1.59 mm)	0.73	278	9
Carpet #1 (wool, stock)	0.11	465	23
Carpet #2 (wool, untreated)	0.25	435	20
Carpet #2 (wool, treated)	0.24	455	22
Carpet (nylon/wool blend)	0.68	412	18
Carpet (acrylic)	0.42	300	10
Gypsum board, common (1.27 mm)	0.45	565	35
Gypsum board, FR (1.27 cm)	0.40	510	28
Gypsum board, wallpaper	0.57	412	18
Asphalt shingle	0.70	378	15
Fiberglass shingle	0.50	445	21
Glass-reinforced polyester (2.24 mm)	0.32	390	16
Glass-reinforced polyester (1.14 mm)	0.72	400	17
Aircraft panel, epoxy fiberite	0.24	505	28

Source: Derived from Quintiere and Harkleroad 1984.

long-term heating (months to years), for which the self-heating properties of wood and some other materials will have to be recognized.

autoignition ■ Ignition due to sufficient surrounding temperature in the absence of an external source of ignition; nonpiloted ignition.

The temperature of an object is a measure of the relative amount of heat (energy) contained within it. This heat is determined by the mass of the object and its heat capacity or specific heat. The heat capacity, c_p, of an object is a property that determines how much heat must be added to the object to increase its temperature.

The ignition of an object can be calculated when it is exposed to radiant or convective heat flux, $\dot{q}''$, with no initial contact with flames (**autoignition**). The calculation assumes that a first burning object (ignition source) provides sufficient radiant heat flux to ignite a second, nonburning fuel object nearby. Experimental fire testing by Babrauskas has established three levels of ignitability: *easy, normally resistant,* and *difficult* (SFPE 2002b). To establish these relationships, this fire testing included paper, wood, polyurethane, and polyethylene fuels.

Thin materials such as paper and curtains will easily be ignited when exposed to a threshold radiant flux of 10 kW/m^2 or greater. Thicker or more massive items with low thermal inertia values such as upholstered furniture will ignite when exposed to a radiant flux of 20 kW/m^2 or greater. Solid materials with thicknesses greater than 13 mm (0.5 in.) such as plastics or dense woods having large thermal inertia values will ignite when exposed to a radiant flux of 40 kW/m^2 or greater (SFPE 2008). However, there are exceptions to this basic yardstick.

Equations (2.17), (2.18), and (2.19) are derived from data plots from Babrauskas (1982) published in *SFPE Engineering Guide to* **Piloted Ignition** *of Solid Materials under Radiant Exposure* (SFPE 2002b). The heat release rates, $\dot{Q}$, needed for a fire source to ignite another fuel package a distance D (measured in meters) away, can be calculated for fuels classified as *easy* ($\dot{q}''_{crit} \geq 10$ kW/m^2), *normally resistant* ($\dot{q}''_{crit} \geq 20$ kW/m^2), and *difficult* ($\dot{q}''_{crit} \geq 40$ kW/m^2), as shown, respectively, in the following equations:

piloted ignition ▪ The initiation of combustion by means of contact of gaseous material with an external high-energy source, such as a flame, spark, electrical arc, or glowing wire.

$$\dot{Q} = 30 \times 10^{\left(\frac{D+0.8}{0.89}\right)}, \quad \dot{q}'' \geq 10 \text{ kW/m}^2, \tag{2.17}$$

$$\dot{Q} = 30\left(\frac{D+0.05}{0.019}\right), \quad \dot{q}'' \geq 20 \text{ kW/m}^2, \tag{2.18}$$

$$\dot{Q} = 30\left(\frac{D+0.02}{0.0092}\right), \quad \dot{q}'' \geq 40 \text{ kW/m}^2, \tag{2.19}$$

where

  $\dot{Q}$ = heat release rate of the source fire (kW), and
  D = distance from the radiation source to the second object (m).

These relationships were derived for ignition of a second item from an initially burning piece of furniture. They should not be applied to heat release rates from sources that can burn indefinitely (e.g., gas jets).

The time to ignition for *thermally thin* and *thermally thick* materials can be estimated by several equations contained in engineering practice guides developed by the SFPE (2002a, 2008). The generally accepted equations for *ignition time* of thermally thin and thermally thick materials are (2.18) and (2.19), respectively. Under thermally thin conditions, $l_p \leq 1$ mm (l = thickness of the material),

$$t_{ig} = \rho c_p l_p \left(\frac{T_{ig} - T_\infty}{\dot{q}''}\right). \tag{2.20}$$

In thermally thick conditions, $l_p \geq 1$ mm, the thickness term disappears:

$$t_{ig} = \frac{\pi}{4} k\rho c_p \left(\frac{T_{ig} - T_\infty}{\dot{q}''}\right)^2, \tag{2.21}$$

where

  t_{ig} = time to ignition (s),
  k = thermal conductivity [W/(m · K)],
  ρ = density (kg/m^3),
  c_p = specific heat capacity (kJ/kg · K),
  l_p = thickness of material (m),

T_{ig} = piloted fuel ignition temperature (K),

T_{∞} = initial temperature (K), and

$\dot{q}''$ = radiant heat flux (kW/m^2).

The ignition behavior of liquids is different, and the preceding equations for solids should not be used.

PHASE 2: FIRE GROWTH

Fire growth rates, as shown in Figure 2-10, can sometimes be modeled using mathematical relationships that show the flame spread and fire growth estimations using heat release rates. The growth of the flame front across a horizontal fuel surface of a solid fuel is known as the *lateral flame spread* and can be expressed by the equation (Quintiere and Harkelroad 1984)

$$V = \frac{\dot{q}''}{\rho c_p A \left(T_{ig} - T_S\right)^2} \tag{2.22}$$

or as

$$V = \frac{\phi}{k\rho c_p \left(T_{ig} - T_S\right)^2}, \tag{2.23}$$

where

V = lateral velocity of flame spread (m/s),

$\dot{q}''$ = radiant heat flux (kW/m^2),

ρ = density (kg/m^3),

c_p = specific heat capacity [kJ/(kg · K)],

A = cross-sectional area affected by the heating of $\dot{q}''$,

ϕ = ignition factor from flame spread data (kW^2/m^3) (experimentally derived),

k = thermal conductivity [W/(m · K)],

T_{ig} = piloted fuel ignition temperature (°C), and

T_S = unignited, ambient surface temperature (°C).

The values for ϕ and other variables needed for calculating lateral flame spread can be obtained from *ASTM E1321-09* (2009). Table 2-7 lists the horizontal flame spread data for several common materials. Obviously, an external draft pushes the flame down against the

TABLE 2-7	Example Lateral Flame Spread Data		
MATERIAL	**PILOTED IGNITION TEMPERATURE** T_{ig} **(°C)**	**AMBIENT SURFACE TEMPERATURE** T_S **(°C)**	**IGNITION FACTOR** ϕ **(kW^2/m^3)**
Wool carpet			
Treated	435	335	7.3
Untreated	455	365	0.89
Plywood	390	120	13.0
Foam plastic	390	120	11.7
PMMA	378	<90	14.4
Asphalt shingle	356	140	5.4
Acrylic carpet	300	165	9.9

Source: Derived from *ASTM E1321* (ASTM 2009).

downwind side and increases its radiant heat effect on the fuel in front of its path, so the flame front will grow faster if wind aided.

Lateral or downward flame spread on thick solids such as wood is typically slow, on the order of 1×10^{-3} m/s (1 mm/s). In contrast, upward spread on solid fuels is typically 1×10^{-2} to 1 m/s (10–1000 mm/s). Lateral flame spread rates across liquid pools are typically 0.01–1 m/s, depending on the ambient temperature and the **flash point** of the liquid. (Figure 2-11) Premixed vapor/air mixtures can have flame spread rates of 0.1–0.5 m/s for laminar flames. Turbulent flames cause higher flame spreads (1–2 m/s over pools of gasoline are typical). Some fuel/air mixtures can produce pressure-driven flames that accelerate to **detonation** velocities (over 1000 m/s) (Drysdale 2011, 278–84).

Another example of lateral surface flame spread over a solid fuel is a grass or wildland fire. Lateral flame spread in this type of fire depends on several key variables: the type of fuel and its configuration, wind velocity and direction, humidity, and terrain. The brush or duff along the forest floor is usually involved in the initial phases of the fire.

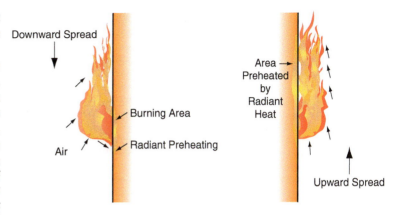

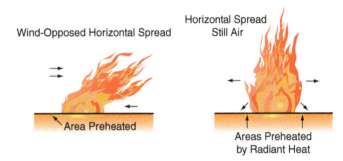

FIGURE 2-11 Radiant heat ahead of an existing flame determines its rate of spread.

The terrain, radiant heat transfer, and wind-induced convective heat transfer can contribute significantly to the flame spread along the forest floor. Flame spread through the canopy of trees or brush is a different process and is part of the complexity being studied in ongoing research relating to wildland–urban interface fires.

The *Thomas formula* (1971), which can be used to approximate wind-aided horizontal flame spread in grass fires on flat ground, is

$$V = \frac{k(1+V_\infty)}{\rho_b}, \tag{2.24}$$

where

V = velocity of flame spread (m/s),
k = 0.07 wildland fire, 0.05 wood crib (kg/m^3),
V_∞ = wind speed (concurrent) (m/s), and
ρ_b = bulk density of the fuel (kg/m^3).

flash point ■ The minimum temperature at which an ignitable vapor is first produced by a liquid fuel.

detonation ■ An extremely rapid reaction that generates very high temperatures and an intense pressure/shock wave that produces violently disruptive effects. It propagates through the material at supersonic speeds.

EXAMPLE 2-4 ■ Estimating Lateral Flame Travel Velocity

Problem. An investigator determines that a fire ignited at the edge of a wool carpet (untreated). What was the estimated velocity of the lateral flame spread, assuming a high radiant heat impact that produced an initial ambient surface temperature of 365°C?

Suggested Solution. Use the Quintiere and Harkelroad equation (2.23) and *ASTM E1321-09* data, from Table 2-7.

Ignition factor ϕ = 0.89 kW^2/m^3
Fuel thermal inertia $k\rho c_p$ = 0.25 $\left[\text{kW}^2 \cdot \text{s} / \left(\text{m}^4 \cdot \text{K}^2 \right) \right]$

Piloted fuel ignition temperature	$T_{ig} = 455°C$
Ambient surface temperature	$T_S = 365°C$
Velocity of flame spread	$V = \phi/k\rho c_p (T_{ig} - T_S)^2 = 0.89/0.25(455 - 365)^2$
	$= 0.33 \times 10^{-3}$ m/s $= 26$ mm/min

Notice that the lower the initial temperature (T_S) is, the larger the denominator of equation (2.23) is, and the slower V will be. If the fuel surface is not preheated, the ambient surface temperature becomes $T_S = T_a = 25°C$, where $T_a =$ ambient air temperature, and the velocity of the flame spread V is calculated as follows:

$$V = \phi/k\rho c_p (T_{ig} - T_S)^2 = 0.89/0.25(455 - 25)^2$$

$$= 0.019 \times 10^{-3} \text{ m/s} = 1.14 \text{ mm/min}$$

EXAMPLE 2-5 ■ Wildland Flame-Spread Velocity

Problem. Children playing with matches set a fire during a dry season of the year in a flat-terrain wildland area. The wind is blowing at a velocity of 2 m/s, and the bulk density of the low brush along the forest floor is 0.04 g/cm³. Determine the velocity of the flame spread from this fire with and without considering the effect of the wind.

Suggested Solution. Use the Thomas equation (2.24).

Velocity of flame spread	$V = k(1 + V_\infty)/\rho_b$
Wind speed	$V_\infty = 2$ m/s
Bulk density	$\rho_b = 0.04$ g/cm³ $= 40$ kg/m³
Equation constant	$k = 0.07$ kg/m³
Flame spread velocity (with $V_\infty = 0$ m/s)	$= (0.07)(1 + 0)/(40)$
	$= 0.00175$ m/s $= 1.75$ mm/s $= 0.1$ m/min
Flame spread velocity (with $V_\infty = 2$ m/s)	$= (0.07)(1 + 2)/(40)$
	$= 0.00525$ m/s $= 5.25$ mm/s
	$= 0.315$ m/min

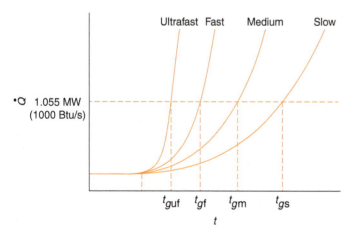

FIGURE 2-12 The fire growth factor α is derived from the graph of Q versus time, where $\alpha = \left(1055/t_g^2\right)$, and t_g is the time required for the heat release rate to grow from its baseline (incipient fire) value to 1.055 MW (1000 Btu/s).

Fire growth rates during the initial phases of development can sometimes be modeled using mathematical relationships. One common relationship assumes that the initial growth rate geometrically approximates the square of the time that a fire has burned (also called a t^2 fire), as shown in Figure 2-12. A perfect fire, unlimited by fuel or ventilation factors, would have an exponential growth rate. This behavior applies only to the growth phase.

This accepted growth rate formula is

$$\dot{Q} = \left(1055/t_g^2\right)t^2 = \alpha t^2, \qquad (2.25)$$

where

t = time (s),
t_g = time required for fire to grow from ignition to 1.055 MW (1000 BTU/s),
$\dot{Q}$ = heat release rate (MW) at time t, and
α = fire growth factor (kW/s²) for the material being burned.

TABLE 2-8	Four Growth Rates in t^2 Fires		
TYPE OF FIRE	**EXAMPLE OBJECTS**	**TIME t_g (s)**	**FIRE GROWTH FACTOR α (kW/s^2)**
Slow	Thick wooden objects (tables, cabinets, dressers)	600	0.003
Medium	Lower-density objects (furniture)	300	0.012
Fast	Combustible objects (paper, cardboard boxes, drapes)	150	0.047
Ultrafast	Volatile fuels (flammable liquids, synthetic mattresses)	75	0.190

Sources: Derived from *NFPA 92* (NFPA 2012) and *NFPA 72* (NFPA 2010).

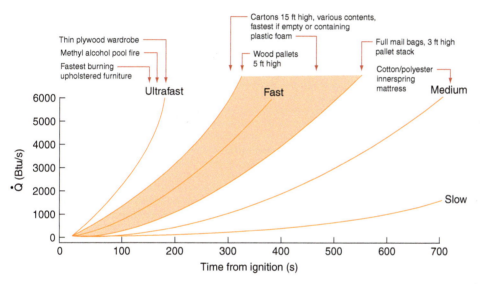

FIGURE 2-13 Relation of t^2 fires to several individual fire tests. The heat release rate is (1000 BTU/s = 1.055 MW). *Courtesy of NIST, from Nelson and Tu 1991.*

According to an accepted principle used by *NFPA 72* (detectors) (NFPA 2010) and *NFPA 92* (NFPA 2012) (smoke control systems), Table 2-8 and Figure 2-13 illustrate four types of growth times used to define fires by the time required to reach 1.055 MW (1000 BTU/s). The time t_g is obtained from repeated calorimetry tests of specific fuel packages.

Table 2-9 shows typical ranges for the heat release rates per unit area and characteristic times for a fire to reach 1.055 MW (1000 Btu/s) in size. These data were obtained from actual fire testing of common materials stored in warehouses.

The *slow growth curve* ($t_g = 600$ s) usually involves thick solid objects such as solidwood tables, cabinets, and dressers. The *medium growth curve* ($t_g = 300$ s) applies to lower-density solid fuels such as upholstered and lightweight furniture. The *fast growth curve* ($t_g = 150$ s) involves thin combustible items such as paper, cardboard boxes, and drapes. The *ultrafast growth curve* ($t_g = 75$ s) is for some flammable liquids, some older types of upholstered furniture and mattresses, or materials containing other **volatile** fuels. (NFPA 2008, 3–127). Of course, the geometry of the objects also has an impact on growth rates.

volatile ■ A liquid having a low boiling point; one that is readily evaporated into the vapor state.

TABLE 2-9	Typical Ranges for Fire Behavior of Materials Stored in Warehouses	
MATERIAL	TYPICAL HEAT RELEASE RATE PER FLOOR AREA COVERED (MW/m^2)	CHARACTERISTIC TIME FOR t^2 FIRES TO REACH 1 MW (s)
Wood pallets		
Stacked 0.46 m (1.5 ft) high	1.3	155–310
Stacked 1.5 m (5 ft) high	3.7	92–187
Stacked 3.1 m (10 ft) high	6.6	77–115
Stacked 4.6 m (15 ft) high	9.9	72–115
Polyethylene bottles in cartons, stacked 4.6 m (15 ft) high	1.9	72
Polyethylene letter trays, stacked 1.5 m (5 ft) high	8.2	189
Mail bags, filled, stored 4.6 m (15 ft) high	0.39	187
Polystyrene jars in cartons, 4.6 m (15 ft) high	14	53

Source: Derived from Quintiere 1998, tables 6-5 and 6-6.

The t^2 growth rate applies after the initial incipient fire phase, during which the growth rate is nearly zero. In some cases involving polyurethane upholstered furniture, the steady-state phase does not exist, and the approximation for the heat release rate resembles a triangular shape. The approach of fitting triangular shape to model the heat release rate signatures of furniture can represent up to 91 percent of the total heat released (Babrauskas and Walton 1986).

EXAMPLE 2-6 ■ Loading Dock Fire

Problem. A fire, which was extinguished by an automatic sprinkler system, occurred on the back loading dock of a supply distribution center. A nearby security camera captured the silhouette of a young male running from the loading dock prior to the activation of the sprinkler system.

On examination of the fire scene, the investigator determines that the fire was intentionally set in a cart of polyethylene letter trays on two pallets stacked 1.56 m (5 ft) high on the loading dock. Determine for the investigator the heat release rate 30 s and 120 s after the arsonist set the fire.

Suggested Solution. Use equation (2.25) for t^2 fire growth and data from Table 2-9 on typical ranges for fire behavior of polyethylene letter trays.

Time to reach 1 MW $\quad t_g = 189$ s
Growth time of fire $\quad t = 30$ s
Heat release rate
$$\dot{Q} = \left(1055/t_g^2\right) t^2$$
$$= \left(1055/189^2\right)\left(30^2\right) \cong 25 \text{ kW}$$

and for $\quad t = 120$ s
$$\dot{Q} = \left(1055/189^2\right)\left(120^2\right) = 425 \text{ kW}$$

PHASE 3: FULLY DEVELOPED FIRE

The fully developed phase is also referred to as the *steady-state phase,* either when the fire reaches its maximum burning rate (based on the amount of fuel available) or when there is insufficient oxygen to continue the growth process (Karlsson and Quintiere 2000, 43). Fuel-controlled fires are associated with steady-state fires, particularly when sufficient oxygen is supplied. The term *fully developed* does not distinguish postflashover fires from those that die out due to a lack of oxygen.

When there is insufficient oxygen, the fire becomes ventilation controlled. In this case, the fire is usually in an enclosed room, and temperatures may become higher than in fuel-controlled fires. Fully developed fires in compartments are often postflashover fires in which all the fuel is involved, and the size of the fire becomes ventilation controlled.

PHASE 4: DECAY

The fire decay during phase 4 usually occurs when approximately 20 percent of the original fuel remains (Bukowski 1995a). Although most fire service and fire protection specialists are interested in the first three phases of a fire, there are significant problems associated with the decay phase.

In high-rise buildings, for example, even after the fire is extinguished, trapped occupants may need to be rescued. Furthermore, fire investigators entering the building may also be exposed to residual amounts of toxic by-products of combustion, since high concentrations of CO and other toxic gases are produced during the decay phase (often dominated by smoldering).

EXAMPLE 2-7 ■ Reliable Test Data

Problem. A fire investigator is trying to develop reliable test data on the length of time a fire may have burned and an estimate of the corresponding heat release rate. The fire was reportedly started by a child playing with a cigarette lighter while sitting at the center of a mattress. The mattress was the only piece of furniture involved in the fire in the room, and a smoke detector in the hallway reportedly activated shortly after the fire began.

Suggested Solution. The information needed by the investigator is found at the "Fire on the Web" Internet site of the NIST (fire.nist.gov/). Collections of actual test data for commonly found items in residential and commercial settings can be found in the Fire Tests/Data section. The data files include still photos, video, graphs, miscellaneous data, and setup data for NIST fire modeling software programs.

Figure 2-14 shows the information needed by the investigator to compare this case with a similar fire under laboratory testing conditions. The fire test data revealed that the peak heat release rate occurred at approximately 150 seconds (2.5 minutes) after ignition, resulting in a 750 kW fire. When forming the hypothesis for the fire, the investigator could evaluate whether the fire's timeline is reasonably consistent with witness accounts, damage indicators, and heat release data. The investigator needs to remember that the test mattress may not be the same construction as the one involved in the fire and that even the same mattress will behave

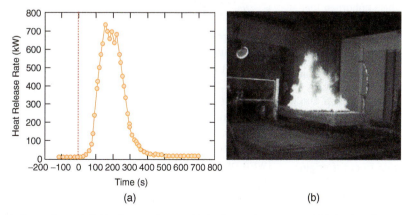

FIGURE 2-14 Reliable fire test data on (a) the heat release rate and (b) the fire after 120 seconds to support an investigation of a mattress fire. *Courtesy of NIST, Fire on the Web, fire.nist.gov/.*

differently if ignited at a corner or from underneath. The test was conducted in a large enclosure so as to minimize heat radiation effects (which would increase the rate of burning) and ventilation effects (which could limit the maximum rate). These variables must be recognized by the investigator when testing the relevant hypothesis. Good documentation of the fire scene will be the key to success.

Enclosure Fires

The growth of a fire confined to a room is usually constrained by the ventilation-limited flow of air, smoke, and hot gases into and out of the enclosure. Confining variables may include the ceiling height, ventilation openings formed by windows and doors, room volume, and location of the fire in the room or compartment. While the fire gases are constrained, a notable event called *flashover* may occur, as a consequence of the heat release rate of the combustible materials contained within the room.

MINIMUM HEAT RELEASE RATE FOR FLASHOVER

In the growth of a compartment (room) fire, *flashover* is defined (per *NFPA 921*) as the transition phase in the development of a compartment fire in which surfaces exposed to thermal radiation reach ignition temperature more or less simultaneously, and fire spreads rapidly throughout the space, resulting in full room involvement or total involvement of the compartment or enclosed space (NFPA 2011). This transition may occur in just a few seconds in small, heavily fueled rooms, or over a period of minutes in large rooms, or may not take place at all due to insufficient heat release rate of the fire. A useful reference on this subject is found in *NFPA 555: Guide on Methods for Evaluating Potential for Room Flashover* (NFPA 2009a).

Fire researchers have documented other observations during this transition phase. Defining observations often include one or more of the following criteria that address the effects of flashover: ignition of floor targets, high heat flux (>20 kW/m^2) to the floor, and flame extension out of the compartment's vents (Babrauskas, Peacock, and Reneke 2003; Milke and Mowrer 2001; NFPA 2009a; Peacock et al. 1999).

The following may be observed during flashover:

- Flames are emitted from openings in the compartment.
- Upper-layer gas temperature rises to ≥600°C.
- Heat flux at floor level reaches ≥20 kW/m^2.
- Oxygen level in the upper portions of the room decreases to approximately 0–5 percent.
- There is a small, short-term pressure rise of approximately 25 Pa (0.0036 psi).

There are two fundamental definitions for the occurrence of flashover. The first looks at flashover as a thermal balance in which critical conditions are produced when the compartment exceeds its ability to lose heat. The second definition views the compartment as being in a mechanical fluid-filling process. In this approach, the point at which the compartment's cool air layer is replaced with hot fire gases is flashover.

Because flashover represents a transition phase, defining the exact moment of occurrence is often a problem. Prior to flashover, the highest average temperatures are found in the upper or hot gas layer. If the average temperature of the hot gas layer exceeds 600°C (with or without the flaming ignition of the hot smoke called **flameover** or **rollover**, the radiant heat from the layer that is reaching all other exposed fuel in the room exceeds the minimum ignition radiant heat flux for exposed fuels, and those fuels ignite. If flashover occurs, temperatures throughout the room reach their maximum (1000°C is not uncommon) as the two-layer environment of the room breaks down and the entire room becomes a turbulently mixed combustion zone—from floor to ceiling.

flameover ■ The flaming ignition of the hot gas layer in a developing compartment fire. The condition in which unburned fuel (pyrolysate) from the originating fire has accumulated in the ceiling layer to a sufficient concentration (i.e., at or above the lower flammable limit) that it ignites and burns. Can occur without ignition of, or prior to the ignition of, other fuels separate from the origin. Also known as *rollover* (*NFPA 921*, 2011 ed. pt. 3.3.71).

rollover ■ *See* flameover.

The active mixing promotes very effective combustion, with oxygen concentrations dropping below 3 percent and producing very high temperatures. This environment in turn produces radiant heat fluxes of 120 kW/m² or higher, ensuring rapid ignition and burning of all exposed fuel surfaces, including walls, carpets, flooring, and low-lying fuels like baseboards. Also, combustion near ventilation points is enhanced.

Ignition of carpets can produce floor-level flames that sweep under chairs, tables, and other surfaces that were initially protected from the downward radiant heat from the hot gas layer alone. Burning proceeds throughout the room until the fuel supply is exhausted or some attempt is made at extinguishment.

Several variables can influence the minimum heat release rates necessary for flashover to occur. First, the size of the compartment may influence the impact of radiation from the fire to the surfaces of the walls, floor, and ceiling. For small compartments, radiation from the fire enhances rapid temperature increase of exposed surfaces. Also, small compartments may negate the two-zone model.

Vent openings have an impact on flashover, as large vent openings make it necessary to have a very large fire in a room to produce flashover and may produce inaccurate calculations for vent flows. The surface materials on the walls can influence the minimum flashover energy, and some calculations take the heat transfer characteristics of the materials into account (Peacock et al. 1999).

Recent work on flashover calculations along with comparison with experimental data demonstrates that the wall surfaces play a role in heat transfer. Researchers report a trend that the shorter exposure times increase the needed minimum heat release rate for flashover (Babrauskas, Peacock, and Reneke 2003).

An approximation for the minimum heat release rate required for flashover in a room, $\dot{Q}_{fo}$, can be found using the following equation, called the *Thomas correlation* (1981):

FIGURE 2-15 Example room with doorway opening for calculation of the minimum heat release rate needed for flashover.

$$\dot{Q}_{fo} = 378A_o\sqrt{h_o} + 7.8A_w, \qquad (2.26)$$

where

$\dot{Q}_{fo}$ = heat release rate for flashover (kW),
A_o = area of the opening to the compartment (m²),
h_o = height of the compartment opening (m), and
A_w = area of the walls, ceiling, and floor minus the opening (m²).

EXAMPLE 2-8 ■ Calculating Flashover

Problem. Given a 10 m × 10 m room, with a 3-m-high ceiling and a 2.5 m × 1 m opening (see Figure 2-15), determine the minimum heat release rate needed to cause flashover.

Suggested Solution. Use equation (2.26).

Area of compartment opening	A_o	$= (2.5 \text{ m})(1 \text{ m}) = 2.5 \text{ m}^2$
Height of compartment opening	h_o	$= 2.5 \text{ m}$
Area calculations	A_w	$= \text{floor} + \text{walls} + \text{ceiling} - \text{compartment opening}$

One floor $= (10 \text{ m})(10 \text{ m}) = 100 \text{ m}^2$

Four walls $= (4)(10 \text{ m})(3 \text{ m}) = 120 \text{ m}^2$

One ceiling $= (10 \text{ m})(10 \text{ m}) = 100 \text{ m}^2$

$= 320 \text{ m}^2 - 2.5 \text{ m}^2 = 317.5 \text{ m}^2$

Flashover heat release rate
$$\dot{Q}_{fo} = 378 A_o \sqrt{h_o} + 7.8 A_w$$
$$= (378)(2.5)\sqrt{2.5} + (7.8)(317.5)$$
$$= 1494 + 2477 = 3971 \text{ kW} = 3.97 \text{ MW}$$

A realistic answer would be about 4 MW.

ALTERNATIVE METHODS FOR ESTIMATING FLASHOVER

There are several alternative methods for estimating the minimum heat release rate necessary for flashover. The use of several estimates is necessary when applying the scientific method. Consider alternative approaches, evaluate the results of fire models, and compare them with eyewitness accounts of the fire.

The first alternative approach assumes a flashover criterion temperature of a suitable increase in temperature, $\Delta T = 575°C$ (Babrauskas 1980), where

$$\Delta T = T_{fo} - T_{ambient} = 600 - 25 = 575°C. \tag{2.27}$$

$$\dot{Q}_{fo} = 0.6 A_v \sqrt{H_v}, \tag{2.28}$$

where

$\dot{Q}_{fo}$ = heat release rate for flashover (MW),
A_v = area of the door (m^2), and
H_v = door height (m).

The term $A_v \sqrt{H_v}$ is often referred to as the *ventilation factor* and frequently appears in similar equations. In actual experiments, two-thirds of the cases agree and are bounded by the following equations (Babrauskas 1984):

$$\dot{Q}_{fo} = 0.45 A_v \sqrt{H_v}, \tag{2.29}$$

$$\dot{Q}_{fo} = 1.05 A_v \sqrt{H_v}. \tag{2.30}$$

EXAMPLE 2-9 ■ Calculating Flashover: Alternative Method 1

Problem. Repeat the calculation for minimum heat release rate needed to cause flashover as previously presented in Example 2-8 using an alternative method.

Suggested Solution. Use equations (2.28), (2.29), and (2.30), also known as the *Babrauskas correlation formulas.*

Area of the door opening	$A_v = (2.5 \text{ m})(1 \text{ m}) = 2.5 \text{ m}^2$
Height of door opening	$H_v = 2.5 \text{ m}$
Flashover heat release rate	$\dot{Q}_{fo} = 0.6 A_v \sqrt{H_v}$
	$= (0.6)(2.5)\sqrt{2.5} = 2.37 \text{ MW}$
Minimum heat release rate	$\dot{Q}_{fo} = 0.45 A_v \sqrt{H_v} = 1.78 \text{ MW}$
Maximum heat release rate	$\dot{Q}_{fo} = 1.05 A_v \sqrt{H_v} = 4.15 \text{ MW}$

The second alternative approach is based on test data and a flashover criterion temperature difference of $\Delta T = 500°C$ (Lawson and Quintiere 1985; McCaffrey, Quintiere, and Harkleroad 1981):

$$\dot{Q}_{fo} = 610 \sqrt{h_k A_s A_v \sqrt{H_v}}, \tag{2.31}$$

where

h_k = enclosure conductance to ceiling and walls $\left[kW/(m^2 \cdot K) \text{ or } \left(kW/m^2 \right)/K \right]$,

A_s = thermal enclosure surface area, excluding vent or door area (m^2),

A_v = total area of vent or door opening (m^2), and

H_v = height of vent or door opening (m).

The *McCaffrey, Quintiere, and Harkleroad (MQH) approach* requires an estimate for the effective heat transfer enclosure *conductance coefficient*, h_k, of the enclosure. Assuming uniform heating of the enclosure with adequate thermal penetration, the conductance coefficient h_k can be estimated as

$$h_k = \frac{k}{\delta}, \tag{2.32}$$

where

h_k = enclosure conductance to ceiling and walls $[(kW/m^2)/K]$,

k = thermal conductivity of enclosure material $[(kW/m)/K]$, and

δ = enclosure material thickness (m).

EXAMPLE 2-10 ■ Calculating Flashover: Alternative Method 2

Problem. Repeat the calculation for the minimum heat release rate needed to cause flashover given an enclosure lined with 16-mm $\left(\frac{5}{8}\text{-in.} \right)$-thick gypsum board. Also calculate the estimated enclosure conductance.

Suggested Solution. Assuming uniform heating of the enclosure with adequate thermal penetration of the enclosure material, the typical values are as follows:

Heat release rate	$\dot{Q}_{fo} = 610\sqrt{h_k A_s A_v \sqrt{h_v}}$	
Thermal conductivity	k	$= 0.00017 \, (kW/m)/K$
Enclosure thickness	δ	$= 0.016 \, m$
Enclosure conductance	h_k	$= k/\delta$
		$= 0.00017/0.016 = \left(1.062 \times 10^{-2} \right) \left[\left(kW/m^2 \right)/K \right]$
Total surface area	A_s	$= 317.5 \, m^2$
Total vent area	A_v	$= 2.5 \, m^2$
Height of vent	H_v	$= 2.5 \, m$
Flashover heat release rate	$\dot{Q}_{fo}$	$= 610\sqrt{(0.01062)(317.5)(2.5)\sqrt{2.5}}$
		$= 2226.9 \, kW = 2.23 \, MW$

A realistic answer would be about 2.2 MW.

The results 4.0 MW (Example 2-8) and 2.23 MW (Example 2-10) fall within the minimum and maximum calculated range of 1.78 to 4.15 MW, from equations (2.29) and (2.30). Taking a numerical average of the three $\dot{Q}_{fo}$ predictions is an inappropriate calculation.

IMPORTANCE OF RECOGNIZING FLASHOVER IN ROOM FIRES

Those who examine a fire scene after extinguishment or burnout should be aware of the importance of recognizing whether flashover did actually occur and the impact it may have had on the burning of a room's contents. Chapter 3 addresses the interpretation of fire burn patterns, some of which may be created during the *postflashover* burning period.

Postflashover burning in a room produces numerous effects that were once thought to be produced only by accelerated (arson) fires involving flammable liquids such as gasoline. A fireball of burning gasoline vapors can sometimes produce floor-to-ceiling damage whose shallow penetration will be the result of the brief duration of such fires. Walls scorched or charred from floor to ceiling can be due to postflashover burning no matter how the fire began, as well as from a fireball of burning gasoline vapors (DeHaan and Icove 2012).

The combustion of carpets and underlying pads, once considered by fire investigators to be fairly unusual in accidental fires, is quite common in rooms filled with today's high-heat-release materials such as polyurethane foam, and synthetic (thermoplastic) fabrics in upholstery, draperies, and carpets. This problem is exacerbated by the increasing use of highly combustible fibers like polypropylene for the face yarns of economical carpets as well as in the backings of nearly all wall-to-wall carpets. These carpets can melt, shrink, and ignite under moderate radiant heat flux, exposing the combustible urethane foam pad underneath to the radiant heat and flames. This combustion produces intense fires at floor level that in turn can create deep irregular burn patterns on the floor beneath. The involvement of common carpet in corridor fires was the reason that the flooring radiant panel test was developed. The intense radiant heat effects in postflashover fires are not uniform owing to the extremely turbulent combustion, so the effects on floors are not uniform (DeHaan 2001).

Other published experiments report the presence and absence of burn-throughs with gasoline poured and ignited on flooring along with various tile and carpet arrangements (Sanderson 1995). The experiments reported that without the presence of carpet padding, burn-throughs did not occur. However, burn-throughs sometimes were reported with carpet padding, but not at the location where the gasoline was poured. Babrauskas (2005) concurs with DeHaan and Icove (2012) that the most common reason for burn-throughs of flooring is radiant heat from above and not the burning of combustible liquid on the floor. Extended slow combustion of collapsed furniture and bedding can also produce localized burn-throughs of flooring.

Fire damage under tables and chairs (once thought to be the result of the burning of flammable liquids at floor level) can be produced by the flaming combustion of the carpet and pad as they ignite during flashover. High temperatures and total room involvement, at one time thought to be linked to flammable liquid accelerants, are produced during postflashover burning without accelerants. The extremely high heat fluxes produced in postflashover fires can char wood or burn away other material at rates up to 10 times the rate at lower heat fluxes (Babrauskas 2005; Butler 1971).

Localized patterns of smoke and fire damage that can help locate fuel packages, characterize their flame heights, and reveal the direction of flame spread can be obliterated with prolonged postflashover burning, making the reconstruction of the fire very difficult. The fire investigator should be aware of the possibility of flashover in modern room fires and the effects it can produce. Postflashover or full room involvement fires typically exhibit ignition of all exposed fuel surfaces (but not necessarily complete combustion) from floor to ceiling with room corners sometimes spared. Larger rooms have been observed by the authors sometimes to exhibit progressive flashover, with floor-to-ceiling ignition at one end and undamaged floor covering at the other. Brief exposure (under 5 minutes) to postflashover fires does not necessarily compromise fire patterns on walls (see Chapter 8; Hopkins, Gorbett, and Kennedy 2009).

In preflashover burning, the most intense thermal damage will be in areas immediately around (or above) burning fuel packages. The higher temperatures of the hot gas layer will produce more thermal damage in the upper half or upper third of the room. In postflashover fires, all fuels are involved, the fires may become ventilation limited, and the most efficient (highest temperature) combustion occurs in turbulent mixing around the ventilation openings, where the oxygen supply is best. Oxygen-depleted burning occurs in many postflashover rooms where the available fuel exceeds the air supply. See the work by Carman (2008) on improving the understanding of postflashover fire behavior. He has done extensive testing on the production of clean burn patterns by ventilation effects in ventilation-controlled fies (Carman 2010).

Interviews of first-in firefighters may reveal whether a room was fully involved, floor to ceiling, when entry was first made. Investigators must be aware of the conditions that can lead to flashover and how to recognize whether it has occurred. A thorough understanding of heat release rates and fire spread characteristics of modern furniture and the dynamics of flashover will be of great assistance (Babrauskas and Peacock 1992).

POSTFLASHOVER CONDITIONS

As previously stated, the average hot gas temperature in a room or compartment at flashover rises to $\geq 600°C$. At postflashover conditions, the entire compartment is treated as a single homogenous (turbulent and well-mixed) volume sharing a common temperature and fuel/oxygen concentrations of oxygen and other by-products of combustion.

For *ventilation-controlled fires* during postflashover, the heat release rate of the fuel within the compartment is regulated by the available ventilation. More specifically, the rate of inflow of fresh air and the outflow of gases is determined by a combination of the temperature differential across the vent, and the ventilation factor $(A_v \sqrt{H_v})$.

Understanding this insight, we can use the relationship of mass flow into an opening to estimate the maximum heat release rate of fuels burning within a compartment. The mass flow rate of air through an opening is approximately (Karlsson and Quintiere 2000)

$$\dot{m}_{air} = 0.5\, A_v \sqrt{H_v}. \tag{2.33}$$

Because the heat release rate is regulated by the amount of available air, the following relationship can be used to estimate this rate:

$$\dot{Q} = \dot{m}_{air} \frac{\Delta h_c}{r}, \tag{2.34}$$

$$\dot{Q} = 0.5 A_v \sqrt{H_v} \frac{\Delta h_c}{r}, \tag{2.35}$$

where

Δh_c = heat of combustion of the fuel (kJ/kg),
$\dot{m}_{air}$ = mass of air required to burn a mass of fuel [kg (air)/kg (fuel)],
$\dot{Q}$ = heat release rate (kJ/s or kW), and
r = air-to-fuel ratio 5.7 kg (air)/kg (wood).

The term is approximately constant for most fuels. For example, for wood,

$$\Delta h_c = 15,000 \text{ kJ/kg}$$
$$r = 5.7 \text{ kg (air)/kg (wood)}$$
$$\frac{\Delta h_c}{r} = 2630 \text{ kJ/kg (air)}$$

For a wood-fueled fire in a room, the maximum-sized fire can be estimated by assuming 100 percent efficiency. Substituting this value into equation (2.35), we obtain

$$\dot{Q} = 1370\, A_v \sqrt{H_v} \text{ (kJ/s = kW).} \tag{2.36}$$

In postflashover fires, the temperature of the compartment, which may reach 1100°C (2012°F), can also be estimated. The combined work of Thomas (1974) and Law (1978) produced the following correlation for predicting the maximum postflashover temperature, assuming natural ventilation:

$$T_{fo(\max)} = 6000 \frac{\left(1 - e^{-0.1\Omega}\right)}{\sqrt{\Omega}} \tag{2.37}$$

$$\Omega = \frac{A_T - A_v}{A_v\sqrt{H_v}} \tag{2.38}$$

where

$T_{fo(\max)}$ = maximum compartment temperature at flashover,
Ω = ventilation factor
A_T = total area of the compartment-enclosing surfaces, excluding the area of vent openings (m^2),
A_V = total area of the vent openings (m^2), and
H_V = height of the vent openings (m).

EXAMPLE 2-11 ■ Calculating Maximum Heat Release Rate and Temperature in a Ventilation-Controlled Compartment Fire

Problem. Given a ventilation-controlled fire consisting of burning wooden furniture in a 10 m × 10 m room, with a 3-m-high ceiling and a 2.5 m × 1 m opening (see Figure 2-15), determine the minimum heat release rate and maximum compartment temperature, assuming natural ventilation.

Suggested Solution. The minimum heat release rate for a fire to progress in this room with the given opening can be calculated from equation (2.36).

Minimum heat release rate	$\dot{Q}$	$= 1370\, A_v\sqrt{H_v}$
Height of opening	H_v	$= 2.5$ m
Area of opening	A_v	$= (1)(2.5) = 2.5\ m^2$
Minimum heat release rate	$\dot{Q}$	$= 1370 A_v\sqrt{H_v}$
		$= 1370(2.5)\sqrt{2.5} = 5415$ kW $= 5.415$ MW

Maximum compartment temp. at flashover	$T_{fo(\max)} = 6000\dfrac{\left(1 - e^{-0.1\Omega}\right)}{\sqrt{\Omega}}$	
Ventilation factor	Ω	$= \dfrac{A_T - A_v}{A_v\sqrt{H_v}}$
Total area of the vent opening	Av	$= (1)(2.5) = 2.5\ m^2$

Total area of the compartment-enclosing surfaces, excluding area of vent opening

$$
\begin{aligned}
&\text{One floor} = (10\ \text{m})(10\ \text{m}) = 100\ m^2\\
A_T = \ &\text{Four walls} = (4)(10\ \text{m})(3\ \text{m}) = 120\ m^2\\
&\text{One ceiling} = (10\ \text{m})(10\ \text{m}) = 100\ m^2\\
&\qquad\qquad = 320\ m^2 - 2.5\ m^2 = 317.5\ m^2
\end{aligned}
$$

Height of the vent opening $H_v = 2.5$ m

| Ventilation factor | Ω | $= \dfrac{A_T - A_v}{A_v \sqrt{H_v}} = \dfrac{317.5 - 2.5}{2.5\sqrt{2.5}} = \dfrac{315}{3.95} = 79.75$ |

Maximum compartment
temperature at flashover

$$T_{fo(\max)} = 6000 \frac{\left(1 - e^{-0.1\Omega}\right)}{\sqrt{\Omega}}$$

$$= 6000 \frac{\left(1 - e^{-7.975}\right)}{\sqrt{79.75}}$$

$$= 6000 \frac{0.999}{8.93} = 671.6 \text{ °C}$$

Discussion. Since we earlier predicted that a fire of 3–4 MW would be at the verge of triggering flashover in this large room, and the ventilation opening will support a fire of 5 MW or better, there is agreement that the maximum predicted temperature 671°C will exceed the threshold for flashover (i.e., 600°C).

Other Enclosure Fire Events

DURATION

As part of forensic fire scene reconstruction and analysis, other features of fire behavior in enclosures also become important to investigators. The following are commonly cited features (1) smoke detector activation, (2) heat and sprinkler system operation, and (3) limited-ventilation fires. Each of these features is discussed in turn.

Enclosure fire features can be used to answer several questions that arise during an investigation, for example, How long did the fire effectively burn? and, When during this time period did the smoke detector or sprinkler head activate? Answers to these questions can be placed on a timeline of information needed to fill the voids, particularly when fire deaths occur.

In tightly closed rooms, a fire consumes available oxygen and pumps oxygen-depleted air into the hot smoke layer. This layer descends until it reaches the level of the fire. The fire then experiences vitiated burning, existing on what oxygen is available to it in the smoke layer, and reducing, accordingly, its heat release rate. If there is a leak or vent up high in the room, the smoke layer may rise to where the fire is back in normal room air, and the fire resumes its full heat release rate, until the layer descends again. This sequence may produce a cyclical behavior of open flame/smolder/open flame/smolder until the fuel is depleted.

NFPA 555 offers two useful relationships for estimating the duration of burning in a tightly closed compartment—one for fires with a steady heat release rate of $\dot{Q}$, the second for unsteady t^2 fires (NFPA 2009a).

For steady fires:

$$t = \frac{V_{O2}}{\dot{Q}\left(\Delta h_c \rho_{O2}\right)}. \tag{2.39}$$

For growing fires:

$$t = \left[\frac{3V_{O2}}{\alpha\left(\Delta h_c \rho_{O2}\right)}\right]^{1/3} \tag{2.40}$$

where

t = time (s),

V_{O2} = volume of oxygen available to be consumed in the combustion process (m³),

$\dot{Q}$ = heat release rate from steady fire (kW),

$\Delta h_c \rho_{O_2}$ = heat release rate per unit volume of oxygen consumed (kJ/m³), and

α = constant governing the speed of fire growth (kJ/s³) (slow 2.93×10^{-3}, medium 11.72×10^{-3}, fast 46.88×10^{-3})

The quantity V_{O_2} is generally considered to be half the total available oxygen in the room $\left[\text{so } V_{O_2} = 0.5(0.21\, V_{\text{room}}) \right]$, since flaming combustion will generally not be sustained once O_2 levels are in the range 8–12 percent, as cited in *NFPA 555* (NFPA 2009a, 9). This relationship also assumes that the fire is at floor level in the room; if the fire is elevated and not so large that it will induce a great deal of localized turbulent mixing, V_{O_2} will be controlled by the volume of the room at and above the level of the fire.

SMOKE DETECTOR ACTIVATION

A common problem faced by fire investigators is determining when on a timeline of events a thermal fire detector, smoke detector, or sprinkler system was activated. Historically, fire protection engineers have done much work on this area, specifically the DETACT-QS program (Evans and Stroup 1986). These milestone events are usually noticed by a witness or electronically reported by alarm systems and often serve as critical markers in an investigation.

Figure 2-16 shows the measurements needed to calculate these activations. The dimension H is the distance to the ceiling above the fuel surface, and r is the radial distance from the plume centerline to the heat detector or sprinkler head. The centerline is the imaginary vertical line emerging from the center of the fuel source to the ceiling. R is the equivalent radius of the burning fuel package. Virtual origins are discussed in detail in Chapter 3.

Three common methods are used in estimating smoke detector response time, namely, Alpert (1972), Milke (1990), and Mowrer (1990). In some approaches, such as Alpert and Milke, the practice is to use the convective portion of the heat release rate, $\dot{Q}_c$, in calculations. The convective heat release rate fraction, X_c (typically 0.70), is used, where $\dot{Q}_c = X_c \dot{Q}$.

In the Mowrer method, two calculations are necessary to estimate smoke detector activations for steady-state fires. The first calculates the time for the fire gases to reach the ceiling at the plume centerline, or the *plume lag time*. The second calculates the time for the gases to reach the detector from the plume centerline, or the *ceiling jet lag*.

For a steady-state fire of heat release rate $\dot{Q}$, these equations are expressed as

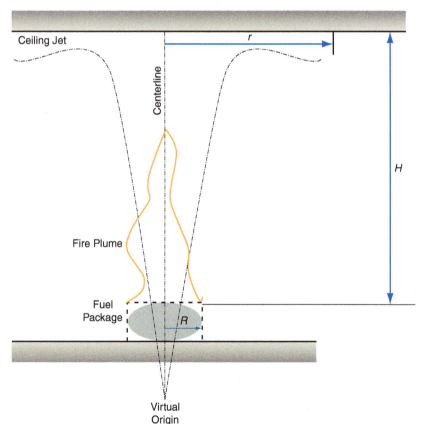

FIGURE 2-16 Fire plume measurements needed in fire detector, smoke detector, or sprinkler system calculations, consisting of fuel surface to ceiling along centerline (H), radius of burning fuel surface (R), and radial distance of ceiling jet from centerline (r).

$$t_{pl} = C_{pl}\, \frac{H^{4/3}}{\dot{Q}^{1/3}}, \qquad (2.41)$$

$$t_{cj} = \frac{1}{C_{cj}} \frac{r^{11/6}}{\dot{Q}^{1/3} H^{1/2}}, \qquad (2.42)$$

where

t_{pl} = transport lag time of plume (s),
t_{cj} = transport lag time of ceiling jet (s),
C_{pl} = plume lag time constant = 0.67 (experimentally determined),
C_{cj} = ceiling jet lag time constant = 1.2 (experimentally determined),
r = radius of distance from plume centerline to the detector (m),
H = height of ceiling above top of fuel (m), and
$\dot{Q}$ = heat release rate of the fire (kW).

EXAMPLE 2-12 ■ Smoke Detector Activation by Wastepaper Basket Fire

Problem. A business owner closed his shop and later claimed that he may have caused a fire by carelessly discarding a lit match into a wastepaper basket before leaving. The fire ignited a small 0.3-m (0.98-ft)-diameter basket filled with ordinary papers. The fire developed quickly to reach a steady-state heat release rate of 100 kW.

The initial temperature of the room was 20°C (293 K, 68°F). The distance from the top of the wastepaper basket to the ceiling is 4 m (13.12 ft). A smoke detector attached to the ceiling is located radially 2 m (6.56 ft) from the fire plume's centerline. Assuming that there are negligible radiation losses in a smooth, level ceiling, determine the time of the smoke detector's activation by the heated ceiling jet gases.

Suggested Solution. Assume that steady-state conditions exist, along with immediate activation.

Height of ceiling above fuel	H	= 4 m
Radius from plume centerline	r	= 2 m
Plume lag time constant	C_{pl}	= 0.67
Ceiling jet lag time constant	C_{cj}	= 1.2
Heat release rate	$\dot{Q}$	= 100 kW
Plume transport time lag	t_{pl}	$= C_{pl} \dfrac{H^{4/3}}{\dot{Q}^{1/3}} = 0.92$ s
Ceiling jet transport lag time	t_{cj}	$= \dfrac{1}{C_{cj}} \dfrac{r^{11/6}}{\dot{Q}^{1/3} H^{1/2}} = 0.32$ s
Detector activation time	t	$= t_{pl} + t_{cj} = 0.92 + 0.32 = 1.24$ s

Discussion. Owing to variables such as fire size, room geometry, the type of detector (photoelectric or ionization), and environmental interactions, the time to activate a smoke detector may vary, especially with ceiling-mounted units. Therefore, the preceding solution may not be accurate, because it does not account for the lag time for the smoke to enter the detector's sensing chamber. The solution assumes that an instantaneous fire of 100 kW occurs, something that takes a real-world fire 30 or more seconds to develop. A suggested practice for determining the activation time for ceiling-mounted smoke detectors is to consider the temperature of the smoke layer (Heskestad and Delichatsios 1977).

Testing by Collier (1996) showed that, approximately, a 4°C rise at the detector location is sufficient for an activation. In several approaches to predicting the time for smoke, heat, and sprinkler systems to activate, the engineering calculations typically use only the convective heating of the sensing elements by the hot fire gases and do not account for any direct heating by radiation from the flames. In heated rooms, the smoke layer from low-energy fires may not be buoyant enough to reach the detector through the fire's static hot layer.

For comparison, the methods of Alpert and Milke provide estimates of 5.89 and 12.48 seconds, respectively, for this problem. Experimental reenactment of a fire condition may reveal critical variables (e.g., placement/location on a wall, ceiling vent location, or time required for development of the fire to steady state).

SPRINKLER HEAD AND HEAT DETECTOR ACTIVATION

In estimating the time to activate a sprinkler head or heat detector, both the ceiling jet temperature and its velocity must be calculated along with the ratio r/H. The equations for these estimates are based on data from a series of actual tests for 670 kW to 100 MW fires (Alpert 1972; NFPA 2008, sec. 4, chap. 1).

The first calculation necessary is the centerline temperature directly above the plume produced by the burning fuel source:

$$T_m = 16.9 \frac{\dot{Q}^{2/3}}{H^{5/3}} + T_\infty \quad \text{for} \quad r/H \leq 0.18. \quad (2.43)$$

For r/H ratios greater than 0.18, the detector or sprinkler falls within the ceiling jet portion of the plume.

$$T_{m_{jet}} = 5.38 \frac{\left(\dot{Q}/r\right)^{2/3}}{H} + T_\infty \quad \text{for} \quad r/H > 0.18, \quad (2.44)$$

where

T_m = plume gas temperature above fire (K),
$T_{m_{jet}}$ = temperature of ceiling jet (K),
T_∞ = ambient room temperature (K),
$\dot{Q}$ = heat release rate from fire (kW),
$\dot{Q}_c$ = convective heat release rate (kW), where $\dot{Q}_c = X_c \dot{Q}$
r = radial distance from plume centerline to device (m), and
H = distance above fuel surface (m).

For calculating the time to activation of a sprinkler head, the generally accepted engineering practice is to consider only the convective heating of the sensing elements by the hot fire gases (Iqbal and Salley 2004, 10–12). This practice does not account for direct heating by radiation and assumes that conduction of heat from the sprinkler to the piping is negligible. Therefore, the value for the convective heat release rate for the fire, $\dot{Q}_c$, is often used in place of the total heat release rate, $\dot{Q}$.

To study detector and sprinkler activation problems further, the maximum velocity of the ceiling jet, U_m, needs to be calculated. The following correlations depend on the value of the r/H ratio.

For maximum jet velocity close to the centerline,

$$U_m = 0.96 \left(\frac{\dot{Q}}{H}\right)^{1/3} \quad \text{for} \quad r/H \leq 0.15. \quad (2.45)$$

For jet velocities farther away from the centerline,

$$U_m = 0.195 \left(\frac{\dot{Q}^{1/3} H^{1/2}}{r^{5/6}}\right)^{1/3} \quad \text{for} \quad r/H > 0.15, \quad (2.46)$$

where

U_m = gas velocity (m/s),
$\dot{Q}$ = heat release rate (kW),
H = distance above fuel surface (m), and
r = radial distance from plume centerline to device (m).

Determination of the time to activation of the heat detector or sprinkler during steady-state fires relies on a term called the *response time index* (RTI). This index assesses the ability of a heat detector to activate from an initial condition of ambient room temperature. Because the detector of a sprinkler has a finite mass, the RTI takes into account the time lag before the temperature of the detector rises.

$$t_{operation} = \frac{RTI}{\sqrt{U_m}} \log_e \left(\frac{T_m - T_\infty}{T_m - T_{operation}} \right), \qquad (2.47)$$

where

RTI	= response time index $\left(m^{1/2} \cdot s^{1/2} \right)$,
U_m	= gas velocity (m/s),
T_m	= plume gas temperature above fire (K),
T_∞	= ambient room temperature (K), and
$T_{operation}$	= operation temperature (K).

RTI values are specified by the manufacturer for each style or model of sprinkler.

EXAMPLE 2-13 ■ Sprinkler Head Activation to Wastepaper Basket Fire

Problem. A closer examination of the debris in Example 2-12 shows that the wastebasket actually contains both paper and plastic, producing a steady-state fire of 500 kW. A standard-response bulb sprinkler head, whose operation temperature is 74°C (165°F, 347 K) and RTI is 235 $m^{1/2} \cdot s^{1/2}$, is located directly above the basket. The ambient temperature of the room is 20°C (68°F, 293 K). Estimate the time to activation of the sprinkler head. When calculating the ceiling jet temperature, use the convective heat release rate, $\dot{Q}_c$.

Suggested Solution. Since the area of interest is directly above the fire plume, the radius from the centerline is zero, which gives a ratio $r/H < 0.18$. Also, the conservative assumption is that the convective heat release rate fraction (X_c) of the fire is 0.70. Therefore, use the appropriate equations to calculate the temperature and velocity.

Heat release rate	$\dot{Q}$	= 500 kW
Convective heat release rate	$\dot{Q}_c$	= 350 kW
Distance above fuel surface	H	= 4 m
Ambient room temperature	T_∞	= 20°C + 273 = 293 K
Response time index	RTI	= 235 $m^{1/2} \cdot s^{1/2}$
Temperature of operation	$T_{operation}$	= 74°C + 273 = 347 K
Temperature of ceiling jet	T_m	$= 16.9 \left(\dot{Q}^{2/3} / H^{5/3} \right) + T_\infty$
		$= (16.9)\left(350^{2/3} / 4^{5/3} \right) + 293$
		$= 83 + 293 = 376 \text{ K} = 103°C$
Ceiling jet velocity	U_m	$= 0.96 \left(\frac{\dot{Q}}{H} \right)^{1/3}$
		$= (0.96)(500/4)^{1/3} = 4.8 \text{ m/s}$
Time to operation	$t_{operation}$	$= \left(RTI/\sqrt{U_m} \right) \log_e \left[(T_m - T_\infty) / \left(T_m - T_{operation} \right) \right]$
		$= \left(235/\sqrt{4.45} \right) \log_e \left[(376 - 293)/(376 - 347) \right]$
		$= 61.8 \text{ s} \approx 1 \text{ min}$

The calculations for ceiling jet behaviors are valid only for smooth, level ceilings. Open joists, ceiling ducts, or pitched surfaces will dramatically affect the movement of gases.

EXAMPLE 2-14 ■ Nursing Home Case Study

Problem. An elderly woman seated in a wheelchair in a nursing home died when her clothing and lap blanket ignited while she was sitting smoking cigarettes on a covered exterior patio of the facility. When staff members first noticed a problem, the flames were readily visible, and by the time they reached her, the flames had set off a sprinkler immediately above her position. The sprinkler doused the flames but not before they had consumed the plastic seat back of the wheelchair and much of the clothing and lap blanket.

The woman succumbed to inhalation injuries (soot and edema in airway) and shock from 70 percent total body surface area (TBSA) burns some 8 hours after the incident. (No postmortem or toxicology tests were conducted.) Witnesses reported they observed no movement and no outcry from the victim who "sat like a mannequin." She had had a stroke some months prior, and possibly had suffered another. On the table next to her were found a partial pack of cigarettes, an ashtray, and a partial box of matches.

The woman, a lifelong smoker, always had access to cigarettes and lighters or matches. Exemplars of a dressing gown and lap blanket fabric were provided. These fabrics were tested using *NFPA 705* methods (NFPA 2009b) and were characterized as a cotton/synthetic blend dressing gown and a combustible lightweight quilt with a thin polyester fiberfill pad covered with a combustible cotton/polyester fabric.

Proposed Solution. Data from various databases and published sources indicated that a fire in such synthetic materials as the lap blanket or wheelchair seat back could be ignited only by an open flame. The chair was a manual one with no power supply or electrical fittings. A dropped match on such materials could provide competent ignition to ignite a rapidly growing, flaming fire. Data from the NIST website (fire.nist.gov) on synthetic-upholstered chairs indicated a fire of a maximum heat release rate of 300–700 kW would be established within 1 minute of flame ignition.

At the location of the fire, the ceiling was finished with painted wood and recessed lighting fixtures. It was sloped with a height of 2.36 m (93 in.) at the fire location, as shown in Figure 2-17. The sprinkler head was identified as a 165°F normal pendant fixture whose deflector head was set 2.3 m (90 in.) from the floor. Sprinklers were located on 3-m (10-ft) centers well away from support beams (also on 3-m centers). There was a circular charred area on the ceiling surrounding the sprinkler head approximately 1 m (3 ft) in diameter with a halo of soot deposit outside it. The seat of an exemplar wheelchair was found to be approximately 0.5 m (19 in.) above the floor. Laboratory analysis of samples of clothing and debris found no residues of ignitable liquid.

The Zukoski (1978) relationship for plume height (plume heights are discussed in Chapter 3) to reach the ceiling at 1.86 m (73 in.) above a fire surface at 0.5 m (19 in.) height is

$$Z = 0.175k\dot{Q}^{0.4}, \tag{2.48}$$

where

Z = fire plume height (m),
k = constant (1), and
$\dot{Q}$ = heat release rate (kW).

Therefore,

$$\begin{aligned} Z &= 0.175(k)\dot{Q}^{0.4} \\ 1.86 &= 0.175(1)\dot{Q}^{0.4} \\ \dot{Q}^{0.4} &= 10.63 \end{aligned}$$

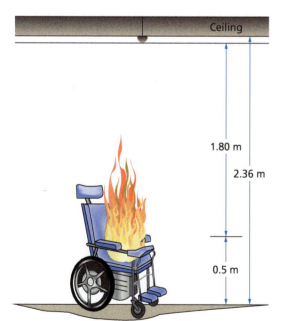

FIGURE 2-17 A fire in a wheelchair under a sprinkler causes rapid activation of the sprinkler (less than 100 s).

1.80 m

2.36 m

0.5 m

Ceiling

$$\left(\dot{Q}^{0.4}\right)^{2.5} = (10.63)^{2.5}$$
$$\dot{Q} \qquad = 368 \text{ kW}$$

Thus, a typical fire from the combustion of a synthetic-upholstered chair (or its equivalent) could easily reach the ceiling and cause ceiling jet damage to the surrounding finish.

Time to activation of a standard-response link sprinkler (RTI = 130 $m^{1/2} \cdot s^{1/2}$, temperature rating = 165°F (74°C) at a height of 1.80 m above the seat of the chair (immediately below the sprinkler head), would be approximately 12 seconds, calculated using NRC Fire Dynamics Tools spreadsheet 10 (Iqbal and Salley 2004, chap. 10) for sprinkler response time (later discussed in detail) and assuming a steady-state fire of 368 kW. Since the fire would have to grow from ignition to 368 kW, a fast t^2 fire, for which $\alpha = 0.047$ (given the properties of synthetic blankets and clothing), could be assumed. Then,

$$\dot{Q} = \alpha t^2,$$

and

$$368 = 0.047 \, t^2$$
$$t^2 = \frac{368}{0.047} = 7830$$

$$t = 88.5 \text{ s}$$

So, realistically, the sprinkler would have activated in less than 100 seconds from the time of ignition.

It was concluded that the decedent had dropped a match while attempting to light a cigarette (probably owing to a stroke or other medical event), and the match ignited a very rapidly growing fire whose flames reached ceiling height. The pattern of burn injuries to the victim and to the chair was entirely consistent with the finding that her clothing and lap blanket were the major fuel packages. The combustion and collapse of the wheelchair were the result of fire spread from there.

Case study courtesy of John D. DeHaan

Summary

The body of fire dynamics knowledge as it applies to fire scene reconstruction and analysis is based on the combined disciplines of thermodynamics, chemistry, heat transfer, and fluid mechanics. We have seen that the growth and development of fires are influenced by a number of variables such as available fuel load, ventilation, and physical configurations of the room. To estimate accurately a fire's origin, intensity, growth, direction of travel, and duration, investigators must rely on and understand the principles of fire dynamics.

Fire investigators can also benefit by applying the wealth of knowledge of fire dynamics contained in textbooks, expert treatises, and authoritatively conducted fire research. Case examples along with discussions on sound fire protection engineering calculations can assist investigators in interpreting fire behavior.

Problems

2.1. Recalculate the heat release rate needed for flashover using several methods for a room measuring 5×5 m, with a 3-m ceiling and two 2.5-m-high $\times$ 1-m-wide openings.

2.2. Photograph or review a recent fire scene and estimate the heat release rate of the first article ignited.

2.3. Search for references to fires in warehouses in which the fire protection system detected and extinguished the incipient fire with the activation of only one sprinkler head. Obtain the floor plan to a warehouse and estimate the time to sprinkler head activation.

2.4. Visit the fire.nist.gov/ website and examine the fire data availability there. What are the maximum heat release rates for upholstered chairs, mattresses, and Christmas trees?

2.5. Test the sensitivity of some of the mathematical calculations described in this chapter by changing a room dimension or ventilation opening by 0.2 m. Compare the results.

Suggested Readings

DeHaan, J. D., and D. J. Icove. 2012. *Kirk's Fire Investigation,* 7th ed. Upper Saddle River, NJ: Prentice Hall, chap. 3.

Drysdale, D. 2011. *An Introduction to Fire Dynamics,* 3rd ed. Hoboken, N.J., Wiley.

Gorbett, G. E., and J. L. Pharr. 2011. *Fire Dynamics.* Upper Saddle River, N.J., Pearson.

Karlsson, B., and J. G. Quintiere. 2000. *Enclosure Fire Dynamics.* Boca Raton, FL: CRC Press.

Quintiere, J. G. 1998. *Principles of Fire Behavior.* Albany, NY: Delmar, chaps. 3–9.

Iqbal, N., and M. H. Salley. 2004. Fire Dynamics Tools (FDTs): Quantitative fire hazard analysis methods for the U.S. Nuclear Regulatory Commission Fire Protection Inspection Program.

References

Alpert, R. 1972. Calculation of response time of ceiling-mounted fire detectors. *Fire Technology* 8 (3): 181–95, doi: 10.1007/bf02590543.

ASTM. 2009. *ASTM E1321-09: Standard test method for determining material ignition and flame spread properties.* West Conshohocken, PA: ASTM International.

Babrauskas, V.1980. Estimating room flashover potential. *Fire Technology* 16 (2): 94–103, doi: 10.1007/bf02481843.

———. 1984. Upholstered furniture room fires: Measurements, comparison with furniture calorimeter data, and flashover predictions. *Journal of Fire Sciences* 2 (1): 5–19, doi: 10.1177/073490418400200103.

———. 2003. *Ignition handbook: Principles and applications to fire safety engineering, fire investigation, risk management and forensic science.* Issaquah, WA: Fire Science Publishers, Society of Fire Protection Engineers.

———. 2005. Charring rate of wood as a tool for fire investigations. *Fire Safety Journal* 40 (6): 528–54, doi: 10.1016/j.firesaf.2005.05.006.

Babrauskas, V., & Peacock, R. D. 1992. Heat release rate: The single most important variable in fire hazard. *Fire Safety Journal* 18 (3): 255–72, doi: 10.1016/0379-7112(92)90019-9.

Babrauskas, V., Peacock, R. D., & Reneke, P. A. 2003. Defining flashover for fire hazard calculations: Part II. *Fire Safety Journal* 38:613–22.

Babrauskas, V., and Walton, W. D. 1986. A simplified characterization for upholstered furniture heat release rates. *Fire Safety Journal* 11:181–92.

Bukowski, R. W. 1995a. How to evaluate alternative designs based on fire modeling. *Fire Journal* 89 (2): 68–74.

———. 1995b. Predicting the fire performance of buildings: Establishing appropriate calculation methods for regulatory applications. In *Proceedings of ASIAFLAM '95, 1st International Conference on Fire Science and Engineering,* March 15–16, Kowloon, Hong Kong, Interscience Communications Limited, 9–18.

Butler, C. P. 1971. Notes on charring rates in wood. *Fire Research Notes 896.* London: Joint Fire Research Organization. Fire Research Station, Borehamwood, England.

Carman, S. W. 2008. Improving the understanding of post-flashover fire behavior. *Proceedings of the international symposium on fire investigation science and technology.* Cincinnati, OH, May 19–21.

———. 2010. Clean burn fire patterns: A new perspective for interpretation. Paper presented at Interflam, July 5–7, Nottingham, UK.

Collier, P.C.R. 1996. Fire in a residential building: Comparisons between experimental data and a fire zone model. *Fire Technology* 32 (3): 195–218, doi: 10.1007/bf01040214.

DeHaan, J. D. 2001. Full-scale compartment fire tests. *CAC News* (Second Quarter): 14–21.

DeHaan, J. D., & Icove, D. J. 2012. *Kirk's fire investigation,* 7th ed. Upper Saddle River, NJ: Pearson-Prentice Hall.

Drysdale, D. 2011. *An introduction to fire dynamics,* 3rd ed. Chichester, West Sussex, UK: Wiley.

Evans, D., & Stroup, D. 1986. Methods to calculate the response time of heat and smoke detectors installed below large unobstructed ceilings. *Fire Technology* 22 (1): 54–65, doi: 10.1007/bf01040244.

Gorbett, G. E., & Pharr, J. L. 2011. *Fire dynamics.* Upper Saddle River, NJ: Pearson.

Gratkowski, M. T., Dembsey, N. A., & Beyler, C. L. 2006. Radiant smoldering ignition of plywood. *Fire Safety Journal* 41 (6): 427–43, doi: 10.1016/j.firesaf.2006.03.006.

Heskestad, G. H., & Delichatsios, M. A. 1977. *Environments of fire detectors—Phase 1: Effects of fire size, ceiling height, and material.* Gaithersburg, MD: National Bureau of Standards.

Hopkins, R. L., Gorbett, G. E., & Kennedy, P. M. 2009. Fire pattern persistence and predictability during full-scale comparison fire tests and the use for comparison of post fire analysis. Paper presented at Fire and Materials 2009, 11th International Conference, January 26–28, San Francisco, CA.

Iqbal, N., & Salley, M. H. 2004. *Fire dynamics tools (FDTs): Quantitative fire hazard analysis methods for the U.S. Nuclear Regulatory Commission Fire Protection Inspection Program.* Washington, DC: Nuclear Regulatory Commission.

Karlsson, B., & Quintiere, J. G. 2000. *Enclosure fire dynamics.* Boca Raton, FL: CRC Press.

Law, M. 1978. Fire safety of external building elements: The design approach. *AISC Engineering Journal* (Second Quarter).

Lawson, J. R., & Quintiere, J. G. 1985. Slide-rule estimates of fire growth. *Fire Technology* 21 (4): 267–92.

McCaffrey, B., Quintiere, J., & Harkleroad, M. 1981. Estimating room temperatures and the likelihood of flashover using fire test data correlations. *Fire Technology* 17 (2): 98–119, doi: 10.1007/bf02479583.

Milke, J. A. 1990. Smoke management for covered malls and atria. *Fire Technology* 26 (3): 223–43, doi: 10.1007/bf01040110.

Milke, J. A., & Mowrer, F. W. 2001. Application of fire behavior and compartment fire models seminar. Paper presented at the Tennessee Valley Society of Fire Protection Engineers (TVSFPE), September 27–28, Oak Ridge, TN.

Mowrer, F. W. 1990. Lag times associated with fire detection and suppression. *Fire Technology* 26 (3): 244–65, doi: 10.1007/bf01040111.

NFPA. 2008. *Fire protection handbook.* Quincy, MA: National Fire Protection Association.

NFPA. 2009a. *NFPA 555: Guide on methods for evaluating potential for room flashover.* Quincy, MA: National Fire Protection Association.

———. 2009b. *NFPA 705: Recommended practice for a field flame test for textiles and films.* Quincy, MA: National Fire Protection Association.

———. 2010. *NFPA 72: National fire alarm and signaling code.* Quincy, MA: National Fire Protection Association.

———. 2011. *NFPA 921: Guide for fire and explosion investigations.* Quincy, MA: National Fire Protection Association.

———. 2012. *NFPA 92: Standard for smoke control systems.* Quincy, MA: National Fire Protection Association.

Peacock, R. D., Reneke, P. A., Bukowski, R. W., & Babrauskas, V. 1999. Defining flashover for fire hazard calculations. *Fire Safety Journal* 32 (4): 331–45, doi: 10.1016/s0379-7112(98)00048-4.

Putorti, Jr., A. D., McElroy, J. A., & Madrzykowski, D. 2001. Flammable and combustible liquid spill/burn patterns. Rockville, MD: National Institute of Standards and Technology.

Quintiere, J. G. 1998. *Principles of fire behavior.* Albany, N.Y.: Delmar.

———. 2006. *Fundamentals of fire phenomena.* West Sussex, England, UK: Wiley.

Quintiere, J. G., and Harkleroad, M. 1984. New concepts for measuring flame spread properties. NBSIR-84-2943. Gaithersburg, MD: National Bureau of Standards.

Sanderson, J. L. 1995. Test results add further doubt to the reliability of concrete spalling as an indicator. *Fire Findings* 3 (4): 1–3.

SFPE. 2002a. *SFPE Engineering guide to predicting 1st and 2nd degree skin burns.* Bethesda, MD: Society of Fire Protection Engineers, Task Group on Engineering Practices.

———. 2002b. *SFPE Engineering guide to piloted ignition of solid materials under radiant exposure.* Bethesda, MD: Society of Fire Protection Engineers, Task Group on Engineering Practices.

———. 2008. *SFPE handbook of fire protection engineering,* 4th ed. Quincy, MA: National Fire Protection Association, Society of Fire Protection Engineers.

Thomas, P. H. 1971. Rates of spread of some wind-driven fires. *Forestry* 44:155–75.

———. 1974. Fires in model rooms: CIB Research Programmes. Building Research Establishment Current Paper CP 32/74. BRE, Borehamwood, UK.

———. 1981. Testing products and materials for their contribution to flashover in rooms. *Fire and Materials* 5 (3): 103–11, doi: 10.1002/fam.810050305.

Zukoski, E. E. 1978. Development of a stratified ceiling layer in the early stages of a closed-room fire. *Fire and Materials* 2 (2): 54–62, doi: 10.1002/fam.810020203.

3

Fire Pattern Analysis

There is no branch of detective science which is so important and so much neglected as the art of tracing footsteps.

—Sir Arthur Conan Doyle
A Study in Scarlet

KEY TERMS

annealing, *p. 128*
backdraft, *p. 125*
calcination, *p. 119*
char, *p. 120*
clean burn, *p. 126*
crazing, *p. 126*

entrainment, *p. 98*
eutectic, *p. 127*
fire patterns, *p. 98*
flame plume, *p. 104*
ghost marks, *p. 124*
isochar, *p. 121*

organic, *p. 120*
overhaul, *p. 135*
rekindle, *p. 135*
spalling, *p. 121*

OBJECTIVES

After reading this chapter, the student should be able to:

- Recognize fire patterns and their causes.
- Translate the pattern's relation to fire dynamics.
- Correlate fire plume calculations to assist in explaining fire patterns.

The ability to document and interpret **fire patterns** accurately is essential to investigators reconstructing fire scenes. Fire patterns are often the only visible evidence remaining after a fire is extinguished. This chapter describes how investigators analyze and use fire patterns in assessing fire damage and determining a fire's origin.

Smoke deposits, heat transfer, and flame spread are the major causes of change to the exposed surface and appearance of materials during a fire. Fire patterns are formed by the heating effects of fire plumes on exposed solid surfaces such as floors, ceilings, and walls. These burn patterns are influenced by a number of variables including the available fuel load, ventilation, and the physical configurations of the room. Many common combustible materials and ignitable liquids can produce these fire plumes and their resulting damage.

Case examples are offered here along with discussions on validated fire science and engineering calculations to assist investigators in interpreting fire patterns. Also addressed in this and subsequent chapters are various documentation techniques that can be helpful in fire scene reconstruction using the concepts of fire pattern analysis.

Fire Plumes

The single most important factor in fire scene reconstruction is the fire plume, which is a buoyant column of hot gases produced by the combustion of a fuel source emitting a vertical column of flames and hot products of combustion (Cox and Chitty 1982; McCaffrey 1979; Beyler 1986; Drysdale 2011). Fire plumes can originate from any fuel combusting that has a sufficient heat release rate to generate a buoyant column of flames and smoke. Fire plumes result from any significant fire. They can be produced by floor-level pools of ignitable liquids; however, many combustible solids, including certain types of foam mattresses and plastics, melt and collapse while burning and behave like liquid fuels. Fire plumes can also result from fuel geometries having multiple vertical and horizontal fuel surfaces, complex internal structures, and varying fuel types. For the purpose of the following discussion, we shall treat the effects of fire plumes from a simple flat fuel array as behaving analogously to a pool fire.

The shape of burning pools that produce fire plumes depends on several variables including the geometry of the containment of the pool, the type of substrate on which the pool is resting, and, in some cases, the external winds. Because fire plumes are three-dimensional, their location can often be determined and documented by evaluating the patterns of heat transfer, flame spread, and smoke damage they cause to adjacent flooring, wall, and ceiling surfaces, as illustrated earlier in Figure 1-5. A fire investigator can gain additional insight through a fundamental understanding of the nature, physics, and heat transfer characteristics of fire plumes. The buoyancy of hot gases is the driving force behind the vertical movement and horizontal spread of fire plumes as they encounter obstacles.

Research with simulated pool fires in compartments shows that plumes lean with dominant air flow from their vertical position when influenced by air movement (Steckler, Quintiere, and Rinkinen 1982). With 55 full-scale steady-state experiments, except for the smallest openings in the compartment, disturbances caused by the incoming air flow behave much like a fire plume exposed to a cross wind.

V PATTERNS

The impingement, intensity, and direction of the fire plume's travel form lines or areas of demarcation on walls, ceilings, floors, and other materials. As the hot gases and smoke rise from a fire, they mix with surrounding air, and the mixing zone becomes wider as hot gases rise above the fuel. **Entrainment** mixes and spreads the rising column, so it forms a V approximately 30° in width (i.e., half-angle of 15°) if unconfined in still air from a turbulent diffusion fire (You 1984). Therefore, the total width of the *unconfined* plume in still air is approximately half its height above the fuel surface. As the fire gases mix, they are diluted and cooled. As a result, the diameter of the hottest part of the plume along the centerline becomes *smaller* with increasing height, as seen in Figure 3-1. The rapid cooling of fire gases by mixing with the air limits their thermal damage on noncombustible walls, so the thermal pattern on adjacent walls will not reflect the entire 30° angle width described for the plume.

Analyzing the shape of fire patterns caused by plumes can provide valuable information. For example, assuming again that fuel is directly beneath the fire pattern, the most pronounced ceiling damage is often in the plume impingement area directly above the fuel source (as in Figure 1-5), a point from which a gas-movement vector points directly back to the fire's origin. The temperature of the gases determines the extent of thermal effects on surfaces they encounter. If their temperature is too low, the heat transfer will be insufficient and there may be no thermal effect; but combustion products will still condense on cooler surfaces.

Damage patterns forming lines of demarcation frequently appear in simple two-dimensional (cross-sectional) views of the location where the fire plume comes into contact with and damages wall surfaces. These lines are often referred to as *V-shaped fire patterns,* based on their characteristic upward-sweeping curves.

NFPA 921, 2011 ed. (NFPA 2011) and *Kirk's Fire Investigation* (DeHaan and Icove 2012) dispel past misconceptions regarding the shape and geometry of V patterns, once thought to relate to the rapidity of fire growth. The shape of a V pattern is actually related to the heat release rate, geometry of the fuel, ventilation effects, ignitability and combustibility of affected surfaces, and intersection with vertical and horizontal surfaces *NFPA 921,* 2011 ed., pt. 6.3.2.1(NFPA 2011).

If the V-shaped fire pattern's angle of demarcation is followed downward to its base, the area closest to the virtual origin of the plume can be located. Several mathematical relationships aid in documenting and analyzing the features of fire plume height,

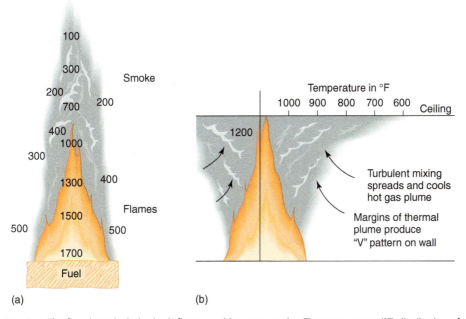

FIGURE 3-1 The fire plume includes both flames and buoyant smoke. The temperature (°F) distribution of gases in the buoyant plume: (a) unrestricted, (b) under ceiling.

temperature, velocity, vortex shedding frequency, and virtual origin (Heskestad 2008; Quintiere 1998).

HOURGLASS PATTERNS

In certain cases, when a fuel package burns in a room adjacent to a wall or corner, the damage appears in the shape of an hourglass burn pattern instead of a V, as shown in Figure 3-2 (Icove and DeHaan 2006). Both fire testing and mathematical analysis by the authors have shown that the formation of hourglass burn patterns is a direct function of the fire plume's virtual origin, which is mathematically tied to the heat release rate and surface area of the fuel package. A later section in this chapter will illustrate in full the formation of these hourglass patterns.

FIRE PLUME DAMAGE CORRELATIONS

Testing by the Factory Mutual Research Corporation (FMRC) provided additional insight into the formation of V patterns. In FMRC's testing, a closer examination of a fire plume's lines of demarcation reveals distinct areas of damage to the wall surfaces, also known as the *fire propagation boundary.* This boundary is a visually distinguishable line where the heavy pyrolysis of the surface ends.

FIGURE 3-2 Hourglass pattern from fire in a corner. Note demarcation between flame contact area (burned paper on drywall) and radiant heat (scorched paper). *Courtesy of J. D. DeHaan.*

Fire testing by FMRC has documented the close correlation of the fire propagation boundary with the *critical heat flux boundary,* which is where the minimum heat flux is at or below the point at which a flammable vapor/air mixture is produced by pyrolysis at the surface of the solid fuel (Tewarson 2008). Critical heat fluxes have also been experimentally determined by successive exposures of material samples to progressively decreasing incident heat fluxes until ignition no longer takes place (Spearpoint and Quintiere 2001).

Figure 3-3 illustrates this close correlation in FMRC's 8-m (25-ft) corner tests of a growing fire peaking at a 3 MW heat release rate (Newman 1993). This test is designed for evaluating fires involving low-density, high–char forming wall and ceiling insulation materials. In Figure 3-3 the critical heat flux boundary as measured by radiometers attached to the surface is denoted by the dashed line, and the visual damage evaluation by the dashed-dotted line. Note how closely these observations correlate. The position and impact of the fire plume on the walls, ceilings, and surface areas are important to interpreting these correlations correctly when evaluating the various types of fire patterns.

At first, the soot and pyrolysis products of the smoke condense on the cooler surfaces, with no chemical or thermal effect. As hot gases from a plume come into contact with the surface, heat is transferred by convective and radiative processes. As the heat is transferred, the temperature of the surface increases. The heat may be conducted into the surface. At some point, the temperature increases to the point at which the surface coating begins to scorch, melt, or pyrolyze. At higher temperatures it may actually ignite. These temperatures are reached when the critical heat flux is reached for that material and sustained for a sufficient time. These observations allow us to characterize patterns as surface deposit only, thermal effect to surface only, charring and ignition, penetration, and consumption.

Types of Fire Plumes

When a fire scene is examined, often the only evidence remaining is the fire burn pattern indicators reflecting their thermal impact. In cases where a fire was extinguished quickly by fire suppression or self-extinguished, clearly defined burn patterns from the plume are left on the walls, floors, ceilings, and exterior surfaces of the structure. Many of these patterns are formed by the intersection of fire plumes with structure and other target surfaces.

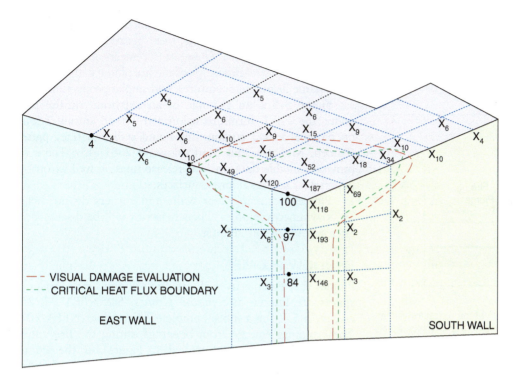

FIGURE 3-3 The correlation between critical heat flux boundary (dashed line) and visual lines of demarcation (dashed-dotted line) in an FMRC 25-ft corner fire test provides a scientific explanation for the formation of V patterns on walls. *From SFPE Handbook of Fire Protection Engineering, 3rd ed., 2002, by permission of Society of Fire Protection Engineers.*

There are four major types of plumes of interest in fire investigation and reconstruction (Milke and Mowrer 2001):

- axisymmetric
- window
- balcony
- line

Understanding these plumes provides a keener insight into the dynamics of more complex fires. The investigator examining the damage of these plumes must have a working knowledge of their origin, type of fuel package, and ventilation. Furthermore, empirical testing has established mathematical relationships describing the behavior, shape, and impact of these classes of fire plumes.

For a summary of physical-scale fire testing and computer modeling of thermal spill plumes, see the work by Harrison (2004) and later Harrison and Spearpoint (2007) at the University of Canterbury, Christchurch, New Zealand. These reports expand the initial fundamental research by Morgan and Marshall (1975) regarding experimentally based theories on thermal spill plumes, which are used to interpret conditions of smoke production and hazards in covered multilevel shopping malls.

AXISYMMETRIC PLUMES

Axisymmetric plumes have uniform radial distribution from a common vertical plume centerline. Generally, they occur in open areas or near centers of rooms where there is no nearby wall. Much research has been conducted on the correlations of the plume height, temperatures generated, gas velocities, and entrainment. Figure 3-4 shows the typical descriptions

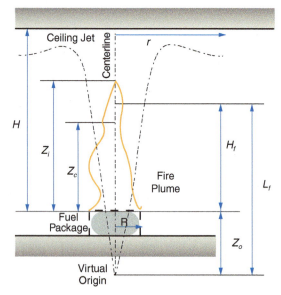

FIGURE 3-4 Schematic diagram of the typical axisymmetric fire plume and its interaction with a ceiling, where Z_0 is the distance from the fuel surface to the virtual origin, H_f is the mean flame height above the fuel surface, Z_c is the continuous flame height above the fuel surface, Z_i is the intermittent flame height above the fuel surface, H is the distance to the ceiling from the fuel surface, R is the radius of the fuel package, and r is the distance of the ceiling jet from the plume centerline. The statistical mean flame height is the portion of luminous flame that is visible during 50 percent of its total exposure time.

and measurements used for flame and plume characteristics describing the plume centerline, mean flame height, and virtual origin (Heskestad 2008). These measurements are used in calculations that show relationships of the fire plume with heat release rates, flame height, temperatures, and smoke production.

Figure 3-5 is an example of an axisymmetric fire plume under the NIST furniture calorimetry hood during research on flammable and combustible liquid spill and burn patterns (Putorti, McElroy, and Madrzykowski 2001, 15). These tests provide valuable data on heat release rates as well as on burn patterns on various flooring surfaces.

Plumes occurring near walls and corners do not behave as axisymmetric plumes. Their behavior will be discussed in a later section of this chapter.

WINDOW PLUMES

Plumes that emerge from doors and windows into large open spaces are known as *window plumes*. See *NFPA 92*, 2012 ed., pt. 5.5.3, for a more complete description (NFPA 2012). These plumes emerge from openings during the fire and are commonly ventilation controlled. In Figure 3-6, the plume is emerging from the opening and measures Z_w above the soffit.

Window plumes result when the rate at which combustible vapors and gases are produced by pyrolysis inside the room exceeds the rate at which those products can be burned by combining with air entering the room and flow out openings to burn outside the compartment. Window plumes are often, but not exclusively, associated with postflashover burning in that room. As such, they may be a visible indicator of a fire's development.

Window plumes can be a major factor in vertical fire spread in multistory buildings, since entrainment can push the plume against combustible siding or spandrel panels or windows of stories above. The building facades above the window sometimes experience very high heat fluxes under such conditions, resulting in rapid failure (DeHaan and Icove 2012, 45–50).

FIGURE 3-5 An axisymmetric fire plume under a NIST furniture calorimetry hood during research on flammable and combustible liquid spill and burn patterns. *Courtesy of Putorti, McElroy, and Madrzykowski (2001,15).*

WINDOW PLUME CALCULATIONS

The area and height of the ventilation openings of window plumes can be used to estimate the maximum heat release rate of the fire in a compartment that is being ventilated by that opening. When a smoke plume is viewed as a window plume, a mathematical relationship based on actual experimental data for wood and polyurethane can be used to predict the maximum heat release rate, assuming that the single opening is ventilating the fire (Orloff, Modale, and Alpert 1977; Tewarson 2008).

Similarly, in the case of window plumes, the mass loss rate can be also determined. *NFPA 92*, 2012 ed., pt. 5.5.3 (NFPA 2012), uses the following equation for making this determination:

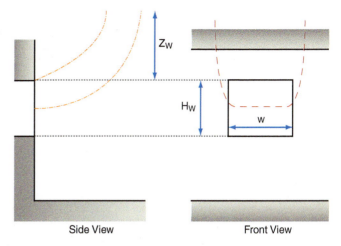

FIGURE 3-6 A schematic diagram of a window plume showing side and front orientations.

$$\dot{m} = \left[0.68 \left(A_w \sqrt{H_w} \right)^{1/3} \left(Z_w + a \right)^{5/3} \right] + 1.59 A_w \sqrt{H_w}, \qquad (3.1)$$

where

$\dot{m}$ = mass flow rate at height Z_w (kg/s),
A_w = area of ventilation opening (m²),
H_w = height of ventilation opening (m),
Z_w = height above the top of the window (m), and
a = $\left(2.40 A_w^{2/5} H_w^{1/5} \right) - 2.1 H_w$ (m).

This relationship is not valid if the temperature rise above ambient temperature is less than 2.2°C (4°F).

This simple relationship for estimating the mass flow rate may be worthwhile when considering the observations of on-scene eyewitnesses and responding firefighters. For example, the description of the window plume may be sufficient to estimate the extent of the involvement of the fire based on the observed area and height of the ventilation opening.

BALCONY PLUMES

Plumes that emerge under overhangs and from doors are known as *balcony spill plumes* as described in *NFPA 92*, 2012 ed., pt. 5.5.2 (NFPA 2012). These plumes (Figure 3-7) are characteristic of fires occurring in enclosed rooms and spreading through patio doors or windows to covered porches, patios, or balconies. The buoyancy of these gases causes them to flow along the underside of horizontal surfaces until they can rise vertically.

The pattern of thermal damage or smoke damage to the underside of horizontal surfaces may be used to estimate the dimensions of the plume. The plume spreads laterally as it flows. The width of the plume is estimated by re-creating the horizontal and vertical dimensions of its contact area such that

$$W = w + b, \qquad (3.2)$$

where

W = maximum width of the plume (m),
w = width of the opening from the area of origin to the balcony (m), and
b = distance from the opening to the balcony edge (m).

This calculation has been used in estimating the width and height of the fire plume to assess its smoke production. Locations of balcony plumes include exterior doors leading

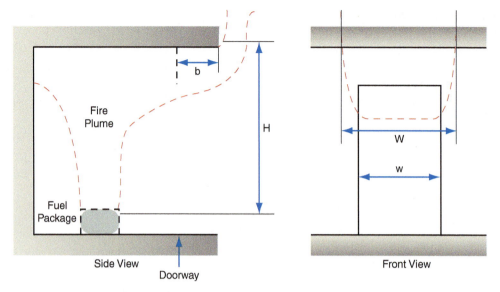

FIGURE 3-7 A schematic diagram of a balcony spill plume showing side and front orientations.

to garden apartments, internal hotel rooms in atrium-style buildings, multistory shopping malls, and multiple-level prison cells that face catwalks.

The characteristic of a balcony plume is that flames are deflected away from the upper floor by the external overhang, or by an open porch or deck in an apartment or condominium. The overhang is an example of the use of passive fire engineering in building construction to reduce the chance that external plumes will spread fires to floors above the floor of fire origin by preventing re-entrainment against the building.

Figure 3-8 shows an example of a window plume coming into exterior contact with a porch overhang, creating a situation similar to a balcony spill. As the plume exits the window, it travels along the ceiling of the porch overhang, and flames are projected from the edge of the overhang by their buoyant momentum. This situation is common in residential fires, particularly when the dwelling has a porch or a deck.

flame plume ■ The buoyant column of incandescently hot gases from a fire visible to the unaided eye.

FIGURE 3-8 Window plume that comes into exterior contact with a porch overhang creates a situation similar to a balcony spill. *Courtesy of D. J. Icove.*

LINE PLUMES

Line plumes (Figure 3-9) have elongated geometric shapes that produce narrow, thin, shallow plumes. The relationship of the heat release rate to the flame height is described in a later section. A line plume can be used to describe a **flame plume** whose length is many time greater than its width.

Possible scenarios for line plumes include exterior fires in ditches where spilled ignitable liquids collect and ignite, a row of townhouses, a long sofa, the advancing front of a forest fire, flame spread over flammable wall linings, and even a balcony spill plume (Quintiere and Grove 1998). Line plumes can also be used to approximate elongated fires in open areas of atria or warehouses.

FLAME HEIGHTS OF LINE PLUMES

The relationship of the heat release rate to the flame height for line plumes (Figure 3-9), where $B > 3A$, is described next. For flame plume heights greater than five times the B measurement in Figure 3-9, the line plume more closely resembles an axisymmetric plume (Hasemi and Nishihata 1989):

$$L_f = 0.035 \left(\frac{\dot{Q}}{B} \right)^{2/3} \quad \text{for} \quad B > 3A; \tag{3.3}$$

$$\dot{m} = 0.21z \left(\frac{\dot{Q}}{B} \right)^{2/3} \quad \text{for} \quad B < 3A \quad \text{and} \quad L < z < 5B, \tag{3.4}$$

where

    L_f = flame height (m),
    $\dot{Q}$ = heat release rate (kW),
    A = shorter side of line plume base (m),
    B = longer side of line plume base (m),
    $\dot{m}$ = mass flow rate (kg/s), and
    z = ceiling height above fuel (m).

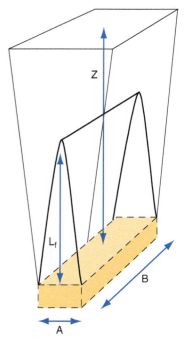

FIGURE 3-9 Illustration of a line source plume.

ALTERNATIVE METHOD

The entrainment of air around a long line plume can be envisioned as air coming from two opposite sides of an axisymmetric plume. It has been modeled in a theoretical study by Quintiere and Grove (1998). For a wall fire where the flame is against a vertical, noncombustible wall, the entrainment is from one side only. If the $\dot{Q}$ for an equivalent fuel surface can be calculated (from the equations in Chapter 2), the $\dot{Q}$ (heat release rate per unit length of line) can be calculated from the ratio $\dot{Q}/B$.

The *Fire Protection Handbook* (SFPE 2008) gives the relationships

$$L_f = 0.017\dot{Q}^{2/3} \quad \text{for line fire plumes;}$$
$$L_f = 0.034\dot{Q}^{3/4} \quad \text{for wall fire plumes.}$$

Axisymmetric Fire Plume Calculations

As indicated in Chapter 2, investigators can benefit from information on fire plumes, which are merely the physical manifestations of the combustion process. Sound scientific and engineering principles, based on a combination of theoretical and actual fire testing data, describe fire plume behavior.

Information on fire plume behavior is expressed in terms of five basic calculated measurements often used in hazard and risk analysis (SFPE 2008). They model the following fire plume characteristics:

- Equivalent fire diameter
- Virtual origin
- Flame height
- Plume centerline temperatures and velocities
- Plume air entrainment

These basic characteristics constitute the bare minimum information needed for fire scene reconstruction. Several common equations for performing a fire engineering analysis of fire plumes rely on the energy or heat release rate and provide answers to the questions: How tall was the fire plume? What was the placement of the virtual source in relation to the floor? What was the depth of the burning flammable liquid pool?

A fire investigator must be able to use these calculations in applying basic fire science and engineering concepts. These calculations are essential in evaluating the various working hypotheses, such as the amount of fuel burned, the fire's time duration, and the impact of ventilation of the room of origin.

Many of these calculations are found in the various fire modeling software tools, such as FPETool and CFAST. The calculations are briefly explained here to demonstrate their underlying concepts and potential applications.

EQUIVALENT FIRE DIAMETER

In calculations of the flame height for fire plumes, the base of the fire is assumed to be circular. This is normally not the case when ignitable liquids are spilled onto a floor or when the fuel package is rectangular in shape. In these cases, the *equivalent diameter* must be calculated. From the basic relationship for circular areas ($A = \pi r^2$, where $r = D/2$), the relationship is

$$D = \sqrt{\frac{4A}{\pi}}, \qquad (3.5)$$

where

D = equivalent diameter (m),
A = total area of burning fuel package (m^2), and
π = 3.1416.

Areas of irregular pools can be calculated by re-creating the shape and size of pools from fire scene photos and obtaining dimensions from fire scene notes and duplicating the shapes on paper cutouts. The cutouts can then be weighed using a postal scale and the area determined from the weight of a reference paper sample of known area. Areas can also be re-created from photogrammetry of the fire scene photos (see Chapter 4 on scene documentation).

Investigators should be aware that the heat release rate of a smoke spill plume is much less than that of the plume emitted by a burning liquid pool of significant depth. This aspect will affect both the heat release rate and flame heights. For a detailed discussion on the fire forensics aspects and analysis of liquid fuel fires, see research by Putorti, McElroy, and Madrzykowski (2001) and Mealy, Benfer, and Gottuk (2011).

Rectangular contained spills include those produced when an oil-filled electric transformer or storage tank is compromised and spills its contents into a rectangular diked area and ignites. It is then necessary to determine the diameter of a circle containing the same area as the rectangular spill, using the aspect ratio of the spill (Figure 3-10).

The *aspect ratio* is the width-to-length ratio of the spill area. The equivalent diameter relationship generally holds true for pools with an aspect ratio ≤2.5. Aspect ratios >2.5, such as would be seen with trench or line fires, utilize other models (Beyler 2008).

In cases of short-duration fires, the area of a liquid fuel spill can be measured directly from the burned areas of the floor or covering. In postflashover or prolonged fires where the floor covering is readily combustible, the fire-damaged areas may extend well beyond the margins of the original pool and are therefore not a reliable guideline.

DeHaan (1995, 2002, and 2004) has shown a simple relationship among surface area, volume of liquid, and type of surface that controls the equivalent depth of the pool. Putorti, McElroy, and Madrzykowski

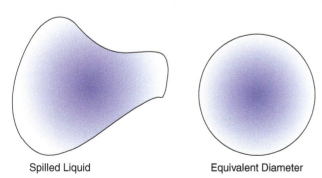

Spilled Liquid   Equivalent Diameter

FIGURE 3-10 The equivalent-diameter concept used in estimating the diameter of a circle that represents the same area as a noncircular spill pattern.

(2001) have confirmed this general relationship. Mealy, Benfer, and Gottuk (2011) explored the forensic aspects through tests that provided insight into the differences in fire dynamics between pool and spill fires (i.e., thick and thin fuel depths). They provided a methodology by which liquid fuel fire events can be assessed and identified forensic indicators that can be used in the analysis of liquid fuel fire events.

Mealy, Benfer, and Gottuk (2011) confirmed that for spills on nonporous surfaces such as sealed concrete, the depth of a freestanding pool of gasoline is about 1 mm (1×10^{-3} m). The area (m^2) is then calculated by dividing the volume of the spilled liquid (m^3) by the depth (m), where

$$\text{Area} \left(\text{m}^2\right) = \frac{\text{Volume} \left(\text{m}^3\right)}{\text{Depth} \left(\text{m}\right)}. \qquad (3.6)$$

For conversion of units, 1 liter (L) = 10^{-3} m^3.

On semiporous surfaces such as wood, the equivalent depth is 2–3 mm ($2–3 \times 10^{-3}$ m). On porous surfaces such as carpet, the maximum depth will be the thickness of the carpet, assuming full saturation of the carpet.

EXAMPLE 3-1 ■ Equivalent Fire Diameter

Problem. A container filled with 3.8 L (1 gal or 3800 cm^3) of gasoline is accidentally tipped over onto an enclosed patio of sealed concrete. The depth of the spilled pool is 1 mm. Calculate the (1) coverage area and (2) equivalent diameter of a circle of the same area.

Suggested Solution.

$$\text{Area of spill} = \frac{\text{Volume of liquid}}{\text{Depth of pool}} = \frac{3800 \text{ cm}^3}{0.1 \text{ cm}} = 38,000 \text{ cm}^2 = 3.8 \text{ m}^2$$

The equivalent diameter D then is

Equivalent diameter	$D = \sqrt{4A/\pi}$
Total surface area	$A = 3.8 \text{ m}^2$
Constant	$\pi = 3.1416$
Equivalent diameter	$D = \sqrt{(4)(3.8)/3.1416} = 2.2 \text{ m}$

VIRTUAL ORIGIN

As shown in Figures 2-7, 2-16, and 3-4, the *virtual origin* represents the actual fire as a point source. Its location is a point along the fire plume centerline where flames appear to originate, measured from the top burning surface of the fuel package. Identifying the virtual origin is helpful when evaluating the exposure of other targets fuels in the room to the fire. The investigator may need to make several assumptions depending on the distance to the target and the size of the burning fuel surface.

The virtual origin can be mathematically calculated, and its location can be helpful in reconstructing and documenting the fire's virtual source, point of origin, area, and direction of travel. The virtual origin is also used in some calculations for the measurement of the fire plume flame height.

The virtual origin, Z_0, is calculated from the *Heskestad equation* (1982, 1983):

$$Z_0 = 0.083\dot{Q}^{2/5} - 1.02D, \qquad (3.7)$$

where

Z_0 = virtual origin (m),
$\dot{Q}$ = total heat release rate (kW), and
D = equivalent diameter.

for the conditions

$$T_{amb} = 293 \text{ K}, (20°C, 68°F),$$
$$P_{atm} = 101.325 \text{ kPa (standard temperature and pressure), and}$$
$$D \quad \leq 100 \text{ m (328 ft)}.$$

The virtual origin, depending on the equivalent diameter and the total heat release rate, may fall above or below the fuel's surface. This placement depends primarily on the heat release rate and effective diameter of the fuel. The resulting value can then be used in equations and models of other fire plume relationships.

Small-diameter ignitable liquid spills are more likely to exhibit virtual origins above the fuel surface. Inspection of equation (3.7) reveals that the smaller the value of the fire's equivalent diameter, D, and/or the larger the value of the heat release rate, $\dot{Q}$, the greater the chance for a positive virtual origin, Z_0, that is, above the fire surface. Because not all fires burn at floor level, the concept of Z_0 is useful in estimating the effective origin of the fire plume in relation to the fuel surface and the compartment.

EXAMPLE 3-2 ■ Virtual Origin

Problem. A fire starts in a wastepaper basket filled with papers and quickly reaches a steady-state heat release rate of 100 kW. The basket measures 0.305 m (1 ft) in diameter. Determine the virtual origin of this plume using the Heskestad equation.

Suggested Solution. Use the equation for the virtual origin.

Virtual origin	$Z_0 = 0.083\dot{Q}^{2/5} - 1.02D$
Equivalent diameter	$D = 0.305$ m
Total heat release rate	$\dot{Q} = 100$ kW
Virtual origin	$Z_0 = (0.083)(100)^{2/5} - (1.02)(0.305)$
	$= 0.524 - 0.311 = 0.213$ m (0.698 ft, or 8.9 in.)

In this case the virtual origin is a positive number, indicating that it is above the surface of the fuel. If the same fuel was spread out over the floor such that its equivalent diameter was 1.0 m while its $\dot{Q}$ remained 100 kW, its virtual origin would be $Z_0 = -0.5$ m (1.64 ft, or 19.7 in., below the floor). This means that the impact of the fire on the room would be as if the fire were burning farther away from the ceiling.

Discussion. If the trash can was filled with gasoline and $\dot{Q} = 500$ kW, recalculate the virtual origin. What can you assume about the virtual origin when the heat release rate is raised?

FLAME HEIGHT

Flames from plumes represent the visible portion of a fire's combustion process. They are visible owing to the luminosity of heated soot particles and pyrolysis products. The buoyancy of these fire gases allows a fire plume to rise vertically. The height of the flames is described by several terms, including *continuous, intermittent,* and *average.*

Knowing the *flame height* and the approximate duration of the fire allows the investigator to examine and confirm what potential damage can be expected based on heat transfer to the ceiling, walls, flooring, and nearby objects. Several equations, based on fitting regression formulas to actual fire testing, are used to estimate the flame height. As with all these equations, users are cautioned that the equations are bounded by the limitations of the testing.

Flames from burning fuels, especially pool fires, fluctuate in height. Researchers typically define flame height as the distance above the burning fuel surface at which the flame tip is observed at least 50 percent of the time. Zukoski, Cetegen, and Kubota (1985), in defining the mean flame height, examined the intermittent flame, the fraction of the time the flame is 0.5 times above a certain height. Audouin et al. (1995) defined in an analysis similar to Zukoski's that the continuous and intermittent flame heights were greater than 0.95 and less than 0.05 times, respectively. Stratton (2005) described a method to measure these heights, as well as the flame's pulsation frequency, from a three-dimensional cloud captured on video.

Within the flaming region, the two *McCaffrey flame height measurements* (1979) along the centerline are depicted as continuous, Z_c, and intermittent, Z_i:

$$Z_c = 0.08 \, \dot{Q}^{2/5}, \tag{3.8}$$

$$Z_i = 0.20 \, \dot{Q}^{2/5}, \tag{3.9}$$

where

Z_c = continuous flame height (m),
Z_i = intermittent flame height (m), and
$\dot{Q}$ = heat release rate (kW).

Crude estimates of heat release rates based upon estimates of the flame heights made by witnesses and responding firefighters can be drawn by algebraically rearranging the McCaffrey flame height equation for intermittent flame heights:

$$\dot{Q} = 56.0 \, Z_i^{5/2}. \tag{3.10}$$

EXAMPLE 3-3 ■ Flame Height: McCaffrey Method

Problem. For Example 3-2 involving the 100 kW wastepaper basket fire, determine the continuous and intermittent flame heights using the McCaffrey method.

Suggested Solution. Use the equation for the McCaffrey method.

Heat release rate	$\dot{Q} = 100 \text{ kW}$
Continuous flame height	$Z_c = 0.08 \, \dot{Q}^{2/5} = (0.08)(100)^{2/5}$
	$= 0.505 \text{ m } (1.66 \text{ ft})$
Intermittent flame height	$Z_i = 0.20 \, \dot{Q}^{2/5} = (0.20)(100)^{2/5}$
	$= 1.26 \text{ m } (4.1 \text{ ft}) \text{ above the fuel surface}$

EXAMPLE 3-4 ■ Heat Release Rate from Flame Height: McCaffrey Method

Problem. A witness to an accident involving an overturned lawn mower sees the gasoline ignite and burn, forming an axisymmetric fire plume. He observes intermittent flames reaching 3 m (9.84 ft) into the air from the ground. Estimate the heat release rate, $\dot{Q}$, given off by the burning fuel spill.

Suggested Solution. Use the equation for the McCaffrey method.

General equation	$Z_i = 0.20 \, \dot{Q}^{2/5}$
Intermittent flame height	$Z_i = 3.0 \text{ m}$
Estimated heat release rate	$\dot{Q} = 56.0 \, Z_i^{2/5}$
	$= (56.0)(3.0)^{2/5} = 873 \text{ kW}$

A practical estimate would be 800–900 kW.

EXAMPLE 3-5 ■ Calculating the Area of a Pool Fire

Problem. From the data in Table 2-4, calculate the area of the pool fire at 873 kW from Example 3-4.

Suggested Solution. Use the equation for the heat release rate.

General equation	$\dot{Q} = \dot{m}'' \Delta h_c A$
Heat release rate	$\dot{Q} = 873$ kW
Mass burning rate per unit area $\dot{m}''$	$= 53$ g/$(m^2 \cdot s)$
Heat of combustion	$\Delta h_{eff} = 43.7$ kJ/g
Area of pool fire	$A = \dot{Q}/(\dot{m}'' \Delta h_c) = 873/[(53)(43.7)] = 0.38$ m^2 (4.1 ft^2)

One assessment of the flame height is its statistical *mean flame height*, which is the portion of luminous flame that is visible during 50 percent of its total exposure time. When the human eye observes a pulsating flame, it tends to interpolate the variations into a rough approximation of the mean flame height.

A good estimate of the mean or average flame height, L_f, uses the *Heskestad* equation:

$$L_f = 0.235 \dot{Q}^{2/5} - 1.02 D, \qquad (3.11)$$

where

D = effective diameter (m),
$\dot{Q}$ = total heat release rate (kW), and
$L_f = Z + Z_0$ = mean flame height above the virtual origin (m).

This equation includes a diameter term, accounting for the effects of large-diameter fuels on the visible flame. The larger the size of the burning fuel area, the lower the flames will be for the same heat release rate.

EXAMPLE 3-6 ■ Flame Height: Heskestad Method

Problem. From Example 3-2 involving the 100 kW wastepaper basket fire, determine the mean flame height, L_f, above the virtual origin, Z_0.

Suggested Solution. Use equation (3.11) for calculating the mean or average flame height.

Mean flame height	$L_f = 0.235 \dot{Q}^{2/5} - 1.02 D$
Diameter of basket	$D = 0.305$ m
Total heat release rate	$\dot{Q} = 100$ kW
Mean flame height	$L_f = 0.235 \dot{Q}^{2/5} - 1.02 D$
	$= (0.235)(100)^{2/5} - (1.0)(0.305)$
	$= 1.17$ m (3.84 ft) above the virtual origin (3–4 ft)

The virtual origin from Example 3-2 was about 0.2 m above the fuel surface.

ENTRAINMENT EFFECTS

When a fire is burning in the center of a room, the buoyant upward flow of the fire gases drags fresh room air into contact with it. This process, called *entrainment* because surrounding cooler air is "mixed" or "entrained" into the upward plume flow, changes the plume's temperature, velocity, and diameter above the fuel source. Entrainment reduces the plume temperature and increases its width.

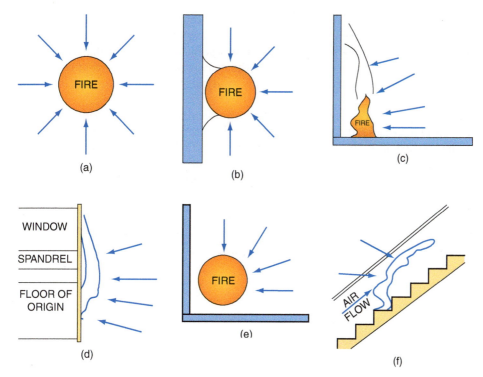

FIGURE 3-11 (a) Entrained air flow for an axisymmetric plume, from above; (b) entrainment in wall configuration; (c) side view; (d) typical high-rise fire plume; (e) entrainment in corner fire; (f) trench effect, side view.

If there is a horizontal ceiling and the fire plume reaches this height, the plume hits the ceiling and flows horizontally, forming what is called a *ceiling jet*. In this case ceiling entrainment also takes place (Babrauskas 1980).

When some of the heat is transferred to the air and is removed by convective processes, there is usually turbulent mixing that dilutes the rising column of hot gases and cools it further. Looking down on the fire plume from above, as in Figure 3-11(a), we can see that the inward air currents are roughly equal around the entire perimeter, so the plume remains symmetrical and upright.

If the fire is against a wall, it cannot draw room air into itself equally from all sides, so the air being drawn from the "free" side acts like a draft to force the plume sideways against the wall (shown in Figures 3-11b and c). The reduced entrainment by 50 percent means that less cooling air is brought in, slowing the cooling process. Thus, the gases have more time to rise farther, stretching out the flame. This process can occur on nearby wall surfaces even if they are not vertical. This is the mechanism responsible for external flame spread between floors of modern high-rise buildings. When the windows of the room first involved fail, the plume exits the building and then entrains itself against the side of the building (as in Figure 3-11d). The windows and spandrel panels of the floors above are then exposed to the radiant and convective heat transfer (>50 kW/m^2) of the plume and quickly fail, allowing the flames to enter the floor above and ignite the contents of that floor. If there is not sufficient fire protection to suppress this newly ignited fire, then fire can leapfrog up the building. In buildings with a ledge or balcony at each floor level, the plume is directed away from the side of the building, so it never gets a chance to attach to the facade. The chances of ignition by window failure of the floors above are much reduced by a ledge flame-guard effect, compared with a (sheer) curtain wall design.

If the fire occurs in a corner, the perimeter through which cooling air can be drawn is reduced to one-fourth its original size (there is still plenty of air reaching the combustion zone, so the fire continues to burn at the same rate as represented in Figure 3-11e). In a corner, then, the gases cool even more slowly and have even more time to rise before cooling to the point at which they are no longer incandescent (500°C), and the visible flame becomes even taller. This effect is enhanced by the increased radiant heat that reaches the fuel surface after being reflected off the nearby wall or corner surfaces, thus increasing the heat release rate.

Because a nearby wall or corner reduces the cooling entrainment of room air into the rising plume and increases radiant heat feedback to the burning fuel, the height of the visible flame can be affected. In its 2004 edition, *NFPA 921* listed relationships that are very useful to investigators and have been subjected to numerous studies and tests for determining the mean flame height near walls (Alpert and Ward 1984). These relationships are as follows:

$$H_f = 0.174 \left(k\dot{Q} \right)^{2/5}, \tag{3.12}$$

$$\dot{Q} = \frac{79.18 \, H_f^{5/2}}{k}, \tag{3.13}$$

where

H_f = flame height (m),
$\dot{Q}$ = heat release rate (kW), and
k = wall effect factor, with
k = 1: no nearby walls,
k = 2: fuel package at wall, and
k = 4: fuel package in corner.

EXAMPLE 3-7 ■ Flame Height: *NFPA 921* Method

Problem. Using Example 3-2 involving the 100 kW wastepaper basket fire, estimate the height of the flame in three configurations—in the middle of the room, near a wall, and near a corner. Assume that this is an unconstrained fire burning in a large room with no ventilation constraints.

Suggested Solution. From equation (3.12), the flame height H_f measured from the top of the fuel surface is as follows:

Total heat release rate $\dot{Q} = 100 \text{ kW}$
Flame height $H_f = 0.174 \left(k\dot{Q} \right)^{2/5}$
No nearby walls ($k = 1$) $H_f = (0.174)(1 \times 100)^{2/5}$

$$= (0.174)(100)^{0.4} = 1.10 \text{ m } (3.61 \text{ ft})$$

Near a wall ($k = 2$) $H_f = (0.174)(2 \times 100)^{2/5}$

$$= 1.45 \text{ m } (4.76 \text{ ft})$$

In a corner ($k = 4$) $H_f = (0.174)(4 \times 100)^{2/5}$

$$= 1.91 \text{ m } (6.27 \text{ ft})$$

A special case of this phenomenon is called the *trench effect,* also known as the *Coanda effect,* and is the tendency of a fast-moving stream of air to deflect itself toward and along nearby surfaces. The trench effect can occur wherever a fire starts on the floor of an inclined surface with walls on both sides, such as a stairway or escalator (shown in Figure 3-11f). In this case, the flow of the air drawn into the fire from below is aided by the confinement of the buoyant flow. When the fire reaches a critical size (i.e., a critical airflow) the plume lies down in the trench and is stretched out by the directional entrainment. If the floor and/or walls of the trench are combustible, they ignite, and the fire can spread the length of the stairway very quickly.

The trench effect was the mechanism that drove a small fire on the floor of an all-wood escalator in the King's Cross underground station in London in 1987 to develop quickly into a massive fireball that engulfed the large ticket hall at the top of the escalator (Moodie and Jagger 1992). Further research by Edinburgh University identified four factors relating to the conditions that produce the trench effect: (1) the slope of the trench, (2) its geometric profile, (3) the combustible construction material, and (4) the fire ignition source (Wu and Drysdale 1996). Researchers determined that the critical angle above horizontal is 21° for a large trench and 26° for a small trench.

PLUME CENTERLINE TEMPERATURES AND VELOCITIES

The temperature distribution along the centerline of a typical plume is represented in Figure 3-12(a), and the temperature distribution along a horizontal cross section through the flames is represented in Figure 3-12(b). Estimates of the centerline maximum temperature rise, velocity, and mass flow rate of a plume are useful in determining the impact of the fire plume in compartment fires. This impact includes the maximum effect of the plume as it is rising to the ceiling and may come into contact with other fuel packages. There are several formulas for estimating these values.

The *Heskestad method* (1982, 1983) used for estimating the centerline maximum temperature rise, velocity, and mass flow rate of a plume is

$$T_0 - T_\infty = 25 \left[\frac{\dot{Q}_c^{2/5}}{(Z - Z_0)} \right]^{5/3}, \tag{3.14}$$

$$U_0 = 1.0 \left[\frac{\dot{Q}_c}{(Z - Z_0)} \right]^{1/3}, \tag{3.15}$$

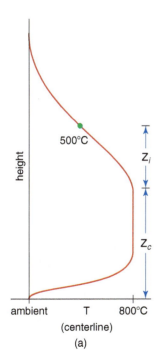

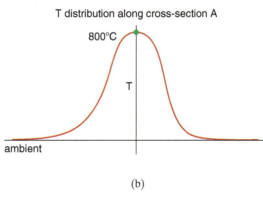

T distribution along cross-section A

(a)

(b)

FIGURE 3-12 (a) Plot of average flame temperature versus height shows the uniform maximum temperature in the center of the continuous flame zone and the decreasing average flame temperature in the intermittent region. (b) Plot of average flame temperature versus horizontal position shows the maximum temperature at the axis of the flame plume.

$$\dot{m} = 0.0056 \, \dot{Q}_c \frac{Z}{L_f} \quad \text{for} \quad Z < L_f \tag{3.16}$$

(i.e., for points within the flame plume),
where

$$\dot{Q}_c = 0.6 \, \dot{Q} \quad \text{to} \quad \dot{Q}_c = 0.8 \dot{Q}. \tag{3.17}$$

The *McCaffrey method* (1979) for estimating the centerline maximum temperature rise, velocity, and mass flow rate of a plume is

$$T_0 - T_\infty = 21.6 \, \dot{Q}^{2/3} \, Z^{-5/2}, \tag{3.18}$$

$$U_0 = 1.17 \, \dot{Q}^{1/3} \, Z^{-1/3}, \tag{3.19}$$

$$\dot{m}_p = 0.076 \, \dot{Q}^{0.24} \, Z^{1.895}. \tag{3.20}$$

An alternative method for estimating the maximum temperature rise and velocity of a plume is (Alpert 1972)

$$T_0 - T_\infty = \frac{16.9 \, \dot{Q}^{2/3}}{H^{5/3}} \quad \text{for} \quad \frac{r}{H} \leq 0.18, \tag{3.21}$$

$$T_0 - T_\infty = \frac{5.38 \, (\dot{Q}/r)^{2/3}}{H^{5/3}} \quad \text{for} \quad \frac{r}{H} > 0.18, \tag{3.22}$$

$$U = 0.96 \frac{\dot{Q}^{1/3}}{H} \quad \text{for} \quad \frac{r}{H} \leq 0.15, \tag{3.23}$$

$$U = \frac{0.195 \, \dot{Q}^{1/3} \, H^{1/2}}{r^{5/6}} \quad \text{for} \quad \frac{r}{H} > 0.15, \tag{3.24}$$

where

T_0 = maximum ceiling temperature (°C),
T_∞ = ambient air temperature (°C),
$\dot{Q}$ = total heat release rate (kW),
$\dot{Q}_c$ = convective heat release rate (kW),
Z = flame height along the centerline (m),
Z_0 = distance to the virtual origin (m),
$L_f = Z + Z_0$,
$\dot{m}_p$ = total plume mass flow rate (kg/s),
r = radial distance from the plume centerline (m),
H = ceiling height (m),
U = maximum ceiling jet velocity (m/s), and
U_0 = centerline plume velocity (m/s).

EXAMPLE 3-8 ■ Plume Temperature, Velocity, and Mass Flow Rate at Ceiling Level: McCaffrey Method

Problem. A fire is discovered in a wastepaper basket in a room that has a ceiling height of 2.44 m (8 ft). The heat release rate estimate is 100 kW. Assume that the ambient air temperature is 20°C (68°F). Estimate the plume temperature, velocity, and mass flow rate at the ceiling using the McCaffrey method. (Note that the proper approach is to measure the distance from the top of the fuel source or the lowest point where the fuel can entrain air. You may want to consider subtracting the height of the wastepaper basket from the ceiling height.)

Suggested Solution. Use the McCaffrey equation.

Maximum plume ceiling temperature $\quad T_0 = 21.6\,\dot{Q}^{2/3}\,Z^{-5/2} + T_\infty$

$$= (21.6)(100)^{2/3}(2.44)^{-5/2} + 20 = 68.5°C$$

Centerline plume velocity $\quad U_0 = 1.17\,\dot{Q}^{1/3}\,Z^{-1/3}$

$$= (1.17)(100)^{1/3}(2.44)^{-1/3} = 4.033 \text{ m/s}$$

Plume mass flow rate $\quad \dot{m}_p = 0.076\,\dot{Q}^{0.24}\,Z^{1.895}$

$$= (0.076)(100)^{0.24}(2.44)^{1.895} = 1.244 \text{ kg/s}$$

Discussion. In Example 3-2, the calculated virtual origin, Z_0, was 0.213 m (0.698 ft) above the burning fuel surface. Solve this same problem using the methods of Heskestad and Alpert.

SMOKE PRODUCTION RATES

During a fire, the smoke production within a building can expose people and property to smoke, toxic gases, and flames. The spread of fire and smoke can also cause a significant amount of property damage, including the disruption of electronic equipment by deposits of soot on circuitry (Tanaka, Nowlen, and Anderson 1996.

The *smoke production rate* is estimated to be close to twice that of the mass flow rate of gas produced by a fire plume. This assumption is based on the theory that the amount of air entrained into the rising plume equals the amount of smoke-filled gas (NFPA 2000, sec. 11, chap. 10).

In the *Zukoski method* (1978) for determining smoke production rates, the equation for estimating the rate of the plume mass flow above the flame at 20°C (68°F) is

$$\dot{m}_s = 0.065\,\dot{Q}^{1/3}\,Y^{5/3}, \tag{3.25}$$

where

$\dot{m}_s$ = plume mass flow rate (kg/s),
$\dot{Q}$ = total heat release rate (kW), and
Y = distance from virtual point source to bottom of smoke layer (m).

The best correlations to this theory are found in cases involving simple circular equivalent pool fires and in which the enclosure has ventilation in the lower layer.

ENCLOSURE SMOKE FILLING

When smoke from a fire accumulates within an enclosure or compartment, it tends to rise to the ceiling. This accumulation or *enclosure smoke-filling rate* depends on the amount of smoke being produced by the fire and its ability to escape through vents within the compartment. As the smoke accumulates, a layer forms and descends toward the floor of the compartment.

The enclosure smoke-filling rate is important when estimating how much smoke accumulated below the ceiling. Factors contributing to how fast smoke stratifies and descends from the ceiling relate primarily to the amount of smoke produced and the size and location of smoke vents. Fire investigators finding a room with smoke stratification deposits evenly distributed in an enclosed room may find this evidence important when later determining the fire duration or the smoke exposure a victim might have encountered at a particular time. Note that this research supports that at the time of the fire the level of the smoke in the room corresponds to smoke stratification deposits. It would not be unusual for smoke to fill the room, but the stratification deposits might be limited to 1.22 m (4 ft) down from the ceiling.

The relationship for the descent of this upper layer in a closed room is expressed as

$$U_t = \frac{\dot{m}_s}{\rho_l A_r},$$
(3.26)

where

U_t = rate of layer descent (m/s),

$\dot{m}_s$ = plume mass flow rate (kg/s),

ρ_l = density of the upper gas layer (kg/m^3), and

A_r = enclosure floor area (m^2).

Note that the density of smoke varies inversely with temperature but is conservatively estimated to be the same as air at standard temperature and pressure (STP; 1.22 kg/m^3).

EXAMPLE 3-9 ■ Smoke Filling

Problem. The 100 kW fire discovered in a wastepaper basket in the previous example is in a closed room that has a ceiling height of 2.44 m (8 ft). The room measures 2.44 m (8 ft) square. Estimate the enclosure smoke-filling rate for this fire.

Suggested Solution. Use the data previously calculated using the McCaffrey method. Assume the lower limit of the velocity of descent by using the ambient density of air as 1.22 kg/m^3, and 17°C.

Rate of layer descent $\qquad U_t = \dfrac{\dot{m}_s}{\rho_l A_p}$

Mass rate of smoke production $\qquad \dot{m}_s = 1.244 \text{ kg/s}$
Density of upper gas layer $\qquad \rho_l = 1.22 \text{ kg/m}^3$
Enclosure floor area $\qquad A = (2.44)(2.44) = 5.95 \text{m}^2$
Rate of layer descent $\qquad U_t = (1.244)/(1.22)(5.95) = 0.171 \text{ m/s}$

Fire Patterns

TYPOLOGY

When reconstructing fire scenes, skilled fire investigators use numerous indicators to estimate the areas and points of origin, distribution, and behavior of a fire such as direction of spread. These indicators, called fire patterns, are characteristically broken down into the following five major groupings summarized in Table 3-1. Some specific patterns are illustrated in Figure 3-13.

- Demarcations
- Surface effects
- Penetrations
- Loss of material
- Victim injuries

Note that the first four of these correlate to the categories of the fire effects discussed previously. Victim injuries can be considered a special case of applying those categories to human skin and tissue.

DEMARCATIONS

Demarcations occur in locations where a combination of smoke, heat, and flames impinge on materials, forming intersections between affected and unaffected areas. Examples include

TABLE 3-1	Typology of Common Fire Patterns		
TYPE OF PATTERN	**DESCRIPTION**	**VARIABLES**	**EXAMPLES**
Demarcations	The intersection of affected and unaffected areas on materials where smoke and heat impinge	Type of material Fire temperature, duration, and rate of heat release Fire suppression Ventilation Heat flux reaching surface	V patterns on walls Discoloration patterns Smoke stains
Surface effects	Determined by the boundary shape, areas of demarcation, and nature of heat application to various surface types	Surface texture (rougher surfaces sustain more damage) Surface-to-mass ratio Surface coverings Combustible and noncombustible surfaces Thermal conductivity Temperature of surface	Spalling of floor surfaces Combustible surfaces scorched or charred by pyrolysis and oxidation Dehydration Noncombustible surfaces changing color, oxidizing, melting, or burning cleanly Smoke stains Heat shadowing
Penetrations	Penetration of horizontal and vertical surfaces	Preexisting openings in floors, ceilings, and walls Direct flame impingement Heat flux Duration of fire Suppression	Downward penetrations under furniture Saddle burns on exposed floor joists Failure of walls, ceilings, and floors Internal damage to walls and ceilings Calcination
Loss of material	Combustible surfaces with loss of material and mass	Type of materials Construction methods Room contents and fuel loads Duration of fire	Tops of wooden wall studs Fall-down of debris Isolated areas of consumed carpet Beveling of corners and edges
Victim injuries	Areas, depths, and degrees of burns on victim's clothing and body	Location of victim in relation to other objects or victims Actions taken prior to, during, and after the fire	Burns to face and hands Absence of burn injuries protected by clothing or furniture Burns on lower torso Heat shadowing

Source: Expanded from Icove 1995.

smoke layers deposited on walls, and plume patterns on walls. See item 1 in Figure 3-13. Demarcations can vary depending on the type of exposed material, gas temperature, rate of heat release, and ventilation. Demarcations can identify deposits on the surface where temperatures have not been hot enough to thermally change the surface's covering. Areas of material loss are sometimes helpful in determining demarcations.

Heat and *smoke levels* (also referred to as *horizons*) refer to the height at which smoke or heat has stained or marked the walls and windows of a room. In fire modeling programs, these levels can be predicted throughout the structure being modeled and correlated with the duration and size of the fire. Heat and smoke levels are also important when evaluating

FIGURE 3-13 Fire pattern indicators found at fire scenes consisting of (1) demarcation, (2) calcination, (3) loss of material from wooden wainscoting and baseboard, (4) fractured glass, (5) ignitable liquid burn pattern on carpet, (6) penetration into the ceiling, and (7) area of clean burn where paper and pyrolysis products burned completely off wall. *Courtesy of D. J. Icove.*

victim accounts of whether they could see an exit if standing in a room or whether they would be affected by smoke or heat as they navigated through the building. In cases involving victims injured or killed in fires, the smoke and heat levels become important when evaluating whether victims stood up in a smoke-filled room and sustained their injuries from the cloud of heated toxic gases and smoky by-products of combustion.

As the ceiling jet plume extends throughout the room, demarcations are formed by stratified heat and smoke damage. In Figure 3-13, the ceiling plume extends to the right along the wall, forming the start of a stratified line. Notice the sharp demarcation in the degree of damage between the affected and the unaffected areas on the wall surface. Note also the acute angle between the demarcation line and the ceiling. This angle is the result of buoyancy-driven flow along the ceiling and is a very reliable indicator of direction of spread.

Fire suppression efforts such as the application of hose spray to the walls, ceilings, and floors can also affect demarcations. Ventilation efforts during fire suppression, such as the venting of the roof and the opening of doors and windows, also can affect the demarcations formed by heat and smoke throughout the structure. Figure 3-14 illustrates the formation of an upward pattern on the wall surface by the impact of a hose spray, which scoured the heat-affected paper and peeled it off the wall.

The cooler the surface, the faster soot and pyrolysates will condense. Airflow will also control depositions. Figure 3-15 shows the effects of airflow on soot deposition and how it can reveal the direction of hot gas spread.

SURFACE EFFECTS

Surface effects occur when heat transfer is sufficient to cause discoloration of masonry surfaces, scorching or charring of combustible surfaces, melting or changes in color, or oxidizing of noncombustible surfaces. Areas not damaged by the heat and flames are called *protected areas* and are significant in cases where objects or bodies have been found or have been removed prior to inspection. Surface effects can also include nonthermal deposits of soot or pyrolysis products that condense on the cooler surfaces. The cooler the surface, the faster these deposits occur. Variables influencing surface effects include the type and smoothness of the surface, the thickness and nature of the coverings, and even the object's surface-to-mass ratio.

Extended postflashover burning in a room exposes surfaces to high temperatures and heat fluxes. This "erases" the demarcations and surface effects (DeHaan and Icove 2012). Research at Eastern Kentucky University on burn patterns showed that brief (2–9 min) exposure to postflashover fires did not always totally obliterate the fire patterns needed by investigators (Hopkins, Gorbett, and Kennedy 2007). Examples are included in Chapter 8.

An example of a surface effect is the discoloration of paint on the external metallic surfaces of a burned automobile. Heat transfer through conduction often produces discernible fire patterns on the exterior and interior surfaces of the vehicle (as in Figures 2-3 and 2-5).

Discoloration of metallic surfaces can also be caused by *oxidation*, which is a chemical process associated with combustion. Particularly with metal surfaces, the boundaries between higher and lower temperatures can be observed and often can be documented through photographs and/or drawings.

Gypsum plasterboard or gypsum board is a material used in residential and commercial construction primarily for interior walls, in a process sometimes referred to as *drywall construction*. The drywall board contains gypsum (calcium sulfate dihydrate) bonded between two layers of paper. Fire-rated or X-type drywall is reinforced with glass fibers. The *ASTM C1396/C1396M-11* standard is the international standard for its manufacture (ASTM 2011).

Gypsum is a rock mineral that can be mixed with water and then dried into a hardened paste, resulting in a product that is 21 percent water by weight. The **calcination** of gypsum wallboard is a process by which heat transfer causes the calcium sulfate to dehydrate in two discrete steps at characteristic temperatures, typically around 100°C (212°F) and 180°C (356°F).

As gypsum dehydrates, it shrinks and loses the water of hydration, and its mechanical strength degrades. It can fail, exposing combustible wall structure, when it loses so much strength from dehydration, shrinkage, or combustion of paper coverings that it can no

FIGURE 3-14 The impact of a hose spray on a wall during fire suppression efforts formed an upward pattern on the wall surface demarcations previously formed by heat and smoke. *Courtesy of D. J. Icove.*

calcination ■ Loss of water of crystallization caused by heating.

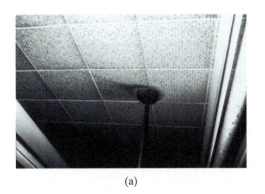

(a)

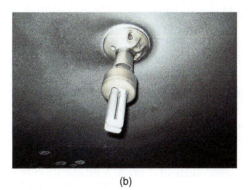

(b)

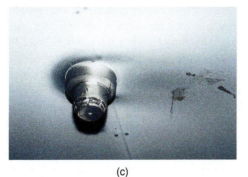

(c)

FIGURE 3-15 Deposits of soot (a and b) around light fixtures and (c) around a heat detector indicate direction of air movement. *(a) Courtesy of Jim Allen, Nipomo, California; (b and c) Courtesy of Ross Brogan, NSW Fire Brigades, Greenacre, NSW, Australia.*

longer support its own weight. Eventually, the fully dehydrated gypsum turns powdery and loses all its mechanical strength, causing the wall to collapse. This process of calcination can also form lines of demarcation on the wall surface (DeHaan and Icove 2012). The effect of calcination on a gypsum board wall is shown as items 2 and 7 in Figure 3-13.

A change in color from white to gray to white is seen in a cross section of affected gypsum. Dehydration alone probably does not cause the gray-to-white change, since this is a colorless transition. During fire exposure other carbonaceous pyrolysis products accumulate that turn the absorbent gypsum gray. Subsequent heating burns away or evaporates these deposits, causing whitening to progress through the thickness of the gypsum.

During fire exposure the exposed paper and painted surfaces of the gypsum wallboard burn away. This reduces the overall mechanical strength of the gypsum wallboard. The process in which dehydrated gypsum falls or flakes off the surface during heat exposure is called *ablation,* which occurs at 600°C (1112°F) for ceilings and 800°C (1472°F) for walls (König and Walleij 2000). Ablation reduces the thickness of the board (Buchanan 2001, 341). Several techniques are available for documenting the measured postfire loss of gypsum from the wall surface and estimating the radiant heat flux exposure time (Kennedy 2004; Schroeder and Williamson 2003). The measurements of depth of calcination of gypsum surfaces are often recorded on a plot of the room.

Research by Schroeder and Williamson (2001) indicates that gypsum wallboard is a useful source for documenting a fire's intensity, spread, and direction. In a series of experiments using an *ASTM E1354* cone calorimeter, they concluded that thermally induced changes in gypsum wallboard, along with X-ray diffraction analysis to document the crystalline structure change, provided a quantifiable method for assessing time, temperature, and heat flux exposure. Additional cited references on the thermal properties of gypsum board can be found in Lawson (1977), Thomas (2002), and McGraw and Mowrer (1999). Also see additional references on calcination of gypsum board during fire in Mowrer (2001), Chu Nguong (2004), and Mann and Putaansuu (2006).

organic ■ Compounds based on carbon.

char ■ Carbonaceous material resulting from pyrolysis, often appearing to be blackened blisters on the surface of cellulose and other solid organic fuels.

Charring is the pyrolytic action of heat that converts the **organic** material, most often paper or wood, into its volatile fractions, leaving behind carbonaceous **char**. Depending on the type of wood, this char layer may burn off or flake off as the fire exposure continues. The linear depth of the pyrolyzed material affected, including the char layer and lost material, is referred to as the *char depth*.

The rate at which wood chars is not linear. Wood chars quickly at first and then at a much slower rate owing to various factors such as the insulation effect of the new char, reduced air–fuel interactions, ventilation, heat transfer, and physical arrangement. The reported charring behavior of large-section structural lumber has been modeled. A review of 55 specimens of 2 in. × 4 in. spruce-pine-fir lumber subjected to a constant-temperature exposure of 500°C (932°F) produced a linear model. The model showed a constant rate of charring of 1.628 mm²/s and 0.45 mm/min (Lau, White, and Van Zeeland 1999). Real-world fire exposures do not produce constant temperature or heat flux conditions.

Research was undertaken to numerically simulate pyrolysis. In this research, the movement of the char/virgin front was modeled with good agreement of mass loss with experimental testing (Jia, Galea, and Patel 1999). Babrauskas offers a comprehensive analysis of charring rate of wood as a tool for fire investigators (2005; also DeHaan and Icove 2012, chap. 7). Babrauskas includes data illustrating the variability of char rates with type of wood and nature of fire exposure. The char rate is highly dependent on applied heat flux (2005; also Butler 1971). During a fire, heat fluxes on a surface may vary from near zero to 50 kW/m² for contact by small flames to 120–150 kW/m² under postflashover conditions, resulting in a char rate that can vary from zero to several millimeters per minute.

To identify conductive heat transfer through walls, investigators often document the charring of wood studs within walls covered by materials such as wooden paneling or gypsum board. Figure 3-16 shows varying degrees of damage to a gypsum board–covered wall and previously protected wood studs within the wall after exposure to a test fire. The test results agreed with the varying degrees of charring damage predicted by a fire model known as WALL2D (Figure 3-17), which predicts heat transfer and char damage to wood studs protected by gypsum board and exposed to fire. The WALL2D developers reported good agreement of their heat transfer model with the results of both small- and full-scale fire tests (Takeda and Mehaffey 1998). WALL2D was the focus of an engineering study that identified it as a viable fire model for users in both the regulatory and fire investigation communities (Richardson et al. 2000).

Char depth is usually documented using a penetrating tool and survey grid diagrams (*NFPA 921*, 2011 ed., sec. 17.4.3) (NFPA 2011). In these diagrams, the data from the measurements are converted to lines of equal depth, called **isochar** lines. Each line connects depths of the same value, and investigators should consider recording elevation (height above the floor) measurements for greater precision (Sanderson 2002 Because variations in char rate can be caused by the differences in affected woods, coatings, and flame-retardant treatments, investigators are cautioned to realize that char depths should be compared on similar types of wood (baseboards, door frames, etc.).

FIGURE 3-16 Varying degrees of damage to a paneled wall and previously protected wood studs within the wall after being exposed to a test fire. *Courtesy of D. J. Icove.*

Owing to the nonlinear nature of charring, the measurement of char depth cannot establish precise times of fire exposure, but deeper relative depths of char may indicate areas of more intense or prolonged burning. Also, the appearance of the char surface often depends on the individual characteristics of the wood and ventilation conditions (DeHaan and Icove 2012). However, char depth relationships assuming a constant 50 kW/m^2 incident flux have been suggested for fire protection engineering applications (Silcock and Shields 2001).

Spalling is the chipping or crumbling of a concrete or masonry surface when it is exposed to varying heat, cold, or mechanical pressures. As a fire pattern damage indicator,

isochar ■ Line drawn on a scene diagram that connects points of similar char depth on wood surfaces.

spall ■ Crumbling or fracturing of concrete or brick surface as a result of exposure to thermal or mechanical stress.

FIGURE 3-17 The progression of a char front upward through a wood ceiling joist after exposure to a test fire. *Courtesy of D. J. Icove.*

spalling can indicate steep temperature gradients caused by radiated heat, a significant fuel load of ordinary combustibles, or other sources of localized heating that last long enough for heat to penetrate the concrete.

Spalling occurs during rapidly rising temperatures, typically 20°C–30°C/min (70°F–86°F/min) (Khoury 2000). Several factors influence spalling, but the moisture content and nature of the aggregate used (limestone versus granite) are the leading variables. Spalling can also be caused by suddenly cooling a very hot concrete surface with a hose stream. High-strength concrete tends to spall more than normal concrete owing to its higher density. The pores of the high-density concrete become filled more quickly with the high-pressure water vapor (Buchanan 2001, 228). Spalling can be reduced if the concrete is made with fine plastic fibers mixed into the slurry. As the fibers melt at low temperatures they provide pores to vent the internally produced steam.

Steel reinforcements expand much more rapidly when heated than does concrete. Thus, if reinforcements are not protected deeply in the concrete, they expand and cause tensile failure of the concrete. Lightweight concrete spalls more readily because of the vermiculite used as an aggregate.

Testing by the SP Swedish National Testing and Research Institute on spalling reinforced the two classic explanations for concrete spalling. The most common cause is thermal stress, which forces water vapor out of the hotter side of the concrete. Experiments by Jansson (2006) documented that water vapor is also forced out of the cold side of the concrete during fire tests. During these fire tests, the internal pressures also started to fluctuate severely after 15 minutes. New material test methods have been proposed for determining whether an actual exposed concrete may suffer from explosive spalling at a specified moisture level. A specimen cylinder is used, which is a cost-effective alternative to full-scale testing (Hertz and Sorensen 2005).

Studies by the UK Building Research Establishment (BRE) in fire damage on natural stone masonry in buildings showed that natural stone can be seriously affected in building fires, with damage concentrated around window openings and doorways (Chakrabarti, Yates, and Lewry 1996). Color changes at 200°C–300°C (392°F–572°F) were followed by localized disintegration at 600°C–800°C (1112°F–1472°F). Color changes included the reddening of stones containing iron, which is irreversible and causes significant damage to historical buildings. Other damage noted by BRE included costly smoke staining and salt efflorescence resulting from fire hose streams.

BRE also reported that brown or light-colored limestone containing hydrated iron oxide changes to reddish brown at 250°C–300°C (482°F–572°F), more reddish at 400°C (752°F), and a gray-white powder at 800°C–1000°C (1472°F–1832°F). The depth of calcination of limestone is seldom less than 20 mm, which begins at 600°C (1112°F), reducing the stone's strength. Magnesium limestone is white or buff colored and changes to pale pink at 250°C (482°F) and pink at 300°C (572°F).

Sandstone changes to a brown color owing to the dehydration of iron compounds starting at 250°C–300°C (482°F–572°F) and to reddish brown above 400°C (752°F). Structural weakening of sandstone starts at 573°C (1063°F), when the quartz grains internally rupture. Loss of strength of such stone is particularly critical when solid stone lintels are used above windows, doors, and gateways, as massive structural collapse can occur without warning.

Further research by BRE indicated that granite will show no color changes but will crack or shatter at temperatures above 573°C (1063°F) through quartz expansion, and marble's internal calcite crystals will undergo thermal hysteresis, causing a reduction in flexural strength. Marble has been known to crumble into a powder when exposed to extreme heat (>600°C, 1150°F). This is particularly dangerous for marble staircases in fires.

The UK Building Research Station published a historical review of the visible changes in concrete or mortar when exposed to high temperatures (Bessey 1950). The study examined aggregates, concrete, and mortar, and the reproducibility and interpretation of observed changes in terms of temperature. Research by Short, Guise, and Purkiss (1996) at the Department of Civil Engineering, Aston University, UK, using optical microscopy documents their use of color analysis in the assessment of fire-damaged concrete.

Flammable liquids very rarely produce significant spalling on bare concrete because a typical flammable liquid pool fire on bare concrete lasts only 1–2 min, and sometimes often much less (DeHaan 1995). This leaves little time for heat to penetrate into the concrete, which has poor conductivity. In addition, the radiant heat from the flame is, in part, absorbed by the liquid fuel during combustion, and the maximum temperature of the surface cannot significantly exceed the boiling point of the liquid in contact with it. Because the boiling point range of gasoline is 40°C–150°C (104°F–302°F), the concrete will not usually be heated to the point at which it will be affected (DeHaan and Icove 2012, chap. 7). Extensive experiments by Novak (unpublished data) have demonstrated the lack of spalling beneath flammable liquid pool fires on a variety of concrete surfaces that are readily spalled by a wood fire (see Figure 3-18).

If tile or carpet is present on the concrete, the gasoline flames can trigger localized charring, melting, and combustion of the floor covering. Because this combustion is more

(a)

(b)

(c)

(d)

FIGURE 3-18 Spalling concrete tests: (a) gasoline pool fire, (b) staining but no spalling of concrete after gasoline fire self-extinguished, (c) wood pallet fire on same type concrete, (d) extensive spalling under wood fire. *Courtesy of Jamie Novak, Novak Investigations, Inc.*

FIGURE 3-19 Surface effects due to a flammable liquid pour pattern (center of the photograph) that charred the floor tiles. Two parallel pour patterns are documented, characteristic of a back-and-forth spilling of flammable liquids as the arsonist backed out of the building. *Courtesy of D. J. Icove.*

ghost marks ■ Stained outlines of floor tiles produced on noncombustible floors by the dissolution and combustion of tile adhesive.

prolonged than the gasoline flame itself, the molten mass has a very high boiling point, and there is no freestanding liquid to protect the concrete, therefore, there is more opportunity for heat to affect the concrete and cause it to spall if it is susceptible. Figure 3-19 shows an example of localized burning due to a flammable liquid pour pattern on floor tiles. The collapse of molten burning plastic or roofing tar onto a concrete surface will produce significant and prolonged heat transfer that will induce spalling. Prolonged burning of structural wood collapsed onto a concrete floor surface is especially likely to cause spalling.

Additional cited references on spalling of concrete include Canfield (1984), Smith (1991), Sanderson (1995), and Bostrom (2005).

Localized combustion of flammable liquids on vinyl- or asphalt-tiled concrete surfaces can produce a type of floor damage called **ghost marks**. Ghost marks are subsurface staining of the concrete beneath the joints of tile floors that appears to be caused by combustion of the cement/mastik dissolved by the applied flammable liquid. Radiant heat in postflashover fires and long-term heating by *fall-down* of burning materials can produce spalling but do not apparently produce ghost marks (DeHaan and Icove 2012, chap. 7)). Charring of wood floors along tile seams is not the same.

Fractured glass is another surface effect that usually occurs when glass is stressed by nonuniform heating. Common variables that affect glass breakage include its thickness, the presence of any defects, the rate at which it was heated or cooled, and the method used to secure the glass to the window frame (Shields, Silcock, and Flood 2001).

The potential for glass to fracture and break out has been the topic of several research initiatives. The prediction of the probability of breakage fits a Gaussian statistical correlation, as shown in Figure 3-20 (Babrauskas 2004). This relationship shows that the mean temperature for the breakage of 3-mm-thick (0.12 in.) window glass is 340°C (644°F), with a standard deviation of 50°C (122°F).

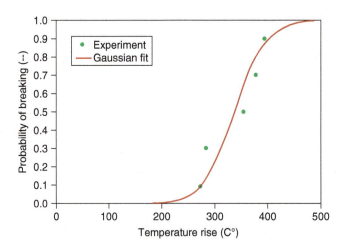

FIGURE 3-20 The probability of breakage of 3-mm-thick window glass fits a Gaussian statistical correlation with a mean temperature of 340°C (644°F) and a standard deviation of 50°C (122°F). *Courtesy of Babrauskas (2004); www.doctorfire.com.*

The fire modeling program BREAK1 (Berkeley Algorithm for Breaking Window Glass in a Compartment Fire) can calculate the temperature history of a glass window given several fire parameters combined with a physical composition of the glass (NIST 1991). The model also predicts the temperature gradients of the glass normal to the glass surface and the window breakage time. Large-scale glass fracturing by heat is often observed as a room approaches flashover and the temperatures and heat fluxes dramatically increase. Note however, that this program computes only the cracking of glass, whereas the investigator is often interested in knowing when the glass failed enough so that pieces fell out, rather than the time when cracks first appeared. It is the physical loss of glass from windows that can affect the development of the fire in the room, sometimes dramatically.

Hietaniemi (2005) at the VTT Building and Transport, Finland, has produced a probabilistic simulation model for glass fracture and fallout in a fire. This approach uses two steps: the model first predicts the simulation time for the initial occurrence of a crack using the BREAK1 program, followed by a thermal response model of the glass.

Heat-induced failure of dual-glazed windows is considerably less predictable. Often, the inside pane fails, but its fragments protect the exterior glass pane. Dual-glazed windows have been observed to survive postflashover room fires only to break when struck by hose streams. Dual glazing with heat-absorbing film is even more resistant to heat-induced failure.

One common mechanism for thermal fracturing of common window glass is the stress induced by nonuniform heating as a result of "thermal shadow" cast by framing or trim. When a window is exposed to radiant or convected heat, the exposed portions increase in temperature and begin to expand, but the areas shaded by the framing or window putty are not heated. Because glass is a relatively poor conductor of heat, those "shaded" areas remain cool and do not expand, as shown in Figure 3-21(a). This differentiation induces stress in the glass that can cause the glass to fracture, often in patterns roughly parallel to the edges, as shown in Figure 3-21(b). The fractures radiating from a single point on the right often are caused by a preexisting nick in the edge of the glass or by stress induced by a glazier point or frame nail (DeHaan and Icove 2012, chap. 7).

It should be noted that the observation that the windows "blew out" during a fire is rarely associated with an overpressure event but, rather, with thermally induced massive fracturing. The pressures produced by a flashover transition in a room are very rarely adequate to cause mechanical breakage (Fang and Breese 1980; Mitler and Rockett 1987; Mowrer 1998). However, the development of a **backdraft** condition, in which the fire becomes severely underventilated and is then given additional air, can produce overpressures sufficient to cause windows to break.

The edges of ordinary glass fractured by mechanical stress are characterized by curved, conchoidal ridge lines caused by the propagation of the fracture, as in Figure 3-22(a). The fracture pattern on the face of glass when broken by mechanical shock or pressure is typically a spiderweb pattern of mostly straight fractures, as in Figure 3.22(b). The pattern usually consists of a number of radial fractures from the point of contact and concentric fractures connecting them.

Mechanical impacts of very brief duration cause minimal radial fractures, because the glass does not have time to flex and may produce a domed plug of glass to be ejected from the side opposite a localized impact. Thermal stress usually results in more random

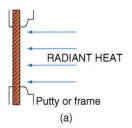

RADIANT HEAT

Putty or frame

(a)

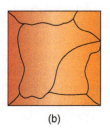

(b)

FIGURE 3-21 Heat fractures of a window pane caused by "shadowing" of radiant heat. Fractures radiating out from a point on the right often are caused by a preexisting nick in the edge of the glass or by stress induced by a glazier point or frame nail.

Chapter 3 Fire Pattern Analysis **125**

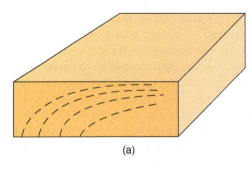

(a)

(b)

FIGURE 3-22 (a) Edge of glass fragment showing conchoidal fracture lines. (b) Face of glass sheet broken by mechanical impact near the center, with radial fractures and concentric fractures.

clean burn ■ An area of wall or ceiling or other surface where the charred organic residues have been burned away by direct flame contact or other source of high temperature.

crazing ■ Stress cracks in glass as the result of rapid cooling.

FIGURE 3-23 Mechanical fractures in tempered glass door (dual glazed) retain the radial fracture pattern indicating point of breakage (by firefighters). *Courtesy of J. D. DeHaan.*

patterns with wavy fractures. These fractures usually have mirror-smooth edges with minimal conchoidal ridges, since they are usually produced at lower speeds than mechanical fractures.

If enough glass can be recovered to jigsaw-fit pieces together for at least part of the pattern, and the inside or outside face can be identified by soil, paint, decals, or lettering, the direction of force can be established by the 3R rule: the conchoidal lines on radial fractures start at right angles on the reverse (side away from the force). The direction is reversed on concentric fractures. Toughened (safety) glass used in the side and rear windows of vehicles and in structural doors and shower doors fractures into small, rectangular blocks. As a general rule, the cause and direction of failure of safety glass cannot be established, but under some conditions fragments trapped in the frame may reveal the approximate location of a mechanical impact point, as seen in Figure 3-23.

The sequence of impacts can be established from intersecting and ending fractures. Sequence of failure versus time of exposure to fire can be established by the presence or absence of soot on the broken edges. Direct contact with flame can heat glass to the point at which all accumulated soot and smoke deposits are oxidized off, a so-called **clean burn**. Glass breakage patterns can also reveal a great deal of information about an explosion (DeHaan and Icove 2012, chap. 12). The distance to which the glass is blown is related to the thickness of the glass, the size of the window, and the pressures developed (Harris, Marshall, and Moppett 1977; Harris 1983). As a general rule, the thinner the glass or the larger the surface area of the pane, the lower the failure pressure will be.

A distinctive form of glass fracturing called **crazing** is usually produced when fire suppression water spray strikes a heated glass surface. Crazing is characterized by a complex pattern (road map–like) of partial-thickness fractures and concave pitting (Lentini 1992). Field experiments during test fires confirmed that when a water spray was applied to the hot glass panes on one side of a heated window, crazing occurred, as in Figure 3-24. Notice the melted glass in the upper right-hand pane. Common window glass softens when its temperature reaches 700°C–760°C (1300°F–1400°F).

Another descriptive form of surface burn pattern is called clean burn (see Kennedy and Kennedy 1985, 451; *NFPA 921*, 2011 ed., pt. 2004a, sec. 6.7.6 [NFPA 2011]; DeHaan 2007, 275). When a solid surface is exposed to a fire environment, products of combustion—water vapor, soot, and pyrolysis products—will condense on it. The cooler the surface, the faster products will condense, giving rise to the shadowy outlines of studs, nail heads, and other hidden features of a wall structure that cause a temperature differential. When a sooted surface is exposed to direct flame, it can reach high enough temperatures that the accumulated products can burn away, leaving behind a clean surface. This pattern can be used to identify areas of direct contact between a surface and the flaming portion of the plume. Surfaces not in contact with the flame will rarely reach sufficiently high temperatures to allow the complete combustion of condensed soot or pyrolysates. An unpublished study by Ingolf Kotthoff indicates that for a clean burn to occur, the material must reach approximately 700°C (1300°F). A clean burn can also result where charred organic materials are burned away, leaving a bare noncombustible substrate exposed.

In ATF Fire Research Laboratory test burns in 2008 involving single-room compartments in ventilation-controlled conditions, each fire was started in the same manner and location (Carman 2010). The main variable in each of the tests was the length of time the fires burned. The results showed both similarities and differences in the ensuing burn patterns, particularly in clean-burn indicators. One hypothesis is that the mechanism for clean-burn pattern development may be different from the popular notion of combustion of previously deposited soot on that surface. The patterns may be due in part to high thermal gradients on surfaces that prevent localized deposition. Accordingly, further study of such mechanisms may improve the interpretation of such patterns and the timing of their creation.

FIGURE 3-24 Crazing of glass caused by rapid cooling of glass, often during fire suppression, as seen in this field experiment in which a water spray was applied to the window panes. *Courtesy of Lamont "Monty" McGill, Gardnerville, NV.*

Melted materials can also reveal much about the temperatures to which surfaces were exposed during the fire. Historical studies by the UK Building Research Station provided an estimation of the maximum temperatures attained in building fires from an examination of the debris. Researchers examined lead and zinc plumbing fixtures, aluminum and alloys used in small machinery, molded glass from windows and jars, sheet glass in window panes, silver jewelry and eating utensils, brass doorknobs and locks, bronze window frames and bells, copper electric wiring, and cast iron pipes and machinery (Parker and Nurse 1950). The findings of this study demonstrated that the analysis of melted materials in a fire could identify and document areas of varying temperature exposure.

Plastics, glass, copper, aluminum, and tin are the most common melted materials found at fire scenes that can document the wide range of temperatures produced by the fire (DeHaan and Icove 2012). Melting points of thermoplastics can vary depending on the nature of their internal molecular structure. Glass melts, softens, or becomes molten over a range of temperatures. Table 3-2 lists approximate melting temperatures of common materials. It must be remembered that interactions between melting materials can cause unusual effects. Zinc or aluminum melting onto steel can produce alloying, causing the steel to melt at ordinary fire temperatures. Zinc can come from pot-metal fittings or galvanized coatings. Aluminum is well known for creating a **eutectic** alloy with copper with a very low melting point. Calcium sulfate (gypsum or plaster) can cause localized melting of black iron under fire conditions.

eutectic ■ An alloy of two materials having special physical or chemical properties, typically having the lowest melting point of any combination of the two.

MATERIAL	°C	°F
Aluminum alloys[a]	570–660	1060–1220
Aluminum (pure)[c]	660	1220
Copper[c]	1083	1981
Cast iron (gray)[a]	1150–1200	2100–2200
50/50 solder[a]	183–216	360–420
Carbon steel[a]	1520	2770
Tin[c]	232	450
Zinc[c]	420	787
Magnesium alloy[a]	589–651	1092–1204
Stainless steel[a]	1400–1530	2550–2790
Iron (pure)[c]	1535	2795
Lead (pure)[c]	327	621
Gold (pure)[c]	1065	1950
Paraffin (wax)[d]	50–57	122–135
Human fat[e]	30–50	86–122
Polystyrene[b]	240	465
Polypropylene[d]	165	330
Polyester (Dacron)[d]	dec 250	480
Polyethylene[b,c]	130* (85–110)	185–230
PMMA (acrylic plastic)[b]	160*	320
PVC[b]	dec 250*	480
Nylon 66[b]	250–260	480–500
Polyurethane foam[c]	dec 200	390

TABLE 3-2 Approximate Melting-Point Temperatures of Common Materials

Sources: [a]Perry and Green 1984, Table 23-6, 23–40; [b]Drysdale 2011; [c]DeHaan 2007, 274–86; [d]*The Merck Index;* [e]DeHaan, Campbell, and Nurbakhsh 1999.

Note: dec = decomposes.

*Varies with cross-linking.

As we have seen, localized temperatures in any flame can readily reach 1200°C (2200°F), but the actual values recorded depend greatly on the measuring technique. Temperatures in a growing room fire can vary from ambient room temperature to over 800°C (1470°F). Postflashover fires can produce temperatures of over 1000°C (1800°F) throughout a room. Burning plastics can produce localized flame temperatures of 1100°C–1200°C (1980°F–2200°F). Most other materials, whether ignitable liquids or ordinary combustibles, produce maximum flame temperatures in air of 800°C (1470°F). Evidence of high temperatures (800°C [1470°F] or higher) is not proof of the use of accelerants. Plastics can indicate the intensity and distribution of heat exposure as they soften and distort. Figure 3-25 shows the thermal plastic face of a wall clock that verifies the depth of the hot layer in a room. Polyethylene laundry bags on the floor near it were not affected.

Annealed furniture springs result when their temperature exceeds the typical or **annealing** point (538°C, 1000°F), and the coiled steel springs lose their temper (Tobin and Monson 1989). Collapsed springs are indicators of this phenomenon, and an investigator should compare the relative degree of collapse to try to ascertain the direction and intensity

annealing ■ Loss of temper in metal caused by heating.

of the fire (DeHaan and Icove 2012). Such collapse is not an indicator that accelerants were used, since smoldering fires of accidental cause can readily produce such effects. The investigator should also examine the framework (usually metal or wood) for other signs of localized thermal damage, *NFPA 921,* 2011 ed., sec. 6.12.14 (NFPA 2011). Modern urethane foam furniture often burns so quickly that even the high-temperature flames cannot bring the steel springs to their annealing temperatures.

Steel springs can also corrode to the point of failure from fire exposure due to stress-corrosion cracking, which occurs when normally ductile metals are subjected to a tensile stress at elevated temperature (NPL 2000).

PENETRATIONS

Another common fire pattern is the occurrence of *penetrations* through horizontal and vertical surfaces as direct flame impingement or intense radiant heat on walls, ceilings, and floors is sustained long enough to affect deeper layers of the material. Ceiling areas directly overhead of fire plumes may burn through or collapse, allowing flames to extend into confined spaces. In some cases, fall-down of combustible debris followed by extended smoldering can lead to localized penetration of wooden floors, sills, and toe plates in wood-framed structures.

FIGURE 3-25 Softening of plastic clock face indicates temperature of layer (face should be saved and its height from floor measured). *Courtesy of J. D. DeHaan.*

Variables influencing penetrations include preexisting openings in floors, ceilings, and walls. Also, direct flame impingement coupled with high heat fluxes plays an important role in creating the conditions necessary for penetrations. For example, the area where a fire plume intersects a ceiling, such as item 6 in Figure 3-13, is an example of a penetration into the ceiling directly above the fire plume due to the high heat flux and temperatures that caused the failure of the ceiling materials.

Penetrations are not always above the fire plume. During flashover conditions, downward penetrations are sometimes found. As discussed previously, the most common reason for penetrations through wooden flooring is radiant heat from above and not the combustion of ignitable liquids (DeHaan 1987; Babrauskas 2005). Sustained burning of collapsed furniture, bedding, or clothing also can result in localized penetrations. Fire testing involving no ignitable liquids in furnished rooms has demonstrated that the resulting postflashover conditions can produce penetrations through carpet, pad, and plywood floors, as illustrated in Figure 3-26.

In postflashover fires where ignitable liquids were poured onto carpeted floors, burning penetrations can char supporting joists (from postflashover fire, not the liquid fuel fire). In fire tests conducted by the U.S. Fire Administration, 1.75 L (0.46 gal) of gasoline was poured onto carpeted floor (FEMA 1997). Flashover occurred at 40 seconds, with extinguishment at 10.3 minutes. An examination of the scene after extinguishment showed the fire to have burned through the carpet, carpet underpadding, and 9.5-mm (0.37-in.) plywood floor. As shown in Figure 3-27, there was widespread charring on the 2 in. × 8 in. supporting joists. The test involved nearly 10 minutes of postflashover burning, which caused most of the observed damage long after the gasoline burned away.

Sometimes, burning foam mattresses drop enough burning molten liquid from pyrolysis onto the floor directly under the bed to cause burns through the floor. Burning thermoplastics such as polyethylene trash containers or thermoplastic appliance liners can do the same. Traditional cotton-stuffed upholstered furniture such as sofas, chairs, and mattresses can support combustion long enough to burn through wooden floors beneath them after they collapse.

Failure of walls, floors, and ceilings during a fire may also contribute to formation of penetrations through which the fire extends throughout a structure. These failures have been known to transport heat, flames, and smoke to other areas of the structure, leading inexperienced investigators to conclude improperly that two or more fires were set within the structure. Such penetrations can occur as gypsum wallboard fails, lath chars, plaster collapses, and metal ceilings or other noncombustible ceiling coverings peel away. Ventilation drafts up stairwells can cause very intense fires with rapid failure of doors and ceilings.

LOSS OF MATERIAL

Loss of material occurs in any combustible material located in the path of a fire's development and spread. This loss of material is helpful in determining the fire's intensity and direction of travel. *Heat shadowing* can affect the loss of materials by shielding the heat transfer and masking the potential lines of demarcation of other fire patterns (Kennedy and Kennedy 1985, 450; *NFPA 921*, 2011 ed., sec 6.3.4.1 [NFPA 2011]). Loss of material can be driven by direct flame impingement or radiant heat alone. The rapid charring caused by postflashover radiant heat can be enhanced by ventilation from doors or windows to produce penetrations. These can occur through floors near doors or windows, in the center of rooms, or across the room from vents where air flows, where postflashover mixing is most efficient (DeHaan and Icove 2012) (see Figure 3-26). Extensive destruction around heater vents has been observed in rooms where there was prolonged postflashover burning because of the extra oxygen the vents provided to fuels already burning. The results of full-scale fire tests reported by Carman (2008, 2009, and 2010) and supplemented by postfire computer modeling further demonstrated the impact of open doorways on fire patterns.

FIGURE 3-26 The penetration (burn-through) of carpet, pad, and 12-mm (0.5-in.) plywood floor in a furnished room fire test involving no ignitable liquids. Postflashover fire lasted less than 5 minutes but produced extensive destruction to exposed areas, especially at the door (foreground). *Courtesy of J. D. DeHaan.*

Items 3 and 7 in Figure 3-13 are two examples of loss of materials due to direct impingement of the fire plume. The most significant is the relative loss of the wooden wainscoting material. The closer to the plume centerline, assume the deeper the charring. Localized damage to the wooden baseboard documents the radiant heat from above.

Loss of material may occur on the tops of wooden studs, on corners and edges of furniture and moldings, and floors and carpeted surfaces. This gives rise to beveling of solid fuels or extended V patterns that can reliably indicate direction of fire spread. Other forms of loss of materials come from fall-down of debris from above. Typical

FIGURE 3-27 The burn-through of carpet, underpadding, and 12-mm (0.5-in.) plywood floor in a room fire test involving 1.75 L (0.5 gal) of gasoline and approximately 10 minutes of postflashover burning. *Courtesy of the U.S. Fire Administration.*

examples include burning drapery or curtains, which ignite other objects in the room and sometimes mislead novice investigators into thinking that multiple fires may have been set.

Loss of material may play an important role in determining the nature and extent of fire damage to upholstered furniture exposed to smoldering versus flaming combustion. Tests published by Ogle and Schumacher (1998) demonstrated that smoldering fire patterns often tended to consist of char zones with a thickness equal to that of the fuel element (i.e., the entire thickness). If the source of the smoldering fire was, for example, a cigarette, they found that the smoldering patterns persisted as long as the cigarette continued to smolder. Fire patterns created by smoldering fires were destroyed if a transition to flaming combustion occurred. In the case of cigarette-initiated fires, the cigarette itself was consumed in the flaming phase.

Flaming fire patterns had thin char zones with thicknesses smaller than the thickness of the fuel element. Ogle and Schumacher (1998) observed that the flaming fire consumed the cover fabric and underlying padding. When examining flame spread rates with polyurethane foam (PUF) seat cushions, they observed that horizontal burning was faster than vertical downward burning. Also, since the cushions were thermally thick, the depth of char into the foam tended to be thin when compared with the overall cushion thickness.

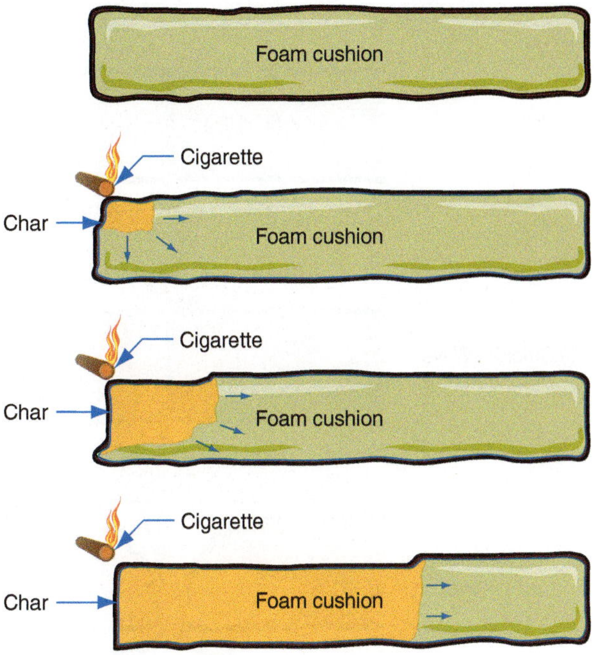

FIGURE 3-28 Cross-sectional view of the development of smoldering fire patterns in a cotton-upholstered foam cushion exposed to a smoldering cigarette at its corner.

Figures 3-28, 3-29, and 3-30 show graphic representations of the burn patterns observed by Ogle and Schumacher in fires involving upholstered PUF cushions. These

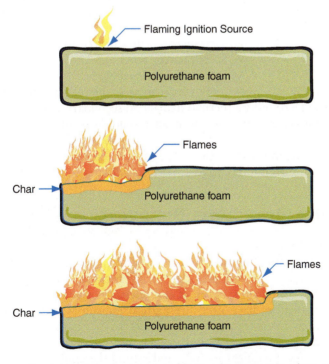

FIGURE 3-29 Cross-sectional view of the development of flaming fire patterns in a foam cushion exposed to a flaming ignition source at its surface.

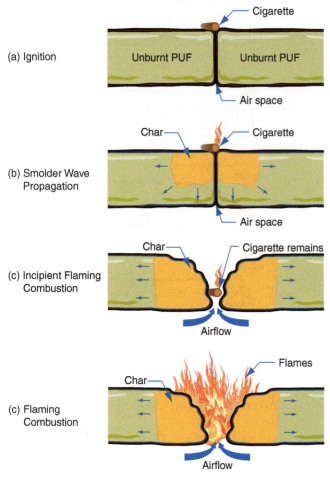

(a) Ignition

Cigarette

Unburnt PUF | Unburnt PUF

Air space

(b) Smolder Wave Propagation

Char — Cigarette

Air space

(c) Incipient Flaming Combustion

Char — Cigarette remains

Airflow

(c) Flaming Combustion

Flames

Char

Airflow

FIGURE 3-30 Cross-sectional view of the development of a burn-through and the transition from smoldering to flaming combustion in a pair of upholstered foam cushions exposed to a smoldering cigarette at their junction.

figures show the cross-sectional views for smoldering, flaming, and transition, respectively. Although polyurethane foam by itself is not usually ignitable by a smoldering cigarette, PUF cushions are more likely to be ignited in combination with cellulosic fabrics and paddings (Holleyhead 1999; Babrauskas and Krasny 1985).

Hot surfaces or a sustained heat source such as a lighted incandescent bulb in contact with polyurethane foam can cause the formation of a cakelike, rigid char, whereas the flaming combustion of the same foam causes the production of viscous yellow polyol. Postfire examination may reveal the different residues, as seen in Figure 3-31.

VICTIM INJURIES

Victim injuries may occur during discovery of, interaction with, or escape from a fire and may reveal actions taken during critical time periods. This is especially true in situations resulting in fire deaths where the only remaining fire pattern indicators are the areas and degrees of burns on the victim's body and clothing. Human tissues exhibit a variety of behaviors, much like the layers of a tree trunk. Body fat can also contribute to the combustion process in prolonged-exposure fires (DeHaan, Campbell, and Nurbakhsh 1999; DeHaan and Pope 2007; DeHaan 2012).

Radiant heat (infrared radiation) travels in straight lines and is absorbed to some extent by most materials. Some thin synthetic fabrics are sufficiently transparent to infrared that first- and second-degree burns can be induced through the fabric. Most materials to some degree absorb or reflect radiant heat, but its energy can be reflected from metallic surfaces with sufficient energy to induce thermal damage to secondary target surfaces. The bodies of victims incapacitated during fires may block radiant heat from reaching the surfaces on which they are lying, resulting in a "heat or smoke shadowing" pattern. In these instances, a silhouette of the victim is left on the bed, carpet, or chair, as illustrated in Figure 3-32. The mechanisms of burn injuries are discussed in full later in this text.

Interpreting Fire Plume Behavior

FIRE VECTORING

A *vector* is a mathematical pointer that has a direction and a magnitude. An applied technique known as *fire vectoring* provides the investigator with a tool for understanding and interpreting the combined *movement* and *intensity* of plumes that create fire pattern damage. Fire investigators have used fire pattern indicators of various types since the earliest investigations.

Kennedy and Kennedy (1985) introduced the concept of heat and flame vector analysis. However, few investigators fully understood the application of the concept and systematically applied it. Scientific research into fire scene investigations has addressed the topic only in the last decade.

(a)

(b)

(c)

FIGURE 3-31 (a) A polyurethane foam cushion placed over a high-wattage lamp. (b) The foam around the lamp formed a dry, rigid, crusty char. When the cushion was ignited by open flame, the residue was the typical sticky liquid and semimolten combustion product. (c) Postfire residues—dry and rigid where charred near lamp, brown sticky mass where burned in flame. *Test courtesy of Jack Malooly. Photos (a) and (b) courtesy of John Galvin, London Fire Brigade. Photo (c) courtesy of J. D. DeHaan.*

Heat and flame vectoring, to which we refer simply as fire vectoring, is becoming a standard technique for using fire patterns to help determine a fire's area and point of origin (Kennedy 2004; *NFPA 921*, 2011 ed., sec. 17.4.2 [NFPA 2011]). Several examples throughout this text use this simple yet effective principle for tracing a fire's development, which was fully explained and used to document the 1997 USFA Burn Pattern Study (FEMA 1997).

Although the underlying concept of fire vectors has two components—direction of movement and relative intensity—some investigators successfully use only the vector's directional component. However, in postflashover burning, heat fluxes due to turbulent fuel/air mixing may be high enough to result in the most severe burn damage to a room, irrespective of the fire's actual area or point of origin.

Movement patterns are caused by flame, heat, and by-products of combustion produced when the fire spreads away from its initial source. Evaluating areas of relative damage from the least to the most damaged areas can reveal the fire's movement, a technique

FIGURE 3-32 The bodies of incapacitated victims during fires may block heat from reaching the surfaces on which they are lying, resulting in a "heat or smoke shadowing" fire pattern. In these instances, a silhouette of the victim is left on the bed, carpet, or chair. *Courtesy of D. J. Icove.*

for tracking a fire back to its possible room, area, and point of origin. Beveled edges to burned materials, smoke levels in a series of connected rooms, or protected areas on one side of an object are all examples of movement patterns. As long as the target material's thermal properties are considered, a trained investigator may make some interpretation of approximate duration.

Intensity patterns are the result of the fire's sustained impingement on exposed surfaces, such as walls, ceilings, furnishings, wall coverings, and floors. Various intensities of heat transfer result in thermal gradients along these surfaces and sometimes form lines of demarcation dividing burned and unburned areas.

The fire vector's position and length quantitatively indicates the intensity of the particular burn pattern, and its direction typically indicates the apparent flow from hottest to coolest. Once the vectors are documented and reviewed, their cumulative effect assists in determining the source of the fire. Each vector is assigned a number that is used for notations in the report. A fire vector diagram used in USFA test 5 is shown in Figure 3-33 (FEMA 1997, 47). This technique is more acceptable and accurate in assessing preflashover fires.

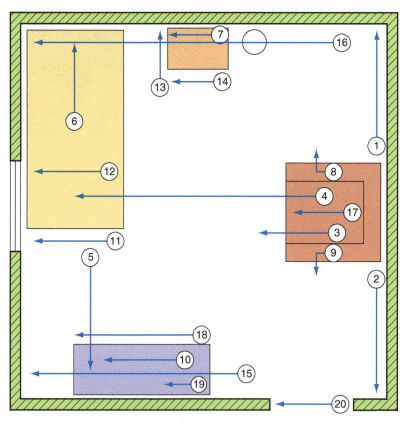

FIGURE 3-33 The fire vectoring technique uses the length of a drawn vector to indicate quantitatively the direction and intensity of the particular burn pattern. *Courtesy of FEMA* (1997, 47).

VIRTUAL ORIGIN

When a majority of the vectors point to the fire plume's base, it is often appropriate to consider this to be the location of the initial fuel package. Vectoring is helpful in estimating the location of the virtual origin or source of thermal energy of the fire. As discussed previously, the virtual origin is a point located along the plume's centerline that would give the same radiative output as the actual fire. This location is sometimes lower than the floor level. If the physical and thermal properties of the fuel package can be determined from the scene investigation, the virtual origin can be mathematically calculated, and its location can be helpful in reconstructing and documenting the fire's virtual source, point of origin, area, and direction of travel. However, if the pattern is clear enough, simple geometric extension of the V sides suffices to establish the virtual origin.

As we saw in Chapter 2, the virtual origin of a fire is located below the floor level when the area of the fuel source is large compared with the energy released by the fuel, as illustrated in previous examples. Likewise, a fuel releasing a high energy over a small area will produce a virtual origin above the floor level (Karlsson and Quintiere 2000). An example of a high-energy release over a small area is a small gasoline pool fire.

TRACING THE FIRE

The fire investigator's task would be quite easy if he or she were present for the duration of the incident. Unfortunately, fire scenes are often cold and in disarray after the extinguishment and subsequent **overhaul** to assure that the fire will not **rekindle**. The investigator's role then becomes one of reconstructing the sequence of the fire backward to its area and point of origin. A firm understanding of fire behavior becomes imperative during this examination and documentation of the fire scene.

The ability to trace the fire's behavior and growth is important to the investigation. Like an experienced tracker, the investigator relies on pattern damage as trail markers of the fire's path. Based on experience, the following rules or characteristics of fire behavior are offered to help the investigator understand and interpret this pattern damage (DeHaan and Icove 2012, chap. 7).

- *Fire's Tendency to Burn Upward:* A fire plume's hot gases, including flames, are much lighter than the surrounding air and therefore will rise. In the absence of strong winds or physical barriers, such as noncombustible ceilings, that divert flames, fire will tend to burn upward. Radiation from the plume will cause some downward and outward travel.

- *Ignition of Combustible Materials:* Combustible materials in the path of the plume's flames will be ignited, thereby increasing the extent and intensity of the fire by increasing the heat release rate. The more intense the fire grows, the faster it will rise and spread.

- *Potential for Flashover:* A flame plume that is large enough to reach the ceiling of a compartment is likely to trigger full involvement of a room and increase the chances for flashover to occur. If there is not more fuel above or beside the initial plume's flame to be ignited by convected or radiated heat, or if the initial fire is too small to create the necessary heat flux on those fuels, the fire will be self-limiting and often will burn itself out.

- *Location and Distribution of Fuel Load:* In evaluating a fire's progress through a room, the investigator must establish what fuels were present and where they were located. This fuel load includes not only the structure itself but its furnishings; contents; and wall, floor, and ceiling coverings; as well as combustible roofing materials, which feed a fire and offer it paths and directions of travel.

- *Lateral Flame Spread:* Variations on the upward spread of the fire plume will occur when air currents deflect the flame, when horizontal surfaces block the vertical

overhaul ■ The firefighting operation of eliminating hidden flames, glowing embers, or sparks that may rekindle the fire, usually accompanied by the removal of structural contents. A firefighting term involving the process of final extinguishment after the main body of the fire has been knocked down. All traces of fire must be extinguished at this time (*NFPA 921*, 2011 ed., pt. 3.3.121).

rekindle ■ Reignition of a fire by latent heat, sparks, or embers after apparent but incomplete extinguishment.

travel, or when radiation from established flames ignites nearby surfaces. If fuel is present in these new areas, it will ignite and spread the flames laterally.

- **Vertical Flame Spread:** Natural (buoyancy-driven) upward, vertical spread is enhanced when the fire plume finds chimneylike configurations. Stairways, elevators, utility shafts, air ducts, and interiors of walls all offer openings for carrying flames generated elsewhere. Fires may burn more intensely because of the enhanced draft.

- **Downward Flame Spread:** Flames may spread downward whenever there is suitable fuel in the area. Combustible wall coverings, particularly paneling, encourage the travel of fire downward as well as outward. Fire plumes may ignite portions of ceilings, roof coverings, draperies, and lighting fixtures, which can fall onto ignitable fuels below and start new fires that quickly join the main fire overhead, a process called fall-down or drop-down. Radiant heat and momentum flow of hot gases can also cause downward spread under appropriate conditions. An energetic fire can produce radiant heat sufficient to ignite adjacent floor surfaces. The momentum flow of ceiling jets from such fires can extend downward on nearby walls.

- **Impact of Radiation:** Fire plumes that are large enough to intersect with ceilings will usually form ceiling jets that extend radially along the ceiling surface. Radiation from overhead ceiling jets or hot gas layers can ignite floor coverings, furniture, and walls even at some distance, creating new points of fire origin. The investigator is cautioned to take into account what fuel packages were present in the room from the standpoint of their potential ignitability and heat release rate contributions.

- **Impact of Suppression Efforts:** Suppression efforts can also greatly influence fire spread, so the investigator must remember to confer with the fire suppression personnel present about direct consequences of their actions in extinguishing the fire. Positive-pressure ventilation (PPV) or an active attack on one face of a fire may force it back into other areas that may or may not already have been involved and push the fire down and even under obstructions such as doors and cabinets. The investigator should interview the firefighters and obtain details on how water was applied, since this important information is rarely entered into a fire department's narrative of the fire.

- **Behavior of Heat and Smoke Plumes:** Because of their buoyancy, heat and smoke plumes tend to flow through a room or structure much like a liquid—upward in relatively straight paths and outward around barriers.

- **Impact of Fire Vectors:** The total fire damage to an object observed after a fire is the result of both the intensity of the heat applied to that object and the duration of that exposure. Both factors may vary considerably during the fire.

- **Impact of Plume's Radiant Heat Flux:** The highest-temperature area of a plume will produce the highest radiant heat flux and will, therefore, affect a surface faster and more deeply than will cooler areas. This pattern can show the investigator where a flame plume contacted a surface or in which direction it was moving (since it will lose heat to the surface and cool as it moves across).

- **Impact of the Plume's Placement:** The contribution that fire plume placement makes to the growth process in a room depends not only on its size (heat release rate) and direction of travel but also on its location in the room—in the center, against a wall, in a rear corner, away from ventilation sources, or close to a ventilation opening.

Fire investigation and reconstruction of a fire's growth pattern back to its origin are based on the fact that fire plumes form patterns of damage that are, to a large extent, predictable. With an understanding of fire plume behavior, combustion properties, and general fire behavior, the investigator can prepare to examine a fire scene for these particular indicators. As with the application of the scientific method, each indicator is an independent test for direction of travel, intensity, duration of heat application, or point of origin. There is no one indicator that proves the origin or cause of a fire. All indicators must be evaluated together, and yet they may not all agree. Finally, the application of fire vectoring can also enrich and buttress the opinion rendered as to the original location of the fire plume.

Fire Burn Pattern Tests

Fire testing can produce all types of fire pattern damage phenomena (Table 3-1) including demarcations, surface effects, penetrations, and loss of materials. Fire testing is one method for expanding the range of knowledge about recognizing and interpreting these fire patterns.

An example of validation testing to explore fire pattern generation was conducted with the assistance of NIST, the Federal Emergency Management Agency's U.S. Fire Administration (FEMA/USFA), and the U.S. Tennessee Valley Authority (TVA) Police in Florence, Alabama. The tests used two forms of structures, a vacant single-family dwelling and a multiple-use assembly building (FEMA 1997; Icove 1995; NIST 1997). While the U.S. TVA Police study concentrated on extending the results of fire pattern burn testing on larger-scale structures, NIST and FEMA worked down the street on the single-family dwelling scenarios. Subsequent to those tests, additional full-scale fire tests were reported by Carman (2008, 2009, and 2010), supplemented by postfire computer modeling.

The primary goals of the project test included demonstrating the production of fire burn patterns, exploring fire scene evidence and documentation techniques, and validating basic fire modeling—all concepts that form the basics of forensic fire scene reconstruction. Owing to the immense amount of diverse information gleaned from the Florence test burns, various aspects of this fire testing appear in examples and problems throughout this text.

FIRE TESTING GOALS

The objectives for conducting the test burns on a multiple-use facility included the simulation of arson fires using plastic containers of ignitable liquids, since arsonists often use these containers when setting fires to structures (Icove et al. 1992). Often, when a burned, self-extinguished firebomb is found, investigators ask the questions: How long did the device burn? What damage can be expected? and Where is the best place to collect evidence of the device once it is damaged?

To simulate the arson, a typical firebomb consisting of a 3.8-L (1-gal) plastic container filled with unleaded gasoline was placed in the corner of the multiple-use facility's main assembly room, as shown in the floor plan in Figure 3-34. The main room was carpeted with institutional-grade synthetic carpeting.

The firebomb was remotely ignited with an electric match. Thermocouple data for the test showed that the resulting fire lasted approximately

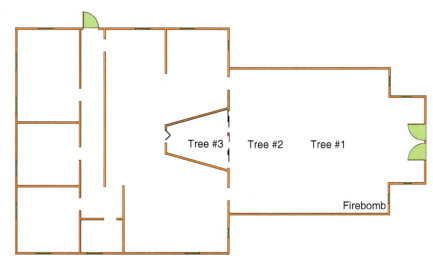

FIGURE 3-34 Floor plan of structure used to simulate the use of a typical firebomb placed in the corner of the room as indicated. *Courtesy of D. J. Icove.*

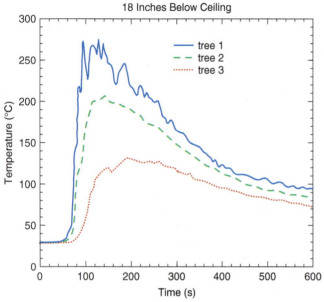

FIGURE 3-35 Three thermocouple trees recorded the room temperature profiles after the firebomb was remotely ignited with an electric match.

300 seconds and then self-extinguished, leaving a 0.46-m (1.5-ft)-diameter circular burned area. Three thermocouple trees recorded the room-temperature profiles, as shown in Figure 3-35.

The pre- and postfire damage patterns are depicted schematically and photographically in Figures 3-36(a), (b), and (c). The exact location of the firebomb and its subsequent burning pool diameter are indicated. The plastic jug melted quickly, so the majority of the fire occurred when its contents were released to burn as a pool fire on the floor. Figure 3-36(b) illustrates the fire patterns from demarcations, surface effects, penetrations, and loss of material. The schematic uses the relative damages revealed on the exposed surfaces to form isochar lines corresponding to nearly equal char depths. These isochar lines are useful in assisting in the placement and interpretation of fire vectors. The lines also document the ignitable liquid burn patterns found on carpeted and other floor surfaces.

ESTIMATED HEAT RELEASE RATE

The peak estimated heat release rate, $\dot{Q}$, for the gasoline pool fire can be calculated using equation (2.7):

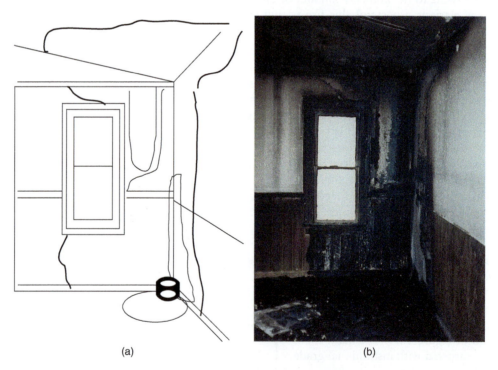

(a) (b)

FIGURE 3-36 The postfire pattern damage from the firebomb test depicted both (a) schematically and (b) photographically. (a) Dark lines indicate major demarcations of burned/unburned areas. Light lines indicate isochar demarcations. (b) The wall finish above the wainscoting was painted drywall and played no significant role in vertical flame spread. Colored lines can be used to delineate areas of surface deposit, penetration, and consumption. *Courtesy of D. J. Icove.*

Heat release rate	$\dot{Q} = \dot{m}'' \Delta h_c A$
Area of the burning pool	$A = (3.1415/4)(0.457^2) = 0.164 \ \mathrm{m}^2$
Mass flux for gasoline	$\dot{m}'' = 0.036 \ \mathrm{kg/(m^2 \cdot s)}$
Heat of combustion for gasoline	$\Delta h_c = 43.7 \ \mathrm{MJ/kg}$
Heat release rate	$\dot{Q} = (0.036)(43700)(0.164) = 258 \ \mathrm{kW}$

The real-world estimate for the heat release rate would be 200–300 kW.

VIRTUAL ORIGIN

The investigator can use a fire vector analysis of this firebomb test case to better understand and interpret the combined movement and intensity of plumes that created the fire pattern damage shown in Figure 3-37. The vectors originated from an area above where the firebomb was located on the floor, which corresponded to the theoretical virtual origin of the fire plume.

The calculation for the location of the virtual origin confirms this observation. As previously calculated in Example 2-3, the heat release rate for a 0.46-m (1.5-ft)-diameter

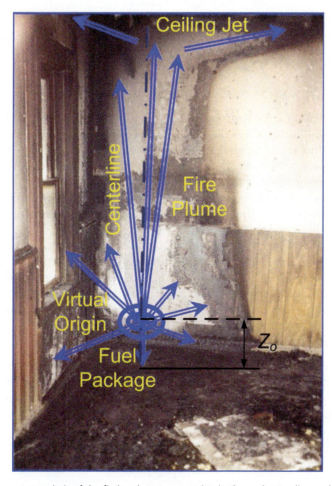

FIGURE 3-37 Fire vector analysis of the firebomb test case assists in the understanding and interpreting the fire plume and points to an area above the floor that corresponds to the plume's theoretical virtual origin. *Courtesy of D. J. Icove.*

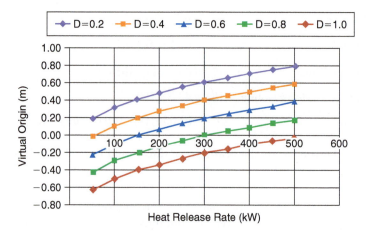

FIGURE 3-38 Plot of the relationship of the virtual origin with changes to the heat release rate and effective diameter of the burning fuel. *Courtesy of Icove and DeHaan (2006).*

gasoline fire is approximately 258 kW. The calculation for virtual origin, as taken from the Heskestad equation, is as follows:

Virtual origin	$Z_0 = 0.083\,\dot{Q}^{2/5} - 1.02\,D$
Equivalent diameter	$D = 0.457$ m
Total heat release rate	$\dot{Q} = 258$ kW
Virtual origin	$Z_0 = (0.083)(258)^{2/5} - (1.02)(0.457)$
	$= 0.766 - 0.466 = 0.3$ m $(1$ ft$)$

For this case example, Z_0 is determined to be 0.3 m (1 ft), which is above the floor level, as documented in the vector diagram in Figure 3-37. As stated previously, research has shown that a fuel releasing a high energy over a small area, as in this example, may produce a virtual origin above the floor level (Karlsson and Quintiere 2000).

A more detailed analysis of the relationship of the virtual origin with changes to the heat release rate and effective diameter of the burning fuel is plotted in Figure 3-38. This plot explores the prediction of the Heskestad equation when comparing fire plumes whose virtual origin may be above or below the burning fuel surface.

For example, a combustible fire with a high heat release rate and small-diameter fuel package will typically have a positive value for the virtual origin, indicating it is located along the centerline above the burning surface of the fuel package. This relationship can be further explored in the graph plotting the relationship of the virtual origin with changes in the heat release rate and effective diameter of the burning fuel.

FLAME HEIGHT

The *NFPA 921* flame height calculation takes into account the placement of the fire within the compartment, particularly in corners. By observation of the damage caused by the fire plume as well as its positive virtual origin above the floor, we can assume that the flame height at least touched the ceiling, which, based upon measurements taken at the scene, is 3.18 m (10.43 ft) high.

Total heat release rate	$\dot{Q} = 258$ kW
Flame height	$H_f = 0.174\left(k\dot{Q}\right)^{2/5}$
In a corner $(k = 4)$	$H_f = (0.174)(4 \times 258)^{2/5}$
	$= 2.79$ m $(9.15$ ft$)$

Including the virtual origin $H_{\text{total}} = Z_0 + H_f$

$$= 0.299 + 2.79 = 3.09 \text{ m} (10.14 \text{ ft})$$

(about 3 m, or 10 ft)

This value corresponds to a greater-than-ceiling height, thus creating the conditions for a ceiling jet across the ceiling. Therefore, the fire plume was high enough to reach the ceiling and radially disperse ceiling jets extending out from the centerline, a behavior that was observed during the test.

FIRE DURATION

Determining the initial growth period is useful in fire scene reconstruction. In validating estimates of fire duration, it is always fortunate to have historical testing, actual loss history data, and realistic fire models.

When comparing the development of real-life fires with models, the investigator relies on observations, documentation, analysis of historical fire test data, and similar cases that relate to the fire under study. A wealth of information is available from handbooks available in the fire protection engineering field (SFPE 2008, sec. 3-1).

Fire duration was one of the key questions to be answered in the firebomb test case. After the firebomb ignited, the plastic jug melted, releasing its contents of gasoline into a circular pool of burning liquid. The question then arose: How long could the pool of gasoline be expected to burn? The question of whether the surface under the spill was nonporous concrete or a carpeted material is addressed in a later section.

The burning duration of pool fires can usually be estimated based on the steady-state mass burning rate, assuming rapid growth and a large fuel supply. Gasoline has a density of 740 kg/m^3(0.74 kg/L). The steady-state mass loss rate (mass flux) of gasoline is 0.036 $\text{kg/(m}^2 \cdot \text{s)}$ (SFPE 2008, fig. 3-1.2[a]).

The following calculation is the working hypothesis for the estimated burning duration of the gasoline, assuming that it would self-extinguish after all the available fuel was consumed.

Mass of gasoline	$m = (1 \text{ gal})(3.785 \text{ L/gal})(740 \text{ g/L}) = 2.7 \text{ kg}$
Mass burning rate for gasoline	$\dot{m}'' = 0.036 \text{ kg/} (\text{m}^2 \cdot \text{s})$
Burning area	$A = \pi r^2 = (3.1416)(0.23 \text{ m})^2 = 0.166 \text{ m}^2$
Burn rate of gasoline	$\dot{m} = A\dot{m}'' = (0.166 \text{ m}^2) \left[0.036 \text{ kg/} (\text{m}^2 \cdot \text{s}) \right]$
	$= 0.00598 \text{ kg/s}$
Approximate burning duration	$= \dot{m}/m = 2.7 \text{ kg} /0.00598 \text{ kg/s} = 451 \text{ s}$

The real-world estimate for the average burning duration should be 7 to 8 minutes. Comparison of this result with the time–temperature curve (Figure 3-35), confirms that this time frame is consistent with the actual test results.

REGRESSION RATES

A flammable liquid pool will burn from the top down at a fairly predictable rate, called its *regression rate,* depending on its chemical composition and physical properties if all other factors (e.g., pool diameter and depth) are constant. A thin layer of liquid will burn away very quickly, often so quickly that only the most transient thermal effects will be observed. The depth of a pool is determined by the quantity of liquid, its physical properties (viscosity), and the nature of the surface.

On a level, smooth nonporous surface, a low-viscosity liquid like gasoline will form a pool approximately 1 mm or less deep if not otherwise limited. On a porous surface the liquid will tend to penetrate as far as it can and then spread horizontally by gravity flow and capillary action. The amount of liquid and the depth of the porous material, the porosity of the substrate, and the rate at which the liquid is poured will then control the size of the pool. On carpeted floors, as a rough guideline, the thickness of carpet can be used as a maximum pool depth, since saturated carpet represents a pool whose depth equals the thickness of the carpet and of any porous pad beneath (DeHaan 1995).

The surface area of the liquid pool is an important variable. A large quantity of gasoline dripping slowly from a leaking gas tank may produce a visible pool on soil or sand not much larger than the diameter of the individual drops but many feet deep. The same quantity dumped quickly onto the same ground may produce a much shallower pool of great diameter. A similar volume of liquid can be broadcast in an arc to generate a shallow layer (film) over a large area. The viscosity and surface tension of the liquid and the speed with which the material is ejected will determine the nature (thickness) of this film. Testing may be needed to establish the bounding conditions of maximum and minimum areas.

Once the liquid is ignited, the rate of burning per unit area (mass flux) is controlled by the amount of heat that can reach the surface of the pool and the size of the perimeter through which air can be entrained. The mass flux depends on the fuel and the size of the pool. For gasoline pools of 0.05–0.2 m in diameter the minimum regression rate is 1–2 mm/min. (Blinov, in Drysdale 2011). For very small pools the rate is much higher (owing to the laminar flame structure). The heat effects (scorching) beneath the burning pool of gasoline are superficial because there is insufficient time and heat flux to produce them. Because the fire size is driven by the evaporation rate per area (mass flux), the larger the pool in surface area, the larger the fire (the greater the heat release rate, $\dot{Q}$). Gasoline pools over 1 m in diameter have a maximum regression rate of 3–4 mm/min owing to the limitations of turbulent mixing in the plume (Blinov, in Drysdale 2011).

The flame produces a variety of effects on nearby surfaces depending on the geometry and nature of heat application. Because the fuel-covered substrate is being cooled by evaporation of the liquid fuel in contact with it, its temperature cannot be more than a few degrees hotter than the boiling point of the liquid and so will not do much more than scorch the surface directly under the pool. Low-melting-point materials like synthetics can melt as well as scorch. The areas of carpet or floor immediately outside the pool will be exposed to the radiant heat of the plume during the fire. Without the protection of the evaporating liquid, materials will melt and scorch, particularly if small or thin (carpet fibers). Materials in direct contact with the flame front will be scorched and ignited or at least charred.

Thermal effects of materials near the flame front often produce what is called a *halo* or *ring effect* around the outside of the pool, as shown in tests (Figure 3-39) by DeHaan (2007) and Putorti, McElroy, and Madrzykowski (2001). If the substrate is ignitable, a ring of damage may extend some distance from the pool. Eventually, the center of the protected area may be burned as the protective layer of liquid fuel evaporates. In traditional carpets (wool or nylon) this area will be noticeable, and sufficient residues of the liquid fuel may survive in the center of the burned area to be recovered and identified. If the fire burns itself out or is quickly extinguished by a sprinkler system or oxygen depletion, the pool area can be roughly estimated from the dimensions of the burned margins. If the fire is sustained by the carpet, or if the fire (externally) reaches flashover, the area of burned carpet can grow well beyond the area of the original liquid pool. This is particularly true for the new generation of synthetic fiber carpets with polypropylene backing and polypropylene face yarn pile (DeHaan and Icove 2012).

FIGURE 3-39 "Halo" or ring pattern formed at outside of burning methanol pool on carpet. Center of pool is protected by evaporating fuel. *Courtesy of J. D. DeHaan.*

Wool or nylon carpets with jute backing will tend to self-extinguish, but polypropylene will not. Combustion tests have shown that flames on such carpets can propagate at a rate of approximately 0.5–1 m^2/hr (5–11 ft^2/hr), generating very small flames at the margins of the burning area. Such carpets will pass the methenamine tablet test (*ASTM D2859-06(2011)*) (ASTM 2011b), since its igniting flame is about 50 W and of brief duration (similar to that of a dropped match), but if there is any significant additional heat flux from a larger ignition source (wad of paper) or a continuously burning object like a piece of furniture, these can propagate a fire over a large area if given enough time (DeHaan and Icove 2012).

NIJ-FUNDED RESEARCH

The National Institute of Justice (NIJ) funded research (Mealy, Benfer, and Gottuk 2011) on the fire dynamics and forensic analysis of liquid fuel fires that explored fluid depths, substrates, burning rates, and impact on ignition delay times. The researchers documented the statistical bounds of spill sizes and depths caused by flammable and combustible fuels including gasoline, kerosene, and denatured alcohol. They also examined spill depths affected by the type of substrates, including vinyl, carpet, plywood, concrete (smooth, brushed, and coated), and oriented strand board.

For spill depth, the researchers conducted a statistical analysis of fluid depths of common liquid types (except lube oil) on smooth substrates, reporting an average depth of 0.72 mm (0.028 in.) with a standard deviation of 0.34 mm (0.013 in.). Spill depths ranged from 0.22 mm (0.0087 in.) up to 2.4 mm (0.094 in.) depending on the specific liquid/substrate scenario.

Research confirmed two factors governing the spread of a fuel's spill depth—the surface tension of the liquid and the surface characteristics of the substrate on which it is spilled. Because the surface tensions of most fuels are relatively similar, the dominant variable is the substrate's surface topography, which can have an impact on the spill's overall spread and equilibrium spill depth.

The researchers confirmed that the mass burning rates of a fuel are different when it is spilled onto surfaces of differing thermal properties. For each fire, the highest burning rates occurred on the vinyl flooring, and the lowest on concrete. No correlation could be found between burning rates and any specific thermal property of the substrates (i.e., thermal conductivity, thermal inertia, thermal effusivity, and thermal diffusivity).

Finally, these researchers found that the ignition delay time—the time between the initial spill and its ignition—affects the peak mass burning rate. In general, the 300-second ignition delay tests resulted in larger areas with reduced peak mass burning rates per unit area. However, the differences in the average increase in spill area from the 30-second to the 300-second ignition delay times ranged from 8 to 76 percent, with an average value of 36 percent. The percent decrease in the average peak heat release rates ranged from 25 to 74 percent, with an average of 52 percent, for the 30-second and 300-second ignition delay times, respectively. The effects resulting from delayed ignition could not be fully explained from the data, but substrate cooling and evaporative fuel losses were shown to account for some of the changes in peak mass burning rate.

ADJUSTMENTS TO FIRE DURATION

A comparison of the calculated burning duration of 451 seconds in the previously described firebomb test (Icove 1995) with the actual burning duration shown in Figure 3-35 showed that the actual duration was less than the estimated duration. This working hypothesis had to be modified to account for the carpet present rather than the concrete floor assumed in the initial calculation.

Experiments on carpet saturated with pentane (a component of gasoline) showed that its rate of evaporation (nonburning) is 1.5 times higher than that for a freestanding pool at the same temperature (DeHaan 1995). Thus, the steady-state mass loss rate by evaporation of gasoline saturated into carpet is approximately 1.5 times the freestanding value of 0.036, or 0.054 kg/(m² · s), assuming in this case that the thermal degradation of the carpet does not affect the mass transfer. Because the combustion processes will progress in the same manner as evaporation on this carpet, the recalculated burning duration for a carpeted surface gave a corrected estimate of 316 seconds and a heat release rate of 387 kW.

Knowledge of the burning characteristics of common flammable and combustible liquids on various surfaces, including those of pool fires such as the one in this test case, is crucial in fire testing and analysis. Synthetic carpets may melt and reduce the mass flux. Also, heat release rate contributions from the molten plastic container pool and carpet were not included. These would be expected to add to the heat release rate and possibly to reduce the duration of the fire.

NIST research on flammable and combustible liquid spill and burn patterns reported that the peak spill fire heat release rates for thin layers of burning liquids on nonporous surfaces were found to be only 12.5 percent to 25 percent of those from equivalent deep-pool fires. The peak heat release rates for fires on carpeted surfaces were found to be approximately equal to those for equivalent pool fires (Putorti, McElroy, and Madrzykowski 2001). Since the nature of the carpet substrates was not reported in these tests, the carpet's contribution cannot be assessed.

In small liquid pool fires, researchers have reported that after the ignition of the fuel, the mass burning rate will increase until a steady rate is reached (Hayasaka 1997). Research and validation experiments by Ma et al. (2004) on burning rates of liquid fuels on carpet indicate that the phenomenon is possibly more complex, particularly regarding the role the carpet plays in supplying fuel to the fire. Ma notes that several factors affect carpet fires: the capillary or "wick" effect of its fibers; evaporation, combustion, and heat transfer; and mass transfer–limited combustion. He surmises that the carpet has two conflicting roles on the mass burning rate: (1) as an insulating material that blocks heat transfer losses to the depth of the pool and increases the mass burning rate along with resulting higher temperatures and (2) as a porous medium that decreases the mass burning rate owing to an insufficient capillary effect. Note that a thin fuel layer on a hard surface such as concrete or wood, whose conductivity is greater than that of the liquid fuel, will result in greater conductive heat transfer losses. Ma points out that the reports of observations that liquid fires burn more severely on carpets than on smooth uncarpeted floors or ground can be explained. It turns out that carpet enhances the mass burning rate at the earlier stages of burning because it insulates the fabric pile, making the burning liquid fuel appear to behave as a steady deep-pool fire.

FLAME HEIGHT ADJUSTMENT FOR FIRE LOCATION

The flame height is a function of the fire location within the room. As shown in equation (3.12), the flame height depends on the fire's placement either in the center ($k = 1$), against the wall ($k = 2$), or in the corner ($k = 4$) of the compartment.

In the initial approach, the corner of the test in Figure 3-36 was assumed to have full impact on the flame height, and a total flame height of 3.1 m (10.1 ft) was predicted. With the new heat release rate of 387 kW and the corner configuration, the new flame height estimate would be recalculated to be 3.28 m plus the virtual origin of 0.43 m, or 3.71 m (12.2 ft). The damage to the corner of the ceiling above the fire demonstrated that a sufficient flame plume was entrained in the corner to cause such penetration. The additional fire from the combustible wall covering (wainscoting) added an unknown factor to the plume generation.

POOL FIRES AND DAMAGE TO SUBSTRATES

When a flammable liquid burns, it does so by evaporating vapors from the liquid, creating a layer of vapor denser than air. Brownian motion at elevated temperatures causes this layer to have a finite thickness and to diffuse into overlying air to form a steep concentration gradient with distance from the liquid surface (DeHaan 1995). Wherever this gradient is within the flammable range of the vapors, a flame can be supported. A *diffusion flame* is one supported by fuel vapors diffusing from the fuel surface into the surrounding air/oxygen.

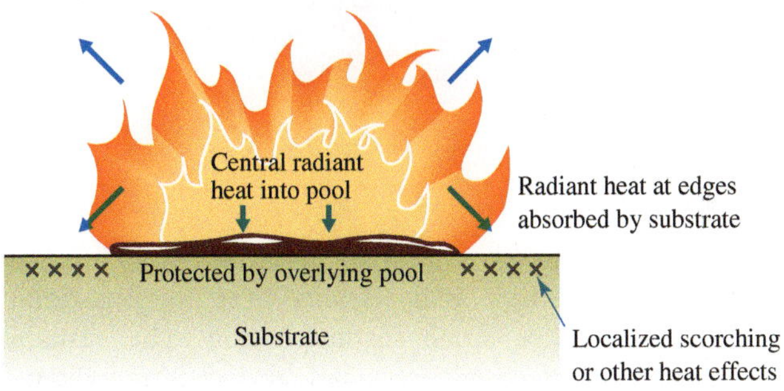

The distance between the fuel surface and the flame front varies with the temperature (and thereby the vapor pressure) of the fuel. Radiant heat from the layer of flame travels in all directions. Some of the heat radiated downward is absorbed by the fuel, keeping the temperature high enough that there is a

FIGURE 3-40 Cross section of burning pool of fuel. *Courtesy of J. D. DeHaan.*

continual supply of vapors to support a flame. Some heat is transferred through the fuel into the substrate beneath and is absorbed, increasing the temperature of the substrate. Owing to intimate contact between the substrate and the liquid fuel, heat is transferred to the fuel and then distributed through the fuel (if it is a deep enough layer) by convective circulation.

The temperature of the surface under the pool will not be more than a few degrees above the boiling point of the liquid overlying it. If the fuel's boiling-point temperature is low enough (<200°C, 390°F), the liquid can burn off without any visible effect if the surface is smooth, with no pores, joints, or seams, and has a relatively high decomposition temperature. The higher the boiling point of the liquid, the greater the chances of thermal damage to the floor, such as pyrolysis (scorching), melting, or both. In a burning pool of fuel, the radiant heat at the edges absorbed by the substrate may produce localized scorching or other heat effects, as illustrated in Figures 3-39 and 3-40. If a fuel contains a mixture of compounds with different boiling points, the low-boiling liquid will tend to burn off first, leaving the higher-boiling compounds to continue to heat. A mixture like gasoline covers the range of boiling points 40°C–150°C (100°F–300°F). As the mixture burns, the boiling point of the residue increases, and therefore the limiting temperature factor increases. At 250°C (450°F), wood surfaces will be only scorched, but some synthetic floor coverings can be significantly damaged. Synthetic carpet fibers can also melt and reduce the mass transfer rate of the liquid fuel as the fire progresses. Field fire tests demonstrating the absence of significant thermal effects on wood with gasoline pools have been published (DeHaan 2007).

FINAL HYPOTHESIS

The final hypothesis noted for this test, which Figure 3-35 reflects, is the consumption of the gasoline in the time–temperature curve, leveling out approximately 400 seconds after ignition, which is consistent with the estimate. This result demonstrates the effects of the use of flammable liquids in arson and other fire-related crimes—a very rapid increase in localized temperature followed by a rapid consumption to self-extinguishment.

Expert conclusions or opinions on the strength of these hypotheses rest partially on the use of accepted and historically proven fire testing techniques that validate this methodology and can easily be replicated. Additionally, this methodology and similar studies have been peer reviewed and published (DeHaan 1995; Ma et al. 2004; Icove and DeHaan 2006). There are established error rates for applying this methodology specifically to known variables such as room size and fire development times. See Mealy, Benfer, and Gottuk (2011) and Wolfe, Mealy, Gottuk (2009) for research into the application of fire dynamics to the forensic analysis of liquid-fuel and limited-ventilation compartment fires. Standardized methods are also maintained by independent and unbiased organizations for applying and interpreting these relationships. Finally, this methodology is generally accepted in the scientific community.

Summary

Fire pattern damage analysis is a vital investigation technique for fire scene reconstruction. The visual interpretation of damage created by fire plumes can isolate and accurately identify the area of fire origin.

Locating and identifying the first fuel package ignited is a critical step in the accurate reconstruction of a fire incident. Careful analysis of fire patterns can significantly aid the scene investigator in this effort. Because effects like charring, melting, ignition, and protection are predictable, their location and distribution, which can be confirmed by interviews or prefire photos, offer a sound basis for locating fuel packages.

Fire investigators can use systematic steps to evaluate and document thermal damage patterns, identify the fire's direction and intensity, confirm significant witness observations, and verify the results of fire modeling. These systematic steps invoke the scientific method to test and evaluate various hypotheses of the fire's origin and spread.

From these observations, the authors conclude the following:

- As long as there is not too much damage, ample fire pattern heat indicators exist and can be documented.
- Evidence of fire pattern damage at scenes can be found at most scenes for fire plumes that can indicate a fire's source, area and point of origin, and direction of travel.

- There are only a limited number of scientifically validated fire-related patterns, and they should be evaluated for their intensity, direction, and duration.
- Fire engineering analysis using the scientific method is necessary to perform a comprehensive analysis and may be the major usable tool if the building has been substantially destroyed.
- Investigators should not overlook signs of physical evidence documenting human activity (hand- or fingerprints, shoeprints, blood spatters, broken glass, discarded hoses/fire extinguishers, and burn injuries).
- Fire analysis and computer-assisted modeling can provide further insights into fire scene behavior.
- Peer review is a healthy and prudent step to ensure that all hypotheses are addressed.

Recommendations for continuing education on fire pattern analysis include the ongoing use of the latest editions of *Kirk's Fire Investigation* and *NFPA 921* as the universally accepted guides to documenting all fire and explosion investigations. Research into fire pattern analysis and reconstruction technologies, and international conferences on the forensic aspects of fire investigation, are important means of increasing the body of knowledge in the field.

Problems

3.1. Find a photograph of a fire plume in a recent news story or article. Estimate the dimensions of the plume and attempt to calculate its heat release rate. What information on the fire can you infer from these calculations that relate to the story or article?

3.2. Photograph a recent fire scene and illustrate as many fire patterns as you can find.

3.3. From the plume information in the photos and the identified first fuel package, estimate the heat release rate and virtual origin of the initial fire.

3.4. Use the equations provided in this chapter to determine the radiant heat fluxes at various distances from the fire's virtual origin.

3.5. A wooden box open on one side can be used to create a small fire with a wad of paper or a small block of urethane foam. Using this reduced-scale test device, observe and describe wall and corner effects, ceiling jets, and rollover events.

References

Alpert, R. 1972. Calculation of response time of ceiling-mounted fire detectors. *Fire Technology* 8 (3): 181–95, doi: 10.1007/bf02590543.

ASTM. 2011a. *ASTM C1396/C1396M-11: Standard specification for gypsum board*. West Conshohocken, PA: ASTM International.

———. 2011b. *ASTM D2859-06(2011): Standard test method for ignition characteristics of finished textile floor covering materials*. West Conshohocken, PA: ASTM International.

Audouin, L., Kolb, G., Torero, J. L., & Most, J. M. 1995. Average centreline temperatures of a buoyant pool fire obtained by image processing of video recordings. *Fire Safety Journal* 24 (2): 167–87, doi: 10.1016/0379-7112(95)00021-k.

Babrauskas, V. 1980. Flame lengths under ceilings. *Fire and Materials* 4 (3): 119–26, doi: 10.1002/fam.810040304.

———. 2004. *Glass breakage in fires*. Issaquah, WA: Fire Science and Technology Inc.

———. 2005. Charring rate of wood as a tool for fire investigations. *Fire Safety Journal* 40 (6): 528–54, doi: 10.1016/j.firesaf.2005.05.006.

Bessey, G. E. 1950. Investigations on building fires. Part II: The visible changes in concrete or mortar exposed to high temperatures. *Technical Paper No. 4*. Garston, England: Department of Scientific and Industrial Research, Building Research Station.

Beyler, C. L. 1986. Fire plumes and ceiling jets. *Fire Safety Journal* 11 (1–2): 53–75, doi: 10.1016/0379-7112(86)90052-4.

Beyler, C. L. 2008. Fire hazard calculations for large, open hydrocarbon fires. In *SFPE Handbook of Fire Protection Engineering,* 4th ed., ed. P. J. DiNenno. Quincy, MA: National Fire Protection Association.

Bostrom, L. 2005. Methodology for measurement of spalling of concrete. Paper presented at Fire and Materials 2005, 9th International Conference, January 31–February 1, San Francisco, CA.

Buchanan, A. H. 2001. *Structural design for fire safety*. Chichester, England: Wiley.

Butler, C. P. 1971. Notes on charring rates in wood. *Fire Research Note No. 896*. London: Department of the Environment and Fire Offices' Committee, Joint Fire Research Organisation.

Canfield, D. 1984. Causes of spalling of concrete at elevated temperatures *Fire and Arson Investigator* 34 (4): 22–23.

Carman, S. W. 2008. Improving the understanding of post-flashover fire behavior. Paper presented at the International Symposium on Fire Investigation Science and Technology, May 19–21, Cincinnati, OH.

———. 2009. Progressive burn pattern developments in post-flashover fires. Paper presented at Fire and Materials 2009, 11th International Conference, January 26–28, San Francisco, CA.

———. 2010. Clean burn fire patterns: A new perspective for interpretation. Paper presented at Interflam, July 5–7, Nottingham, UK.

Chakrabarti, B., Yates, T., & Lewry, A. 1996. Effects of fire damage on natural stonework in buildings. *Construction and Building Materials* 10 (7): 539–44.

Chu Nguong, N. 2004. Calcination of gypsum plasterboard under fire exposure. Master's thesis, University of Canterbury, Christchurch, New Zealand (*Fire Engineering Research Report 04/6*).

Cox, G., & Chitty, R. 1982. Some stochastic properties of fire plumes. *Fire and Materials* 6 (3–4): 127–34, doi: 10.1002/fam.810060306.

DeHaan, J. D. 1987. Are localized burns proof of flammable liquid accelerants? *Fire and Arson Investigator* 38 (1): 45–49.

———. 1995. The reconstruction of fires involving highly flammable hydrocarbon liquids. PhD diss., University of Strathclyde, Glasgow, Scotland, UK.

———. 2001. Full-scale compartment fire tests. *CAC News* (Second Quarter): 14–21.

———. 2002. Our changing world. Part 2: Ignitable liquids: Petroleum distillates, petroleum products and other stuff. *Fire and Arson Investigator* (July): 20–23.

———. 2004. Advanced tools for use in forensic fire scene investigation, reconstruction and documentation. Paper presented at the 10th International Fire Science and Engineering Conference, Interflam 2004, July 5–7, Edinburgh, UK.

———. 2007. *Kirk's fire investigation,* 6th ed. Upper Saddle River, N.J.: Pearson-Prentice Hall.

———. 2012. Sustained combustion of bodies: Some observations. *Journal of Forensic Science*, May 4, doi: 10.1111/j.1556-4029.2012.02190.x.

DeHaan, J. D., Campbell, S. J., & Nurbakhsh, S. 1999. Combustion of animal fat and its implications for the consumption of human bodies in fires. *Science & Justice* 39 (1): 27–38, doi: 10.1016/s1355-0306(99)72011-3.

DeHaan, J. D., & Icove, D. J. 2012. *Kirk's fire investigation,* 7th ed. Upper Saddle River, NJ: Pearson-Prentice Hall.

DeHaan, J. D., & Pope, E. J. 2007. Combustion properties of human and large animal remains. Paper presented at 11th International Fire Science and Engineering Conference, Interflam 2007, September 3–5, London, UK.

Drysdale, D. 2011. *An introduction to fire dynamics,* 3rd ed. Chichester, England: Wiley.

Fang, J. B., & Breese, J. N. 1980. Fire development in residential basement rooms. Gaithersburg, MD: National Bureau of Standards.

FEMA. 1997. USFA fire burn pattern tests. *Report FA 178*. Emmitsburg, MD: Federal Emergency Management Agency, U.S. Fire Administration.

Harris, R. J. 1983. The investigation and control of gas explosions in buildings and heating plants, 96–100. London: British Gas Corp.

Harris, R. J., Marshall, M. R., & Moppett, D. J. 1977. The response of glass windows to explosion pressures. Paper presented at the I. Chem E. Symposium Series No. 49, April 5–7.

Harrison, R. 2004. Smoke control in atrium buildings: A study of the thermal spill plume. In *Fire Engineering Research Report 04/1*, ed. M. Spearpoint. Christchurch, New Zealand: University of Canterbury.

Harrison, R. and Spearpoint, M. 2007. The balcony spill plume: Entrainment of air into a flow from a compartment opening to a higher projecting balcony. *Fire Technology* 43 (4):301–17, doi: 10.1007/s10694-007-0019-3.

Hasemi, Y., & Nishihata, M. 1989. Fuel shape effects on the deterministic properties of turbulent diffusion flames. Paper presented at Fire Safety Science, Second International Symposium, Washington, DC.

Hayasaka, H. 1997. Unsteady burning rates of small pool fires. In *Proceedings of 5th Symposium on Fire Safety Science*, ed. Y. Hasemi. Tsukuba, Japan.

Hertz, K. D., & Sorensen, L. S. 2005. Test method for spalling of fire-exposed concrete. *Fire Safety Journal* 40:466–76.

Heskestad, G. H. 1982. Engineering relations for fire plumes. SFPE Technology Report 82-8. Boston: Society of Fire Protection Engineers, 6.

———. 1983. Virtual origins of fire plumes. *Fire Safety Journal* 5 (2): 109–14, doi: 10.1016/0379-7112(83)90003-6.

———. 2008. Fire plumes, flame height, and air entrainment. Chapter 2-1 in *SFPE Handbook of Fire Protection Engineering*, ed. P. J. DiNenno. Quincy, MA: National Fire Protection Association, Society of Fire Protection Engineers.

Hietaniemi, J. 2005. Probabilistic simulation of glass fracture and fallout in fire. *VTT Working Papers 41*, ESPOO 2005. Finland: VTT Building and Transport.

Holleyhead, R. 1999. Ignition of solid materials and furniture by lighted cigarettes. A review. *Science & Justice* 39 (2): 75–102.

Hopkins, R. L., Gorbett, G. E., & Kennedy, P. M. 2007. Fire pattern persistence and predictability on interior finish and construction materials during pre- and post-flashover compartment fires. Paper presented at Fire and Materials 2007, 10th International Conference, January 29–31, San Francisco, CA.

Icove, D. J. 1995. Fire scene reconstruction. Paper presented at the First International Symposium on the Forensic Aspects of Arson Investigations, July 31, Fairfax, VA.

Icove, D. J., & DeHaan, J. D. 2006. Hourglass burn patterns: A scientific explanation for their formation. Paper presented at the International Symposium on Fire Investigation Science and Technology, June 26–28, Cincinnati, OH.

Icove, D. J., Douglas, J. E., Gary, G., Huff, T. G., & Smerick, P. A. 1992. Arson. In *Crime classification manual*, ed. J. E. Douglas, A. W. Burgess, A. G. Burgess, & R. K. Ressler. New York: Macmillan.

Jansson, R. 2006. Thermal stresses cause spalling. *Brand Posten SP, Swedish National Testing and Research Institute* 33:24–25.

Jia, F., Galea, E. R., & Patel, M. K. 1999. Numerical simulation of the mass loss process in pyrolyzing char materials. *Fire and Materials* 23:71–78.

Karlsson, B., & Quintiere, J. G. 2000. *Enclosure fire dynamics*. Boca Raton, FL: CRC Press.

Kennedy, J., & Kennedy, P. 1985. *Fires and explosions: Determining the cause and origin*. Chicago, IL: Investigations Institute.

Kennedy, P. M. 2004. Fire pattern analysis in origin determination. Paper presented at the International Symposium on Fire Investigation Science and Technology, Cincinnati, OH.

Khoury, G. A. 2000. Effect of fire on concrete and concrete structures. *Progress in Structural Engineering Materials* 2:429–42.

König, J., & Walleij, L. 2000. Timber frame assemblies exposed to standard and parametric fires. Part 2: A design model for standard fire exposure. *Report No. 100010001*. Stockholm, Sweden: Swedish Institute for Wood Technology Research.

Lau, P. W., White, C. R., & Van Zeeland, I. 1999. Modelling the charring behavior of structural lumber. *Fire and Materials* 23:209–16.

Lawson, J. R. 1977. *An evaluation of fire properties of generic gypsum board products*, Gaithersburg, MD: National Bureau of Standards.

Lentini, J. J. 1992. Behavior of glass at elevated temperatures. *Journal of Forensic Sciences* 37 (5): 1358–62.

Ma, T. S., Olenick, M., Klassen, M. S., Roby, R. J., & Torero, L. J. 2004. Burning rate of liquid fuel on carpet (porous media). *Fire Technology* 40 (3): 227–46.

Mann, D. C., & Putaansuu, N. D. 2006. Alternative sampling methods to collect ignitable liquid residues from non-porous areas such as concrete. *Fire and Arson Investigator* 57 (1): 43–46.

McCaffrey, B. J. 1979. Purely buoyant diffusion flames: Some experimental results (final report). Washington, DC: National Bureau of Standards.

McGraw, J. R., & Mowrer, F. W. 1999. Flammability of painted gypsum wallboard subjected to fire heat fluxes. Paper presented at Interflam 1999, June 29–July 1, Edinburgh, Scotland, UK.

Mealy, C. L., Benfer, M. E., & Gottuk, D. T. 2011. Fire dynamics and forensic analysis of liquid fuel fires. Baltimore, MD: Hughes Associates, Inc.

Merck index, 11th ed. 1989. Rahway, NJ: Merck Co.

Milke, J. A., & Mowrer, F. W. 2001. Application of fire behavior and compartment fire models seminar. Paper presented at the Tennessee Valley Society of Fire Protection Engineers (TVSFPE), September 27–28, Oak Ridge, TN.

Mitler, H. E., & Rockett, J. A. 1987. *Users' guide to FIRST, a comprehensive single-room fire model*, Gaithersburg, MD: National Bureau of Standards.

Moodie, K., & Jagger, S. F. 1992. The King's Cross fire: Results and analysis from the scale model tests. *Fire Safety Journal* 18 (1): 83–103, doi: 10.1016/0379-7112(92)90049-i.

Morgan, H. P., & Marshall, N. R. 1975. Smoke hazards in covered multi-level shopping malls: An experimentally based theory for smoke production. *BRE Current Paper* 48/75: 23. Garston, England, UK: Building Research Establishment.

Mowrer, F. W. 1998. Window breakage induced by exterior fires. Washington, DC: U.S. Department of Commerce.

———. 2001. Calcination of gypsum wallboard in fire. Paper presented at the NFPA World Fire Safety Congress, May 13–17, Anaheim, CA.

Newman, J. S. 1993. Integrated approach to flammability evaluation of polyurethane wall/ceiling materials. *Journal of Cellular Plastics* 29 (5), doi: 10.1177/0021955X9302900535.

NFPA. 2000. *Fire protection handbook,* 18th ed. Quincy, MA: National Fire Protection Association.

———. 2004. *NFPA 921: Guide for fire and explosion investigations,* 2004 ed. Quincy, MA: National Fire Protection Association.

———. 2011. *NFPA 921: Guide for fire and explosion investigations,* 2011 ed. Quincy, MA: National Fire Protection Association.

———. 2012. *NFPA 92: Standard for smoke control systems.* Quincy, MA: National Fire Protection Association.

NIST. 1991. *Users' guide to BREAK1, the Berkeley algorithm for breaking window glass in a compartment fire.* Gaithersburg, MD: National Institute of Standards and Technology.

NPL. 2000. *Guides to good practices in corrosion control.* National Physical Laboratory, Queens Road, Teddington, Middlesex TW11 0LW.

Ogle, R. A., & Schumacher, J. L. 1998. Fire patterns on upholstered furniture: Smoldering versus flaming combustion. *Fire Technology* 34 (3): 247–65.

Orloff, L., Modak, A. T., & Alpert, R. L. 1977. Burning of large-scale vertical surfaces. *Symposium (International) on Combustion* 16 (1): 1345–54, doi: 10.1016/s0082-0784(77)80420-7.

Parker, T. W., & Nurse, R. W. 1950. Investigations on building fires. Part I: The estimation of the maximum temperature attained in building fires from examination of the debris. *National Building Studies,* Technical Paper no. 4: 1–5. Building Research Station, Garston, England: Department of Scientific and Industrial Research.

Perry, R. H., & Green, D. W., eds. 1984. *Perry's chemical engineers' handbook,* 6th ed. New York: McGraw-Hill.

Putorti Jr., A. D., McElroy, J. A., & Madrzykowski, D. 2001. *Flammable and combustible liquid spill/burn patterns,* Rockville, MD: National Institute of Standards and Technology.

Quintiere, J. G. 1998. *Principles of fire behavior.* Albany, NY: Delmar.

Quintiere, J. G., & Grove, B. S. 1998. *Correlations for fire plumes.* (NIST-GCR-98-744). Gaithersburg, MD: National Institute of Standards and Technology.

Richardson, J. K., Richardson, L. R., Mehaffey, J. R., & Richardson, C. A. 2000. What users want fire model developers to address. *Fire Protection Engineering* (Spring): 22–25.

Sanderson, J. L. 1995. Tests results add further doubt to the reliability of concrete spalling as an indicator. *Fire Finding,* 3 (4): 1–3.

———. 2002. Depth of char: Consider elevation measurements for greater precision. *Fire Findings* 10, no. 2 (Spring): 6.

Schroeder, R. A., & Williamson, R. B. 2001. Application of materials science to fire investigation. Paper presented at Fire and Materials 2001, 7th International Conference, January 22–24, San Francisco, CA.

———. 2003. Post-fire analysis of construction materials: Gypsum wallboard. Paper presented at Fire and Materials 2001, 8th International Conference, January 28–29, San Francisco, CA.

SFPE. 2008. *SFPE handbook of fire protection engineering,* 4th ed. Quincy, MA: National Fire Protection Association, Society of Fire Protection Engineers.

Shields, T. J., Silcock, G. W. H., & Flood, M. F. 2001. Performance of a single glazing assembly exposed to enclosure corner fires of increasing severity. *Fire and Materials* 22:123–52.

Short, N. R., Guise, S. E., & Purkiss, J. A. 1996. Assessment of fire-damaged concrete using color analysis. *Inter-Flam '96 Proceedings.* London: Interscience.

Silcock, G. W. H., & Shields, T. J. 2001. Relating char depth to fire severity conditions. *Fire and Materials* 25:9–11.

Smith, F. P. 1991. Concrete spalling: Controlled fire test and review. *Journal of Forensic Science* 31 (1): 67–75.

Spearpoint, M. J., & Quintiere, J. G. 2001. Predicting the piloted ignition of wood in the cone calorimeter using an integral model: Effect of species, grain orientation and heat flux. *Fire Safety Journal* 36 (4): 391–415, doi: 10.1016/s0379-7112(00)00055-2.

Steckler, K. D., Quintiere, J. G., & Rinkinen, W. J. 1982. Flow induced by fire in a compartment. *Symposium (International) on Combustion* 19 (1): 913–20, doi: 10.1016/s0082-0784(82)80267-1.

Stratton, B. J. 2005. Determining flame height and flame pulsation frequency and estimating heat release rate from 3D flame reconstruction. Master's thesis, University of Canterbury, Christchurch, New Zealand (*Fire Engineering Research Report 05/2*).

Takeda, H., & Mehaffey, J. R. 1998. WALL2D: A model for predicting heat transfer through wood-stud walls exposed to fire. *Fire and Materials* 22:133–40.

Tanaka, T. J., Nowlen, S. P., & Anderson, D. J. 1996. Circuit Bridging of Components by Smoke. S. N. Laboratory. Albuquerque, New Mexico, U.S. Nuclear Regulatory Commission.

Tewarson, A. 2008. Generation of heat and gaseous, liquid, and solid products in fires. Chapter 3-4 in *SFPE Handbook of Fire Protection Engineering*, ed. P. J. DiNenno. Quincy, MA: National Fire Protection Association, Society of Fire Protection Engineers.

Thomas, G. 2002. Thermal properties of gypsum plasterboard at high temperatures. *Fire and Materials* 26 (1): 37–45, doi: 10.1002/fam.786.

Tobin, W. A., & Monson, K. L. 1989. Collapsed spring observations in arson investigations: A critical metallurgical evaluation. *Fire Technology* 25 (4): 317–35.

Wolfe, A. J., Mealy, C. L., & Gottuk D. 2009. Fire Dynamics and Forensic Analysis of Limited Ventilation Compartment Fires. National Institute of Justice: 194.

Wu, Y., & Drysdale, D. D. 1996. *Study of upward flame spread on inclined surfaces,*. Edinburgh, UK: Health & Safety Executive.

You, H-Z. 1984. An investigation of fire plume impingement on a horizontal ceiling. 1: Plume region. *Fire and Materials* 8 (1): 28–39.

Zukoski, E. E. 1978. Development of a stratified ceiling layer in the early stages of a closed-room fire. *Fire and Materials* 2 (2): 54–62, doi: 10.1002/fam.810020203.

Zukoski, E. E., Cetegen, B. M., & Kubota, T. 1985. Visible structure of buoyant diffusion flames. *Symposium (International) on Combustion* 20 (1): 361–66, doi: 10.1016/s0082-0784(85)80522-1.

4

Fire Scene Documentation

❝When you have eliminated the impossible, whatever remains, however improbable, must be the truth. **❞**

—Sir Arthur Conan Doyle
The Sign of the Four

Courtesy of D. J. Icove.

OBJECTIVES

After reading this chapter, the student should be able to:

- Define and outline proper fire scene documentation.
- Illustrate the fire scene and its reconstruction.
- Employ proper tools to record the scene.
- Recognize and minimize the possibility of spoliation when processing a fire scene.

Fire scenes often contain complex information that must be thoroughly documented, a task of paramount importance to the investigator. A single photograph and scene diagram are not sufficient to capture vital information on fire dynamics, building construction, fire indicators, evidence collection, and avenues of escape for the building's occupants.

Thorough documentation is a comprehensive effort comprising forensic photography, measurements, sketches, drawings, and analysis. The purposes of forensic fire scene documentation are to record visual observations, to authenticate physical evidence found at the scene, and to ensure the integrity of the scene delayering process.

This chapter provides the fundamental concepts of forensic fire scene documentation, with emphasis on a systematic set of guidelines. Concepts and techniques for producing accurate and legally acceptable documentation for both investigative reports and courtroom presentations are reviewed. Various computer-assisted photographic and sketching technologies that help ensure accurate and representational diagramming are also explored.

Report writing guidelines are not included in this textbook, only the documentation needed to assemble authoritative reports. For a detailed discussion and examples, see *Kirk's Fire Investigation,* 7th ed. (DeHaan and Icove 2012), *Combating Arson-for-Profit: Advanced Techniques for Investigators,* 2nd ed. (Icove, Wherry, and Schroeder 1998), and several articles on the subject (Icove and Henry 2010; Icove and Dalton 2008).

National Protocols

The *systematic documentation* of a fire scene from its initial stages plays an essential role in the effort to record and preserve the events and evidence for all parties involved, particularly for forensic and other experts who are retained later and may be asked to render their opinion in the case. This systematic approach captures all available information needed for later use in criminal, civil, or administrative matters. Systematic documentation ensures that an independent, qualified investigator examining the evidence can arrive at the same opinions as the documenting investigator. Additionally, a failure to systematically document a scene can have severe consequences in litigation when *Daubert* (1993 and 1995) or similar types of challenges are raised. Several national protocols cite the need for systematic documentation of fire scenes. Two recent protocols include those by the U.S. Department of Justice and the National Fire Protection Association (NFPA).

U.S. DEPARTMENT OF JUSTICE

One example of a peer-reviewed national protocol for fire investigation was developed by the National Institute of Justice (NIJ), which serves as the applied research and technology agency of the U.S. Department of Justice. The NIJ's (2000) *Fire and Arson Scene Evidence: A Guide for Public Safety Personnel* was the product of their Technical Working Group on Fire/Arson Investigation, which consisted of 31 national experts from law enforcement, prosecution, defense, and fire investigation communities. Two of the authors (DeHaan and Icove) participated in the preparation and editing of the NIJ guide.

The intent of the NIJ guide is to expose as many public sector personnel (mainly fire, police, and emergency medical) as possible to the process of identifying, documenting, collecting, and preserving critical physical evidence at fire scenes. This document has become the most widely distributed public guide on fire scene processing since the 1980 historic publication by the National Bureau of Standards (NBS;

TABLE 4-1	**Forms Used in Assuring the Collection of Uniform and Complete Field Data for Constructing Written Reports**	

FORM	NAME	DESCRIPTION
921-1	Fire Incident Field Notes	Collects general identification and contact information
921-2	Structure Fires	Used for documenting structure fires
921-3	Motor Vehicle Fires	Used for documenting vehicle fires inspections
921-4	Wildland Fires	Used for documenting grass, brush, or wildland fires
921-5	Casualties	Records information on persons injured or killed in the fire
921-6	Evidence	Documents and records recovered, seized, and released evidence
921-7	Photographs	Logs the descriptions of all photographs taken by the investigators
921-8	Electrical Panel	Records information as to the position and identification of circuit breakers in an electrical distribution panel
FFSR/Kirk's	Fire Modeling	Data for compartment fire modeling

Sources: Derived from NFPA 1998; NFPA 2011; and *Kirk's Fire Investigation,* 7th ed. (DeHaan and Icove 2012).

forerunner of the National Institute of Standards and Technology), *Fire Investigation Handbook* (NBS 1980)).

NATIONAL FIRE PROTECTION ASSOCIATION

The NFPA's *Guide for Fire Incident Field Notes, NFPA 906,* first published in 1988, and republished in 1998, served for a long time as a standardized protocol for recording field notes of fire scenes (NFPA 1998). The NIJ's 2000 guide cites *NFPA 906* and includes the data collection forms in its appendix. These forms have been merged into *NFPA 921,* and a synopsis appears in Table 4-1. Similar forms are found in the appendixes to *Kirk's Fire Investigation* (DeHaan and Icove 2012), and the NFPA forms are reprinted in this chapter with their permission.

The intended users of these forms include all persons having responsibility for investigating fires—fire company officer, the incident commander, the fire marshal, and any private investigators. These data collection forms constitute an organized investigative protocol. The purpose of these forms is to serve as a protocol for collecting and recording preliminary information needed to prepare a formal incident or comprehensive investigative report and for constructing a compartment fire model.

The reports cover structure, vehicle, and wildland fires; information on casualties, witnesses, evidence, photographs, and sketches; and documentary data on insurance and public records. A cover case management form is used to track the progress of the investigation. The maintenance of the updated and expanded *NFPA 906* forms is now under the responsibility of NFPA's Technical Committee on Fire Investigations, which also oversees *NFPA 921* (NFPA 2011).

Systematic Documentation

Fire investigators should adopt a *systematic procedure* or *protocol* for documenting through sketches, photographs, witness statements the examination of a fire scene as well as recording the chain of custody of all evidence removed. A systematic four-phase process (Icove and Gohar 1980) was first introduced with the publication of the NBS (1980) *Fire Investigation Handbook.*

This process recommends a careful documentation of the fire scene in the following four phases.

- ■ *Phase 1:* Exterior
- ■ *Phase 2:* Interior
- ■ *Phase 3:* Investigative
- ■ *Phase 4:* Panoramic or specialized photography

Table 4-2 shows the recommended updated guidance and purpose for this systematic documentation philosophy. This approach covers fire investigations without any preconception as to whether the fire was accidental, natural, or incendiary in cause.

TABLE 4-2	Systematic Documentation Techniques in Fire Investigation		
STEP	**TECHNIQUE**	**GUIDANCE**	**PURPOSE**
1	Exterior	Photograph perimeter and exterior of property.	Establishes venue and location of fire scene in relation to surrounding visual landmarks
		Sketch exterior.	Documents exposure damage to adjacent properties
		Use GPS to obtain location.	Reveals structural conditions, failures, violations, or deficiencies
			Establishes extent of fire damage to exterior of scene
			Establishes egress and condition of doors and windows
			Documents and preserves possible physical evidence remote from the fire
2	Interior	Record and sketch extent of fire damages, potential ignition sources, and data needed for fire reconstruction and modeling.	Traces fire travel and development from exterior to suspected point(s) of fire origin
			Documents heat transfer damage, heat and smoke stratification levels, and breaches of structural element
			Documents condition of power utility and distribution, furnace, water heater, and heat-producing appliances
			Documents position and damage to windows, doors, stairwells, and crawl space access points
			Documents fire protection equipment locations and operation (sprinklers, heat and smoke detectors, extinguishers)
			Documents readings on clocks and utility equipment
			Documents alarm and arc fault information
3	Investigative	Document clearing of debris and evidence prior to removal, fire burn and charring patterns, packaged evidence.	Assists in recording fire pattern and plume damage, isochar lines
			Establishes condition of physical evidence, utilities, distribution, and protection equipment (breakers, relief valves)
			Documents integrity of the chain of custody of evidence
4	Panoramic specialized	Produce multidimensional sketches.	Provides clearer peripheral views of exterior and interiors
		Capture evidence using forensic light sources or special recovery techniques.	Establishes photograph viewpoints of witnesses
			Documents and preserves critical evidence

Source: Updated from Icove and Gohar 1980.

Agencies that use the *NFPA 921* field or similar-styled notes often copy the blank forms directly from the standard and bind them into packets that are placed into case jackets. Several additional copies of the witness statement form are often included for multiple interviews.

Many agencies use a cover sheet for their investigative notes, which is to be used as a working document for both investigator and supervisor. The form typically includes check boxes indicating whether a form is contained in the case file, when it was completed, and remarks as to its status. The disposition of evidence, court-imposed deadlines, and other significant pieces of information may be noted in the activity section of the report.

Another high-level case document is Form 921-1, "Fire Incident Field Notes" (Figure 4-1). This form records how an agency was notified of the incident, the conditions on arrival, the owner/occupant of the property, other agencies involved, and an estimated total financial loss. Also documented are the time of arrival, the basis for an investigator's legal authority to enter the scene, and the time the scene was released.

Form 921-2, "Structure Fire Field Notes" (Figure 4-2), assists in the documentation of the type of property, geographic area, address, construction techniques, security, alarm protection, and utilities. Documenting the security at the time of the fire is an important consideration in arson cases, where the issue of "exclusive opportunity" is raised, along with the condition of doors, windows, and protection systems. Also important is documentation of the condition of doors, windows, and evidence found as part of the external documentation, noting fire patterns emanating from the doors or windows (or the lack of them) that may later provide critical information relative to fire development, ventilation, and timeline. The location and distances of window glass displaced from the structure or vehicle are documented during the exterior examination.

The "Vehicle Inspection Field Notes" (Form 921-3; Figure 4-3), "Wildfire Notes" (Form 921-4; Figure 4-4), "Casualty Field Notes" (Form 921-5; Figure 4-5), "Evidence Form" (Form 921-6; Figure 4-6), "Photograph Log" (Form 921-7; Figure 4-7), and "Electrical Panel Documentation" (Form 921-8; Figure 4-8) are used to further document the various aspects of the fire investigation. The "Room Fire Data" form" (Figure 4-9) is a specialized one-page data collection form developed by the authors and used to record measurements and information at the scene necessary for conducting a fire hazard analysis or constructing compartment fire models.

The *weather conditions* prior to the fire incident sometimes become important, especially when high winds, temperature fluctuations, or lightning come into play. The National Weather Service, Office of Climate, Water, and Weather Services, has a forensic services program to support investigations, particularly those involved in litigation. Certified climatological records (including radar images, satellite photos, and surface analysis) can be obtained from the National Weather Service headquarters in Silver Spring, Maryland. Alternative free and low-cost weather notification and historical systems available in the United States provide radar and severe-weather notifications. One popular website is Weather Underground (wunderground.com/), which provides historical weather data for many U.S. locations.

The location of *lightning strikes* is important information when addressing the possibility that lightning caused a particular fire. The U.S. National Lightning Detection Network locates strikes across the United States. For a fee, the private firm Vaisala will provide a report on all lightning strikes in a particular area for a given time frame. Vaisala (2012) also supplies data and custom software to its customers at vaisala.com/.

Exterior

The first major phase of the process is to document the exterior of the structure, vehicle, forest, wildland, boat, or object prior to probing into the cause of the fire. While circling the exterior of the structure or vehicle, the investigator can make a cursory field search

FIRE INCIDENT FIELD NOTES

Agency: ______________________ File No: ______________

TYPE OF OCCUPANCY

Location/ Address						
Property Description	Structure	Residential	Commercial	Vehicle	Wildland	Other
Other Relevant Info						

WEATHER CONDITIONS

Indicate Relevant Weather Information					
	Visibility	Rel. humidity	GPS	Elevation	Lightning
	Temperature	Wind direction	Wind speed	Precipitation	

OWNER

Name			DOB
d/b/a (if applicable)			
Address			
Telephone	Home	Business	Cellular

OCCUPANT

Name			DOB
d/b/a (if applicable)			
Permanent Address			
Temporary Address			
Telephone	Home	Business	Cellular

DISCOVERED BY

Incident Discovered by	Name		DOB
Address			
Telephone	Home	Business	Cellular

 NFPA 921 (p. 1 of 2)

FIRE INCIDENT FIELD NOTES (Continued)

File No: ______________

REPORTED BY

Incident Reported by	Name		DOB
Address			
Telephone	Home	Business	Cellular

INVESTIGATION INITIATION

Request Date and Time	Date of request	Time of request
Investigation Requested by	Agency name	Contact person/Telephone no.
Request Received by	Agency name	Contact person/Telephone no.

SCENE INFORMATION

Arrival Information	Date		Time		Comments		
Scene Secured	Yes	No	Securing agency		Manner of security		
Authority to Enter	Contemporaneous to exigency		Consent		Warrant		
			Written	Verbal	Admin.	Crim.	Other
Departure Information	Date		Time		Comments		

OTHER AGENCIES INVOLVED

	Dept. or Agency Name	Incident No.	Contact Person/Phone
Primary Fire Department			
Secondary Fire Department(s)			
Law Enforcement			
Private Investigators			

ADDITIONAL REMARKS

 NFPA 921 (p. 2 of 2)

FIGURE 4-1 The *NFPA 921-1* "Fire Incident Field Notes" form collects information on the type of occupancy, weather conditions, owner and occupant, person discovering the fire, how the investigation was initiated, and scene information. *Reprinted with permission from* NFPA 921: Guide for Fire and Explosion Investigations, *2011 ed. Copyright © 2011, National Fire Protection Association, Quincy, MA 02269. This reprinted material is not the complete and official position of the National Fire Protection Association on the referenced subject, which is represented only by the standard in its entirety.*

STRUCTURE FIRE

Agency: _____ Case number: _____

TYPE OF OCCUPANCY

Residential	Single family	Multifamily	Commercial	Governmental
Church	School	Other:		
Estimated age:	Height (stories):	Length:	Width:	

PROPERTY STATUS

Occupied at time of fire? ☐ Yes ☐ No   Unoccupied at time of fire? ☐ Yes ☐ No   Vacant at time of fire? ☐ Yes ☐ No

Name of person last in structure prior to fire: _____   Time and date in structure: _____   Exited via which door/egress: _____

Remarks: _____

BUILDING CONSTRUCTION

Foundation Type	Basement	Crawl space	Slab	Other:
Material	Masonry	Concrete	Stone	Other:
Exterior Covering	Wood	Brick/Stone	Vinyl	Asphalt	Metal	Concrete	Other:
Roof	Asphalt	Wood	Tile	Metal	Other:
Type of Construction	Wood frame	Balloon	Heavy timber	Ordinary	Fire resistive	Non-combustible	Other:

ALARM/PROTECTION/SECURITY

Sprinklers ☐ Yes ☐ No   Standpipes ☐ Yes ☐ No   Security camera(s) ☐ Yes ☐ No

Smoke detectors ☐ Yes ☐ No   Hardwired ☐ Yes ☐ No   Battery ☐ Yes ☐ No

Were batteries in place? ☐ Yes ☐ No   Location(s): _____

Hidden keys ☐ Yes ☐ No   where: _____   Security bars: Windows? ☐ Yes ☐ No   Doors? ☐ Yes ☐ No

Remarks: _____

© 2010 National Fire Protection Association   NFPA 921 (p. 1 of 2)

STRUCTURE FIRE (Continued)

CONDITION OF DOORS/WINDOWS

	Locked	Unlocked but closed	Open
Doors	Forced entry? ☐ Yes ☐ No	Who forced if known?	
Windows	Secure	Unlocked but closed	Open
	Broken by first responders? ☐ Yes ☐ No	Remarks:	Broken

FIRE DEPARTMENT OBSERVATIONS

Name of first on scene: _____   Department: _____

General observations: _____

Obstacles to extinguishment? _____   First-In Report attached? ☐ Yes ☐ No

UTILITIES

Electric	☐ On	☐ Off	☐ None	☐ Overhead	☐ Underground
	Company:	Contact:		Telephone:	
Gas/Fuel	☐ On	☐ Off	☐ None	☐ Natural	☐ LP	☐ Oil
	Company:	Contact:		Telephone:	
Water	Company:	Contact:		Telephone:	
Telephone	Company:	Contact:		Telephone:	
Other	Company:	Contact:		Telephone:	

COMMENTS: _____

© 2010 National Fire Protection Association   NFPA 921 (p. 2 of 2)

FIGURE 4-2 The *NFPA 921-2* "Structure Fire Field Notes" form collects occupancy, property status, building construction, alarm protection and security, condition of doors and windows, fire department observations, and utilities information. *Reprinted with permission from NFPA 921: Guide for Fire and Explosion Investigations, 2011 Edition. Copyright © 2011, National Fire Protection Association, Quincy, MA 02269. This reprinted material is not the complete and official position of the National Fire Protection Association on the referenced subject, which is represented only by the standard in its entirety.*

VEHICLE INSPECTION FIELD NOTES

Job # ______ File # ______
Insured ______
Address (City, State) ______
Loss Location ______

Date of Occurrence ______
Date of Assignment ______
Date of Receipt ______
Date of Inspection ______
Insp Location ______
Time of Inspection ______
Fire Report ______
Alarm Type ______

Stolen? ❑ Yes ❑ No Recovered by ______
Police Report ______
# of Keys ______ Alarm System? ❑ Yes ❑ No
Hidden Keys? ❑ Yes ❑ No Location ______

VEHICLE
Make ______ Model ______ Year ______
VIN ______ Odometer ______

EXTERIOR

Tires	Tire Type	Wheel Type	Tire Tread Depth	Lugs	Missing
LF					
LR					
RR					
RF					
SP					

Doors	Glass Y/N	Window UP/DOWN	Locked Y/N	Open/Closed	Prior Damage
LF					
LR					
RR					
RF					

Body Panels	Construction	Condition	Prior Damage
F Bumper			
Grill			
LF Fender			
LR Quarter			
R Bumper			
RR Quarter			
RF Fender			
Hood			
Roof			
Trunk			

UNDER HOOD	Intact	Missing	Parts Missing	Condition
Engine				
Battery				
Belts & Hoses				
Wiring				
Accessories				

FLUIDS	Level	Condition	Sample Taken
Oil			
Transmission			
Radiator			
Pwr Steer			
Brake			
Clutch			

ATS 851B, 8/97 NFPA 921 (p. 1 of 2)

VEHICLE INSPECTION FIELD NOTES (Continued)

Job # ______

INTERIOR	Intact	Missing	Parts Missing	Condition
Dash Pod				
Glove Box				
Strg Column				
Ignition				
Front Seat				
Rear Seat				
Rear Deck				
			Make/Model	
Stereo				
Speakers				
Accessories				

FLOOR			Sample Taken	
LF				
LR				
RR				
RL				

PERSONAL EFFECTS IN THE INTERIOR

TRUNK OR CARGO AREA

AFTERMARKET ITEMS NOT PREVIOUSLY DESCRIBED

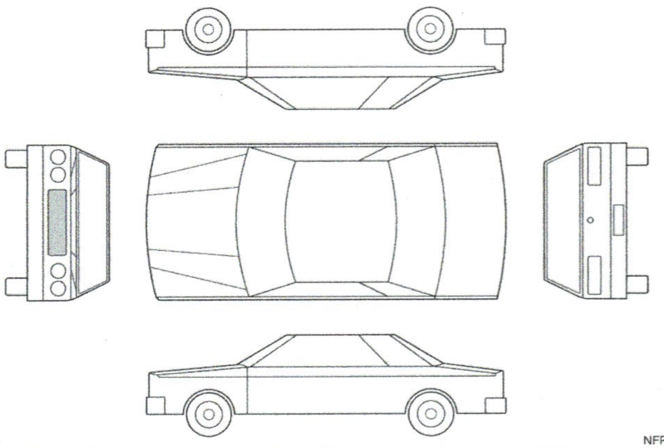

ATS 851B, 8/97 NFPA 921 (p. 2 of 2)

FIGURE 4-3 The *NFPA 921-3* "Vehicle Inspection Field Notes" form collects vehicle description, owner/operator, exterior/interior, security, and area of fire origin information.
Reprinted with permission from NFPA 921: Guide for Fire and Explosion Investigations, *2011 Edition. Copyright © 2011, National Fire Protection Association, Quincy, MA 02269. This reprinted material is not the complete and official position of the National Fire Protection Association on the referenced subject, which is represented only by the standard in its entirety.*

WILDFIRE NOTES

Agency: _____   File number: _____

PROPERTY DESCRIPTION

_____

_____

_____

| Fire damage:
❏ Less than acre _____ No. acres | Other properties involved: |
| Security:
❏ Open   ❏ Fenced   ❏ Locked gate | Comments: |

FIRE SPREAD FACTORS

| Type fire:
❏ Ground ❏ Crown | Factors:
❏ Wind ❏ Terrain | Comments: |

AREA OF ORIGIN

_____

_____

_____

PEOPLE IN AREA

| At time of fire:
❏ Yes ❏ No   ❏ Undetermined | Comments: |

IGNITION SEQUENCE

Heat of ignition: _____

Material ignited: _____

Ignition factor: _____

If equipment involved: Make: _____   Model: _____   Serial no.: _____

Comments: _____

_____

_____

_____

© 2010 National Fire Protection Association   NFPA 921

FIGURE 4-4 The *NFPA 921-4* "Wildfire Notes" form collects property, fire spread factors, area of origin, people in the area, and ignition sequence information. *Reprinted with permission from* NFPA 921: Guide for Fire and Explosion Investigations, *2011 Edition. Copyright © 2011, National Fire Protection Association, Quincy, MA 02269. This reprinted material is not the complete and official position of the National Fire Protection Association on the referenced subject, which is represented only by the standard in its entirety.*

for additional evidence, noting that fire patterns emanating from the doors or windows may provide critical information relative to the fire's development, ventilation, and timeline. This phase establishes the location of the fire scene in relation to surrounding visual landmarks. The exterior inspections should also reveal the extent of fire damage, collapse, structural conditions, failures, code violations, deficiencies, or potential safety concerns. This step also documents exposure damage to adjacent properties from the fire. In large investigations, an overhead crane, aerial ladder truck, or aircraft can be used to obtain overall photographs of the scene.

CASUALTY FIELD NOTES

Agency: _____ Incident date: _____ Case number: _____

DESCRIPTION

Name: _____ DOB: _____ Sex/Race: _____

Address: _____ Phone: _____

Other identifiers: _____

Description of clothing and jewelry: _____

Occupation: _____ Place of employment: _____

Marital status: _____

Victim's doctor: _____ Victim's dentist: _____

Smoker: ☐ Yes ☐ No ☐ Unknown

CASUALTY TREATMENT

Treated at scene: ☐ Yes ☐ No By: _____

Transported to: _____ Remarks: _____

SEVERITY OF INJURY

☐ Minor ☐ Moderate ☐ Severe ☐ Fatal

Describe injury: _____

NEXT OF KIN

Name: _____ Address: _____ Phone: _____

Relationship: _____ Notified on ___/___/___ By: _____

FATALITY INFORMATION

Where was victim initially found: _____

Who located victim: _____

Body position when initially found: _____

Victim's appearance: _____

Body removed by: _____ To: _____

Photographed in place: ☐ Yes ☐ No Significant blood present under/near victim: ☐ Yes ☐ No

MEDICAL EXAMINER/CORONER

Agency: _____

Date of examination: ___/___/___ Location: _____

Autopsy requested: ☐ Yes ☐ No Autopsy completed: ☐ Yes ☐ No Copy attached: ☐ Yes ☐ No

Full body x-rays: ☐ Yes ☐ No Other x-rays: _____

Identification made from: ☐ Physical appearance ☐ Dental records ☐ Fingerprints ☐ Prior injury comparison

☐ Other: _____

Condition of trachea: _____

Evidence of pre-fire injury: ☐ Yes ☐ No Type/location: _____

Blood samples taken: ☐ Yes ☐ No Other specimens collected: _____

CO level: _____ Blood alcohol: _____ Other: _____

Cause of death: _____

COMPLETE BODY DIAGRAM ON REVERSE

© 2010 National Fire Protection Association NFPA 921 (p. 1 of 2)

CASUALTY FIELD NOTES (Continued)

REMARKS

_____

_____

_____

_____

BODY DIAGRAM

Top of Head

Indicate parts of body injured: ☐ None ☐ Blisters (red marker) ☐ Burns (black marker)

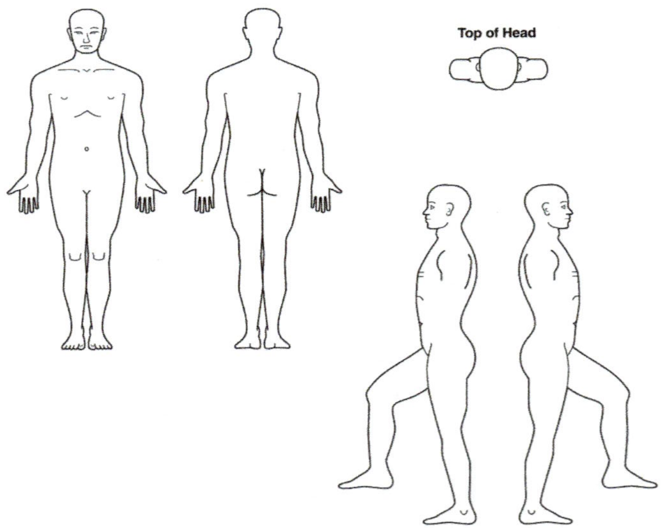

Fire Investigation Data Sheet/Attachment: _____ Initials: _____

Body Diagram

© 2010 National Fire Protection Association NFPA 921 (p. 2 of 2)

FIGURE 4-5 The *NFPA 921-5 "Casualty Field Notes"* form collects the name and description of the injured or deceased person, treatment, next of kin, medical examiner/coroner, and body diagram information. *Reprinted with permission from NFPA 921: Guide for Fire and Explosion Investigations, 2011 Edition. Copyright © 2011, National Fire Protection Association, Quincy, MA 02269. This reprinted material is not the complete and official position of the National Fire Protection Association on the referenced subject, which is represented only by the standard in its entirety.*

EVIDENCE FORM

Date of incident: ___ / ___ / ___   Storage location: _____   Case #: _____

Item No.	Description	Location		
			Destroyed	Released
			Destroyed	Released
			Destroyed	Released
			Destroyed	Released
			Destroyed	Released
			Destroyed	Released
			Destroyed	Released
			Destroyed	Released
			Destroyed	Released
			Destroyed	Released

How was evidence received?   Date received: ___ / ___ / ___   Date stored: ___ / ___ / ___

❑ Removed from scene by investigator.

❑ Received by investigator from: _____ Name, Company, or Dept.

Received via: ❑ UPS ❑ FedEx ❑ Airborne ❑ U.S. Mail ❑ In person ❑ Freight _____ Name of Company

❑ Other: _____ Describe _____

Received by _____   Case Investigator _____

LOCATION EVIDENCE REMOVED

Owner _____   State ___ Zip ___ Phone ___

Company _____   Address 2 _____

Address 1 _____   City _____

City _____   State ___ Zip ___ Phone ___

    NFPA 921 (p. 1 of 2)

EVIDENCE FORM (Continued)

INTERNAL EXAMINATION

	Date Pulled	Date Examined	Date Returned
Investigator			

EVIDENCE DESTRUCTION

Authorized by _____ Date _____

Investigator's Authorization _____ Date _____

Destroyed by _____ Date _____

EVIDENCE RELEASE

Signature of Person Receiving Evidence _____

Person Receiving Evidence (Please Print) _____ Date _____

Company Name _____

Address _____

City _____ State ___ Zip Code ___

Authorized by _____ Date _____

Investigator's Authorization _____ Date _____

Released via _____

EXAMINATION BY OTHERS

Name _____ Date of Examination _____

Company _____

Address _____

City _____ State ___ Zip ___ Phone ___

Authorized by _____

Investigator's Authorization _____ Date _____

Name _____ Date of Examination _____

Company _____

Address _____

City _____ State ___ Zip ___ Phone ___

Authorized by _____

Investigator's Authorization _____ Date _____

Name _____ Date of Examination _____

Company _____

Address _____

City _____ State ___ Zip ___ Phone ___

Authorized by _____

Investigator's Authorization _____ Date _____

REMARKS

    NFPA 921 (p. 2 of 2)

FIGURE 4-6 The *NFPA 921-6* "Evidence Form" collects the information on the disposition of items collected, released, and destroyed from a fire scene. *Reprinted with permission from NFPA 921: Guide for Fire and Explosion Investigations, 2011 Edition. Copyright © 2011, National Fire Protection Association, Quincy, MA 02269. This reprinted material is not the complete and official position of the National Fire Protection Association on the referenced subject, which is represented only by the standard in its entirety.*

PHOTOGRAPH LOG

Roll #: _____   Exposures: _____

Case #: _____   Date: _____

Camera make/type: _____   Film type: _____   Film speed: _____   ASA: _____

Number	Description	Location
1)		
2)		
3)		
4)		
5)		
6)		
7)		
8)		
9)		
10)		
11)		
12)		
13)		
14)		
15)		
16)		
17)		
18)		
19)		
20)		
21)		
22)		
23)		
24)		
25)		
26)		
27)		
28)		
29)		
30)		
31)		
32)		
33)		
34)		
35)		
36)		

Photos taken by: _____   Initials: _____

© 2010 National Fire Protection Association   NFPA 921

FIGURE 4-7 The *NFPA 921-7* "Photograph Log" collects the information on the description and location of which each photograph is taken and by whom at a fire scene. The form should be modified to include digital photo data such as medium identification and image number. *Reprinted with permission from NFPA 921: Guide for Fire and Explosion Investigations, 2011 Edition. Copyright © 2011, National Fire Protection Association, Quincy, MA 02269. This reprinted material is not the complete and official position of the National Fire Protection Association on the referenced subject, which is represented only by the standard in its entirety.*

If there has been any explosion, the investigator must measure and document via both diagramming and photography the points to which fragments of glass or structure were thrown. These concepts are discussed in later sections.

When considering all other sources of ignition, the condition of the utilities at the time of the fire is particularly important. The status of the utilities may indicate whether the owner/occupant(s) was (were) living in the structure. If instructions to disconnect the utilities were given prior to the fire, it is important to identify and document who gave those instructions and why they were given. Determining when gas and electric services were cut off (and by whom) during fire suppression is also important. If electrical power interruptions or surges are suspected, neighbors who share the same power service should be canvassed.

ELECTRICAL PANEL DOCUMENTATION

Fire location:		Date:		Case #:

Panel location:		Main size:		Fuses: ☐
				Circuit breakers: ☐

LEFT BANK

#	Rating Amps	Labeled Circuit	Status
1	—		—
3	—		—
5	—		—
7	—		—
9	—		—
11	—		—
13	—		—
15	—		—
17	—		—
19	—		—
21	—		—
23	—		—
25	—		—
27	—		—
29	—		—

Notes:

RIGHT BANK

#	Rating Amps	Labeled Circuit	Status
2	—		—
4	—		—
6	—		—
8	—		—
10	—		—
12	—		—
14	—		—
16	—		—
18	—		—
20	—		—
22	—		—
24	—		—
26	—		—
28	—		—
30	—		—

Notes:

Documented by: _____

　　　　　　　　　NFPA 921

FIGURE 4-8 The *NFPA 921-8* "Electrical Panel Documentation" form collects the information on the location of the electrical panel and the condition of circuit breakers along with their amperage rating and status. *Reprinted with permission from NFPA 921: Guide for Fire and Explosion Investigations, 2011 Edition. Copyright © 2011, National Fire Protection Association, Quincy, MA 02269. This reprinted material is not the complete and official position of the National Fire Protection Association on the referenced subject, which is represented only by the standard in its entirety.*

Interior

The second phase involves the documentation of interior damage by showing the extent and progress of the fire through the room(s), area(s), and suspected point(s) of fire origin. These pictures and sketches are made prior to excavating the fire debris and serve to document the conditions of the scene found on the investigator's arrival. All undamaged areas of the building should also be examined and documented for comparison and hypothesis formulation (as long as there is legal authority for entry).

DOCUMENTATION OF DAMAGE

While conducting the interior examination the investigator documents damage (or lack of damage) to *all* rooms, including heat and smoke stratification levels, smoke deposits, heat

ROOM FIRE DATA

Room _____   Room # _____

Length _____

Width   _____   **Floor Plan**

Height _____

    Note ceiling height changes _____

Walls: Structure/material _____ Thickness _____ Covering _____ Sample? Y/N

          Structure/material _____ Thickness _____ Covering _____ Sample? Y/N

Ceiling: Structure/material _____ Thickness _____ Covering _____ Sample? Y/N

Floor: Structure/material _____ Thickness _____ Covering _____ Sample? Y/N

Openings (door, window, other vents) into room (number on plan above):

Height (bottom to top of opening)	Sill Height	Soffit Depth (above opening)	Width	Open or Closed? Changes During Fire?
1.				
2.				
3.				
4.				
5.				
6.				
7.				

HVAC System:

    Description _____

    On/off prior to/during fire? _____

Furnishings (descriptions of major fuel items, including floor and wall coverings, draperies):

_____

_____

_____

_____

Time Line: Alarm time _____ FD arr. time _____ Control time _____

Detection: _____

Pre-Fire Events: _____

FIGURE 4-9 The one-page "Room Fire Data" form captures critical information about each room at a fire scene, which is often necessary when conducting a fire scene analysis or constructing a computer fire model. *Courtesy of J. D. DeHaan.*

transfer effects, and breaches of structural elements (walls, floors, ceilings, and doors). The investigator may find it useful to map out and delineate the areas of damage corresponding to the pattern types described in Table 3-1, with color-coded chalk or tape on the surfaces themselves, followed by photography. Areas of surface deposit demarcations (sometimes referred to as smoke horizons), thermal effects, penetrations, and loss of material can also be outlined using colored markers on sketches or photographs. One system might be to use yellow lines to outline areas of surface deposits, green lines for thermal effects, blue lines for penetrations, and red lines to outline areas where the material has been consumed completely.

    This phase includes examination and documentation of the condition of power utility and distribution, furnace, water heater, and heat-producing appliances, as well as the contents and conditions of the rooms. The investigator should document the location of the HVAC system and the condition of the unit, ductwork, and filters. The thickness of window glass, the sizes of windows, and whether they are single or double-glazed must also be noted and documented. The sill height and soffit depth of each door and window in the rooms involved must be noted, as well as their opening dimensions and whether they were open during the fire or broken sometime later.

STRUCTURAL FEATURES

The design and construction of the ceiling may play a key role in spread of smoke or fire, or detection. A smooth, sloping ceiling may direct smoke away from a detector, whereas one with exposed beams or one with deep decorative bays may channel smoke or fire in a preferred direction or prevent it from spreading at all (see Figure 4-10). Because ceilings often suffer major damage during a fire or during overhaul, it is important to document similar undamaged rooms in the building when practical. Otherwise, occupants or maintenance staff should be interviewed about such features. Such features may also be documented using prefire photos, videos, or building plans.

FIGURE 4-10 Complex ceiling structures, as in this hotel conference room will affect fire and smoke spread. This photo captures the structural features as well as ventilator, sprinkler, and alarm sensor locations. *Courtesy of John Houde.*

One of the authors participated in a reconstruction of the 1916 fire that destroyed author Jack London's nearly completed home, Wolf House. Descriptions in contemporary interviews and features still visible in the masonry remains were compared with the builder's plans. It was clear that even though the plans were dated just weeks prior to the fire, there were features in the as-built structure that were different. Wall coverings and stairwell and door locations played a significant role in testing various hypotheses about the fire's area of origin. For the detailed case study on the Wolf House, see Icove and DeHaan (2009), chapter 9. The type, location, and operation of HVAC components are also important. In one case, an accidental fire originating in a chair took the life of the sole occupant of a room designated as the smoking room for the facility. The midmorning fire was well advanced before it was detected by a staff member on the floor above who saw smoke rising past the window of her room and went to investigate. The question ultimately was, Why did the fire get so large and yet the smoke detector in the hall on the ceiling just outside the open door to the room failed to function? (No smoke was seen in the hallway as staff rushed to the room, and the detector was hardwired to an alarm system.) It was discovered that an exhaust fan in the room's window was left on and its CFM rating was sufficient to draw enough smoke-laden air from the limited fire in the room away from the door, preventing its detection outside. Even passive elements of HVAC systems can play a role. The void space above many suspended ceilings acts as an open air-return plenum, allowing smoke and hot gases to spread quickly. See Figure 4-11 for an example. The filters and condensers on HVAC systems should be checked for soot or pyrolysis products.

FIRE PROTECTION SYSTEMS

It is also important to document for later evaluation the fire protection equipment (e.g., fire alarms, smoke detectors, automatic suppression systems, fire doors, thermally operated safety devices, and the presence or lack of compartmentation) that would have helped contain the fire as well as alert the occupants. Documentation includes function and location. The time readings on electric and mechanical clocks can serve to document the approximate times that they were damaged and halted by heat or by interruption of power.

The position, type, and operability of smoke, heat, and carbon monoxide (CO) detectors and sprinkler systems should also be documented. This is important, since NFPA studies show that almost all U.S. households have at least one smoke alarm, yet during the period 2000–2004, no smoke alarms were present or none operated in almost half (46%)

FIGURE 4-11 An air-return plenum vent opening in this kitchen allowed very rapid extension into the ceiling space and faster-than-expected collapse of the suspended ceiling. *Courtesy of Jamie Novak, Novak Investigations.*

of the reported home fires (Ahrens 2007). Also during this time period, 43% of all home fire deaths resulted from fires in homes with no smoke alarms, and 22% of such deaths occurred in homes in which smoke alarms were present but simply did not operate. Unfortunately, investigations of fatal fires often found that batteries had been removed for use in toys or remote controls or to prevent false alarms from being triggered by cooking or even bathroom steam. Hardwired detectors can be disconnected (from unmonitored systems) or covered with tape or plastic film to prevent alarms. The condition, fusing temperature, and orifice size of any open sprinkler heads should also be documented and the head(s) preserved as evidence, including exemplars of unfused sprinkler head.

The data systems in modern fire alarm systems can be electronically examined to capture the time, sequence, and zone of alarms or sprinkler activations. The zones and manner of activation should be established for alarm sensors. Someone knowledgeable about the alarm system should be consulted for assistance. Even systems that are not remotely monitored have a battery-powered memory of recent activations that can be downloaded.

ARC-FAULT MAPPING

arc-fault mapping ■ Locating and analyzing the pattern of arc faults in electrical circuits as a means of locating a possible area of origin.

arcing through char ■ Unintended passage of current through a semiconductive degradation product.

arc-tracking ■ Passage of current across a contaminated nonconductor or through a poor conductor that causes localized heating, which can degrade the material and lead to current flow.

When a flame or heat attacks an insulated electrical cable, the insulation begins to pyrolyze and degrade. When rubber, cloth, or some plastic insulation chars, the carbonaceous residue becomes conductive, and current can pass if the conductor is energized and a return path established (sometimes called **arcing through char**). With PVC, a temperature of 120°C (248°F) is sufficient to initiate an **arc-tracking** process and allow a current to begin to flow (Babrauskas 2006). The resistance decreases as the insulation degrades further. In either case, current begins to flow between the hot or energized wire and the neutral or ground (or to a grounded conduit, appliance, or junction box). This increased current creates heat that further chars the insulation, providing even better current flow.

Because of the resistance provided by the charred insulation, this condition is not the same as a direct short or fault (which has nearly zero resistance). The current flow can vary from very low to very high (as high as 150 A has been measured). This can produce a variety of arc-damaged areas along the conductor—ranging from small craters to beads of molten metal. Because the conductors are not bonded together, they are free to move, so these current flows are usually very short in duration, in many cases too short to cause overcurrent protective devices (OCPDs) to function. This means that arcing can occur at many places along a single energized wire and multiple times until the OCPD trips or the wire is severed by melting (sometimes called *fusing*). Once the wire is separated, no current can flow beyond that point (Carey 2002).

These phenomena provide a way of locating a possible area of origin of a fire by tracing the wiring from its power source (such as the breaker or fuse box) and mapping the locations of all failures of the wire (since wires can be caused to fail by fire effects even when nonenergized). The failures are then characterized as arcing (energized) or melting (nonenergized) by their gross appearance. The locations of the arcing phenomena farthest downstream along a circuit from the power source are indicators of the point at which the fire first attacked the wiring, as illustrated in Figure 4-12. These failures, then, may be very useful in indicating a possible area of origin. Dr. Robert J. Svare (1988) pioneered this approach, which has been tested in numerous live-burn structure tests.

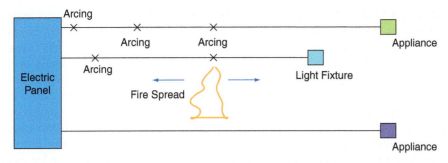

FIGURE 4-12 Mapping the locations of arcing failures farthest downstream along a circuit from the power source, which are indicators of the point at which the fire first attacked the wiring.

This procedure has been shown to be effective even in multistory buildings and installations of ring-mains wiring (as in the United Kingdom) (Carey 2002; Carey and Nic Daéid 2010). It is not effective in three-phase wiring systems, where there are three energized wires in every circuit. The technique requires diligent and careful tracing of wiring and circuits, since the source of the energy has to be established and the nature of the failure accurately evaluated. *NFPA 921* suggests the documentation of the electrical distribution panel using a form similar to that previously shown in Figure 4-8.

In buildings protected by zone alarm systems, arc-fault indicators can be used together with data from the alarm system (alarm, sensor, or sprinkler activation) to estimate areas of possible origin. This approach may not be of use in extensively collapsed or fire-damaged buildings where the relationship and tracing of wires cannot be established accurately.

Investigative

The third phase of the systematic investigation concentrates on the debris clearing operations, fuel loads, char and burn patterns, potential ignition sources, and position of evidence prior to its removal from the fire scene. Evidence of any crimes associated with the fire such as burglary, theft, and homicide should also be documented.

Until photographically recorded and supported by fire scene notes, nothing should be moved, including bodies of victims. Investigative photographs ensure the integrity of the probe and custody of evidence.

CASUALTIES

Information on casualties is recorded on the "Casualty Field Notes" form (Figure 4-5). This report includes a description of the victim, type of injury, circumstances, treatment received, disposition of the body and its examination, next of kin, and other appropriate remarks. This information should be collected on all injured parties and not just on fatalities, although Health Insurance Portability and Accountability Act (HIPAA) regulations (*45 CFR 160* and *164*, subparts A and E) make some information unavailable.

This form does not presently include general information about the victim obtained at autopsies such as burn injuries, tests for blood alcohol, hydrogen cyanide, carboxyhemoglobin levels, and other conditions often documented in fire death investigations. That information is discussed at length in Chapter 7.

WITNESSES

Information from witnesses is documented in field notes taken by the investigator, including identification, home and work addresses, contact information, and expected testimony. *NFPA 921* identifies witness information as one of the primary sources of information to be considered by investigators in determining a fire's origin and cause (see *NFPA 921*,

2011 ed., pts. 17.1.2 and 17.2.5) (NFPA 2011). It is imperative to interview as many witnesses (including fire and police personnel) to a fire as soon as possible while the information is fresh and untainted. *NFPA 921* cautions that the rate of fire growth as solely determined by witness statements may be subjective. It notes that often the time of discovery is considerably later than the time of actual ignition, and witnesses report the fire growth observed from time of discovery not ignition (see *NFPA 921*, 2011 ed., pt. 5.10.1.4) (NFPA 2011).

In cases involving on-scene witnesses to the fire, thorough interviews may reveal additional information relating to the initial stages of the fire and the environmental conditions at the time (rain, wind, extreme cold, etc.). In most cases, survivors can provide information on prefire conditions; fire and smoke development; fuel package location and orientation; activities of victims and suspects before, during, and after discovery of the fire; actions resulting in their survival; and critical fire events such as flashover, structural failure, window breakage, alarm sounding, first observation of smoke, first observation of flame, fire department arrival, and contact with others in the building (see *NFPA 921*, 2011 ed., pt. 10.8.5) (NFPA 2011).

Furthermore, it is important to document details as to the position of the witnesses in relation to the fire scene when recording their visual observations, especially when using witness-viewpoint photographs (see *NFPA 921*, 2011 ed., pt. 15.2.6.10) (NFPA 2011). This process can be enhanced by walking witnesses through the scene (when safe) or back to the location from which they witnessed the fire. This can help establish or confirm lines of sight and prompt more complete statements.

EXAMPLE 4-1 ■ Systematic Analysis of Witness Statements

On January 16, 2007, a fire at the Castle West Apartments, a 129-unit wood-framed three-level apartment building in Colorado Springs, Colorado, resulted in two deaths, 13 injured occupants, six injured firefighters, and an estimated $6 million loss. The building had no automatic sprinklers or a monitored/addressable fire alarm system and had emergency lighting only in hallways and stairs, and only battery-operated smoke detectors in hallways and in some units.

The multiagency investigation team interviewed witnesses from 96 occupied units of the 129 units, 24 of which were vacant. These interviews resulted in coverage of more than 90% of the total occupied units.

To minimize confirmation bias, a study by ATF's Fire Research Laboratory (Geiman and Lord 2011) mapped the mosaics of the individual observations of the witness interviews onto the floor plans of the three levels. The result was an open-ended protocol of essential questions to ask witnesses that increased their independent reliability and corrected for many of the factors that would lead to confirmation bias introduced either conscious or subconsciously by the interviewer.

The results of the analysis demonstrated the survivability of victims. Figure 4-13 shows that almost all the occupants on Level A (first floor) successfully exited their apartment building. The analysis of the individual observations of smoke and fire by occupants exiting the building helped independently isolate the area of fire origin to the north wing of the building on Levels B and C, as seen in Figure 4-14 from the study.

The study concluded that this approach provides a systematic analytical tool for fire investigators to use in testing fire origin hypotheses. This methodology reduces the potential for confirmation bias by both the investigators and witnesses when a consistent protocol of open-ended questions is applied in mass interviews. Encoding the interviews into a text retrieval system allows for later content analysis of the witness information. Finally, the simple graphical presentation of the investigation overlaid on the building plan assists in interpreting the results of the analysis and provides a positive demonstrative evidence for inclusion in reports and potential legal proceedings.

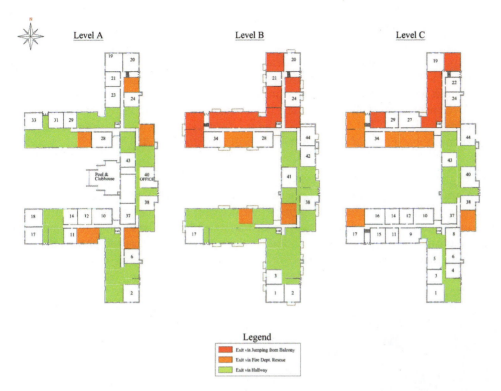

FIGURE 4-13 Three-level plan of the apartment complex showing the egress modes of the occupants via jumping from the balcony, rescue by the fire department, and simple unaided egress. *Courtesy of the Fire Research Laboratory, U.S. Department of Justice, Bureau of Alcohol, Tobacco, Firearms and Explosives.*

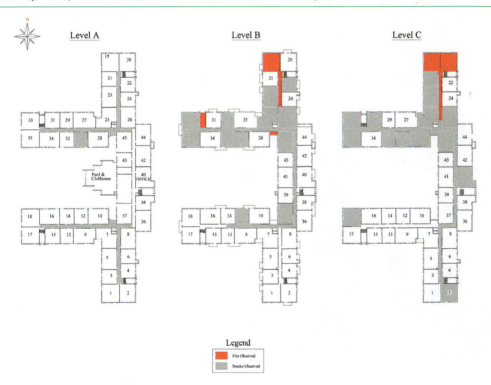

FIGURE 4-14 Three-level plan of the apartment complex showing the observations by escaping occupants of smoke and flame. *Courtesy of the Fire Research Laboratory, U.S. Department of Justice, Bureau of Alcohol, Tobacco, Firearms and Explosives.*

EVIDENCE COLLECTION AND PRESERVATION

The chain of custody is intended to trace the item of evidence from its discovery to court. Its purpose is to authenticate the evidence as it is found as well as to prevent its loss or destruction. The documentation includes photographing the evidence at its discovery and preparing a written list itemizing the transfer of the evidence once it leaves the scene, as on the "Evidence" form (Figure 4-6). Such documentation will demonstrate compliance with guidelines such as *ASTM E1188-05: Standard Practice for Collection and Preservation of Information and Physical Items by a Technical Investigator* (ASTM 2005a).

PHOTOGRAPHY

Photographs should include not only overall views but also close-ups of critical evidence and intermediate "establishing" shots when necessary. Photos should record the scene as found, during debris clearance, after clearance, and with furnishings replaced in their prefire positions. An accurate photo log is nearly as important as the photos themselves to ensure that the investigator can correctly reconstruct the scene days, weeks, or months later when called on.

The "Photograph Log" form (Figure 4-7) is used to record the description, frame, roll, or image storage card, and digital image number of each photograph taken at the fire scene. Investigators should follow a similar systematic documentation scheme for fire scene photographic images taken with video recorders. The form is designed to be filled out as the photographs are taken. The frame and roll (or photo) image numbers are used later on the fire scene sketch to indicate the location and direction from which they were taken. Each roll of film, digital image, and videotape that is preserved by duplicate archived media should be documented on a separate form. The "Remarks" field is used to document the disposition of the film, archived media, and videotapes. Close-up photos of evidence should be accompanied by overall, orientation, or panoramic shots (as in Figure 4-15). Photography is discussed in more detail in a later section of this chapter.

(a)

(b)

FIGURE 4-15 Photographic documentation of critical features such as possible ignition sources should include both (a) overall (orientation) and (b) close-up photos. *Courtesy of J. D. DeHaan.*

SKETCHING

Fire scene sketches, whether or not to scale, are important supplements to photographs. They graphically portray the fire scene and items of evidence as recorded by the investigator. A typical sketch is a simple two-dimensional drawing of the fire scene. It should include all major dimensions of rooms, compartments, buildings, vehicles, and curtilage, where applicable. Square-grid paper is available in many sizes at business and school supply stores. These sketches may range from rough exterior building outlines or room contents to detailed floor plans. Sketches of the distribution of window glass fragments are critical to later reconstruction of explosion events. They must include distances and angular direction data. A more detailed discussion of sketches appears later in this chapter.

DOCUMENTARY RECORDS

Insurance information and documentary records are required to be collected to ensure that a thorough investigation is conducted, particularly when there are multiple policyholders, as in the case of commercial buildings. Care should be taken to catalog accurately and to secure these incident, property, business, and personal documentary records.

COMPARTMENT FIRE MODELING DATA

The documentation needed for compartment fire modeling exceeds the data normally collected by investigators. These additional details are listed on the "Room Fire Data" form shown in Figure 4-9. Included is information on the compartment needed for the model such as room dimensions, construction, surface materials, interior finish, openings in doors and windows, HVAC, event timeline, and fuel packages. The information is critical to hypothesis formation and testing in all fire reconstructions, even if computer modeling is not planned. It is especially important that height information be accurately and fully recorded, not just the floor plan. For example, any places where the ceiling height changes must be clearly identified (because slope, deep beams, and architectural features can affect hot gas flow and layer development).

PHYSICAL EVIDENCE

Physical evidence is sometimes called the silent witness because it can provide reliable answers to questions other investigative techniques cannot address, fill in details, and corroborate other information. Clearly, the investigative phase of systematic investigation combines all the essential elements of forensic evidence collected from the fire scene.

Generally accepted forensic guidelines are designed to prevent contamination, loss, or destruction of the evidence and to provide a reliable chain of custody for that evidence. For instance, when there is an object with dried blood on it, the object itself should be collected whenever possible and allowed to air-dry before packaging. Bloody objects should be placed into individual sealed paper bags, boxes, or envelopes after drying and kept dry and refrigerated if possible. Ordinary plastic bags should not be used, since they will not allow the sample to ventilate. Investigators should be aware of the safety issues involving blood-borne pathogens.

All firearms should be placed into individual manila envelopes, and rifles should be tagged. Firearms should be hand-carried to the laboratory. Special handling instructions include the collecting technique, unloading and noting the cylinder position in revolvers, and recording the serial number, make, and model of the weapon. Fire debris or containers containing volatile liquids must be sealed in appropriate packaging to prevent loss or contamination and kept cool to minimize evaporation. Further details are included in *Kirk's Fire Investigation,* 7th ed. (DeHaan and Icove 2012).

HANDHELD LASER MEASUREMENT TECHNIQUES

Investigators often find themselves documenting the measurements of several rooms within a structure, a cumbersome, time-consuming task that can be inaccurate. Because measuring tapes are most efficient when used by two persons, handheld devices are beneficial when a single person is responsible for measuring distances.

With the introduction of low-cost accurate handheld laser devices, an investigator can now measure not only the distance but also the area and volume of a room. Published specifications for devices sold for less than $100 report an accuracy of ± 6.35 mm (1/4 in.) at 30.48 m (100 ft). The typical range of these devices is 0.6–30 m (2–100 ft), they can be switched to read in English or metric units, and they operate on low-cost 9 V batteries (see Figure 4-16) (Stanley 2012; DeHaan 2004).

FIGURE 4-16 New devices such as this FatMax laser measurement tool make taking room measurements easier and faster. *Courtesy of J. D. DeHaan.*

IGNITION MATRIX

Today, more than ever before, fire investigators face the challenge of identifying all potential ignition sources in the area of origin and then eliminating them, down to one (based on reasonable data or methods). As discussed in Chapter 1, the expected outcome of the scientific method of fire investigation is that an investigator will reach a defensible conclusion by demonstrating not only how this source could have started the fire but also how all other reasonably possible sources have been excluded. If two or more ignition sources are deemed "possible" at the end of the investigation, the cause has to be declared to be "undetermined." Demonstrating these conclusions can be difficult, especially in a concise yet comprehensive manner.

The Ignition Matrix approach forces the investigator to consider a comprehensive range of alternative hypotheses and consider each one on the basis of factors such as heat release rate, heat flux, separation distances, thermal inertia, and routes of fire spread. A completed matrix offers a concise demonstration that all potential ignition sources have been considered and all (but one, presumably) eliminated. Such a matrix is also more easily verified than the traditional item-by-item list, improving the investigator's own review process. Note that this comprehensive approach is not the casual "process of elimination" that *NFPA 921, 2011 ed., pt. 18.6.5,* strongly cautions against using (NFPA 2011); however, it is an exhaustive approach using the true intent of the scientific method.

The process begins with a detailed evaluation and sketch of the suspected area of origin. All identified first fuels and possible ignition sources are documented. The Bilancia Ignition Matrix (as illustrated in Figure 4-48) represents exhaustive pairwise comparisons designed to systematically evaluate numerous ignition sources and also to document how each was or was not competent to ignite a particular first fuel. The suggested ignition matrix is merely a grid on paper. In that layout, each square represents a pairwise assessment of the interaction between an energy source and a first fuel.

Each combination is then evaluated on four primary values:

1. Is this ignition source competent to ignite this fuel? Yes or no.
2. Is this ignition source close enough to this fuel to be capable of igniting it? Yes or no.
3. Is there evidence of ignition? Yes or no.
4. Is there a pathway for a fire ignited in this first fuel to ignite the main fuel? Yes or no.

There are often comments or notes that can be included and applied to each evaluation. These notes might include one or more of the following:

■ This source was not energized (appliance or power cord) or not in use (candle).
■ "Too far away" or "Depends on duration being adequate," or "Unknown—must test exemplar."
■ Physical or visual evidence of burning; ignition was witnessed or recorded in video.
■ Plume developed from first fuel sufficient, or flashover produced ignition.

Finally, the matrix can be color-coded to indicate which combinations are capable, which are excluded, and which need further data. This color legend might be Red—Competent and close, Blue—Not competent, and Yellow— Competent but ruled out. For a more detailed discussion on the application of the Bilancia Ignition Matrix in fire scene documentation, see *Kirk's Fire Investigation*, Chapter 6.

Panoramic Photography

PHOTOGRAPHIC STITCHING

A majority of the photographs taken at fire scenes focus on fire patterns and other evidence. In structure fires, an aerial or perspective view of the building can reveal the overall impact of the fire on the structure, including how the building itself was breached by both the fire and efforts to extinguish the blaze. A simple photo may not be able to capture information around the perimeter of a room. An example of a *panoramic photograph* of a fire scene created by stitching several photographs together is shown in Figure 4-17.

The creation of panoramic images to improve scene review and illustration for court is not new. Panoramic (scanning slit) cameras have been around for more than 100 years, but they were rarely used, because the specialized equipment was heavy, yet delicate. Panoramic cameras, popular in the late 1800s, were sometimes used to capture the ravaging effects of fires. Many present-day cameras simply crop the top and bottom portions of the image, creating a false appearance of a panoramic photograph. Super-wide-angle and 360° rotating cameras are now available, but they are costly (Curtin 2011). Note that human vision typically captures a 170° view of a room, but a typical 50 mm camera lens captures only about one-fifth of that field of view. An investigator attempting to capture fire patterns around a room requires some type of panoramic imagery.

At its simplest, constructing a panoramic photograph can consist of using a steady tripod, taking a series of overlapping photographs, and then physically overlaying prints to form a mosaic view of the fire scene. Overlaying several photographic shots in the form

FIGURE 4-17 Construction of a panoramic view using individual photographs and stitching software. *Courtesy of D. J. Icove.*

FIGURE 4-18 Separate sequential photos stitched together using PTGui to form a 150° panoramic photo of a large industrial fire scene. *Courtesy of D. J. Icove.*

of a mosaic can compensate for the lack of a true panoramic photograph. For best results a lens with 35 to 55 mm focal length is used, since wide-angle (<35 mm) and telephoto lenses introduce unwanted distortions. Once an investigator has become familiar with the results and limitations of the camera, he or she should work on developing an expertise with panoramic photography (see Figure 4-18).

Simple paste-ups of overlapping images are acceptable, but there are always perspective-shift distortions between views that can disorient the viewer. The advent of panoramic view "stitching" as part of various computer photo editing programs has made it possible to create seamless, perspective-corrected panoramic images from a series of simple still photos without specialized cameras such as digital scanning cameras, which are discussed in detail later in this chapter.

EXAMPLE 4-2 ■ The School Fire

Large multiroom school fires with massive structural destruction are difficult to reconstruct. A fire scene reconstruction was requested of a large school fire that occurred some years ago in the Southeast. A majority of the contents and walls had been removed and placed outside the structure prior to examination of the scene.

The fire scene was reconstructed using a combination of eyewitness statements, raw news videos, scene analysis, and knowledge of plume geometry. Investigation revealed that a fire started on the school office floor adjacent to filing cabinets where student records were stored.

A methodical examination of the debris, the dynamic time sequence of the fire's movement, and the damage caused by the fire confirmed the office to be the room of fire origin. A careful fire scene analysis of pattern damage revealed a roughly elliptical penetration at the floor of the office where the fire originated. The apparent point of fire origin was at the center of this elliptical pattern.

Yellow chalk was used to sketch on the floor the location of the desk, cabinets, and walls. These chalk lines were photographed from the second floor and displayed as a mosaic overlay forming a panoramic view of the fire scene. After the panoramic photos were pieced together, the walls were marked using graphic chart tape on a plastic overlay. Plastic overlays are important for courtroom presentation, since they can be removed if objections are raised to their use. A photograph of this exhibit with the overlay present is shown to the left of the schematic floor plans in Figure 4-19. Such "ghost" walls can today be inserted using the graphics packages described previously. Such displays can also be helpful in confirming witness observations.

One of the most persuasive visual exhibits offering a perspective view is a scale model for courtroom presentations. The overall dimensions of the building being re-created must be carefully defined and accurately captured. Architectural firms often have technicians on staff who can prepare these models. A removable roof on the model should be included to provide access to the building's interior layout.

Specialized photographic techniques can involve the use of forensic (alternate) light sources and suitable filters, ultraviolet or infrared films or imaging systems, and close-up (macro) photography or other techniques to preserve physical evidence such as fingerprints. Some of these are discussed in later sections of this chapter.

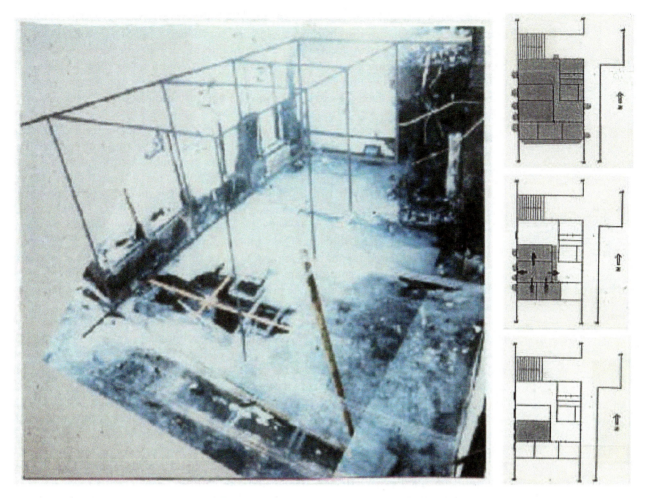

FIGURE 4-19 The panoramic photo overlay from Example 4-1 with graphic chart tape on a plastic overlay that shows missing walls and area of fire origin. Floor plans on right show (bottom to top) the successive fire spread as seen through exterior windows. *Courtesy of D. J. Icove.*

TOTAL STATION SURVEY MAPPING

A new generation of computer-driven surveying technology has come into use for mapping of large interior and exterior scenes. A single reference location is selected and located using a *global positioning system* (GPS) integrated with a *geographic information system* (GIS). The laser sighting unit is then trained on individual features—corners of rooms, curb lines, or evidence locations. The computer notes each feature's direction, angle of elevation, and distance. The program then draws a plan of the scene with extremely accurate dimensions (error of ±1 cm [±0.39 in.] at 150 m [492 ft] is typical) even without reflector posts. Systems by Leica, Sokkia, and Topcon that have been used by traffic accident investigation teams in many police jurisdictions for years are being used by some major arson investigation agencies.

LASER SCANNING SYSTEMS

Laser scanners are gaining ground in documenting fire-related incidents is a technology that combines the accuracy of total station measurements with the completeness of panoramic digital photography. Three-dimensional (3D) laser scanners use LIDAR (LIght Detection And Ranging) technology to quickly capture millions of measurements at a scene. Most scanners are integrated with a high-resolution digital camera (or are coupled

laser scanners ■ The use of technology that combines the accuracy of total station measurements with the completeness of panoramic digital photography. Three-dimensional (3D) laser scanners use LIDAR (LIght Detection And Ranging) technology to quickly capture millions of measurements at a scene.

with an external digital camera) to capture colors and textures of surfaces. This information is then mapped onto the laser measurements to produce a fully navigable 3D virtual reality (VR) (called a *point cloud*) of data of any interior or exterior scene. A scene can be scanned from multiple locations and the images registered (or "stitched") together to produce a comprehensive and visually stunning 3D data collection. Any measurement (in three dimensions) can later be extracted from the data using the software provided with the system.

Laser scanners have become more widely used and lower in cost. Several manufacturers now offer PC-driven scanners that are placed in a room or at an outdoor scene and scan a laser beam vertically and then rotate horizontally (using mirrors) taking distance and angular deflection measurements at the rate of several to many thousand per second. Distance may be measured by time of flight (TOF) (in which the delay time between a pulsed output and its returned reflection determines distance) or phase shift (in which the beam is continuous and the interference between outgoing and returned light waves determines distance). Phase-shift scanners emit a continuous laser beam, and the phase difference between outgoing and returning sine waves determines the distance to an object. Both kinds of systems employ a mirror to direct the beam vertically around a scene while motors drive the scanner to rotate horizontally. Data are typically collected at a rate of tens of thousands of measurements per second.

Several manufacturers now offer high-speed laser scanners, and these systems have become more widely used and are lower in cost. The technology firm 3rd Tech Inc. has developed a laser-scanning device that uses a time-of-flight 5 MW laser range finder that when mounted on a photographer's or surveyor's tripod will scan an entire room, taking 25,000 measurements per second (3rd Tech 2012).

Leica Geosystems (Leica 2012) offers the ScanStation laser scanner, which uses a pulsed green class 3R laser to capture high-accuracy measurements both indoors or outside regardless of lighting conditions (as seen later in Figure 4-23). Their Cyclone software outputs data that are compatible with most third-party computer-aided drafting (CAD) products and can also quickly produce a fully immersive and measurable 3D environment called Leica TruView that is sharable with free viewing software.

The data from such scans can be analyzed and displayed in many ways. Once the registration process is completed in a computer lab, the 3D VR image can be rotated and viewed from any angle or any scanner setup position. Outlines, wireframes, and diagrams can be created. "Hot spot" links to close-up photos, lab reports, notes, or chain-of-custody information can be added. Additional distance and angle measurements can be taken and recorded at any time. Many scanners have already found their way into forensic use via the Department of Justice (DOJ); crime labs; and police, security, and fire agencies. A number of manufacturers offer laser scanners with different capabilities and limitations (Leica 2012; 3rd Tech 2012; and Riegl 2012).

Selection considerations for laser scanners include the following:

Accuracy (resolution): Each time the laser measures a distance, it records a point (scan spot). Once all measurements are merged, the scanner forms a "point cloud," which allows for further viewing and analysis. Point spacing (equivalent to the dpi rating of digital images) can be adjusted by the scanner technician to provide the necessary detail for a given project's needs in relation to distance from the target area. Leica Geosystems has announced the introduction of a new tool that validates the accuracy of forensic 3D laser scans made using the Leica Geosystems ScanStation C10. The new validation tool is designed specifically to help crime laboratories and crime scene units achieve ISO/IEC 17020 and 17025 accreditation. The twin-target pole, placed at the scene in various locations to allow precise calibration and measurements, has a design distance of 1.700 m between the two targets.

To achieve National Institute of Standards and Technology (NIST) traceability, Leica Geosystems has contracted with the Large Scale Coordinate Metrology Group at NIST to perform individual calibrations on specific Leica Geosystems twin-target poles by carefully measuring multiple times between the target centers on each end of the pole (artifact) using a wavelength-compensated helium–neon linear interferometer. NIST then applies serial numbers to each target system and generates a Report of Calibration that is provided to Leica Geosystems customers. "It's the 3D laser scanning equivalent of introducing a scale into a crime scene photograph to provide a control," said Tony Grissim, Leica Geosystems Public Safety and Forensic Account Manager. "The provable accuracy of 3D laser scanner measurements to a known standard is key to forensic credibility in the courtroom, and our new NIST traceable twin-target pole has the added benefit of supporting the ISO/IEC requirements for quality systems and quality control." (Leica 2011).

Court acceptability: Of critical importance to the users of scan data in forensic cases is the technique's successful admissibility in court. As with other forensic tools, the user must be able to explain the device's scientific foundation and validate the accuracy of the measurement. In addition to NIST and ISO, *ASTM E2544-11a: Standard Terminology for Three-Dimensional (3D) Imaging Systems* (ASTM 2011c) offer users assistance with validation should a *Daubert* challenge be executed. Courts are repeatedly ruling in favor of accepting laser-scanned data and animations that have been developed using scanners as the basis for data collection, analysis, and visualization.

Distance range: Some scanners have a minimum range of 2 m (6.6 ft), so their usefulness for small spaces needs to be considered. Others have a maximum range of 16 m (52 ft), so they have limited use in large indoor scenes or exterior scenes. Distance upper limits are dependent on the reflectivity of the target [e.g., 300 m (980 ft) at 90 percent reflectivity versus 134 m (440 ft) at 18 percent reflectivity], so the conditions of the scene and nature of the target are crucial. Ranges are offered up to 280 m (900 ft), but the resolution and accuracy will naturally decline.

Wavelength: Some wavelengths perform better with dark (charred) surfaces, which is a concern to fire investigators. Some wavelengths encounter interference with ambient light and suffer from a condition called "day blindness" and thus require the scanned area to be darkened. Others (e.g., Leica) will work in daylight or full darkness.

Accuracy (distance): Accuracy of distance measurements depends on the type (TOF or phase shift) and range. Accuracy is typically on the order of 4–7 mm (0.16–0.28 in.) at 50 m (160 ft).

Field of view: Some scanners will scan from vertical (90° up to almost 90° down); others can scan only from 45° up to 45° down and require a rescan to capture information from directly overhead.

Scan rate: Most systems will complete a 360° high-resolution scan of a room in 12–30 minutes, taking 4000–50,000 measurements per second for TOF, and 500,000 points per second for phase-based systems.

Safety: All systems use a Class I laser, so eye safety is a concern. Some systems are designed (by scan rate) so that duration of exposure to an unprotected eye will not exceed the federal safety limits. Others require eye protection for users or others in the room.

Ease of "registration" or stitching: Most scenes require scanning from at least two positions. Systems (software) vary in the ease with which integration of these data sets can be done. Some systems produce a data format that can be entered directly in AutoCAD and similar drafting programs.

Such scanners are not inexpensive, but they permit rapid, extremely accurate measurement of all critical features in a fire scene, such as room dimensions, ventilation openings, and structural features (beams, sloping ceilings, and ductwork). For users who do not have the case demands to justify their own system, both public and private sources offer technicians for single-use deployments. The captured images can be revisited at any time to conduct additional measurements. Data from some software systems can be imported into CAD systems for rendering to fixed images or creating a physical model. Readers are encouraged to contact vendors or manufacturer websites (listed at Resource Central) for detailed information. Depending on the wavelength of their lasers, laser scanners can capture variations in surface condition (reflectivity) that may not be apparent in standard color photographs. Compare the "false-color" laser scan of a burned test cubicle with a color photo of the same scene in the following test study.

EXAMPLE 4-3 ▪ Wildland Fire Scene Reconstruction

Problem. Reconstruct a wildland fire scene 2 years postextinguishment. Several factors contributed and need to be considered including a wind event, a damaged tree, and a subsequent wire strike. The "suspected" tree was cut into sections and stored as evidence.

Solution. The steep site was accessed, and the terrain at the scene was laser scanned (Figure 4-20a). An expert arborist assisted with reconstruction of the eight tree sections, via the 3D computer terrain model (Figure 4-20b). The scene and the reconstructed tree were scanned separately with a Leica Geosystems scanner, which produced several million measurements.

Conclusion. A visually and spatially accurate 3D model depicting the tree in position prior to wind damage, wire strike, and fire was produced (Figure 4-20c).

Case study courtesy of Precision Simulations Inc. and Kirk McKinzie CFI.

EXAMPLE 4-4 ▪ Fire Death Documentation

Problem. The San Luis Obispo Fire Investigation Strike Team (SLOFIST) has for several years assembled the nation's finest fire investigators and forensic experts to provide advanced training with test burns specifically designed to promote proper documentation and analysis of scenes involving fire-related human deaths.

Solution. The audience of detectives, coroners, fire and crime investigators, forensic anthropologists, and medical examiners examined several test fires that were also documented via the Leica Geosystems ScanStation 2 scanner. Photo data and laser data were merged to produce a composite (Figure 4-21) that included both the point cloud data from the laser scanner and the color image from the accompanying digital 360° image of the scene immediately after the fire.

The same scene captured by the scanner but with a false-color scale representative of surface reflectivity specific to the laser is shown in Figure 4-22. Note the bright yellow outlines of fire patterns from the most intense fire (highest heat flux), including the pattern above the stove (where the primary fire occurred) and near the floor (from fall-down fires). These patterns can be visually obscured by smoke accumulation in fires that become ventilation limited and are not as well defined in the color photo image.

The "overhead" image or SiteMap in Figure 4-23 was created by combining the data collected from ground level using the Leica ScanStation 2. The image represents the complex "scene" via the merging (registration) of data measured from various points (marked by the black circles).

The virtual reality image/scene can be "toured" by the viewer at any time, and any measurements can be extracted. The Leica Geosystems proprietary TruView software

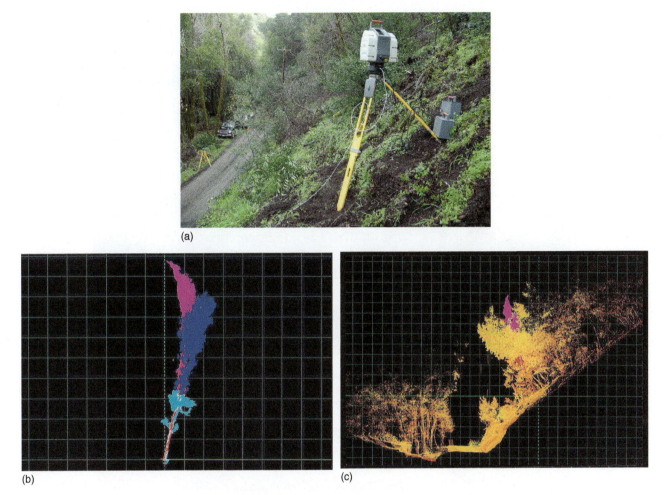

FIGURE 4-20 (a) The scanner and steep terrain near the area of fire origin, 2 years postfire. (b) The main trunk, limbs, and branch structure reassembled in the elevation model. (c) Once the tree was accurately reassembled, it was rotated and properly positioned in elevation, plan, and orientation in the 3D forensic computer model/scene. *Data collection, registration, and images courtesy Precision Simulations Inc. and Kirk McKinzie CFI.*

provides access to scan data via a file transfer protocol (FTP) server from any Internet connection. The free download/plug-in allows investigators, prosecutors, insurance company staff, and other stakeholders to visit, tour, and take measurements of the scene in perpetuity (Figure 4-24). White lines indicate this measuring function. Although the viewer's perspective is slightly skewed owing to the "dome photo" lens, the measurements remain highly accurate (Leica 2012).

One benefit of scanning a scene is that it can complement the use of the NIST Fire Dynamics Simulator (FDS) fire model. Should a particular scene be scanned and a subsequent 3D model constructed, a rapid and exceptionally accurate set of input data can expedite development of models for fire propagation, carbon monoxide, oxygen, heat release, and temperature studies.

An area of future study is the fire patterns associated with calcination. One theory that deserves consideration and study is the possibility that reflectance data from scans of gypsum surfaces that have been exposed to heat energy may provide quantifiable heat flux data. The correlation between laser reflectance at a variety of reflectance levels must be established. Paint, paper, smoke, bare gypsum, and variations in penetrating calcination each have their own reflectance qualities, and there may be a correlative heat flux factor/index relation.

FIGURE 4-21 This composite image includes both the point cloud data from the laser scanner and the color image from the accompanying digital 360° dome photo image and represents the scene appearance immediately after the fire. *Test courtesy of SLOFIST, Inc. Scan data collection courtesy of Tony Grissim, Leica Geosystems. Registration and image analysis courtesy of Robert Gardiner and Kirk McKinzie CFI, both of Precision Simulations, Inc., Grass Valley, CA.*

FIGURE 4-22 The same scanned scene as in Figure 4-20 but with a false-color scale representative of surface reflectivity, specific to the laser. Note the bright yellow outlines of fire patterns from the most intense fire (highest heat flux), including the pattern above the stove (where the primary fire occurred) and near the floor (from fall-down fires). These patterns can be visually obscured by smoke accumulation in fires that become ventilation limited and are not as well defined in the color photo image. *Test courtesy of SLOFIST, Inc. Scan data collection courtesy of Tony Grissim, Leica Geosystems. Registration and image analysis courtesy of Robert Gardiner and Kirk McKinzie CFI, both of Precision Simulations, Inc. Grass Valley, CA.*

FIGURE 4-23 "This "overhead" image was created by combining the data collected from ground level using the Leica Scan Station 2. The image represents the complex "scene" via the merging (registration) of data measured from various points (marked by the black circles). The virtual reality image/scene can be "toured" by the viewer at any time, and any measurements can be extracted. The hot spots designated by the yellow triangles are links to detailed photos and scans of those scenes. *Test courtesy of SLOFIST, Inc. Scan data collection courtesy of Tony Grissim, Leica Geosystems. Registration and image analysis courtesy of Robert Gardiner and Kirk McKinzie CFI, both of Precision Simulations, Inc. Grass Valley, CA.*

FIGURE 4-24 Leica-Geosystems proprietary TruView software allows access to scan data via an FTP server from any Internet connection. The free download/plug-in enables investigators, prosecutors, insurance company staff, and other stakeholders to tour and measure the scene into perpetuity. The white lines indicate measuring function. Although the viewer's perspective is slightly skewed owing to the "dome photo" lens, the measurements remain highly accurate. *Image courtesy of Kirk McKinzie CFI.*

Conclusion. Speed, accuracy, and portability make 3D terrestrial and airborne scanners a natural choice for important scene documentation. With *NFPA 921* and *NFPA 1033* calling for the most complete documentation with the most advanced technology available, it is inevitable that 3D laser scanners will find increased use at fire scenes.

Application of Criminalistics at Fire Scenes

Criminalistics has been defined as the application of the methods and knowledge of the natural sciences (physics, chemistry, biology, botany, etc.) to legal inquiries. Although not commonly in use in the United States until the 1950s, the term derives from the German word *Kriminalistik* (from the 1880s) for forensic evidence analytical techniques. Criminalistics involves the extensive use of physical science to analyze and identify materials but, more important, to compare, classify, and individualize items of physical evidence, often with the intent of establishing or excluding a common origin. This physical evidence can take any form—impressions from shoes, tools, or friction ridge skin (on fingers, palms, or feet), tissue, blood or other physiological fluids, glass, paint, soil, grease, oil, dyes, inks, documents, physical matches, or residues of chemicals associated with fires or explosions. Criminalistics often goes further than simple comparison or identification to include the analysis of human behavior and physical dynamics of force, impact, and transfer, and the reconstruction of scenes in both physical and dynamic aspects.

Criminalistics is the science of examining all these items of physical evidence to link them to a common origin, identify them, or use them in the reconstruction of events. This becomes especially important when the scene is complex or serious (deaths or injuries), and every avenue of information seeking must be explored. Evidence may include physical items such as debris or incendiary devices, witness interviews, casualty injuries or postmortems, and fire scene photographs and sketches.

USE OF CRIMINALISTICS

There is a set of generally accepted forensic guidelines for collecting and preserving physical evidence recovered from fires during a fire scene investigation (DeHaan and Icove 2012). The application of these best practices is derived from procedures used by professional criminalists.

As a simple example, an investigator arrives at a crime scene to find a broken window, an unlatched door, and blood smears—all visible evidence and easily documented (Figure 4-25). The reconstruction of this evidence would determine whether the door was originally locked and its window broken by mechanical force (determined by glass fracture patterns to have come from the outside), and the intruder cut himself/herself on the glass and while fumbling for the lock and latch, left blood smears behind on the door, as shown in Figure 4-25.

Analysis of the blood (or any fingerprints left on the door, glass, or latch) could identify the individual present. Glass found on the clothing could be compared with window glass from the door to confirm a

FIGURE 4-25 The reconstruction shows that the intruder cut himself/herself on the glass and while fumbling for the lock and latch left blood smears behind on the door. Glass fracture patterns document that the entry was made from the outside. *Courtesy of Lamont "Monty" McGill, Gardnerville, NV.*

two-way transfer (suspect blood to scene, scene glass to suspect), and the distribution of cuts or abrasions (and glass/paint fragments) would confirm the method of entry. The services offered by many public and private criminalistics laboratories can be of great help to the fire investigator in reconstructing scenes and events and testing hypotheses about what happened, where, when, and in what sequence. These analyses can possibly link a person with a scene or a victim, and a scene with a person or a vehicle.

GRIDDING

Methodical techniques for assisting in forensic evidence collection and documentation have been widely used by archaeologists when processing field sites. Fire investigators can adopt similar searching and photographic techniques (Bailey 2012). There are several accepted methods for *gridding* and documenting scenes (DeHaan and Icove 2012, chap. 7). Measurements of the location of recovered physical evidence must be reliable and accurate to allow the reconstruction of distances and relationships. Several convenient methods for recording measurements are shown in Figure 4-26.

Coordinate systems aligned perpendicular to major structural walls can aid in the placement of grid lines, especially when the zero point is a corner, usually on the bottom or top left-hand side. *Triangulation* measures distances to an item of interest within a room from two fixed points. *Angular displacement* uses one fixed point and a compass

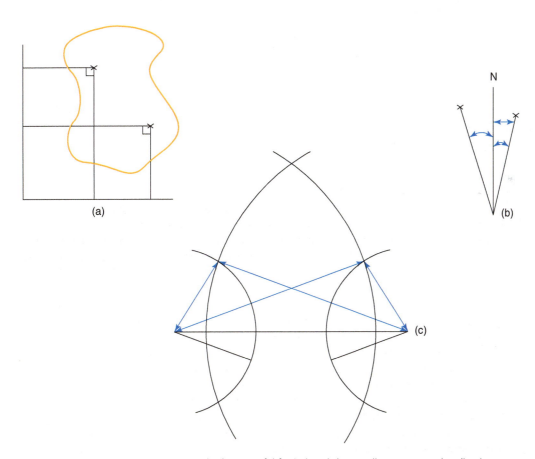

(a)

(b)

(c)

FIGURE 4-26 Several measurement methods are useful for indoor (where walls can serve as baselines) or outdoor scenes. (a) Right-angle transect: baselines at right angles (established with GPS and compass outdoors). (b) Azimuth/baseline: single baseline (in north–south or east–west orientation). Locations are noted by angle and distance from a reference point. (c) Intersecting arcs from two fixed reference points.

FIGURE 4-27 Methodical techniques using gridding assist in forensic evidence collection and documentation.
Courtesy of Lamont "Monty" McGill, McGill Consulting, Gardnerville, NV.

direction and is most appropriate for large outdoor scenes (Wilkinson 2001). A handheld GPS device can be used to establish the location of the fixed reference point as well as compass directions, or for mapping large scenes.

A *grid system* is well suited for large-scale scenes such as explosions where shrapnel and other evidence are hurled away from a central location. A grid system can also be used in small scenes, for example, quadrants within a vehicle. Figure 4-27 illustrates the use of a grid system at a fire scene. A baseline and distance system can also be used for large outdoor scenes, particularly explosions, where there are no natural rectilinear baselines.

Although most scenes can be layered and examined room-by-room, some scenes require more carefully controlled examination. This is particularly true at fire death scenes, where the location of small items is very important, particularly in the vicinity of the body. Using techniques developed in archaeology, the investigator divides the scene into grid squares using the walls as reference base lines. Rope, string, or even chalk can be used to mark out grids Squares within the grid typically are marked using the alphabet along one axis and numerals along the other. Using alphanumeric coordinate identifiers simplifies and reduces the possibility of inadvertently switching the coordinates. These grids may be 0.5–0.8 m (2–3 ft) square in critical areas and as large as 3 × 3 m (10 × 10 ft) in surrounding areas with lighter debris concentrations.

Debris is removed from each grid square, layer by layer, for manual/visual inspection and sieving. Material (including evidence) recovered from each grid is kept in a bag, can, or envelope designated by corresponding number/letter. This ensures that at a later stage the evidence can be placed back to within 0.30 m (1 ft) of its original location. Although time consuming, labor intensive, and costly, this method is the best way of finding and documenting evidence that permits physical reconstruction after the scene has been completely searched.

Iso-damage curves have been used by archaeologists in documenting the impact of fires on historic monuments and have application to the documentation of the thermal impact of modern-day fires. For example, one study evaluated damage sustained by the Parthenon from a fire set by Celts in CE 267 (Tassios 2002). Based on observed burn pattern damage, the study determined that there had been more intense thermal damage within the interior section of the structure than on the exterior. The investigation by Tassios used nondestructive pulse velocity measurements to correlate three quantitative criteria corresponding to the thermal damage.

DOCUMENTATION OF WALLS AND CEILINGS

Although most investigators are diligent about identifying floor coverings and assessing their potential contributions to fire spread, the same diligence is not always utilized in documenting wall and ceiling materials and coverings. As can be seen from the compartment fire data form shown in Figure 4-9, the nature and thickness of walls, ceilings, and their coverings are important for accurate reconstruction of the event.

In some cases, noncombustible walls of concrete, masonry, or stucco will not contribute to the fire spread, but their low thermal conductivity and large thermal mass may affect some stages of the fire's development. Plaster (whether with metal or wood lath) will dehydrate and fail, allowing fire to penetrate into ceilings or wall cavities. Modern gypsum board will resist fire spread for some time if properly installed (typically 15–20 min of direct fire exposure) before it collapses. X- or fire-rated gypsum wallboard is 16 mm ($^5/_8$ in.) or thicker and contains fiberglass fibers to strengthen the gypsum. As a result, it will withstand 30 minutes or longer of direct fire contact (White and Dietenberger 2010).

Walls and ceilings can also be made of solid wood, plywood paneling, fiberboard (high density like Masonite or low density like Celotex), particleboard, OSB (oriented-strand board), or even metal. Each affects fire spread differently. Noncombustible walls can be covered with combustible materials. Thin plywood paneling or low-density cellulose can contribute enormously to fire spread. Karlsson and Quintiere (2000) estimated that lining a room or covering a ceiling with low-density fiberboard would cut the development-to-flashover time in half compared with that for the same room with noncombustible walls or ceilings. These materials will almost guarantee a flashover fire (if adequately ventilated) and can be destroyed so completely as to make their detection difficult.

Fire resistance ratings (1 hr, 2 hr, etc.) are based on laboratory test exposure to a furnace fire whose growth is programmed on a standard time–temperature curve to make the test reproducible, as described in *ASTM E119* (ASTM 2011a) (see Figure 4-28). Such tests do not necessarily replicate real-world compartment fires and are not intended to predict failure times in such fires. However, most fires are less severe (on the average, not at a specific time) than those produced on the standard time–temperature curve. Such fire testing results can often be used as confidence levels for expected performance.

The investigator must be familiar with these materials and how they are installed. Irregular squiggles or zigzag lines of charred adhesive on cement or plasterboard walls are a sure sign that paneling was glued at one time. Remnants of the paneling will usually survive at the toe plate or behind baseboards, plumbing, or electrical fixtures. The small nails used to secure paneling or cellulose tiles are very different from those used to secure gypsum board.

Wood or metal furring strips may be used to install tiles on walls or ceilings. Their presence should be taken as a cue to search out and identify the sometimes fragmentary remains of combustible walls or ceilings. Wall or ceiling tiles are sometimes installed only with dabs of cement on the back side, so patterns of large dark dots on remaining surfaces should be carefully examined. Samples of unburned wall covering should be measured, identified, and retained for later confirmation.

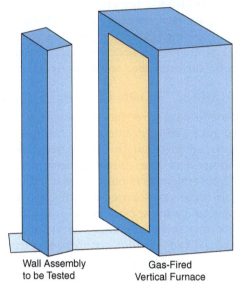

Wall Assembly
to be Tested

Gas-Fired
Vertical Furnace

FIGURE 4-28 The *ASTM E119* test for wall or floor assemblies uses a large vertical or horizontal gas-fired furnace. The test assembly is built against the open face of the furnace, which is operated according to a set time–temperature curve. *Courtesy of NIST.*

In a case of one of the authors' experience, postfire scene photos showed only bare studs, with gypsum board on only one (exterior) face of a common wall. Careful searching revealed that the wall facing the room of origin was covered only in thin plywood veneer paneling that had burned nearly completely and contributed to an extremely intense and fast-growing fire (*Commonwealth of Pennsylvania v. Paul S. Camiolo* 1999).

Although not common in modern structures, the presence of combustible ceilings should not be dismissed. Low-density cellulose ceiling tiles were widely used for many years prior to about 1960. They can be 0.3 m × 0.3 m (12 in. × 12 in.) tiles or 0.61 m × 1.22 m (2 ft × 4 ft) panels. Decorative-edged tongue-and-groove pine or fir boards (sometimes called *beadboard*) were widely used in the late nineteenth to early twentieth centuries for walls and ceilings. Polystyrene ceiling tiles have been widely installed during remodeling of older structures. Such linings will add dramatically to the fuel load and increase the rate of fire development in the room (cutting time to flashover by as much as half, per Karlsson and Quintiere 2000).

The nature and thickness of ceiling materials must be noted. Samples are strongly recommended for later testing. Suspended ceilings are very common in commercial structures, as they can lower ceilings and conceal electrical, plumbing, and HVAC services. These ceilings, although usually noncombustible, allow fire gases to penetrate the large plenum area they offer and spread throughout the concealed space. They typically fail as the steel wires or lightweight steel grid members reach their annealing temperatures and lose their tensile strength. This can occur in as little as 10 minutes once the flames reach the ceiling. For additional examples and discussion, see the NIST experiments as a result of their study into the fire at the Cook County Administration Building (Madrzykowski and Walton 2004).

The interior dimensions of all rooms involved in the fire (by flame or smoke penetration) must be recorded (preferably to the nearest ±50 mm (±2 in.). This includes the heights of various rooms (and not assuming all rooms in a structure to be the same).

Height is often overlooked as a measurement, but is critical when evaluating the size of fire required for flashover or effects of a given initial fire size, estimating smoke-filling rates, visibility, tenability, and the like. Unusual features such as skylights should be measured and photographed.

The nature of the design, shape, and slant of the ceiling itself will affect the spread of smoke and fire. A flat (level), smooth ceiling will allow the smoke and hot gases from the ceiling jet to spread uniformly in all directions; a pitched (slanted) ceiling will direct the majority of the spread upward. Headers above doors, exposed structural beams—even decorative add-on beams—and ceiling-mounted ductwork will limit spread dramatically, in some cases allowing the buildup of a sufficient hot gas layer on the side of the room containing the initial fire to the point of transition to flashover. Open ceiling joists (common in unfinished basements) will direct most of the accumulating hot gases along their length while dramatically reducing (if not preventing) transverse spread into adjacent joist spaces. Such features must be measured and documented.

Efforts should be made to find prefire photos or videos that show the character of the ceiling and walls, as the action of fire and subsequent extinguishment and overhaul activity may have obliterated signs of wall and ceiling coverings. Failing that, interviews should be taken of owners, occupants, guests, customers, visitors, maintenance personnel, and the like, asking them to describe ceiling and wall finish (as well as type and placement of furniture). Examination and documentation of undamaged areas of the building may reveal the original structural features.

LAYERING

At scenes where there has been significant destruction and collapse of furniture, walls, or ceilings, investigators would be well advised to process at least the most critical portions in *layers*. Most evidence of the critical early stages of a fire will be buried beneath the debris from the collapsed ceiling and roof.

Undamaged areas of the building may be surveyed to establish what types of materials to expect. The roof structure can be photographed and its direction of collapse noted and then removed. Roof or ceiling insulation can then be removed noting whether it is loose (blown-in) fiberglass, mineral wool or cellulose, or batting or rolled insulation. Samples should be taken for later identification if ignition or spread through the insulation is considered possible.

The ceiling material can then be identified: gypsum wallboard, lath and plaster (wire or wood lath makes a difference in manner and time of collapse), ceiling tile, plywood, or wood plank. An investigator should never assume that all insulation, ceiling, or lining materials are consistent throughout even a small residence, since repairs, renovations, or additions will usually be made with materials available at the time.

Light fixtures, furnishings, and victims will be found under the ceiling material. Most of the critical evidence will be found between the ceiling and the floor covering, but even here, the sequence of positioning may be of importance.

Window glass will tend to break and collapse when the temperatures or heat flux become high enough during the fire's development. This usually occurs after the smoke products have condensed on the glass, and the fracture pattern would be expected to be thermal rather than mechanical (unless the glass is toughened safety glass) (DeHaan and Icove 2012, chap. 7). Glass falling inward may well fall on furnishings or floors that are already fire damaged. If a window is broken before the fire reaches it, the fracture pattern will be mechanical in appearance, and the glass, bearing few or no soot deposits, may fall and protect unburned materials beneath. (Keep in mind that radiant heat in a subsequent postflashover may melt and char even protected material beneath the glass.) Breakage of glass in locations where fire exposure does not seem to have been sufficiently intense must be examined for other causes.

The location of first collapse of ceiling or wall covering may be estimated by careful examination of the debris at this stage, and should be photographed before further excavation. As the furnishings are documented and removed, the nature of floors and floor coverings can be determined. The distribution of carpet, tile, bare wood, composite floor coverings, and vinyl or asbestos (tile or sheet) should be noted. Comparison samples of carpet, floor covering, and underlayment should be taken in areas of suspected origin.

Testing has shown that carpet fiber content cannot accurately be estimated from observation alone, and even using a match or lighter flame will reveal only whether the face yarn is synthetic or natural fiber (For additional details, see Chapter 8 on fire testing for application of *NFPA 705* and other tests) (NFPA 2009a). Fire tests have further demonstrated that a carpet may not support flame spread alone but if installed over a particular type of pad, may burn readily. It is always best to recover and preserve a small (at least 0.15 m × 0.15 m [6 in. × 6 in.]) sample of unburned carpet and pad for later identification. Such samples can be recovered in most scenes from under large furniture or appliances or in protected corners of the room. Such samples can help forensic testing by being used as a comparison to establish what volatiles they contain or may yield on burning.

Regarding the process of layering, archaeologists use a systematic approach of coordinate systems that account for not only the location but also the depth of where an object is found. Their excavations are carried out layer by layer, so the depth of each is known. This approach can be helpful when layering debris at fire scenes that has fallen onto an area of fire origin. Such cases include multistory buildings, where collapsed floors bury important evidence. In some cases fallen debris often preserves evidence on lower floors.

As previously noted, it is important to place these layers spatially in the documentation process. If forensic evidence collection and documentation are viewed as a scientific endeavor, the use of archaeological techniques will only enhance this effort. In both fire scene investigations and archaeological excavations, it is important to document the physical structures, three-dimensional layers of stratified debris, and the location of critical evidence that is preserved, altered, or destroyed by the excavation process. These surfaces are known as *units of stratification*.

The standard of care in the archaeological community for scene excavations is the use of the *Harris Matrix*, named after British archaeologist Dr. Edward Cecil Harris, who invented it in 1973 (Harris 2012). This matrix documentation technique shows the chronological relationships (temporal sequences) among these stratified sequence of layers excavated at a site in logically presented and abstracted form.

The Harris Matrix has gained wide acceptance in the archaelogical community. For example, in Belgium the Harris Matrix has been added to a set of universal requirements that are to be fulfilled during the processing of every archaelogical site excavation and will soon be required by law (Harris 2012, personal communications).

There are computer packages, developed for both the Windows and Mac OSX operating systems, that can assist in the documentation and construction of a Harris Matrix. Called the Harris Matrix Composer (HMC, Version 1.6b), the program builds and administers a representation of an archaeological stratification in the form of a Harris Matrix (Harris 2012).

SIEVING

In the ashes, all evidence seems to be the same shades of white, gray, or black and not readily distinguishable from fire debris. A manual and visual search may overlook evidence critical to a complete reconstruction. *Wet sieving* of the debris affords a better chance of detecting small items such as glass fragments, projectiles, keys, or jewelry that may go unnoticed in dry ash. Note that bone fragments that are highly calcined by fire exposure can disintegrate if exposed to water, so areas around hands and feet of burned bodies should be manually searched and dry sieved.

Sieving is most effective when done with at least three sieve frames stacked together using 25-, 12.7-, and 6.35-mm (1-, 0.5-, and 0.25-in.) mesh (some examiners will use a fourth sieve with window screen if searching for tiny tooth or bone fragments). Debris is placed in the correct grid and agitated, or a garden hose or hose reel line is used to wash debris through the sieves. The debris should never be pushed through by hand, since that can shatter bones and teeth. Figure 4-29 is an example of a sieve.

Wet sieving for general evidence is preferred, since the water washes off the gray ash and makes small objects visible by color or reflectance. If the objects recovered can be visually identified, they can be placed directly into evidence bags. Items that cannot be readily identified should be kept in a separate container, labeled as to grid of recovery, until they can be properly analyzed.

PRESERVATION

Fire exposure causes materials to become fragile and brittle. This is especially true of copper wiring and electrical insulation and components. Any recovery starts with thorough photographic documentation before an attempt is made to move the object.

Shrink wrap or cling wrap (Saran wrap) is very useful for wrapping large items to keep loose pieces together. Wiring can be preserved on 1 × 4 lumber 4–8 ft long, affixed with cable ties or wraps with the ends numbered and distances from a reference point labeled on the wood. Fragile wiring or components can be protected with cling wrap. Smaller lengths of wiring or wiring from an appliance can be tie-wrapped to a piece of corrugated cardboard. Colored or numbered tape can be used around ends of wires to denote the circuit to which they were connected.

When a scene is gridded off for searching, numbered tarps (plastic) can be used for materials recovered from each grid. If the scene is not gridded off, a separate tarp can be used for each room or each sector of a scene. Large plastic tarps can be marked off with waterproof tape to produce a full-scale reproduction of the room (see DeHaan and Icove 2012 for an example). This technique has been used to reconstruct positions of furniture and other evidence, and even for court display (Rich 2007).

FIGURE 4-29 Sieving assists the investigator in collecting evidence that is not readily distinguishable from fire debris. Searching by hand and visual examination only may cause evidence critical to a complete reconstruction to be overlooked. *Courtesy of Lamont "Monty" McGill, Gardnerville, NV.*

IMPRESSION EVIDENCE

Impression evidence is a general term for the transfer of pattern or contour information from one surface to another. This transfer may take the form of one harder material deforming a softer material on contact. Examples are shoe impressions in soft clay outside a window or a striated rough cut on a piece of wood made by the ragged edge of a hatchet blade. This transfer can also occur when a transfer medium preserves the two-dimensional shape of contoured surfaces that come into contact with it. For example, a dusty shoe can leave an identifiable print on a clean surface, and a clean shoe can remove dust from a dirty surface and leave the same information. Photography is the primary means of documenting most impressions, but care must be taken not to distort the image in the photograph.

Designed by the American Board of Forensic Odontology (ABFO), the ABFO L-shaped No. 2 scale, measuring 10 cm × 10 cm, has become the standard scale for recording through photography fingerprints, wounds, blood spatters, and similar small evidence, as in Figure 4-30. The three circles are useful in compensating for distortion resulting from photographs taken from oblique camera angles.

A larger scale, shown in Figure 4-31, is ideal for shoe prints, since it helps prevent distortion (LeMay 2002). Originally made to FBI specifications, the "Bureau Scale," as it is referred to, contains a 15 cm × 30 cm L-shaped scale. Alternating black and white bands provide a visual reference, and circled crosshairs assist in checking and correcting perspective distortion in photographs.

Fingerprints and shoe prints are commonly found forms of impression evidence. When impression evidence is found, it should be documented in place by photograph. Whenever possible, the object bearing the impression should be recovered and submitted to the lab. The object is always preferable to a photo or cast, but if the object is not removable, photos with scales, or casting or lifting by adhesive or electrostatic lifter will be necessary.

Items bearing tool marks should also be collected and individually wrapped in paper packages to protect their surfaces. Similar collection guidelines apply to the suspected tools. Wrap each tool separately to prevent shifting or damage during submission to the

FIGURE 4-30 The ABFO #2 scale is best suited for photographing small objects, such as fingerprints, tool marks, or matchbooks. *Courtesy of J. D. DeHaan.*

FIGURE 4-31 An example of a forensic photographic scale alongside impression evidence for a shoe by the transfer of pattern or contour information from one surface to another. © *Safariland, Courtesy of Forensics Source™ by Permission.*

laboratory. Place each tool in a separate envelope or box with a folded sheet of paper over the end of the tool to minimize damage and loss of trace evidence adhering to the surfaces and to prevent rusting.

Fingerprints A finger touching a clean surface like glass or metal can leave a reproduction of its friction ridge contours on the surface by the transfer of the nearly invisible oils, fats, and sweat secretions normally found on the skin. Such patterns are often called *latent* because they are not readily visible to the unaided eye and require some sort of physical, chemical, or optical treatment to make them visible and recordable (and comparable to "inked" record prints).

The skin can also be contaminated with blood, food, grease, or paint that leaves behind a visible or *patent* impression. Even though the skin is pliable and deforms on contact with most other materials, it can deform soft materials such as cheese, chocolate, solvent- or heat-softened plastic or paint, and window putty and leave a three-dimensional or molded "plastic" impression of its ridge detail. The same considerations apply to other deformable materials like rubber shoe soles, gloves, tires, or cloth; each may deform softer materials or leave a pattern on harder ones through transfer of some intermediate medium or even residues of itself.

Impressions of friction ridge features from fingers, palms, and feet can and do survive fires. The multitude of chemical, physical, and optical techniques available today to enhance fingerprints (both latent and patent) has made it possible to recover prints from difficult, textured, or contaminated surfaces. Heat alone can cause the constituents in skin oils to darken or even react with the surface beneath, producing a patent impression from a latent one (DeHaan and Icove 2012, 319; Lennard 2007; Deans 2006; Nic Daéid 2004).

Fingerprints are unique to an individual, they are permanent, everyone has them, and chances for transfer on contact are good (with something at the scene, with an incendiary device, or with a container used to transport or distribute ignitable liquids). In addition, there are vast long-term repositories of reference prints (which include noncriminals as well as criminals), and the advent of powerful and fast computers has made it possible to scan millions of record prints in a short time to establish a short list of candidate matches. These databases are designed to run single prints, and even partial prints. Some systems can process palm prints. If fire exposure has not been so severe as to melt or severely char the base material, it is worth considering examination for fingerprints. Handling must be very careful and kept to an absolute minimum. The material must be transported very carefully to the lab (preferably by hand) using containers or devices that minimize contact with potential print-bearing surfaces.

Water contamination from condensation, hose stream, or environmental exposure once meant that prints on paper or cardboard would be impossible to develop (since the amino acids detected by ninhydrin processing are water soluble). The advent of physical developer (which reacts with the fatty constituents that are not water soluble) made it possible to process wet paper or cardboard. Small-particle reagent (SPR) allows processing on nonporous surfaces like metal, glass, and fiberglass even while still wet. Forensic light sources (multiple wavelength, high intensity) and a variety of chemicals and powders make it possible to recover prints from textured or contaminated surfaces. Fingerprint experts such as Jack Deans have documented that modern optical, chemical, and physical methods are capable of developing latents on all manner of surfaces after fire exposure (Deans 2006; Bleay, Bradshaw, and Moore 2006; DeHaan and Icove 2012, chap. 14).

Soot can often be washed off smooth metal or glass surfaces with running water, leaving behind prints "developed" by the soot carbon to be photographed or lifted. A recent Malaysian study assessed the effectiveness of soot removal techniques on glass fire debris without affecting the fingerprints found on the evidence (Ahmad et al. 2011).

Fingerprints in blood can be enhanced with amido black or leucocrystal violet (LCV) sprays, which react chemically with blood to form a dark blue–purple-colored product. The solvent for the LCV has been seen to help rinse overlying soot away from the surface. In one case where two fires had been set in a house, the LCV solution washed soot away, revealing blood spatters on walls where a resident had been bludgeoned to death some two years previously (the fires having been subsequently set to simulate drug gang activity).

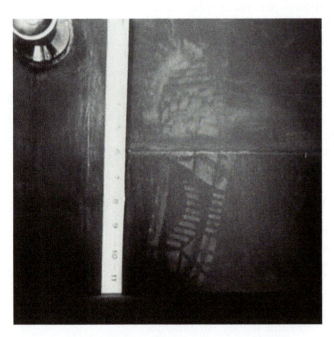

FIGURE 4-32 Shoe prints left behind on doors that are forcibly kicked in may be enhanced by the action of the fire. In this case the dusty shoe print was lightened by heat exposure, and the wood around it scorched, increasing its contrast and readability. *Photo by Joe Konefal, by permission.*

Charred or burned paper documents are important to a fire scene examination, particularly when business records and important documents are involved. Owing to their fragile condition, fire-damaged documents should be placed on soft cotton sheets in rigid containers and hand-carried to the laboratory for examination. Papers should not be treated with any lacquer or coating if they are to be processed for identification or comparison.

Although shoe prints are often compromised by the foot traffic of emergency personnel, tires of vehicles, water, or structural changes, shoe prints can survive fires if they are left on surfaces that have not been destroyed by fire or heat. Shoe prints left behind on doors that are forcibly kicked in may actually be enhanced by the action of the fire. In the example case shown in Figure 4-32, the dusty shoe print was lightened by heat exposure, and the wood around it scorched, increasing its contrast and readability.

Shoe prints are second only to fingerprints in their potential value for linking a person with a scene in that shoes are often (although not always) individualized by the accidental features they acquire through use and damage, they tend not to be changed or discarded for prolonged periods of time, and shoe prints are likely to be left behind at many scenes, because a building is nearly always entered (if not approached) on foot. Paper or cardboard can bear latent shoe impressions (developable by physical developer chemical treatment even after water exposure) or patent impressions in soil, dust, or blood.

In addition to identification, shoe prints can offer information about where someone entered or left a room or building, where he/she went, and in what sequence. If two or more people were present, shoe prints can be used to establish where each of the perpetrators went in the building.

TRACE EVIDENCE FOUND ON CLOTHING AND SHOES

Trace evidence such as soil, glass, paint chips, metal fragments, chemicals, and hairs and fibers may cling to the shoes or clothing of a suspect. Liquids may also soak into the clothing or remain on the bottom of the individual's shoes. Glass, plasterboard, sawdust, and metal shavings have all been found embedded in shoes to offer a link with a scene (as long as comparison samples are recovered). Footwear and clothing can also absorb residues of flammable liquid or chemical accelerants used in arson attacks.

A New Zealand study illustrated the necessity of quickly analyzing clothing and shoes of suspected arsonists for the presence of ignitable liquids. The study found that detectable amounts of petrol were transferred to the clothing and shoes of a person during the action of pouring it around the room. This study also addressed the pouring heights and floor surfaces. Results of the study showed that petrol was always transferred to the shoes and often transferred to both the upper and the lower clothing, but a sample must be seized and properly packaged quickly (Coulson and Morgan-Smith 2000; Coulson et al. 2008). Tests using a carbon strip extraction method found that 10 mL (0.34 oz or 2 tsp) of gasoline on clothing worn continuously, evaporated much more quickly than on the same clothing left on the lab bench at 20°C–25°C (68°C–77°F) and was barely detectable after only 4 hours of wear (Morgan-Smith 2000). Clothing must be recovered and packaged soon after exposure.

The investigator must be aware that the solvents used in glues holding footwear together (particularly athletic shoes) can interfere with identification of ignitable liquid residues, as they can produce false positives (some manufacturers have been known to use gasoline as a substitute glue solvent). Preservation of both shoes separately from other evidence and from each other in vapor-tight containers is essential to prevent cross-contamination (Lentini, Dolan, and Cherry 2000). Owing to the unpredictable presence of gasoline and other ignitable liquids in footwear, many forensic laboratories are reluctant to analyze them.

An Australian transfer study examined the possibility of a natural occurrence of petrol residues tracked inside a vehicle on the shoes of drivers or passengers. The results showed that small quantities of petrol (500 μL) could be detected after 24 hours, but evaporation prevented it from being discovered after one week on the shoes. The researchers concluded that it would be significant if fresh or slightly evaporated petrol is found on the car carpet (Cavanagh-Steer et al. 2005).

A separate study by the same Australian researchers found that car carpets, since they may be manufactured from petroleum feed stocks, might decompose when heated and produce volatile organic compounds that produce background interference when testing for the presence of petrol. The researchers found, however, that these components produced chromatographic patterns that were distinguished from those produced by petrol. Both studies underscore the necessity to obtain reference samples to eliminate or account for background interference (Cavanagh, Pasquier, and Lennard 2002).

Of note is the work by Armstrong et al. (2004). In field experiments, a specified amount of gasoline was poured at designated locations. The subject then stepped into the pour location and walked the length of the testing area. Gasoline on contaminated footwear (fire boots, work boots, and tennis shoes) was found no farther than two steps away from the site of the pour on carpet.

Shoes can also pick up material from scenes. One notable example is a case in which a wad of charred drapery fabric was found melted into the bottom of a heat-damaged athletic shoe. The wearer of the shoe was dumped in the parking lot of a county hospital suffering from extensive second-degree burns from head to foot with his clothing badly burned and still smoldering. His claim of having injured himself when the carburetor of his car engine malfunctioned was not supported by the distribution of burns on him or his clothing. The charred drapery remnant was matched to drapery material recovered from a nearby apartment fire that had been ignited with a substantial quantity of gasoline poured through several rooms. Clearly, the gasoline had ignited prematurely, and the suspect had had to escape through the burning structure. The observation that the synthetic uppers of the shoes were melted and scorched, and the drapery fragment was melted into the bottom of the fire-damaged shoe, precluded the excuse that he was passing by the fire scene and trod in the debris well after the event.

The general forensic guidelines for handling items of clothing stipulate that they should be marked directly on the waistband, pocket, and collar with the investigator's initials and date found. If being preserved for blood or trace evidence, these items must be packed separately into clean paper bags, with each item wrapped separately in clean paper. If they are damp (water or blood), they must be allowed to air-dry before being packaged. Clothing should only be handled by someone wearing clear nitrite or latex gloves to prevent DNA transfer cross contamination.

If flammable liquid residues are suspected to be present, the clothing must be sealed separately in vapor-tight cans or sealed nylon or AMPAC FireDebris bags. AMPAC Fire Debris polymer for fire debris packaging, useful in a wide variety of applications for evidence collection, was reintroduced into the U.S. market in July 2010 and is being evaluated. Preliminary tests showed that heat-sealed AMPAC FireDebris bags will preserve even light hydrocarbons and will not introduce foreign trace volatiles (AAFS 2011). When properly heat-sealed, these airtight and puncture-resistant polyester pouches are ideal for packaging and storing chemicals, controlled substances, and related paraphernalia while preventing cross contamination and protecting individuals from exposure to dangerous substances. (See www.ampaconline.com/.)

When dealing with glass, the investigator should collect as many fragments as possible and place them in paper bags, boxes, or envelopes for later physical reconstruction or reassembly. The fragments should be packaged so that their movement within the containers is kept to a minimum. Paint chips, particularly those at least 1.5 cm² (0.5 in.²) large should be placed in pillboxes, paper envelopes, cellophane, or plastic bags and carefully sealed. Smaller glass or paint chips should be placed in a folded paper bindle and then sealed in a paper envelope. All trace evidence must be kept separate from any control or comparison samples to avoid cross contamination.

Trace evidence sometimes consists of hairs and fibers. Comparison samples from victims or suspects may reveal a direct or indirect relationship. Loose hair combings and clumps of fibers cut from various areas should be packaged separately in pillboxes, paper envelopes, cellophane, or plastic bags. The outside of the container should be sealed and labeled. Hairs from pets in structures may also be transferred to persons, so comparison samples may be suggested.

DEBRIS CONTAINING SUSPECTED VOLATILES

Items containing suspected ignitable liquids should be sealed in clean metal cans, glass jars, or specialized polymeric (AMPAC FireDebris or nylon) bags that have been developed for fire debris preservation, as shown in Figure 4-33. Common polyethylene plastic or paper bags should not be used, since they are porous and allow evaporation or cross contamination and may contain volatile chemicals that may contaminate the evidence, resulting in false-positive or false-negative results when subjected to laboratory analysis. The container should be filled no more than three-fourths full, using clean tools to prevent contamination. Clean disposable plastic, isoprene, or latex gloves should be worn when collecting the debris and then discarded after each sample is taken.

Investigators should seek to collect and preserve comparison samples of uncontaminated flammable and combustible liquids from the scene. Samples of suspected liquids measuring up to 1 pt should be sealed in Teflon-sealed glass vials, glass bottles, or jars with Bakelite or metal tops (preferably with Teflon seals), or metal cans. Such comparison samples should be carefully labeled as to their source if they are not submitted in their original containers. They must be stored and shipped separately from scene samples to avoid any chance of contamination.

FIGURE 4-33 Clean new airtight metal cans, clean glass jars, or special polymeric bags (such as AMPAC FireDebris) are all suitable containers for fire debris evidence. *Courtesy of J. D. DeHaan.*

The collecting officer should label each container with a brief description of the contents and the location from which the sample was collected, per *ASTM E1459* (ASTM 2005b). Collectors are cautioned not to use rubber stoppers or jars with rubber seals, since gasoline, paint thinners, and other volatile liquids compromise the sample by dissolving the seals. Any odor present or reading obtained on a hydrocarbon detector should be noted prior to sealing the container. Plastic bottles are inappropriate containers for debris, liquid, or comparison samples. Various techniques are available for collecting liquids from concrete floors using absorbents such as flour, sweeping compounds, and calcium carbonate or "kitty litter" (Tontarski 1985; Mann and Putaansuu 2006; Nowlan et al. 2007; DeHaan and Icove 2012, chap. 14). Investigators should remember to submit a sample of the type of absorbent material used along with their scene samples for laboratory analysis (Stauffer, Dolan, and Newman 2008, 170).

Samples containing garden soil should be frozen as soon as possible to minimize possible degradation of hydrocarbons by microbial action and kept frozen until submission to the laboratory. All other samples should be kept as cool as possible and submitted to a testing laboratory as soon as possible. Further details are included in *Kirk's Fire Investigation,* 7th ed. (DeHaan and Icove 2012).

UV DETECTION OF PETROLEUM ACCELERANTS

Since the 1950s, *ultraviolet (UV) fluorescence* has been touted as a way of discovering deposits of some hydrocarbon liquids at fire scenes. Unfortunately, simple fluorescence observation (observing the visible light emitted when some materials are exposed to UV) is not reliable for that purpose. Many materials naturally fluoresce, and pyrolysis produces a large number of complex aromatic hydrocarbons as breakdown products that fluoresce and obscure and interfere with any fluorescence from "foreign" accelerants. Some flammable liquids produce no fluorescence at all.

A newly reported innovation promises some success in using UV light, but it involves two electronic manipulations. If the UV source is pulsed rapidly and the electronic detector is gated so it observes only the target after a preset delay, the decay time of most pyrolysis products can be discriminated from those of petroleum products. Another method is to use an image-intensified charge-coupled device (CCD) detector to look at emissions (fluorescence) at wavelengths and intensities not visible to the unaided eye. A time-resolved fluorescence system pulsed laser offers a promising future for the UV detection of stains of ignitable liquid (Saitoh and Takeuchi 2005).

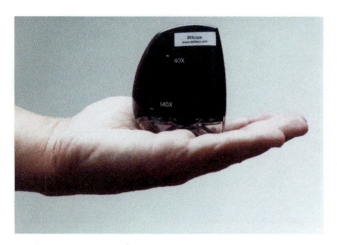

FIGURE 4-34 A handheld digital microscope (magnification 40X–140X) has built-in LED lighting. Its images are displayed on a laptop computer and captured as movie or still images. *Courtesy of Zarbeco LLC.*

DIGITAL MICROSCOPES

Many investigators carry an eye loupe or linen tester magnifier to scenes to enable them to examine small objects, but capturing those images is a problem (which may be critical if the object is likely to change with time owing to evaporation, handling, or storage). A handheld digital microscope (MiScope) is available that combines built-in LED lighting, precision optics, and a digital movie camera (Zarbeco 2012). The microscope offers 40X–140X magnification (field of view 6×8 mm to 1.5×2 mm), and the image is fed directly to a laptop computer (via a USB port) for examination, capture, and labeling (as still, movie, or time-lapse images). It is available for less than $300. A high-resolution version (MiScope MP offering 12X–140X magnification with 2.3-μm resolution is available for under $700 (see Figure 4-34).

PORTABLE X-RAY SYSTEMS

Laboratory X-ray systems such as Faxitron (2012) have been used by crime labs and engineering labs for many years to evaluate evidence of all types after recovery. These are similar to medical X-ray systems using variable voltages and exposure times to suit a wide variety of target materials.

Portable X-ray systems have been used for some years by bomb disposal squads to evaluate suspected devices in situ, but the recent developments in spoliation issues have made on-scene X-ray examination much more desirable for the fire investigator. One such system is a battery-powered X-ray system weighing less than 2.5 kg (5 lb). The XR150 system by Golden Engineering (Golden 2012) uses electrically generated pulses of X-rays of very short duration (50 ns) but very high power (150 kV) (see Figure 4-35). The images are accumulated based on the type and thickness of the material (1–2 pulses for paper envelopes, 100 for steel up to 1 cm (0.4 in.) thick. Images can be produced on a "green screen" fluoroscope, a "live" digital image system, or on a conventional 20 cm × 25 cm (8 in. × 10 in.) Polaroid film sheet. Such devices allow a trained and qualified investigator to assess the internal condition of electrical devices, appliances, or mechanical systems before they are disturbed or removed from the scene (thus minimizing risk of damage or loss). Such systems are currently available for under $5000.

PORTABLE X-RAY FLUORESCENCE

Another valuable technique, X-ray fluorescence elemental analysis, long used in the laboratory, can now be performed on a portable handheld unit the size of a power drill (as shown in Figure 4-36). It uses low-energy X-ray (10–40 kV) beams focused on a nearby surface to identify all but the lightest elements in one scan of a few seconds. It can be used to help detect and identify chemical incendiaries, low explosives, bromine-based flame retardants, chemical treatments, coatings, and fragments of solids of all types (Innovxsys 2012).

ThermoFisher Scientific (2012) also produces a handheld X-ray fluorescence instrument for elemental analysis. The object or surface to be tested is brought into contact with the window of the Niton XL device. A 2 W X-ray tube is triggered, and a spectrum of the elements present is displayed on the built-in screen or (via USB connection or integrated Bluetooth communications) transferred directly to a PC or other storage device. The standard system permits identification of elements from chlorine to uranium. A helium

(a)

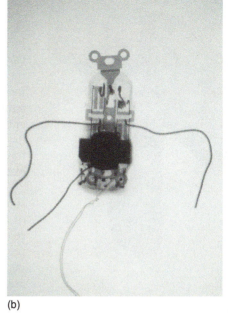

(b)

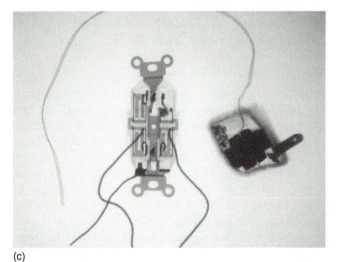

(c)

FIGURE 4-35 (a) A portable X-ray unit can take images of electrical equipment and appliances at a scene using a 150 kV X-ray source. (b) Its image can be captured on film, fluoroscope screen, or digital devices. *Courtesy of Golden Engineering Inc.* (c) A typical X-ray of a household receptacle is shown on the left, plug-in power supply on right. *Courtesy of Golden Engineering Inc.*

purge of the detector extends its sensitivity down to magnesium, making it very useful for characterizing possible residues of incendiary or inorganic explosive materials. The Niton system also permits the choice of full-area analysis or a 3-mm "small-spot" analysis. An optional internal digital camera permits documentation of the sample being analyzed.

INFRARED VIDEO THERMAL IMAGING

Pioneered by Inframetrics, which was acquired by FLIR Systems (FLIR 2012), false-color infrared video recording has been in use in industrial and military applications for more than a decade, to remotely and quickly survey high-tension line insulators for heating from leakage currents, transformers for overheating, and chemical delivery pipes and storage tanks for leaks, and to measure operating temperatures of surfaces of aircraft or vehicles.

As applied to fire research, infrared video thermal imaging allows remote and accurate measurement of the surface temperatures of fuel surfaces or masses of hot gases or of the external surfaces of walls, windows, or other structural elements as they are exposed to fires.

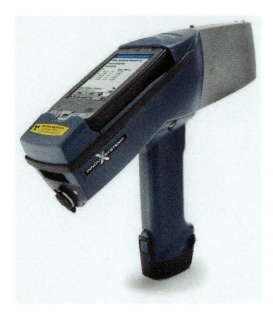

FIGURE 4-36 A handheld energy-dispersive X-ray fluorescence (XRF) analyzer can simultaneously identify up to 25 elements in a sample at a scene, helping identify chemical incendiaries, metal alloys, and other materials. *Courtesy of Innov-X Systems Inc.*

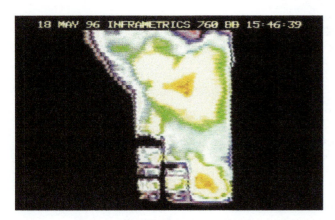

FIGURE 4-37 DeHaan (2001) used Inframetrics infrared (FLIR) video to study the dynamics of chaotic postflashover fires. *Courtesy of J. D. DeHaan.*

A very early version was used by Helmut Brosz in 1991 in monitoring fire development in compartments (Brosz, Posey, and DeHaan, unpublished). Although DeHaan (1995, 1999) used Inframetrics equipment to measure the surface temperatures of carpet during the evaporation of volatile fuels, very few papers on the topic have been published in the fire research literature. Infrared video imaging can distinguish areas of a surface whose temperatures differ by only 0.25°C (0.5°F) in static images. Its applicability to capturing transient events such as the chaotic movement of the high-temperature flames in postflashover fires has been demonstrated (DeHaan 2001). See Figure 4-37 for an example.

The high cost of the necessary equipment has been reduced significantly in recent years with the dramatically increased power of low-cost computer components. It is to be hoped that more researchers will use this type of technology to study surface temperatures that presently are impractical to measure accurately using traditional hard-wire thermocouple devices. Several versions are made today by FLIR Systems (FLIR 2012).

MRI AND CT IMAGING

Progress in magnetic resonance imaging (MRI) and computerized tomography (CT, or multiple-position X-ray imaging) has been very rapid. Recently, MRI and multiple-slice CT were demonstrated in the examination of a very badly charred body to establish the presence or absence of fractured bones and reconstruction of impact injuries as well as to document the presence of surgical implants and the like (Thali et al. 2003). Such techniques have also been suggested for use in examination of melted or charred artifacts from fire scenes where nondestructive testing for pockets of liquid or similar nonradiopaque inclusions would be desirable. Although expensive, such techniques may provide answers where X-rays cannot.

DNA ON MOLOTOV COCKTAILS

Recent work by the Forensic Science Centre at the University of Strathclyde has shown that owing to the STR (short tandem repeat) methods now available for deoxyribonucleic acid (DNA) analysis, identifiable DNA from saliva can be found on the necks of gasoline-filled Molotov cocktails (petrol bombs). Using both sterilized bottles with 50 µL (2 drops) of saliva deposited on the exterior and a random selection of bottles that had been drunk from, researchers recovered identifiable DNA even after gasoline had been poured in and a wick inserted. When the devices were thrown to explode and burn, identifiable (4–7 loci) DNA was found on about 50 percent of the bottle necks, and full DNA profiles were identifiable on another 25 percent of the burned bottles (Mann, Nic Daéid, and Linacre 2003).

DNA RECOVERY IN BLOOD-SPATTERED FIRE SCENES

Sometimes a suspect has set a fire to cover up or destroy possible evidence of a homicide, and bloodstain patterns spattered on surfaces are removed for later DNA analysis.

Research by ATF Fire Reseach Laboratory shows that DNA can be recovered from these fire scenes, even in some cases when the blood spatter or stains have been subjected to elevated temparatures (Tontarski et al. 2009).

The ATF Laboratory found that in similated multiroom fires, bloodstain patterns remained visible and intact inside the structure and on furnishings unless the surface that held the blood was totally burned away. For the most part, the recovered DNA seemed to be unaffected by the heat, until the temperature was 800°C (1500°F) or greater. At this temperature, no DNA profiles were obtained.

ATF cautions that in cases when presumptive chemical tests for the presence of blood were conducted prior to DNA testing, there could be a failure to obtain useful typing results. They recommend that recovery of DNA is most successful in instances where the blood is removed by methods using either swabbing or cutting/scraping.

Photography

Fire investigators should both photograph and sketch carefully even the seemingly least significant fire scene detail. *NPFA 921* notes that photography, whether still or video, is the most efficient reminder of what a fire investigator saw while examining the fire scene. Photography is used to substantiate witness statements, document reports, provide documentary exhibits for courtroom presentations, and ensure the integrity of the debris delayering process. Critical evidence, including complex fire patterns, may also be more visible later than when originally examined (see *NFPA 921*, 2011 ed., pt. 15.2.1) (NFPA 2011).

Other standards of care also mention how photography should be used in documenting fire scenes. *ASTM E1020-06,* pt. 5.1.2 (ASTM 2006), states that investigators must take photographs "which accurately and fairly identify and depict the scene, the items, or systems involved in the incident, and the post-incident conditions." *ASTM E1020* also provides additional guidance in that "photographs should be taken from many directions and should include overall site views, overall item and system views, intermediate views, and close-up views."

Photography, especially digital, is a very inexpensive investigative step in documenting fire scenes. Some benchmark estimates vary as to the actual average direct costs of performing a preliminary scene investigation, yet photography may be the least costly (Icove, Wherry, and Schroeder 1998). In present-day digital cameras (to be discussed later in detail), a 10 gigabyte (GB) photographic image can be stored as a 4.5 megabyte (MB) JPEG image. Therefore, a 1 GB digital storage card will hold about 200 images. With digital storage cards for cameras retailing at about $1.00/GB, the cost per image would be $0.005 (half a cent).

Fire investigators are required by *NFPA 1033* (NFPA 2009b) to be knowledgeable in photographic documentation of fire scenes (see *NFPA 1033*, 2009 ed., pt. 4.3.2). Fire investigators are usually not required to qualify as expert photographers, but they must be able to understand and utilize their equipment to the maximum benefit (Berrin 1977). Thus, the following items should be included as minimum equipment for forensic fire scene photography.

DOCUMENTATION AND STORAGE

The "Photograph Log" form (Figure 4-7) is used to record the description, frame, and unique storage media number (for example, the serial number of digital memory card, CD-ROM, or DVD) of each photograph or video taken at the fire scene. The form is designed to be filled out as the photographs are taken. The frame and storage media numbers are later used on the fire scene sketch or separate photographic log to indicate the location and direction from which they were taken. The documentation should include a brief caption that permits an independent reviewer correctly to identify the object of interest and the position from which the photos were taken. Each storage media number should be documented on a separate form. The "Remarks" field is used to document the disposition of the medium used to store the images or videos.

The fire scene sketch or diagram should contain the image or video number within a circle on the sketch showing the position and viewpoint from which the image was captured. Figure 15.4.2(d) in *NFPA 921* (NFPA 2011) illustrates this technique, which is invaluable for later assessing witness and fire investigator observations (see *NFPA 921*, 2011 ed., pt. 15.4.2).

At the time of the investigation, a descriptive photographic index should be completed using Figure 4-7 or a similar photo log from Appendix H of *Kirk's Fire Investigation*, 7th ed. (DeHaan and Icove 2012). The photographic index or log, complete with narrative, should be included in the investigation report.

FILM CAMERAS

For many years the most versatile cameras traditionally available to fire investigators were 35mm single-lens reflex with focal-plane or between-the-lens shutter systems (Berrin 1977). Single-lens reflex cameras are expensive but allow for close-ups and specialized scene photography and introduce little distortion. Viewfinder cameras are low cost but introduce parallax errors and cannot perform many useful photographic functions.

With the introduction of electronically operated shutter systems, both 35mm film and digital-format cameras have become the recommended standard for fire investigation photography (see *NFPA 921*, 2011 ed., pt. 15.2.3 [NFPA 2011]; Peige and Williams 1977). Competitive pricing has placed the cost of many acceptable units below $100.

Also, many of these versatile film cameras have been updated by outfitting them with digital camera backs, which allows them to take high-resolution digital photographs. Two suppliers of such digital camera backs are Phase One (2012) and Mamiya Leaf (2012).

FILM FORMATS

Economic considerations play a key role in deciding on a film format over less expensive digital images. However if the choice is to use film, especially as a backup camera, the standard 35mm color film is the nearly universal choice for fire scene photography over black-and-white film. Black-and-white photography of fire scenes has been all but abandoned owing to its inability to record important colored fire and burn patterns (Kodak 1968).

Print film requires development and costly printing of photographs. Some investigators may choose to use color slide films having an American National Standards institute (ASA) rating between 100 and 500 to maintain resolution. In some processing laboratories, at development time the customer can choose prints, negatives, slides, or scanned images delivered on CD-ROM.

DIGITAL CAMERAS

Newer and higher-resolution digital cameras are introduced each year. Resolution of image quality is often measured in millions of pixels (megapixels). The present minimum acceptable resolution is upward of 5 megapixels, which is still less than the resolution available using standard film photography. The higher the number of pixels, the better is the quality of the image. Many cameras will capture 6 megapixels per image on a flash card or similar downloadable medium.

Computer technology has made dramatic improvements in the capabilities of even relatively low cost digital cameras. It is common now for cameras to offer image capture quality of 15 MB or higher, with more than 5 GB flash storage memory readily available. Digital cameras are now available with SLR lens-and-shutter assemblies, so a wide variety of traditional photographic tasks can be addressed.

One of the most useful innovations is the incorporation of a microphone and sound recording chip into the camera. With each photo taken, the photographer can dictate a description of the photo up to 20 seconds in length. The chip can then be downloaded and the narration converted to a printed photo log. This eliminates the need to set the camera down and make a log entry with each photo (or to try to later re-create a photo log from memory after viewing the photos). Such features are available on the Nikon D100, Sony 727 and 828, and some Olympus digital cameras. For those who prefer to use a film camera, small voice-activated digital voice recorders are available at low cost from retailers such as Radio Shack. These can record up to 1 hour on a single chip in a device the size of a half-pack of cigarettes without the use of fragile tape cassettes.

The improper use of digital cameras in investigative photography may undermine the viability of a case, owing mainly to evidentiary considerations. Investigators are advised to adhere to a standard protocol for preserving digital images. Some agencies preserve images on their original media (CD or flash cards), whereas others save original images on disk drive memory and make copies for any medium or any digital image processing.

As with all photographs, the courtroom tests for authentication must be met: each image used must be a "true and accurate representation" and "relevant to the testimony" to the fire investigation, particularly under *Federal Rules of Evidence 403* (Lipson 2000). Systematic handling and processing of digital images is the only method for assuring their long-term acceptability (*NFPA 921,* 2011 ed., pt. 15.2.3.4) (NFPA 2011).

Computer operator error and unforeseen crashes can cause digital images to be lost forever, so precautions are necessary. It is often impractical and costly to store high-resolution images permanently on some media or flash cards; they must be copied onto another medium. Two copies of the original electronic/magnetic medium should be made and placed in the case file in the same manner as an audio or video recording. A third copy is used as a working copy. Images should not be compressed, as this causes loss of detail and clarity. Copies should also be placed on external hard drives and stored off site. Operating systems, programs, and file formats change and render some files unreadable. Some investigators use two cameras at critical scenes (print and digital) and take duplicate photos to ensure that some images are always available, even years after the fire.

Digital cameras and their electronic images are very useful in the production of preliminary photographs of fire scenes or final investigative reports, as the images can easily be processed to lighten, darken, or enhance features and easily transferred to printed documents. They are also helpful in producing panoramic views of fire scenes, as discussed previously. However, file compression and other manipulations may reduce the quality of the images and introduce suspicions of manipulation of the photo's content. Digital images can be coded or watermarked while being recorded to demonstrate they are "as taken," an unmodified original. The ease and low per-image cost of digital photography tempts some investigators to take far more photos at scenes than are really necessary or justified. Investigators should consider what information they are attempting to capture with any particular photograph before taking any image.

DIGITAL IMAGE PROCESSING

In addition to easy-to-use computer image software for creating panoramic photos from a series of still photos, digital scanning cameras are also now available. With these technologies, investigators have realized the advantage of panoramic views, especially for briefing clients and making courtroom presentations.

Several packages, both for purchase as well as at no cost (freeware), are available for constructing stitched views of both interior and exteriors of structures. The packages can also be used to photograph large outdoor areas, particularly wildland fires. During this process, a series of photos with overlapping edges are taken from a single viewpoint using a normal focal length lens (35–55mm), and the software easily stitches the images together. Distortion is minimized if the photos are taken using a leveled rotating-head tripod, with 15–20 photos for a full rotation.

Adobe Photoshop is a commercial program with a function called Photomerge that quickly and easily stitches several overlapping photographic images into one (Adobe 2011).

The Roxio Panorama Assistant program is part of Roxio's Creator software that will match features in a series of photos (scanned or digital) to produce a single panoramic scan (Roxio 2012). Photos must be cropped so that all images are exactly the same size, and they are manually overlapped before being merged digitally. The system produces only end-to-end panoramas.

A more advanced system is the Graphical User Interface for Panorama Tools (PTGui), a panoramic stitching software developed for both the Windows and Mac OSX operating systems. PTGui (2012), based on the successful Panorama Tools (PT) program developed by Bernhard Vogl, allows the creation of spherical, cylindrical, or flat interactive panoramas from any number of source images. Refer to Figure 4-18 for an example of a PTGui 150° panorama of a large industrial fire scene. This software supports JPEG, TIFF, PNG, and BMP source images. The computer mouse can be used to move the images to change yaw, roll, and pitch with corrected perspective in real time.

In a typical application, 16 photos taken of a large complex fire scene using a hand-held manual 35mm camera are scanned and stored as separate files then opened in PTGui with the 360° single-row panorama editor selected. The latest version of PTGui is 9.1.3, and a full trial version can be downloaded.

Select Hewlett Packard (HP) digital cameras can also capture panorama-ready images and produce them immediately (Deng and Zhang 2012). HP digital cameras offer both in-camera panorama preview and in-camera panorama stitching. With the in-camera panorama preview, up to five photos can later be combined into one seamless image once they are downloaded to a computer for photo stiching. With in-camera panorama stitching, up to five single shots are stitched together within the camera to create one seamless image without having to be downloaded to a computer.

Another digital scanning approach to panoramic or three-dimensional photography is the iPIX (2012) camera system. Its 180° fish-eye lens and tripod mount are compatible with many film or digital 35mm cameras. Mounted on a tripod, the camera takes a photo, then rotates to take a second picture in the opposite direction. Fish-eye lenses produce severely distorted images when viewed directly, but these images are imported into the iPIX software, which corrects the distortion and seamlessly stitches the two images together. The result is a fully immersive, navigable computer image. The iPIX software then permits interactive examination or illustration of fire patterns or explosion damage around the entire periphery of a room. This process requires that the special equipment be available at the scene. It has been used successfully by a number of U.S. police agencies.

Panoscan (2012) of Van Nuys, California, offers a scanning digital camera that can complete a 360° scan of a room in 8 seconds with high resolution and great dynamic range. Its images are readily viewable as a flat panorama or as a virtual reality movie on QuickTime VR, Flash Panoramas, Immervision JAVA, and other imaging software. Developed for both the Windows and Mac OSX operating systems, Panoscan also offers a photogrammetric capacity using two scans at different heights with specialized software (PanoMatrix), allowing accurate measurements of any scene, indoors or out. The system is fully portable.

QuickTime VR (also known as QuickTime Virtual Reality or QTVR) is an image format for use with the Apple media player QuickTime Player. QTVR is used to both create and view panoramic photos. There are software installable plug-ins for use with both the stand-alone QuickTime Player and a Web browser (QuickTime 2012).

Data are saved in DXF format for use in AutoCAD, Maya, and other CAD programs. Measurements (with an accuracy of fractions of an inch over a 25-ft radius) can be made or added to the file at any time. Some advanced digital image systems allow the user to interactively browse the scene and can even link to views in adjacent rooms. Full 360° immersive images allow viewing of floors, ceilings, and walls in any direction. Images can then be linked together or linked to traditional photographs and renderings as well as to audio or other file types.

Panoramic photographs can be used to establish *viewpoint photographs*. This type of photography can be used to document what a witness might have seen or not seen from a particular location. The peripheral vision or field of view of the normal human eye is not readily duplicated by a normal camera lens, so panoramic imaging is sometimes the only way to replicate the view a witness may have had.

Crime Scene Virtual Tour (CSVT 2012) Version 3.0 uses the Java Virtual Machine (Sun Java) software to produce a fully integrated virtual scanned image with links to floor plans, close-up still photos, and investigative notes. With CSVT, users can immerse themselves in the scene, get accurate measurements of important objects from any angle, measure the distance of any two points, and generate documentation.

The GigaPan System (2012) is a derivative of NASA technology that can capture thousands of digital images and weave them into a single uniform high-resolution picture. The process allows users to share and explore gigapixel panoramas. A gigapixel image is a digital image composed of one billion pixels (1,000 megapixels), which is more than 150 times the graphic information captured by a 6 megapixel digital camera. The technology sold by GigaPan is the product of a collaboration between Carnegie Mellon University and NASA Ames Intelligent Robotics Group, with support from Google. GigaPan sells automatic stepping panning heads for ordinary cameras that make it easy to capture gigapixel panoramas that can later be combined with GigaPan Stitch software.

HIGH DYNAMIC RANGE PHOTOGRAPHY

High dynamic range imaging (HDR) photography is a technique that increases the range between the lightest and darkest areas of an image captured by a digital camera, allowing details to be revealed in areas that are too dark or shaded. It has been suggested that fire scene photography could benefit from the use of HDR (Kimball 2012; Howard 2010).

HDR involves taking three or more images of the same scene at different shutter speeds or apertures and combining them into a single image using a processing algorthm. The final image shows details in both the shadows and highlights. The image processing algorithms are available in many programs, including Adobe Photoshop. The algorithms that produce HDR images use the static metadata stored in each digital image, also known as the *exchangeable image file format* (EXIF) information, that reveal information such as the camera's make and model, orientation, shutter speed, aperture, focal length, metering mode, and ISO settings.

HDR technologies are also used in other visualization programs, such as the Spheron-VR's SceneCenter forensic visual content management software (Spheron 2012). Scene-Center documentation allows virtual access to documentation from a crime scene through its presentation in court. It also allows for 3D photogrammetric measurement.

DIGITAL IMAGING GUIDELINES

The first draft of "Definitions and Guidelines for the Use of Imaging Technologies in the Criminal Justice System" was published in October 1999 in *Forensic Science Communications* (FBI 1999). The guidelines were prepared by the Scientific Working Group on Imaging Technology (SWGIT 2002; FBI 2004) and cover the "documentation of policies and procedures of personnel engaged in the capture, storage, processing, analysis, transmission, or output of imagery in the criminal justice system to ensure that their use of images and imaging technologies are governed by documented policies and procedures."

TABLE 4-3	Recommended Guidelines and Best Practices in Forensic Photography by the Scientific Working Group on Imaging Technology

SECTION	DESCRIPTION
Section 1	Use of Imaging Technologies in the Criminal Justice System
Section 2	Considerations for Managers
Section 3	Guidelines for Field Applications of Imaging Technologies in the Criminal Justice System
Section 4	Recommendations and Guidelines for Using Closed-Circuit Television Security Systems in Commercial Institutions
Section 5	Recommendations and Guidelines for the Use of Digital Image Processing in the Criminal Justice System
Section 6	Guidelines and Recommendations for Training in Imaging Technologies in the Criminal Justice System
Section 7	Recommendations and Guidelines for the Use of Forensic Video Processing in the Criminal Justice System
Section 8	General Guidelines for Capturing Latent Impressions Using a Digital Camera
Section 9	General Guidelines for Photographing Tire Impressions
Section 10	General Guidelines for Photographing Footwear Impressions
Section 11	Best Practices for Documenting Image Enhancement
Section 12	Best Practices for Practitioners of Forensic Image Analysis

Source: *Forensic Science Communications* (FBI 2004).

Table 4-3 lists the present approved guidelines issued by the SWGIT, now known as the "Guidelines for the Use of Imaging Technologies in the Criminal Justice System." These guidelines cover imaging technologies, recommended policies, best practices, and techniques for photographing specific impression evidence. These resource documents are available on the websites for the FBI (fbi.gov/) and the International Association for Identification (theiai.org/).

Agencies or individuals using either simple or complex photography should develop a standard operating procedure and departmental policy. The SWGIT procedures recommend preserving original images by storing and maintaining them in an unaltered state and in their original uncompressed file formats. Duplicates or copies should be used for working images. Original images should be preserved in one of the following durable formats: silver-based (noninstant) film, write-once compact recordable disks (CDRs), or digital versatile recordable disks (DVD-Rs).

If image processing techniques are used on the original image, they should be documented with standard operating procedures. These procedures should be visually verifiable and include cropping, dodging, burning, color balancing, and contrast adjustment. Advanced techniques include those that increase the visibility of the image through multi-image averaging, integration, or Fourier analysis. Other techniques include cropping, overlaying, and creating panoramic views from individual series of photographs. An image processing log is recommended so that the process can later be replicated if required.

A chain of custody should be maintained for the media on which original images are recorded. This chain of custody should document the identity of the personnel who had custody and control of the digital image file from the point of capture to archiving.

When compressing images into formats such as JPEGs, investigators should maintain as high a resolution as possible to avoid degrading the image. Image capture devices should render an accurate representation of the images. Different applications will dictate different standards of accuracy.

Training is important when using imaging technologies. Formal uniform training programs should be established, documented, and maintained. Proficiency testing will ensure that camera equipment, software, and media keep pace with hardware or software updates.

LIGHTING

No matter what camera is used, lighting plays an important role in fire scene photography. Intensely burned areas tend to absorb light from natural or artificial sources. Powerful electronic flashes are useful for lighting char details in dark scenes. Some built-in flashes on film and digital cameras do not have enough power to adequately illuminate large fire scenes, so provisions should be made for external flash connections.

Sometimes, burn patterns can best be photographed using an oblique flash to light the area of interest. Therefore, a camera should be selected that allows for an externally connected flash unit for oblique or remote illumination of the scene (*NFPA 921*, 2011 ed., pt. 15.2.3.7.2) (NFPA 2011).

A major problem for interior fire scenes or exterior scenes at night is the provision of suitable lighting. Quartz-halogen worklights are widely used, but they have limited coverage and produce glare and blindspots. One new innovation uses a halogen–metal halide lamp inside a translucent balloon mounted on an extendable mast. An Airstar (2012) balloon creates uniform white glare-free light over a large area. These balloons range in size from 200 W to 4000 W and with masts from 1.8 to 9 m (6 to 36 ft) in height and can cover areas from 200 m^2 (2100 ft^2) to 10,000 m^2 (108,000 ft^2) .

ACCESSORIES

Many attachments are available for cameras; however, for the sake of simplicity, it is recommended that nonexpert investigators limit their use. The maximum effectiveness of the single lens provided with the camera (typically, 50- to 55-mm focal length) is for normal room and exterior view photos. Close-up (18–35 mm) and telephoto (100–400 mm) are for specialized applications, but they can introduce optical distortions. Furthermore, the use of filters and interchangeable lens configurations might disqualify photographs or slides used in courtroom presentations when investigators cannot authoritatively qualify their uses, advantages, deficiencies, or effects (Icove and Gohar 1980). A clear filter to protect the lens is the only recommended filter for fire scene photography (*NFPA 921*, 2011 ed., pt. 15.2.3.6) (NFPA 2011). Some photographers prefer using polarizing or UV filters (which do not affect the color of the image) in place of the clear filter.

The single most recommended accessory for photographing fire scenes is a sturdy tripod. This simple device enables the photographing of scenes with clear detail when low-lighting conditions command longer shutter exposures. Also, a tripod allows the optimization of depth of field according to available lighting conditions. For example, an aperture setting of f/16 will usually produce a depth of field ranging from 0.91 m (3 ft) to 3.05 m (10 ft) when the camera is focused on a point 1.8 m (6 ft) from the lens. As previously discussed, panoramic scenes are most easily recorded when using a tripod. Longer exposure times may be needed to accommodate smaller aperture settings, and any exposure longer that 1/30 second is subject to tremors if the camera is handheld.

MEASURING AND IMAGE CALIBRATION DEVICES

Measuring and image calibration devices are used to provide additional information in both processing and interpretation. These devices should be used uniformly from case to case. Even though current fire and explosion investigation guidelines in *NFPA 921* recommend the use of an 18 percent grayscale calibration card (*NFPA 921*, 2011 ed., pt. 15.2.7.1.)

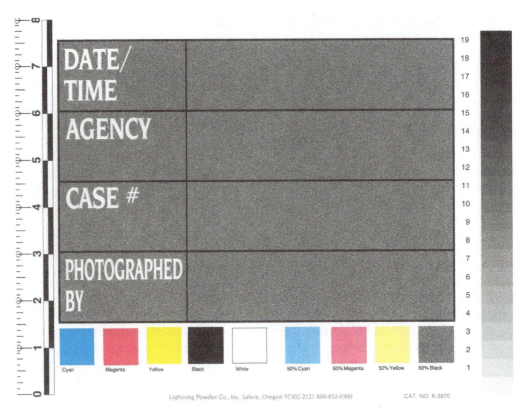

FIGURE 4-38 Commercially available grayscale and color calibration chart suitable for use in forensic fire scene documentation. © *Safariland, Courtesy of Forensics Source™ by Permission.*

FIGURE 4-39 A convenient grayscale ruler is available from several sources. *Courtesy of J. D. DeHaan.*

(NFPA 2011), there are many advantages to using a color calibration card. With the move toward the universal use of color photography for fire scenes, the investigator should consider the use of a combined color and grayscale calibration card.

A *color and grayscale calibration card* bearing the agency name, case number, date, and time should be photographed on the first frame and last frame of each roll of film, digital camera memory card, or video camera recording. With this technique, variations in color accuracy can be detected in both film and digital images, and professional film processing laboratories will calibrate their equipment using these cards to ensure that the best color calibrations and gray scales are met. Also, color printers, such as Xerox (2012), can be calibrated and corrected to ensure the proper colors are used in the graphics display and commercial printing process. Figure 4-38 is an example of a commercially available card with the essential calibration and documentation information.

Forensics Source (2012) produces a 15-cm (6-in.) scale that is printed on 18 percent gray low-reflection plastic so that even close-up digital images can include grayscale calibration. The International Association of Arson Investigators (IAAI) also produces a similar plastic 6-in. ruler as a promotion for their CFITrainer program (Figure 4-39).

(a)

(b)

(c)

FIGURE 4-40 (a) Crime scene measurement devices include yellow numbered tent cards to identify the location of individual points of interest or forensic evidence, such as ignitable liquid containers. (b) Underside of can bears repeated stabbing penetrations from a knife blade, indicating that it was intentionally made to spread flammable liquid contents. Tool marks may link it to the offender. (c) Numbered evidence markers illustrate the position of evidence. Plastic pointers or arrows indicate positions of positive canine alerts at the scene. They may also be used to indicate the direction of fire spread in the fire scene. *Courtesy of J. D. DeHaan.*

Calibration cards also have measuring devices (rulers) on their edges to document distances and sizes of objects photographed. Other measurement devices include longer folding rulers and yellow numbered tent cards to identify the location of individual points of interest or forensic evidence, as illustrated in Figure 4-40. Bright-colored pointers can be used to indicate important features or document direction of fire travel.

AERIAL PHOTOGRAPHY

In rural areas, the U.S. Soil Conservation Service has *aerial photos* (also known as *orthophotoquads*) of a majority of the farmlands within its area of responsibility. These photos are at various resolutions. Aerial photographs do not always need to be taken from aircraft. Perspective photographs, such as in Example 4-1, can be taken from higher floors of adjacent buildings. Photos taken from the elevated platform of a fire truck can be useful, providing a closer look at the overall pattern of damage, as seen in Figure 4-41(b).

Satellite and commercial aerial photos are widely available at a modest cost. Many cities and counties use aerial survey to search for nonpermitted construction. Care must

(a)

(b)

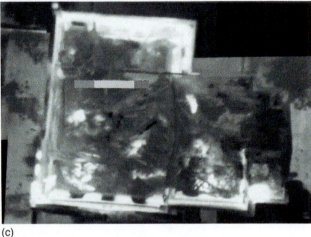

(c)

FIGURE 4-41 (a) Helicopter with mounted camera, (b) aerial image, and (c) FLIR image. *Courtesy of MCSO Unmanned Flight Program.*

be taken in using commercial databases (e.g., Google Earth, Bing), as street addresses are not always up-to-date in data files. Users must also confirm that the image was recent enough that it accurately reflects the prefire conditions.

A developing, low-cost form of aerial photography uses field-deployable remote-controlled drones and helicopters to survey fire scenes at low altitudes. Figures 4-41(a)–(c) show examples of aerial photographic imagery produced by the helicopters courtesy of the Mesa County Sheriff's Office (MCSO) Unmanned Flight Program.

One commercial product is the Draganflyer X8 UAV helicopter (Draganfly 2012), which is battery operated and can be equipped with one of four optional cameras: a high-resolution still camera (with remote zoom, shutter control, and tilt), high-definition video, low-light black-and-white video, or infrared camera.

The helicopter's size—only 25.4 cm (10 in.) high and 99 cm (39 in.) wide including rotors—enables it to move in very close quarters. Because it is battery-powered, the helicopter has a flight time of around 20 minutes. Batteries are easily changed, so that repeat flights are possible. There is an optional capability for live video feed, allowing the operator and/or others on the ground to monitor the imagery simultaneously.

The firm Infinite Jib (2012) is a distributor for the Droidworkx Airframe multirotor helicopter, which can deliver low-altitude imagery suitable for fire scene photography. The current ready-to-fly units consist of the EYE-Droid 4, 6, and 8 models and cost up to $13,000 each.

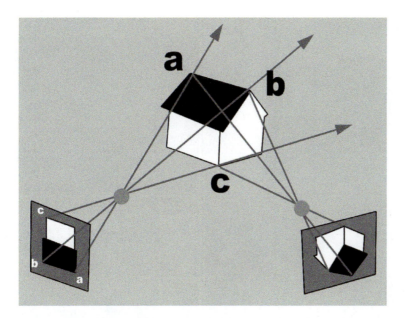

FIGURE 4-42 In photogrammetry, two or more photographs are used in such a way as to extract absolute coordinates and measurements from key points in those images. *PhotoModeler image courtesy of Eos Systems, Inc.*

PHOTOGRAMMETRY

Photogrammetry, the science of extracting measurement data from photographs, was once out of the range of expertise of the typical fire investigator but has become more accessible in recent years through affordable, user-friendly computer software. In photogrammetry, two or more photographs are taken and used to extract absolute coordinates and distance measurements from key features in those images (see Figure 4-42). Close-range photogrammetry programs compensate for the shortening of object lines due to perspective that is present in nearly all photographs, thus allowing users to obtain accurate measurements and create realistic models (Eos 2012).

For example, the walking distance from a deceased victim in a bed to the doorway threshold may be crucial in an investigation. Photogrammetry can capture measurement data in large or complex scenes where tape measure and notepad are inadequate or too time-consuming. It also allows investigators to measure evidence that may have been moved or cleaned up or no longer exists at the scene. Photogrammetry has been in use in North America, Europe, and Australia for many years for accident scenes and major crimes and has been proposed for advanced fire investigations. In one case in the United States, the technique was first used forensically to reconstruct a model of a kitchen area to enhance a V burn pattern on three dimensions (King and Ebert 2002).

Some photogrammetry programs allow users to map two-dimensional images over three-dimensional surfaces that have been created, to provide photorealistic models with perspective-corrected phototextures. There is a wide range of software programs for calculating these measurements and constructing three-dimensional models from two-dimensional photographs. Figure 4-43 is an example of the use of photogrammetry to construct a three-dimensional view of the exterior of a building using two-dimensional photographs.

DIGITAL SCANNING CAMERAS

One very new innovation that combining the capabilities of both panoramic photography and photogrammetry is the digital scanning camera, such as the previously mentioned Panoscan or Spheron VR models. Photogrammetry capability is available on the

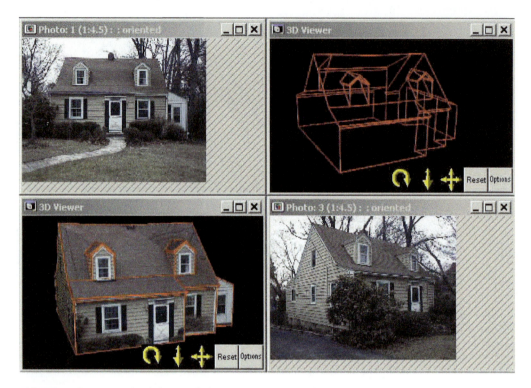

FIGURE 4-43 An example of the use of photogrammetry to combine several two-dimensional images to construct a three-dimensional view. *PhotoModeler image courtesy of Eos Systems, Inc.*

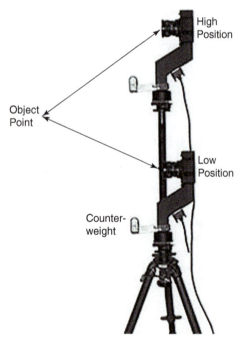

FIGURE 4-44 A digital scanning camera with an extendable mast and specialized software can be used to capture photogrammetric images. *Courtesy of Panoscan Inc.*

Panoscan and the other systems. After the area is captured in a single scan, the camera is elevated on a vertically extendable mast that allows the camera to capture an image from the same position in the room but from a view line some 0.5 m higher, and the area is scanned again. The two images are then combined digitally (the Panoscan system uses proprietary software called PanoMetric) to produce a stereoptic virtual reality in which any measurement can then be taken with millimeter precision.

The core of the system is a specialized digital camera that works much like the moving-slit film cameras of a century ago but uses a high-resolution image converter and a specialized lens-and-shutter system to capture in each vertical "slice" an image of about 170° vertically (see Figure 4-44).

The camera automatically rotates, so that the final image is a spherical high-resolution image. (The maximum vertical resolution is 5200 pixels and the total image capture can be up to 50 megapixels.) The dynamic range of image capture is up to 26 f-stops, so information from very dark areas to those in full sunlight are captured in a single pass. This is an extension of the high dynamic range technology described earlier.

Sketching

Fire scene sketches and diagrams, whether or not to scale, are important supplements to photographs and can assist the investigator in documenting evidence of fire growth, scene conditions, and other details of the fire scene (see *NFPA 921*, 2011 ed., pt. 15.4 [NFPA 2011]; *NFPA 1033*, 2009 ed., pt. 4.3.1 [NFPA 2009]).

A sketch graphically portrays the fire scene and items of evidence as recorded by the investigator. The valuable documentation technique is used by investigators when drawing simple two-dimensional sketches of fire scenes. These sketches may range from rough exterior building outlines to detailed floor plans. Sketches allow the investigator to illustrate relationships between objects that cannot be captured via photography, such as those in separate rooms, under or behind large furniture, or visible only from overhead or via cross section. See the appendix in *Kirk's Fire Investigation,* 7th ed., for additional details on the preparation of sketches.

GENERAL GUIDELINES

Good scene sketches do not require highly artistic skill. Even informal hand-drawn representations can capture locations and relationships between items that cannot be shown in photographs. Forensic fire scene sketches serve many purposes and should be used to illustratate and depict the following elements of information (see DeHaan and Icove 2012; *NFPA 921*, 2011 ed., pts. 15.4.4 (A) through (D) [NFPA 2011]; and *NFPA 170* [NFPA 2012]):

- The investigator's full name, rank, agency, case number, and the date/time when the sketch was prepared along with the full names of other individuals involved in constructing the sketch
- The location and geographic orientation of the fire scene (north arrow). Latitude and longitude using GPS data may be very helpful for rural scenes to determine jurisdiction boundaries (state, county, township, etc.) or to relocate the scene accurately
- The outline and scaled dimensions of the building, room, vehicle, or area(s) of interest, including the compass orientation. Note that if the scale is approximated, this should be indicated by the phrase "Not to Scale" on the sketch or drawing.
- A legend identifying and describing all symbols and their meanings, the scale, and other significant information used on the sketch or diagram (see *NFPA 170: Standard for Fire Safety and Emergency Symbols)* (NFPA 2012).

Sketches are very useful for recording the following:

- The locations of pertinent evidence or critical features, such as fire patterns and plume damage
- The locations and dimensions of all major fuel packages and competent sources of ignition involved in the fire
- Locations and travel distances of possible points of entry and exit of victims and suspects through simple measurement, laser-assisted measurements, and GPS coordinate data
- Conditions, security (locked or unlocked), and dimensions of windows, doors, floors, ceilings, and wall surfaces including sill and soffit heights for later use in fire scene reconstruction and analysis
- The detailed locations of the arc fault sites in arc survey diagrams (see *NFPA 921,* 2011 ed., pt. 17.4.5.2) (NFPA 2011).

In some instances, multiple sketches using the same overall dimensions are necessary. Several separate types of evidence can be preserved in this manner including isochars, plume damage, and sample (evidence collection) sites. A sketch can often reveal characteristics not readily obvious in a photograph.

The areas of surface demarcations, thermal effects, penetrations, and loss of material can also be outlined using colored markers on sketches or photographs. For example, one system uses yellow lines to outline areas of surface deposits, green lines for thermal

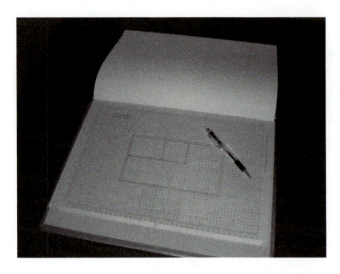

effects, blue lines for penetrations, and red lines for areas where the material has been consumed completely. Background patterns (lines, cross-hatching) can be used in computer diagrams.

A sketch is usually initially drawn by hand and sometimes followed up with a computer-assisted drawing. An effective hand-sketching tool is a grooved grid pad on which the pencil lead follows a slight indentation on a plastic surface (Accu-line 2012) (see Figure 4-45).

A reliable fire scene sketching program, Visio by Microsoft Corporation, is sufficient to produce accurate representations of structures (Visio 2012). Microsoft provides an add-on crime scene package for Visio at no charge through their website. Experienced investigators have developed templates for use in fire scene analysis. Figure 4-46 shows an example of a typical fire scene diagram produced using Visio.

GPS MAPPING

Magellan (2012) has improved on an innovative approach to mapping large scenes using handheld GPS units. Using a GPS device (the Mobile Mapper CE) the size of a cell phone, with GPS Differential software extension, the examiner walks to various points at the scene and keys in a code (ArcPad) denoting a particular set of features. The data are then downloaded and processed by specialized software (Mobile Mapper Office) to produce a plot of the scene accurate to ±0.5 m (depending on the number of satellites in "view" and the stabilization time at each location). Such devices allow the mapping of large and complex scenes where view lines for Total Station surveying would be obscured by terrain

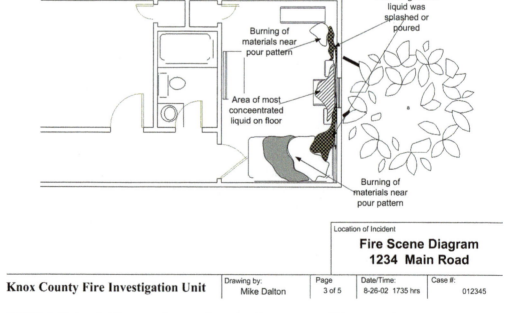

FIGURE 4-46 A typical fire scene diagram using a sketching program sufficient to produce accurate representations of burn patterns, location of evidence, and other details. *Courtesy of Michael Dalton, Knox County Sheriff's Office.*

or features (trees, equipment, etc.). The handheld device contains all the usual features of GIS/GPS location systems

TWO- AND THREE-DIMENSIONAL SKETCHES

With the advent of CAD tools, two- and three-dimensional sketches can now be used for courtroom exhibits. Figure 4-47 shows an example of this technique. Plastic overlays can also be placed over the large sketches, which can be annotated in court using either prepared overlays or impromptu markings on the overlays. Using a convenient coordinate system allows the accurate placement of critical evidence such as the orientation of torsos of victims, fire plumes, and evidence locations.

Google's SketchUp (Google 2012), available in both a free and a for-purchase version, allows the user to construct two- and three-dimensional buildings using simple commands. SketchUp works with the Windows Vista, Windows 7, and Apple Mac OSX operating systems.

A special application of sketching is to copy an overhead or plan view of the scene onto several clear (transparency) sheets. Each sheet can be used to record a different indicator such as fire travel vectors, char depths, calcination, and furniture patterns. The overlays then can be compared with one another or with the original floor plan to illustrate convergence on a possible area of origin. Putting sketches on separate sheets avoids the problem of having too much detail in a single diagram.

Inclement weather can affect the survivability of field notes. HTRI Forensics of Newberg, Oregon (HTRI 2012), distributes two products using special writing paper called "Rite in the Rain." The Field Investigation Field Notes contains blank pages necessary for a full investigation report, including the Bilancia Ignition Matrix (Bilancia 2012), as shown in Figure 4-48. A smaller Fire Incident Report pocket notebook is designed for use by first responders responsible for collecting initial data on fire investigations.

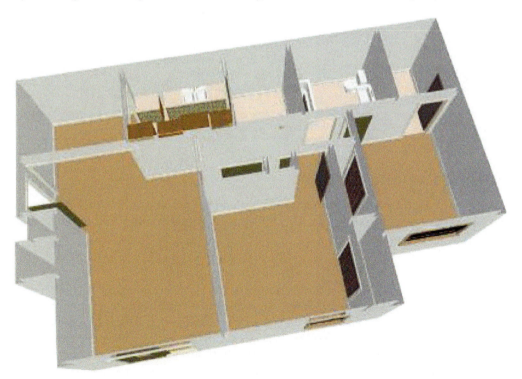

FIGURE 4-47 Construction of a three-dimensional view using an architectural rendering program. *Courtesy of D. J. Icove.*

Source / Fuel	Clock radio	Cell phone charger	Cigarette	Candle	Compact fluorescent lamp	Plug-in room refresher
Lace night table cover	1. yes (a) 2. yes close 3. no (d) 4. yes path	1. yes (a, c) 2. not close 3. 4. yes	1. 2. yes close 3. no evid. 4.	1. yes comp 2. yes 3. maybe 4. yes	1. yes (a, c) 2. yes 3. no (d) 4. yes gravity	1. yes (a) 2. not close 3. no (d) 4.
Sheets or covers	1. no (c) 2. not close 3. no (d) 4. yes gravity	1. yes (a, c) 2. yes close 3. no (d) 4. yes	1. 2. not close 3. no evid. 4.	1. yes comp 2. not close 3. 4.	1. yes (a, c) 2. no 3. no (d) 4. no	1. yes (a) 2. yes close 3. 4.
Curtains	1. no (c) 2. not close 3. no (d) 4. yes path	1. yes (a, c) 2. yes close 3. no (d) 4. yes	1. 2. not close 3. no evid. 4.	1. yes comp 2. not close 3. 4.	1. yes (a, c) 2. no 3. no (d) 4. yes shade	1. yes (a) 2. yes close 3. no (d) 4.
Plastic decorative flowers	1. no (c) 2. not close 3. no (d) 4.	1. yes (a, c) 2. not close 3. no (d) 4. no	1. not comp. 2. 3. no evid. 4.	1. yes comp 2. yes close 3. maybe 4. yes	1. no (a, c) 2. no 3. no (d) 4. yes shade	1. no (a, c) 2. not close 3. no (d) 4. no
Lamp shade	1. yes comp 2. not close 3. no (d) 4. yes path	1. yes (a, c) 2. not close 3. no (d) 4. no	1. 2. not close 3. no evid. 4.	1. yes comp 2. not close 3. indirect 4. yes path	1. no (a, c) 2. yes close 3. no (d) 4. yes	1. no (c) 2. not close 3. no (d) 4. no

1. Competent ignition source Y/N?
2. Proximity, ignition close to fuel Y/N?
3. Evidence of ignition Y/N?
4. Initial fuel path to fuel load Y/N?

Color legend
Red – Competent and close
Blue – Not competent
Yellow – Competent but ruled out

Codes
P – plume or flashover
W – witnessed
F – open flame
N – not energized

Notes
a. if the device failed
b. if fuel was cellulosic with a breeze
c. only with open flame
d. device was intact

FIGURE 4-48 The Bilancia Ignition Matrix. *Courtesy of Lou Bilancia, P.E. Synnovation Engineering and Ignition Matrix Corporation; Copyright 2008–12. All rights reserved. Used with permission.*

Establishment of Time

Several natural processes can be used to establish one of the most difficult factors in scene investigation—the passage of time. These processes include evaporation, warming, cooling, drying, and melting.

EVAPORATION

One of the most critical factors in fire reconstruction is determining the use or presence of accelerants. Estimating the evaporation of accelerants may provide important insight as to the passage of time and the dynamics of the resulting fire. The evaporation rate of a liquid depends on its vapor pressure (which is temperature dependent), the temperature and nature of the surface on which it is deposited, its surface area, and the circulation of air around it.

Vapor pressure is a fundamental physical property of all liquids and solids. It is the pressure that would be generated by the material if it was placed in a vacuum and allowed to come to equilibrium. Vapor pressure is a measure of how volatile the material is (i.e., how easily it vaporizes). Because vapor pressure is temperature-dependent, the higher the temperature, the higher the vapor pressure. It is the temperature of the liquid (or of the surface on which it is spread) that is critical, not necessarily the ambient temperature of the room. A pan of acetone on a hot stove will evaporate very quickly even in a cold room.

Vapor pressure is also highly dependent on molecular weight. The lower the molecular weight of a substance, the higher its vapor pressure, and the faster it will evaporate. The physical form of the material is also critical. A very thin film of volatile liquid on a surface will evaporate more quickly than a deep pool. Volatile liquids on porous surfaces like cloth or carpet will evaporate more quickly than they will from a free-standing pool of the same size. The larger the size of the pool and the more air movement around the pool, the faster the total evaporation (DeHaan 1999).

A complex mixture like gasoline contains more than 200 compounds, some of which, like pentane, are very volatile and evaporate very quickly. Others, like trimethylbenzene, evaporate very slowly and will persist for a long time. When such a liquid is poured out, the lightest, most volatile compounds will represent the entire bulk of the initial vapors generated. Thus, it is the presence of toluene, pentane, and other volatile components that controls the ignitability of the vapors being generated. Components like octane or heavier hydrocarbons slowly evaporate at room temperatures. The vapors created by evaporation are easily distinguished from the residues of the liquid based on their gas chromatographic profiles.

This process means that partially evaporated gasoline will have a very different gas chromatographic profile than fresh gasoline. Simple evaporation and burning of gasoline result in the same disproportionate loss of the lighter, more volatile components. The process is accelerated by the added heat flux from the flames. This property can be used to estimate the relative time of exposure. Gasoline that is present in fire debris as a contaminant from gas-powered firefighting equipment or intentionally added after the fire will have far more volatiles than would be expected from residues after normal evaporation or combustion. Partial evaporation of gasoline will reduce the effective vapor pressure of the remaining liquid by the loss of the most volatile components. This can change the flash point of the remaining liquid and ignitability of the product (Stevick et al. 2011).

Changing the physical form of a liquid by placing it on an absorbent wick or aerosolizing it (changing it to a mist by releasing it under pressure) will increase its vapor pressure and make even its heavier components evaporate more quickly or make them more readily ignitable. If evaporation is a concern, from the standpoint of either creating a risk by evaporation of a liquid fuel or reconstructing time factors (time since release), several factors must be documented. These factors include accurate identification of the material itself; the temperature of ambient air, surfaces, or the liquid itself at the time of release; and the nature of surfaces on which it was spread (liquids evaporate much more quickly from thin materials like clothing with free air movement than from thick porous materials like shoes).

The predominant air conditions are also critical: high temperatures, direct sun, wind, mechanical movement by fans, HVAC systems, and movement of equipment, vehicles, or people all increase the evaporation rate. Evaporation of flammable liquids from evidence that could be significant in establishing the cause of the fire or linking a person with the scene needs to be arrested by sealing the evidence in an appropriate vapor-tight container and keeping it as cool as possible to reduce further losses. On the container, the investigator should note the time at which it was sealed.

DRYING

Drying is usually closely related to evaporation (as in the drying of water from clothing), but when blood or other complex liquids are involved, other processes may be involved. Blood, for instance, changes chemistry, color, and viscosity as it dries, becoming darker and stickier (if it is drying as a pool on a nonporous surface). Documentation of the drying state of blood involves more than simple photography, as the texture needs to be assessed, usually using a sterile cotton swab (which would be used in its eventual collection in any event). The time at which this observation is made must also be recorded. The same considerations apply to paint, adhesives, plastic resins, or other materials that change mechanically as they dry. Also, the ambient temperature must be noted if drying is a critical concern.

COOLING

Cooling and *melting* are time indicators that play roles in crime scene investigation (such as still partly frozen ice cream on the counter) but occasionally are useful in fire reconstruction. The cooling (or lack thereof) of materials directly exposed to fire can be assessed by direct contact, thermal imaging, or temperature probe. Inexpensive digital thermometers that are good from −45°C to 200°C (−50°F to 400°F) are available from kitchen and housewares stores. Once again, ambient conditions of wind, rain, sun, standing or running water, and the like, need to be recorded at the same time.

Cooling of a body is a routine estimate in death investigations but is often ignored in fires. That is unfortunate, because the death and the fire need not have occurred at the same time. An adult human body, once cooled to room temperature, must be exposed to heat (even from a fire) for a significant period of time before its internal core temperature rises. The thermal inertia of human tissue is similar to that of pine wood or polyethylene plastic. The internal temperature (rectal or liver) of a body at a fire scene should be recorded not for estimation of the time of death interval as much as for exclusion of situations in which the victim was dead for many hours prior to the fire. Recent work by DeHaan reports on experiments with the sustained combustion of human bodies (DeHaan 2012; Schmidt and Symes 2008, chap. 1).

Spoliation

The fire scene is often the most important piece of evidence in forensic analysis and reconstruction, particularly when the fire clearly will result in criminal and/or civil litigation. The major concern of fire investigators is the preservation of evidence to prevent its destruction or alteration before it is submitted for competent examination and analysis. *NFPA 921* restates the *ASTM E860* standard that spoliation is the "loss, destruction, or material alteration of an object or document that is evidence or potential evidence in a legal proceeding by one who has the responsibility for its preservation" (see *NFPA 921*, 2011 ed., pt. 3.3.162 [NFPA 2011]; *ASTM E860*, pt. 3.3 [ASTM 2007b]). Both *NFPA 921* and *ASTM E860* caution that **spoliation** may occur during initial and subsequent fire scene processing, particularly when the movement or the alteration of the debris at the fire scene may significantly impair the opportunity of other interested parties to obtain the same evidentiary value from the evidence as did any prior investigator (see *NFPA 921*, 2011 ed., pt. 11.3.5 [NFPA 2011]; *ASTM E860*, pt. 3.3 [ASTM 2007]).

The inherent problem with spoliation is that intentional or negligent destruction or alteration of evidence not only may be the subject of pending litigation but also could be considered in future litigation.

Failure to prevent spoliation can result in disallowance of testimony, sanctions, or other civil or criminal remedies (Burnette 2012). Published standards (*ASTM E860*) establish practices for examining and testing items of evidence that may or may not be involved in product liability litigation. This evidence may include equipment, items, or components that will not be returned to service. In laboratory examinations where the evidence will be altered or destroyed, all persons involved in the present or potential

spoliation ■ The destruction or material alteration of, or failure to save, evidence that could have been used by another in future litigation. Loss, destruction, or material alteration of an object or document that is evidence or potential evidence in a legal proceeding by one who has the responsibility for its preservation (*NFPA 921*, 2011 ed., pt. 3.3.162).

cases should be given the opportunity to express their opinions, to provide input on any proposed examination protocol, and to be present at the testing.

ASTM E860 refers to other ASTM practices, including those concerning the reporting of opinions (*ASTM E620*) (2011b), evaluation of technical data (*ASTM E678*) (2007a), reporting of incidents (*ASTM E1020*) (2006), and collection/preservation of evidence (*ASTM E1188*) (2005a). For example, *ASTM E1188*, sec. 4.1, states that the investigator should "obtain statements as early as feasible from all individuals associated with the event and recovery activity." *ASTM E1188* also guides the investigator in maintaining a **chain of evidence (chain of custody)** by identifying and uniquely labeling evidence as to the location, time and date, and name of collector.

When the evidence is subject to in-place or later laboratory examination by a coordinated and informed group of affected parties, these processes must be documented and photographed. These tests or examinations may consist of disassembly or destructive testing (*ASTM E1188*, sec. 4.3) (ASTM 2005a). Disposal of evidence should also be coordinated with ample notification of all parties and clients.

Both *NFPA 921* and *ASTM E860* provide guidance to fire investigators on their responsibilities in minimizing spoliation of evidence (*NFPA 921*, 2011 ed., pt. 11.3.5 [NFPA 2011]; *ASTM E860*, pt. 5.2 et seq. [ASTM 2007b]):

chain of evidence (chain of custody) ▪ Written documentation of possession of items of physical evidence from their recovery to their submission in court.

- Immediately inform their client that there was a need to secure and preserve crucial evidence;
- Recommend that their client notify all other interested parties which may have a standing or exposure to the potential civil litigation; and
- Cease all destructive disassembly and testing until all other interested parties be given the opportunity to participate in and record the results of the examination and potential destructive testing.

Common testing standards were listed in Table 1-3. Legal experts provide the following practical advice and pointers for avoiding spoliation issues in potential or pending cases (Hewitt 1997):

- Recognize the duty to other interested parties.
- Keep current on your field of expertise and track the law on spoliation.
- Retain only properly qualified experts and seek their guidance.
- Respect industry standards and recommendations and develop standard procedures for the storage of evidence.
- Put those in control of evidence on notice of your rights.
- Raise the opposing party's standard of care if necessary and build a record with the party in control of evidence.
- Notify other interested parties and provide them with an opportunity to examine evidence.
- Seek either a voluntary undertaking or a formal agreement to preserve evidence.
- Consider also obtaining a preservation or protection order.
- If compelled to destroy or damage evidence, first document it thoroughly.
- Consult a lawyer.

Guidelines for avoiding sanctions for spoliation of evidence include conducting an immediate investigation, giving timely notification to potential defendants, and arranging for long-term preservation of the evidence collected (Sweeney and Perdew 2005). The realization that the fire scene cannot be preserved throughout litigation is an incentive to allow interested parties to take steps to identify and preserve other relevant evidence. *NFPA 921* defines an *interested party* as "any person, entity, or organization, including their representatives, with statutory obligations or whose legal rights or interests may be affected by the investigation of a specific incident" (see *NFPA 921*, 2011 ed., pt. 3.3.105) (NFPA 2011).

TABLE 4-4	Summary of Case Citations and Opinions on Spoliation of Evidence	
PRINCIPLE	**CASE CITATION**	**DISCUSSION**
Definition	*County of Solano v. Delancy,* 264 Cal. Rptr. 721, 724 (Cal. Ct. App. 1989)	■ Failure to preserve property for another's use in a pending or future litigation
	Miller v. Montgomery County, 494 A.2d 761 (Md. Ct. Spec. App. 1985)	■ Destruction, mutilation, or alteration of evidence by a party to an action
Dismissal	*Allstate Insurance Co. v. Sunbeam Corp.,* 865 F. Supp. 1267 (N.D. Ill. 1994)	■ Dismissal of action as result of spoliation by deliberate or malicious conduct
	Transamerica Insurance Group v. Maytag Inc., 650 N.E. 3d 169 (Ohio App. 1994)	■ Dismissal of action for failure to preserve all evidence prior to suit
Duty to preserve evidence	*California v. Trombetta,* 479, 488–89 (1984)	■ Duty to preserve exculpatory evidence that will play a role in suspect's defense
Exclusion of expert testimony	*Bright v. Ford Motor Co.,* 578 N.E. 2d 547 (Ohio App. 1990)	■ Exclusion of expert testimony for failure to protect evidence prejudicial to innocent party
	Cincinnati Insurance Co. v. General Motors Corp., 1994 Ohio App. LEXIS 4960 (Ottawa County Oct. 28, 1994)	■ In product liability actions, expert testimony may be precluded as a sanction for spoliation of evidence
	Travelers Insurance Co. v. Dayton Power and Light Co., 663 N.E. 2d 1383 (Ohio Misc, 1996)	■ Negligent or inadvertent destruction of evidence sufficient for sanctions excluding deposition testimony of expert witness
	Travelers Insurance Co. v. Knight Electric Co., 1992 Ohio App. LEXIS 6664 (Stark County Dec. 21, 1992)	■ Opinion evidence of plaintiff's expert struck because physical evidence no longer available
Evidentiary inferences	*State of Ohio v. Strub,* 355 N.E. 2d 819 (Ohio App. 1975)	■ Attempts to suppress evidence indicating consciousness of guilt
	U.S. v Mendez-Ortiz, 810 F.2d 76 (6th Cir. 1986)	■ Inference that evidence unfavorable to spoliator's cause was intentionally spoliated or destroyed
Independent torts	*Continental Insurance Co. v. Herman,* 576 So. 2d 313 (Fla. 3rd DCA 1991)	■ Independent actions for intentional or negligent spoliation of evidence
	Smith v. Howard Johnson Co. Inc., 615 N.E. 2d 1037 (Ohio 1993)	■ Existence of cause of action in tort for interference with or destruction of evidence
Criminal statutes	Ohio Statute, Section 2921.32	■ "Obstructing justice" by destroying or concealing physical evidence of the crime or act

Source: Derived from information given by Burnette (2012).

Table 4-4 summarizes relevant fundamental case citations and opinions. Fire investigators should keep up-to-date on legal issues surrounding the problem of spoliation and how to guard against violating its principles and practices. When questions arise, a legal advisor or prosecutor should be consulted for clarification or advice. Thorough documentation with photography and detailed notes is the best protection against problems arising from accidental damage during storage and transport. A comprehensive review of current spoliation decisions is presented in *Kirk's Fire Investigation*, 7th ed. (DeHaan and Icove 2012, chap. 17).

Summary

Fire scenes often contain complex information that must be thoroughly documented, since a single photograph and scene diagram are not sufficient to capture vital information on fire dynamics, building construction, evidence collection, and avenues of escape for the building's occupants.

Fire scenes are documented through a comprehensive protocol of forensic photography, sketches, drawings, and evidence collection. Various computer-assisted photographic and sketching technologies can help ensure accurate and representational diagramming. Comprehensive documentation aids the primary investigation and any subsequent investigations or inquiries, and minimizes charges of spoliation. It offers support for any opinions offered and makes effective peer or technical review possible. It also demonstrates that the investigator has followed important professional guidelines.

The next chapter discusses accurate fire scene documentation in the analysis of intentionally set fires. From visual observations of the manner of ignition of the fire, the presence or absence of accelerants, and the area of origin, much can be learned about the arsonist. Later chapters demonstrate the need for accurate documentation to construct fire models, to design tests, and to compare case studies.

Problems

4.1. Find an example of a public fire report on the U.S. Fire Administration's website. What is your assessment of the completeness of this report? What information would you add to the report?

4.2. Visit the scene of a fire and take exterior survey photographs of the building without crossing onto the property. What information can you glean from the exterior visual information?

4.3. Practice sketching fire scenes by drawing a plan (overhead) view of your living room, showing the location of furnishings, heat sources, and ventilation openings.

4.4. Review the list of case citations on spoliation of evidence. What guidance would you provide to ensure that your local community investigator meets or exceeds this professional conduct?

Suggested Readings

Cooke, R. A., and R. H. Ide. 1985. *Principles of fire investigation,* chaps. 8 and 9. Leicester, U.K.: Institution of Fire Engineers.

DeHaan, J. D., and D. J. Icove. 2012. *Kirk's fire investigation,* 7th ed., chaps. 6, 7, and 12. Upper Saddle River, N.J.: Pearson-Prentice Hall.

National Institute of Justice. June 2000. *Fire and arson scene evidence: A guide for public safety personnel.* Washington, D.C.: NIJ.

References

AAFS. 2011. Fire debris workshop. Chicago, IL: American Academy of Forensic Science.

Accu-Line. 2012. Accu-line Industries, USA. Retrieved January 31, 2012, from https://www.accu-line.com/.

Adobe. 2011. Adobe Photoshop. Retrieved January 31, 2012, from www.adobe.com/PhotoshopFamily/.

Ahmad, U. K., Mei, Y. S., Bahari, M. S., Huat, N. S., & Paramasivam, V. K. 2011. The effectiveness of soot removal techniques for the recovery of fingerprints on glass fire debris in petrol bomb cases. *Malaysian Journal of Analytical Sciences* 15 (2): 191–201.

Ahrens, M. 2007. U.S. Experience with Smoke Alarms and other Fire Detection/Alarm Equipment. Quincy, MA, National Fire Protection Association.

Airstar. 2012. Airstar Space Lighting USA. Retrieved January 31, 2012, from http://airstar-light.us/.

Armstrong, A., Babrauskas, V., Holmes, D. L., Martin, C., Powell, R., Riggs, S., & Young, L. D. 2004. The evaluation of the extent of transporting or "tracking" an identifiable ignitable liquid (gasoline) throughout fire scenes during the investigative process. *Journal of Forensic Sciences* 49 (4): 741–48

ASTM. 2005a. *ASTM E1188-05: Standard practice for collection and preservation of information and physical items by a technical investigator.* West Conshohocken, PA: ASTM International.

———. 2005b. *ASTM E1459-92: Standard guide for physical evidence labeling and related documentation.* West Conshohocken, PA: ASTM International.

———.2006. *ASTM E1020-96: Standard practice for reporting incidents that may involve criminal or civil litigation.* West Conshohocken, PA: ASTM International.

———. 2007a. *ASTM E678-07: Standard practice for evaluation of scientific or technical data.* West Conshohocken, PA: ASTM International.

———. 2007b. *ASTM E860-07: Standard practice for examining and preparing items that are or may become involved in criminal or civil litigation.* West Conshohocken, PA: ASTM International.

———. 2011a. *ASTM E119-11a: Standard test methods for fire tests of building construction and materials.* West Conshohocken, PA: ASTM International.

———. 2011b. *ASTM E620-11: Standard practice for reporting opinions of scientific or technical experts.* West Conshohocken, PA: ASTM International.

———. 2011c. *ASTM E2544-11a: Standard terminology for three-dimensional (3D) imaging systems.* West Conshohocken, PA: ASTM International.

Babrauskas, V. 2006. Mechanisms and modes for ignition of low-voltage, PVC-insulated electrotechnical products. *Fire and Materials* 30:151–74.

Bailey, C. 2012. One stop shop in structural fire engineering. Retrieved January 31, 2012, from http://www.mace.manchester.ac.uk/project/research/structures/strucfire/.

Berrin, E. R. 1977. Investigative photography. *Technology Report 77-1.* Bethesda, MD: Society of Fire Protection Engineers.

Bilancia. 2012. Bilancia Ignition Matrix. Retrieved January 31, 2012, from http://www.ignition-matrix.com/.

Bleay, S. G., Bradshaw, and J. E. Moore. 2006. Fingerprint development and imaging newsletter: Special edition. Publication No. 26/06. Home Office Scientific Development Branch, UK, April.

Burnette, G. E. 2012. Spoliation of evidence. Tallahassee, FL: Guy E. Burnette, Jr, P.A.

Carey, N. 2002. Powerful techniques: Arc fault mapping. *Fire Engineers Journal,* 47–50.

Carey, N., & Nic Daéid, N. 2010. Arc mapping research. *Fire and Arson Investigator* (October): 34–37.

Cavanagh, K., Pasquier, E. D., & Lennard, C. (2002). Background interference from car carpets: The evidential value of petrol residues in cases of suspected vehicle arson. *Forensic Science International* 125 (1): 22–36, doi: 10.1016/s0379-0738(01)00610-7.

Cavanagh-Steer, K., Du Pasquier, E., Roux, C., & Lennard, C. 2005. The transfer and persistence of petrol on car carpets. *Forensic Science International* 147 (1): 71–79, doi: 10.1016/j.forsciint.2004.04.081.

Coulson, S., Morgan-Smith, R., Mitchell, S., & McBriar, T. 2008. An investigation into the presence of petrol on the clothing and shoes of members of the public. *Forensic Science International* 175 (1): 44–54, doi: 10.1016/j.forsciint.2007.05.005.

Coulson, S. A., & Morgan-Smith, R. K. 2000. The transfer of petrol on to clothing and shoes while pouring petrol around a room. *Forensic Science International* 112 (2–3): 135–41, doi: 10.1016/s0379-0738(00)00179-1.

CSVT. 2012. Crime Scene Virtual Tour. Retrieved January 31, 2012, from http://crime-scene-vr.com/index.html/.

Curtin, D. P. 2011. Curtin's on-line library of digital photography. Retrieved January 30, 2012, from http://www.shortcourses.com/.

Deans, J. 2006. Recovery of fingerprints from fire scenes and associated evidence. *Science and Justice* 46 (3): 153–68.

DeHaan, J. D. 1995. The reconstruction of fires involving highly flammable hydrocarbon liquids. PhD diss., University of Strathclyde, Glasgow, Scotland, UK.

———. 1999. The influence of temperature, pool size, and substrate on the evaporation rates of flammable liquids. *Proceedings of InterFlam 99, June 29–July 1, Edinburgh, UK.* Greenwich, UK: Interscience.

———. 2001. Full-scale compartment fire tests. *CAC News* (Second Quarter): 14–21.

———. 2004. *Advanced tools for use in forensic fire scene investigation, reconstruction and documentation.* Paper presented at the 10th International Fire Science and Engineering Conference, Interflam, July 5–7, Edinburgh, UK.

———.2012. Sustained combustion of bodies: Some observations, *Journal of Forensic Science,* May 4, doi: 10.1111/j.1556-4029.2012.02190.x.

DeHaan, J. D., & Icove, D. J. 2012. *Kirk's fire investigation,* 7th ed. Upper Saddle River, NJ: Pearson-Prentice Hall.

Deng, Y., & Zhang, T. 2012. Generating panorama photos. Palo Alto, CA: Hewlett-Packard Labs.

Draganfly. 2012. DraganFlyerX8: Fire investigation. Retrieved January 31, 2012, from http://www

.draganfly.com/uav-helicopter/draganflyer-x8/
applications/government.php/.

Eos. 2012. Eos Systems: Imaging and measurement technology. Retrieved January 31, 2012, from http://www.eossystems.com/.

Faxitron. 2012. Faxitron bioptics: Forensic analysis. Retrieved January 31, 2012, from http://faxitron.com/scientific-industrial/forensics.html/.

FBI. 1999. Definitions and guidelines for the use of imaging technologies in the criminal justice system. *Forensic Science Communications* 1 (3).

———. 2004. Scientific Working Group on Imaging Technology (SWGIT) references/resources. *Forensic Science Communications* (March).

FLIR. 2012. FLIR: The world leader in thermal imaging. Retrieved January 31, 2012, from http://www.flir.com/US/.

Forensics Source. 2012. Crime scene documentation: Photo ID cards. Retrieved January 31, 2012, from http://www.forensicssource.com/ProductList.aspx?CategoryName=Crime-Scene-Documentation-Photo-ID-Cards/.

Forensic Magazine. 2011. Leica Geosystems introduces NIST traceable targets. *Forensic Magazine*, August 22.

Geiman, J. A., & Lord, J. M. 2011. Systematic analysis of witness statements for fire investigation. *Fire Technology*, 1–13, doi: 10.1007/s10694-010-0208-3.

GigaPan. 2012. GigaPan Systems. Retrieved January 31, 2012, from http://gigapan.org/.

Golden. 2012. Golden Engineering, Inc.: Manufacturer of lightweight portable X-Ray Machines. Retrieved January 31, 2012, from http://www.goldenengineering.com/.

Google. 2012. Google SketchUp: 3D modeling for everyone. Retrieved January 31, 2012, from http://sketchup.google.com/.

Harris, E. C. 2012. Harris Matrix.com: Home of archaeology's premier stratigraphy system, January 31, 2012, from http://www.harrismatrix.com/.

Hewitt, Terry-Dawn. 1997. A primer on the law of spoliation of evidence in Canada. *Fire & Arson Investigator* 48, no. 1 (September):17–21.

HTRI. 2012. HTRI Forensics: The cross-technology integration specialists. Retrieved January 31, 2012, from www.htriforensics.com/.

Howard, J. 2010. *Practical HDRI: High dynamic range imaging for photographers,* 2nd ed. Santa Barbara, CA: Rocky Nook.

Icove, D. J., & Dalton, M. W. 2008. A comprehensive prosecution report format for arson cases. Paper presented at the International Symposium on Fire Investigation Science and Technology, May 19–21, Cincinnati, OH.

Icove, D. J., & DeHaan, J. D. 2009. Forensic fire scene reconstruction, Pearson-Prentice Hall.

Icove, D. J., & Gohar, M. M. 1980. Fire investigation photography. In *Fire investigation handbook*, ed. F. L. Brannigan, R. G. Bright, and N. H. Jason. NBS Handbook 123. Washington, DC: National Bureau of Standards, U.S. Department of Commerce.

Icove, D. J., & Henry, B. P. 2010. Expert report writing: Best practices for producing quality reports. Paper presented at the International Symposium on Fire Investigation Science and Technology, (ISFI 2010) September 27–29, University of Maryland, College Park, MD.

Icove, D. J., Wherry, V. B., & Schroeder, J. D. 1998. *Combating arson-for-profit: Advanced techniques for investigators,* 2nd ed. Columbus, OH: Battelle Press.

Infinite Jib. 2012. Droidworkx airframe multi-rotor helicopter. Retrieved January 31, 2012, from http://www.infinitejib.com/.

Innovxsys. 2012. Olympus inspection and measurement systems. Retrieved January 31, 2012, from http://www.olympus-ims.com/en/innovx-xrf-xrd/.

iPIX. 2012. Minds-Eye-View, Inc. Retrieved January 31, 2012, from http://www.ipix.com/.

Karlsson, B., & Quintiere, J. G. 2000. *Enclosure fire dynamics*. Boca Raton, FL: CRC Press.

Kimball, C. 2012. Fire photographer magazine. Retrieved January 31, 2012, from http://firephotomagazine.com/.

King, C. G., & Ebert, J. I. 2002. Integrating archaeology and photogrammetry with fire investigation. *Fire Engineering* (February): 79.

Kodak. 1968. Basic police photography. Publication no. M77. Rochester, NY: Eastman Kodak Co.

Leica. 2011. Leica Geosystems introduces NIST traceable targets for 3D laser scanning of crime scenes to support ISO accreditation. Retrieved January 31, 2012, from www.leica-geosystems.us/forensic/.

———. 2012. Leica Geosystems. Retrieved January 31, 2012, from http://www.leica-geosystems.us/.

LeMay, J. 2002. Using scales in photography. *Law Enforcement Technology* 29 (10): 142.

Lennard, C. 2007. Fingerprint detection: Current capabilities. *Australian Journal of Forensic Sciences* 39 (2): 55–71, doi: 10.1080/00450610701650021.

Lentini, J. J., Dolan, J. A., & Cherry, C. 2000. The petroleum-laced background. *Journal of Forensic Science* 45 (5): 968–89.

Lipson, A. S. 2000. *Is it admissible?* Costa Mesa, CA: James, 44-2–44-8.

Madrzykowski, D., & Walton, W. D. 2004. Cook County Administration Building fire, October 17, 2003. *NIST SP 1021*. Gaithersburg, MD: National Institute of Standards and Technology.

Magellan. 2012. Magellan Mobile Mapper CE Retrieved January 31, 2012, from http://www.magellangps.com/.

Mamiya Leaf. 2012. Mamiya Leaf Imaging, Ltd. Retrieved January 31, 2012, from http://www.mamiyaleaf.com/.

Mann, D. C., & Putaansuu, N. D. 2006. Alternative sampling methods to collect ignitable liquid resides from non-porous areas such as concrete. *Fire and Arson Investigator,* 57 (1): 43–46.

Mann, J., N. Nic Daèid, A. Linacre. 2003. An investigation into the persistence of DNA on petrol bombs.

Proceedings of the European Academy of Forensic Sciences, Istanbul.

NBS. 1980. Brannigan, F. L., R. G. Bright, and N. H. Jason, eds. *Fire investigation handbook*. NBS Handbook 123. Washington, DC: National Bureau of Standards, U.S. Department of Commerce.

NFPA. 1998. *NFPA 906: guide for fire incident field notes*. Quincy, MA: National Fire Protection Association.

———. 2009a. *NFPA 705: Recommended practice for a field flame test for textiles and films*. Quincy, MA: National Fire Protection Association.

———. 2009b. *NFPA 1033: Standard for professional qualifications for fire investigator*. Quincy, MA: National Fire Protection Association.

———. 2011. *NFPA 921: Guide for fire and explosion investigations*. Quincy, MA: National Fire Protection Association.

———. 2012. *NFPA 170: Standard for fire safety and emergency symbols*. Quincy, MA: National Fire Protection Association.

NIJ. (2000). Fire and arson scene evidence: A guide for public safety personnel (O. o. J. Programs, Trans.) *Technical Working Group on Fire/Arson Scene Investigation (TWGFASI)* (pp. 48). Washington, DC: National Institute of Justice.

Nowlan, M., Stuart, A. W., Basara, G. J., & Sandercock, P.M.L. 2007. Use of a solid absorbent and an accelerant detection canine for the detection of ignitable liquids burned in a structure fire. *Journal of Forensic Sciences* 52 (3): 643-648, doi: 10.1111/j.1556-4029.2007.00408.x.

Panoscan. 2012. Panoscan: A breakthrough in panoramic capture. Retrieved January 31, 2012, from http://www.panoscan.com/.

Peige, J. D., & Williams, C. E. 1977. *Photography for the fire service*. Oklahoma City, OK: Oklahoma State University.

Phase One. 2012. Phase One IQ photography. Retrieved January 31, 2012, from http://www.phaseone.com/.

PTGui. 2012. PTGui: Create high quality panoramic images. Retrieved January 31, 2012, from http://www.ptgui.com/.

QuickTime. 2012. QuickTime 7. Retrieved January 31, 2012, from http://www.apple.com/quicktime/.

Rich, L. 2007. Rooms to go. Personal communication.

Riegl. 2012. Riegl laser measurement systems. Retrieved January 31, 2012, from http://www.riegl.com/.

Roxio. 2012. Roxio Creator 2011 Digital Media Suite. Retrieved January 31, 2012, from http://www.roxio.com/.

Saitoh, N., & Takeuchi, S. 2006. Fluorescence imaging of petroleum accelerants by time-resolved spectroscopy with a pulsed Nd-YAG laser. *Forensic Science International* 163 (1–2): 38–50, doi: 10.1016/j.forsciint.2005.10.025.

Schmidt, C. W., & Symes, S. A., eds. 2008. The analysis of burned human remains. Academic Press. London

Spheron. (2012). Spheron VR Virtual Technologies. Retrieved January 31, 2012, from http://www.spheron.com/.

Stanley. 2012. Stanley FatMax Electronic Distance Measuring Tool. Retrieved January 31, 2012, from http://www.stanleytools.com/.

Stauffer, E., Dolan, J. A., & Newman, R. 2008. *Fire debris analysis*. Boston, MA: Academic Press.

Stevick, G., Zicherman, J., Rondinone, D., & Sagle, A. 2011. Failure analysis and prevention of fires and explosions with plastic gasoline containers. *Journal of Failure Analysis and Prevention* 11(5): 455-465, doi: 10.1007/s11668-011-9462-z.

Svare, R. J. 1988. Determining fire point-of-origin and progression by examination of damage in the single phase, alternating current electrical system. Paper presented at the International Arson Investigation Delegation to the People's Republic of China and Hong Kong.

Sweeney, G. O., & Perdew, P. R. 2005. Spoliation of evidence: Responding to fire scene destruction. *Illinois Bar Journal* 93 (July): 358–67.

SWGIT. 2002. Guidelines for the use of imaging technologies in the criminal justice system. Version 2.3 2002.06.06. Hollywood, FL: International Association for Identification, Scientific Working Group on Imaging Technologies.

Tassios, T. P. 2002. Monumental fires. *Fire Technology* 38 (3): 11–17.

Thali, M. J., Yen, K., Schweitzer, W., Vock, P., Ozdoba, C., & Dirnhofer, R. 2003. Into the decomposed body: Forensic digital autopsy using multislice-computed tomography. *Forensic Science International* 134 (2–3): 109–14, doi: 10.1016/s0379-0738(03)00137-3.

ThermoFisher. 2012. ThermoFisher Scientific. Retrieved January 31, 2012, from http://www.thermofisher.com/.

3rd Tech. 2012. 3rdTech, Inc.: Advanced imaging and 3D products for law enforcement and security applications. Retrieved January 30, 2012, from http://www.3rdtech.com/.

Tontarski, K. L., Hoskins, K. A., Watkins, T. G., Brun-Conti, L., & Michaud, A. L. 2009. Chemical enhancement techniques of bloodstain patterns and DNA recovery after fire exposure. *Journal of Forensic Sciences* 54 (1): 37–48, doi: 10.1111/j.1556-4029.2008.00904.x.

Tontarski, R. E. 1985. Using absorbents to collect hydrocarbon accelerants from concrete. *Journal of Forensic Sciences* 30 (4): 1230–32.

Vaisala. 2012. Vaisala: A global leader in environmental and industrial measurement, especially weather measurement. Retrieved January 31, 2011, from http://www.vaisala.com/.

Visio. 2012. Microsoft Visio 2000 Crime Scenes add-in available for download. Retrieved January 31, 2012, from http://support.microsoft.com/kb/274454/.

White, R. H., & Dietenberger, M. A. 2010. Fire safety of wood construction. Chap. 18 in *Wood handbook*. Madison, WI: U.S. Department of Agriculture, Forest Products Laboratory. Retrieved from http://www.fpl.fs.fed.us/documnts/fplgtr/fplgtr190/.

Wilkinson, P. 2001. Archaeological survey site grids. *Practical Archaeology* 5 (Winter): 21–27.

Xerox. 2012. PhaserMatch 4.0: Xerox professional color matching. Retrieved January 31, 2012, from http://www.office.xerox.com/latest/PMSDS-07.PDF/.

Zarbeco. 2012. Zarbeco: Portable and powerful digital imaging solutions. Retrieved January 31, 2012, from http://www.zarbeco.com/.

Legal References

Commonwealth of Pennsylvania v. Paul S. Camiolo, Montgomery County, No. 1233 of 1999.

Daubert v. Merrell Dow Pharmaceuticals Inc. 509 U.S. 579 (1993); 113 S. Ct. 2756, 215 L. Ed. 2d 469.

Daubert v. Merrell Dow Pharmaceuticals Inc. (Daubert II), 43 F.3d 1311, 1317 (9th Cir. 1995).

5

Arson Crime Scene Analysis

Courtesy of Mike Dalton.

❝You know my methods in such cases, Watson: I put myself in the man's place, and having first gauged his intelligence, I try to imagine how I should myself have proceeded under the same circumstances.**❞**

—Sir Arthur Conan Doyle
The Adventure of the Musgrave Ritual

KEY TERMS

accelerant, *p. 226*

motive, *p. 227*

OBJECTIVES

After reading this chapter, the student should be able to:

- Describe arson crime scene analysis.
- Define the crime of arson.
- Classify motives.
- Identify indicators of motives.

Based on published statistics from the National Fire Protection Association (NFPA), in 2010, U.S. fire departments responded to 1,331,500 fires, or one every approximately 24 seconds. Of these incidents, 36 percent occurred in structures, 16 percent in highway vehicles, and 48 percent in outside properties. Loss of life is always a grave concern in fires. The NFPA estimated that 3,120 civilians died in 2010, with approximately 85 percent while they were at home. These statistics were compiled from fire departments that responded to the NFPA's national fire experience survey (Karter 2011).

A leading cause of fire in the United States, arson continues to be an urgent national problem. Often characterized as a clandestine tool for criminals, the true arson picture is not clearly known. The NFPA statistics for 2010 estimate that there were 27,500 intentionally set structure fires resulting in an estimated 200 civilian deaths and $585 million in property losses. There were also an estimated 14,000 intentionally set vehicle fires resulting in $89 million in property loss (Karter 2011).

Although the NFPA collects and statistically projects fire data trends in the United States, the Federal Bureau of Investigation (FBI) collects actual crime incident data through its Uniform Crime Reporting (UCR) Program. In 2010, 15,475 law enforcement agencies reported 56,825 arsons. From FBI analysis of the reported data, arsons involving structures (e.g., residential, storage, and public) accounted for 45.5 percent of the total number of arson offenses. Mobile property was involved in 26 percent of arsons, and other types of property (such as crops, timber, and fences) accounted for 28.5 percent of reported arsons. The FBI's analysis revealed that the average dollar loss due to arson was $17,612 per incident, with the highest average loss of $133,717 per incident in industrial/manufacturing structures (FBI 2010). With local fire and police resources dwindling, it is nearly certain that many intentionally set fires are never even reported let alone investigated and included in these statistics.

However, insight into arson statistics is not limited just to the United States. In 1999 the UK Home Office provided a view of that country's problems (Home Office 1999). On the other side of the globe, Australia specifically addressed their unique wildland "bushfire" arson problem with a statistics-driven prevention approach (Willis 2004; Anderson 2010).

Economically, arson affects insurance premium rates, removes taxable property assets, and erodes our communities. Historically, the inner areas of our large cities often are the hardest hit, and the result is that much of the cost of this destructive crime falls on those who can least afford it. The correlation between a healthy economy and a decline in business failure equates to fewer arson-for-profit cases. However, historical trends show repeatedly that a decline in the economy results in an increase in arsons (Decker and Ottley 2009).

An understanding of the crime of arson based on its motivations may enhance an investigator's efforts and provide a focus for intervention efforts. Examination of the fire scene and reporting of the findings may facilitate dialogue among the various disciplines and investigative units involved in arson enforcement, investigation, and prevention.

It is intended that the information in this chapter will assist arson investigators in developing skills for reading and interpreting the characteristics of crime scene evidence and applying that evidence to the arsonist's behavior and patterns of motivation.

Arson as a Crime

The definitions for arson range widely, owing primarily to differing statutory terminology. The FBI's UCR program defines arson as "any willful or malicious burning or attempting to burn, with or without intent to defraud, a dwelling house, public building, motor vehicle or aircraft, or personal property of another" (FBI 2010). The most accepted definition is that *arson* is simply the willful and malicious burning of property (Icove et al. 1992).

According to *NFPA 921*, an *incendiary fire* is "a classification of the cause of a fire that is intentionally ignited under circumstances in which the person igniting the fire knows the fire should not be ignited" (see *NFPA 921*, 2011 ed., pt. 3.3.103) (NFPA 2011).

The criminal act of arson is usually divided into three elements (DeHaan and Icove 2012):

1. ***There has been a burning of property.***   This must be shown to the court to be actual destruction, at least in part, not just scorching or sooting (although some states include any physical or visible impairment of any surface). As used here, "burning" includes destruction by explosion.
2. ***The burning is incendiary in origin.***   Proof of the existence of an effective incendiary device, no matter how simple it may be, is adequate. Proof must be established by exhaustively considering all hypotheses using the scientific method.
3. ***The burning is shown to be started with malice.***   The fire is started with the intent of destroying property (i.e., a person starts a fire or causes an explosion with the purpose of destroying a building of another or one's own with fire). The "degree" of the arson charge in most jurisdictions has to do with the occupancy: first degree corresponding to an occupied structure, second to an unoccupied structure, and third to other property.

accelerant ■ A fuel (usually a flammable liquid) that is used to initiate or increase the intensity or speed of spread of fire.

An arsonist is a person apprehended, charged, and convicted of one or more arsons. Arsonists commonly use **accelerants**, which are any type of material or substance added to the targeted materials to enhance the combustion of those materials and to accelerate the burning (Icove et al. 1992). There are many serial arsonists who set fires to available combustibles without using any accelerants.

DEVELOPING THE WORKING HYPOTHESIS

Chapter 1 explored in depth the development of a *working hypothesis* using the *scientific method*. Several factors that may contribute to a working hypothesis for an incendiary fire are discussed in *NFPA 921,* 2011 ed., chap. 22 (NFPA 2011). Displayed in Figure 5-1 is a mosaic of several factors that may contribute to a working hypothesis in an incendiary fire or explosion, derived from *NFPA 921*. These factors should not be considered all-inclusive; other indicators may exist. However, investigators should be cautioned that one or more of these factors are not necessarily sufficient to constitute a finding of an incendiary fire.

MULTIPLE FIRES

When an offender is involved with three or more fires, particular terminology is used to describe that crime (Icove et al. 1992).

■ *Mass arson* involves an offender who sets three or more fires at the same site or location during a limited period of time.
■ *Spree arson* involves an arsonist who sets three or more fires at separate locations with no emotional cooling-off or delay period between the fires.
■ *Serial arson* involves an offender who sets three or more fires with a cooling-off or delay period between the fires.

This chapter includes a detailed discussion of how arson crime scenes are analyzed. Such analysis often leads to the identification of offenders, their method of operation, and their areas of frequent travel.

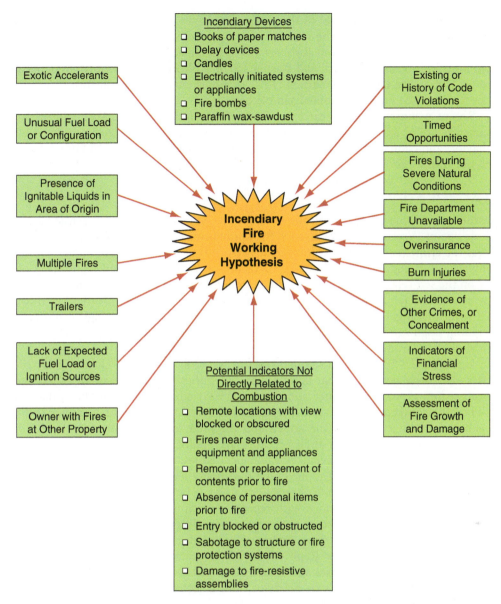

Incendiary Devices
- ❑ Books of paper matches
- ❑ Delay devices
- ❑ Candles
- ❑ Electrically initiated systems or appliances
- ❑ Fire bombs
- ❑ Paraffin wax-sawdust

Exotic Accelerants

Unusual Fuel Load or Configuration

Presence of Ignitable Liquids in Area of Origin

Multiple Fires

Trailers

Lack of Expected Fuel Load or Ignition Sources

Owner with Fires at Other Property

Incendiary Fire Working Hypothesis

Existing or History of Code Violations

Timed Opportunities

Fires During Severe Natural Conditions

Fire Department Unavailable

Overinsurance

Burn Injuries

Evidence of Other Crimes, or Concealment

Indicators of Financial Stress

Assessment of Fire Growth and Damage

Potential Indicators Not Directly Related to Combustion
- ❑ Remote locations with view blocked or obscured
- ❑ Fires near service equipment and appliances
- ❑ Removal or replacement of contents prior to fire
- ❑ Absence of personal items prior to fire
- ❑ Entry blocked or obstructed
- ❑ Sabotage to structure or fire protection systems
- ❑ Damage to fire-resistive assemblies

FIGURE 5-1 Several factors that may contribute to a working hypothesis for an incendiary fire or explosion. *Derived from NFPA 921, 2011 ed., chap. 22 (NFPA 2011).*

Classification of Motive

Juries often want to have a demonstrable **motive** considered with the evidence so that they can justify the verdict to themselves, even though it is not a legal requirement. Although motive is not essential to establish the crime of arson, and need not be demonstrated in court, the development of a motive frequently leads to the identity of the offender. Establishing motive also provides the prosecution with a vital argument when presented to the judge and jury during trial. It is thought that the motive in an arson case often becomes the mortar that holds together the elements of the crime.

Most of the literature on fire setting and arson concentrates on either psychiatric/psychological or criminological studies (Kocsis and Cooksey 2006). The criminological studies on arson deal with either anecdotal case studies or motive-based classification

motive ■ An inner drive or impulse that is the cause, reason, or incentive that induces or prompts a specific behavior (Rider 1980).

schemes and typologies. Several of the earlier typologies contributed significantly to the traditional understanding of the motives and psychological profiles of arsonists (Lewis and Yarnell 1951; Robbins and Robbins 1967; Steinmetz 1966; Hurley and Monahan 1969; Inciardi 1970; Vandersall and Wiener 1970; Wolford 1972; Levin 1976; Icove and Estepp 1987). The Lewis and Yarnell (1951) study is cited only for historical purposes, since it was based on a very limited sample base.

Recent comprehensive research work by Gannon and Pina (2010) addresses the characteristics of adult fire setters, contemporary research on the theories of fire-setting behavior, and clinical interventions. The researchers suggest that clinical knowledge and practice relating to fire setting is extremely underdeveloped.

CLASSIFICATION

Research on arsonists has historically been conducted largely from the forensic psychiatric viewpoint (Vreeland and Waller 1978). Many forensic researchers do not necessarily assess the crime from the law enforcement perspective. They may have limited access to complete adult and juvenile criminal records and investigative case files and must often rely on the self-reported interviews of the offenders as being totally truthful. Often, these researchers do not have the capabilities and time to validate the information through follow-up investigations. Other researchers have cited that methodological difficulties, including small sample sizes of interviews and skewed databases, may have biased some of the previous studies (Harmon, Rosner, and Wiederlight 1985).

Inciardi's (1970) work deals with a typology of adult fire setters. Geller (1992, 2008) offers exhaustive reviews of that literature and identifies 20 or more attempts to classify arsonists into typologies. Canter and Fritzon (1998) advance the hypothesis that there will be behavioral consistencies in the criminal actions of arsonists that characterize them, which include their target of the attack and the offender's motivation. Doley (2003a) examines the various attempts at classification of typology and motive.

Medical studies suggest several physiological issues. Virkkunen et al. (1987) examined cerebrospinal fluid (CSF) monoamine metabolite levels in 20 arsonists and 20 habitually violent offenders, and 10 healthy inpatient volunteers. Although they found no correlation in CSF concentrations with repeat fire-setting behavior, the researchers did find a tendency for the arsonists to suffer from hypoglycemia.

Studies have looked at various aspects of criminal and psychiatric behavior in convicted arsonists. The study by Repo et al. (1997) examined the medical and criminal records of 282 Finnish arsonists and found that recidivist offenders had common traits of alcohol dependence, antisocial behavior, and a history of long-term enuresis (bedwetting) during their childhood.

Stewart and Culver, (1982) and later Yang and Geller (2009), examined the clinical perspective of dealing with juvenile fire setters. The demographics of female arsonists have been studied by Harmon, Rosner, and Wiederlight (1985) and by Miller and Fritzon (2007).

OFFENDER-BASED MOTIVE CLASSIFICATION

For classification purposes, FBI behavioral science research defines *motive* as an inner drive or impulse that is the cause, reason, or incentive that induces or prompts a specific behavior (Rider 1980). A motive-based method of analysis can be used to identify personal traits and characteristics exhibited by an unknown offender (Icove 1979; Icove and Estepp 1987; Douglas et al. 1986; Douglas 1992, 2006; Icove et al. 1992; Allen 1995). For legal purposes, the motive is often helpful in explaining why an offender committed his or her crime. However, motive is not normally a statutory element of a criminal offense. The motivations discussed in this chapter are also outlined and described in *NFPA 921: Guide for Fire and Explosion Investigations* (see NFPA 921, 2011 ed., pt. 22.4.9.3) and the FBI's *Crime Classification Manual* (Douglas et al. 1992, 2006; Icove et al. 1992).

Santtila, Fritzon, and Tamelander (2004) underscored the importance of linking an arsonist to his or her crime. The arsonist's consistency in offense behavior is linked to motive (Fritzon, Canter, and Wilton 2001). The latter group's research examined 248 arson cases that formed 42 series of arsons and examined 45 variables. The research results statistically confirmed earlier studies suggesting that serial and spree arson cases could be linked based on a consistency of behaviors from one crime scene to another. Woodhams, Bull, and Hollin (2007) underscore the importance for investigators to be familiar with case linkage research.

Law enforcement–oriented studies on arson motives are *offender-based*; that is, they look at the relationship between the behavioral and the crime scene characteristics of the offender as it relates to motive. One of the largest offender-based studies consists of 1,016 interviews of both juveniles and adults arrested for arson and fire-related crimes during the years 1980–1984 by the Prince George's County Fire Department (PGFD), Fire Investigations Division (Icove and Estepp 1987). These offenses included 504 arrests for arson, 303 for malicious false alarms, 159 for violations of bombing/explosives/fireworks laws, and 50 for miscellaneous fire-related offenses.

The overall purpose of the PGFD study was to create and promote the use of motive-based offender profiles of individuals who commit incendiary and fire-related crimes to assist in investigations. Prior studies failed to address comprehensively the issues confronting modern law enforcement. Of primary concern were the efforts to provide logical, motive-based investigative leads for incendiary crimes.

The study was conducted primarily because fire and law enforcement professionals were enabled to take on the task of conducting their own independent research into violent incendiary crimes. The PGFD study determined that arrested and incarcerated arsonists most often give the following motives (Icove and Estepp 1987). Since this keynote study these major motive categories have become accepted in the fire investigation field after first being included in *Kirk's Fire Investigation*, then in *NFPA 921*, 2011 ed., pt. 22.4.9:

- Vandalism
- Excitement
- Revenge
- Crime concealment
- Profit
- Extremist beliefs

This system of classification advances the earlier approach by the FBI to divide criminal offenders, including arsonists, into two groups: organized and disorganized (Douglas et al. 1986). Data in a study by Kocsis, Irwin, and Hayes (1998) supported the presence of an organized/disorganized typology. The researchers analyzed and validated hypotheses about the crime of arson using the crime scene characteristics of profit-motivated and vandalism-motivated arson offenses.

In some cases several of the preceding motives may be determined. Such cases can be classified as being mixed motives. For example, a businessperson might murder a partner after work, set a fire at their store to cover up the crime, remove valuable merchandise, and file an inflated insurance claim. This hypothetical case would exhibit overlapping motives of revenge, crime concealment, and profit. See a later section for a more detailed discussion of mixed motives.

VANDALISM-MOTIVATED ARSON

Vandalism-motivated arson is defined as malicious or mischievous fire setting that results in damage to property (see Table 5-1 and Figure 5-2). Some of the most common targets of these usually juvenile arsonists are schools, school property, and educational facilities. Vandals also frequently target abandoned structures and combustible vegetation. The typical vandalism-motivated arsonist will use available materials to set fires with book

TABLE 5-1	Vandalism-Motivated Arson
Characteristics Victimology: targeted property	Educational facilities common target
	Residential areas
	Vegetation (grass, brush, woodland, and timber)
Crime scene indicators frequently noted	Multiple offenders acting spontaneously and impulsively
	Crime scenes reflect spontaneous nature of the offense (disorganized)
	Offenders use available materials at the scene and leave physical evidence behind (shoeprints, fingerprints, etc.)
	Flammable liquids occasionally used
	Entrance may be gained through windows of secured structures
	Matchbooks, cigarettes, and spray-paint cans (graffiti) often present
	Materials missing from scene and general destruction of property
Common forensic findings	Presence of flammable liquids
	Presence of fireworks
	Glass particles on suspect's clothing if entered by breaking a window
Investigative considerations	Typical offender is a juvenile male with 7–9 years of formal education
	Records of poor school performance
	Not employed
	Single and lives with one or both parents
	Alcohol and drug use usually not associated
	Offender may already be known by police and may have arrest record
	Majority of offenders live more than 1 mile from the crime scene
	Most offenders flee from scene immediately and do not return
	If the offenders return, they view the fire from a safe vantage point
	Investigators should solicit assistance from school, fire, and police
Search warrant suggestions	Spray-paint cans
	Items from the scene
	Explosive devices
	Flammable liquids
	Clothing: evidence of flammable liquid, glass particles
	Shoes: shoe prints, flammable liquid traces

Sources: Updated from Icove and Estepp 1987; Icove et al. 1992; Sapp et al. 1995.

matches and cigarette lighters as the ignition devices. Time-delay devices and liquid accelerants are very rarely used. The motive classification for vandalism is referenced in *NFPA 921*, 2011 ed., pt. 22.4.9.3.2 (NFPA 2011).

Usually, vandalism-motivated arsonists will leave the scene and not return to it. Their interest is in setting the fire, not watching it or the firefighting activities it generates. On average, vandalism arsonists will be questioned twice before being arrested and charged.

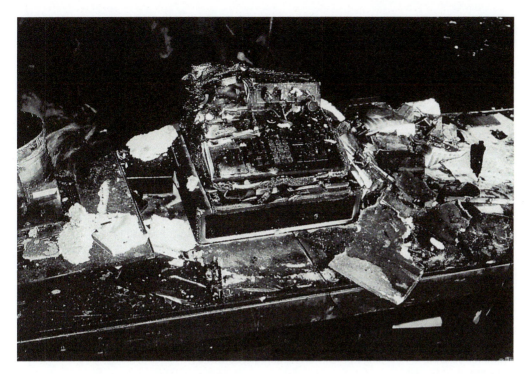

FIGURE 5-2 Vandalism-motivated arson showing a superficially damaged cash register. *Courtesy of D. J. Icove.*

They will offer no resistance when arrested but will qualify and minimize their responsibility. After an initial not-guilty plea, vandalism-motivated arsonists will typically change the plea to guilty before trial.

EXAMPLE 5-1 ■ Vandalism: Case Study of a Vandalism-Motivated Serial Arsonist

A 19-year-old high-school dropout was responsible for a series of arsons in a northeastern city that were set using available materials and lit by a cigarette lighter. He admitted to setting 31 fires in vacant buildings and garages as well as Dumpsters and derelict vehicles. He was questioned twice before being arrested and formally charged with nine of the house fires. When interviewed, he stated, "I just burned them for the hell of it. You know, just to have something to do. Those old houses and that other stuff was not worth anything anyway."

His mother and father had divorced when he was 2 years old. He had lived alternately with his mother and grandmother. He had no contact with his father and said that his mother had remarried several times. After dropping out of school in the 10th grade, he had worked sporadically at unskilled jobs and continued to live with his grandmother.

Several times, he reported just striking a match, tossing it into dry leaves or grass, and walking away without even seeing if it ignited. "I did not care nothing about watching no fire. It was just something to do. There was not much going on most of the time, you know. Just hanging out. It was just kid stuff, just for the hell of it. Half of the guys I knew that hung out would set a fire for the hell of it" (Sapp et al. 1995).

EXCITEMENT-MOTIVATED ARSON

Excitement-motivated offenders include seekers of thrills, attention, recognition, and, rarely but importantly, sexual gratification (see Table 5-2 and Figure 5-3). The arsonist who sets fires for sexual gratification is quite rare. The motive classification for excitement is referenced in *NFPA 921*, 2011 ed., pt. 22.4.9.3.3 (NFPA 2011).

TABLE 5-2	Excitement-Motivated Arson
Characteristics Victimology: targeted property	Dumpsters
	Vegetation (grass, brush, woodland, timber)
	Lumber stacks
	Construction sites
	Residential property
	Unoccupied structures
	Locations that offer a vantage point to observe fire suppression and investigation safely
Crime scene indicators frequently noted	Often adjacent to outdoor "hangouts"
	Often uses available materials on hand to start small fires
	If incendiary devices are used, they usually have time-delayed triggering mechanisms
	Offenders in the 18–30 age group more prone to using accelerants
	Match/cigarette delay device frequently used to ignite vegetation fires
	Small group of offenders is motivated by sexual perversions, leaving ejaculate, fecal deposits, pornographic material
Common forensic findings	Fingerprints, vehicle and bicycle tracks
	Remnants of incendiary devices
	Ejaculate or fecal materials
Investigative considerations	Typical offender is a juvenile or young adult male with 10+ years of formal education
	Offender unemployed, single, and living with one or both parents from middle- to lower-class bracket
	Offender generally is socially inadequate, particularly in heterosexual relationships
	Use of drugs or alcohol limited usually to older offenders
	History of nuisance offenses
	Distance offender lives from the crime scene determined through a cluster analysis
	Some offenders do not leave, mingling in crowds to watch the fire
	Offenders who leave usually return later to assess the damage and their handiwork
Search warrant suggestions	Vehicle: material similar to incendiary devices, floor mats, trunk padding, carpeting, cans, matchbooks, cigarettes, police/fire scanners
	House: material similar to incendiary devices, clothing, shoes, cans, matchbooks, cigarettes, lighter, diaries, journals, notes, logs, records and maps documenting fires, newspaper articles, souvenirs from the crime scene, police/fire scanners

Sources: Updated from Icove and Estepp 1987; Icove et al. 1992; Sapp et al. 1995.

Potential targets of the excitement-motivated arsonist run the full spectrum from so-called nuisance fires to fires in occupied apartment houses at nighttime. A limited number of firefighters have been known to set fires so they can engage in the suppression effort (Huff 1994; USFA 2003). Security guards have set fires to relieve boredom and gain recognition. Research into this motive category further categorizes excitement-motivated arsonists into several subclassifications, including thrill-, sex-, and recognition/attention-motivated arsonists (Icove et al. 1992). Recent reports on dealing with firefighter arson bring new insight into their prevention by the National Volunteer Fire Council by careful pre-employment screening and criminal records checks (NVFC 2011).

EXAMPLE 5-2 ■ Excitement: Case History of an Excitement/Recognition-Motivated Serial Arsonist

A 23-year-old volunteer firefighter was charged with setting a series of fires shortly after joining the fire department. Initially, he set the fires in trashcans and Dumpsters and then progressed to unoccupied and vacant structures.

FIGURE 5-3 Excitement-motivated arsonists sometimes target garages, which contain and have easily accessible all the materials and fuels needed for setting the fire. *Courtesy of D. J. Icove.*

The firefighter became a suspect in the fires when he consistently arrived first on the scene and frequently reported the fire. He stated that he set the fires so that he could get practice and others would view him as a good firefighter. He commented that his father was "real proud of his firefighter son" (Sapp et al. 1995).

REVENGE-MOTIVATED ARSON

Revenge-motivated fires are set in retaliation for some injustice, real or imagined, perceived by the offender (see Table 5-3 and Figure 5-4). Often, revenge is also an element of other motives. The motive classification for revenge is referenced in *NFPA 921,* 2011 ed., pt. 22.4.9.3.4 (NFPA 2011). Mixed motives are discussed later in this chapter.

What may be of concern to investigators is that the event or circumstance that is perceived as unjust may have occurred months or years before the fire-setting activity (Icove

TABLE 5-3	Revenge-Motivated Arson
Characteristics Victimology: targeted property	Victim of revenge fire generally has a history of interpersonal or professional conflict with offender (lovers' triangle, landlord/tenant, employer/employee)
	Tends to be an intraracial offense
	Female offenders usually target something of significance to victim (vehicle, personal effects)
	Ex-lover offender frequently burns clothing, bedding, and/or personal effects
	Societal revenge targets displace aggression to institutions, government facilities, universities, corporations
Crime scene indicators frequently noted	Female offenders usually burn an area of personal significance, using victim's clothing or other personal effects
	Male offenders begin with an area of personal significance, tend to overkill by using more accelerants or incendiary devices than are necessary
Common forensic findings	Laboratory tests for accelerants, pieces of incendiary device, cloth, fingerprints

(continued)

TABLE 5-3	Revenge-Motivated Arson *(continued)*
Investigative considerations	Offender is predominantly an adult male with 10+ years of formal education
	If employed, the offender is usually a blue-collar worker of low socioeconomic status
	Resides in rental property; loner, unstable relationships
	Event happens months or years after precipitating incident
	Most often, has some periodic law enforcement contact for burglary, theft, and/or vandalism
	Use of alcohol more prevalent than drugs, with possible increase after the fire
	Usually alone at scene and seldom returns after fire started to establish an alibi
	Lives in affected community, with mobility an important factor
	Revenge-focused analysis assists in determining true victim
	Investigative investment significant
Search warrant suggestions	If accelerants suspected: shoes, socks, clothing, bottles, flammable liquids, matchbooks

FIGURE 5-4 A revenge-motivated fire set on a bed, which is characteristic of a focused target having personal significance. *Courtesy of D. J. Icove.*

and Horbert 1990). For threat assessment purposes, this time delay may not readily be recognized, and investigators are urged to pursue the historical incidents in which a person or property has been targeted.

Revenge- and spite-motivated fires account for more of the serious arson cases owing to the overkill reflected in the actions of the offender. In these cases, when one container of a flammable liquid would suffice in setting a building on fire or incinerating a body, the offender may use much larger quantities, reflecting the rage or vengeance of the act.

The broad classification of revenge-motivated arsonists is further divided into subgroups based on the target of the retaliation. Studies show that serial revenge-motivated arsonists are more likely to direct their retaliation at institutions and society than at individuals or groups (Icove et al. 1992).

EXAMPLE 5-3 ■ Revenge: Case History of an Institutional Retaliation "Revenge" Serial Arsonist

John (not his real name) is 31 years old and claims to have set more than 60 fires at various local government facilities in the city where he lives. His fire-setting activities have occurred since he was 19 years old.

John started setting fires after he was sentenced to 180 days in the local jail facility for a petty theft. He claims to have set five fires while in the jail and then later 20–25 trash can fires in the local city hall. His method of operation consisted of simply walking through and dropping lighted matches into the trash containers.

His stated motive for setting the fires was "to cost the city some trouble and money. They did not treat me fair and I will get even." When asked when his revenge against the city would be satisfied, his response was "when the whole damn city hall burns down and the jail too" (Sapp et al. 1995).

EXAMPLE 5-4 ■ Revenge: Case History of a Revenge-Motivated Apartment Fire

An apartment fire was reported at 9:09 a.m. by a neighbor who saw a woman leaving the apartment from which he saw smoke coming seconds later. Fire crews arrived at 9:17 a.m. to find the fire in the apartment well involved, with flames emanating from two windows. An interior hose stream attack began at 9:22 a.m., and control of the bulk of the fire was reported at 9:30 a.m. The apartment consisted of two small rooms (as shown in Figure 5-5) each approximately 3.6 m × 3.6 m (12 ft × 12 ft). The living room/kitchen had two large windows approximately 1.2 m × 1 m high—one next to the front door and one over the sink.

The steel-faced entry door was closed at the time of the fire but was burned to failure when fire crews attempted entry. The doorway to the bedroom had no door, and the bedroom window failed during the fire. Both rooms were very heavily damaged by fire, the living room more so than other areas. There were patterns in the doorway and bedroom clearly indicating the fire had extended into that room from the living room. Only the metal frame and springs of the sofa survived, and there was a V pattern of intense damage to the adjoining walls. The top portions of the kitchen cabinets were severely damaged, and they were charred to floor level. There were two large areas where the floor had burned through, allowing debris to fall into the room below. The gypsum board ceiling of the living room had failed, allowing extension of the fire into the attic above.

The registered occupant had left the apartment about half an hour before the fire, after having an argument with his girlfriend (leaving her behind). He reported that no heaters or appliances, no paints or solvents, and no candles were in use prior to his departure. The original scene investigator concluded that the fire was deliberately ignited with an ignitable liquid based on damage to the floor coverings, the floor penetrations, and the alleged rapidity of fire spread (using a 9:09 a.m. alarm time and 9:17 a.m. fire department arrival time as his time frame (not 9:30 a.m. or later as the actual time of suppression). No samples of debris were collected for lab analysis, and no odors of ignitable liquids were detected during the scene examination. An unused tea light candle was recovered from the debris in the room below, as were closed, intact cans of paint that the occupant reportedly had in storage.

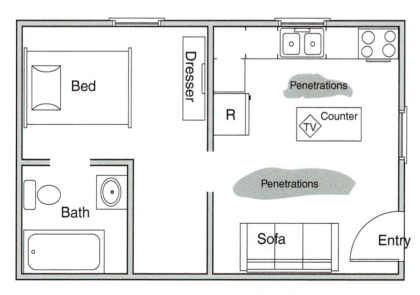

FIGURE 5-5 Floor plan shows sofa and large windows in a small two-room apartment. Sofa was ignited for revenge and drove the living room to flashover, which burned holes through the floor. *Courtesy of Mike Dalton.*

Unfortunately, no comparison samples of sofa or carpet were collected. The photos of the sofa's remains showed that it was a modern all-synthetic upholstered unit that had burned nearly completely away. Such units are not susceptible to smoldering ignition but once ignited by an open flame, will produce a very large fire in less than 5 minutes. In such a small apartment, the ventilation limit would be reached very quickly, but as the large windows failed in the room, more than enough air would have been admitted to support a fire (1500 kW per window) far in excess of that needed to achieve and sustain flashover (700–1700 kW, according to NRC spreadsheet calculations). The fire was (at least) approaching flashover when fire crews arrived and observed fire venting from the windows at 9:17 a.m. Control was reported at approximately 9:30 a.m.; this left at least 13 min (if not more) for a postflashover fire to cause the observed damage without the use of an accelerant. With the elimination of competent accidental ignition sources (by testimony of the occupant), the conclusion was that the girlfriend had set fire to the sofa (in a spite-motivated fire) just before leaving the apartment. She eventually pled guilty.

Case study courtesy of John D. DeHaan

CRIME CONCEALMENT–MOTIVATED ARSON

Arson is the secondary criminal activity in the category of *crime concealment–motivated arson* (see Table 5-4 and Figure 5-6). Examples of crime concealment–motivated arsons

TABLE 5-4	Crime Concealment–Motivated Arson
Characteristics Victimology: targeted property	Dependent on the nature of the concealment
Crime scene indicators frequently noted	Murder: attempts to obliterate forensic evidence of potential lead value or conceal victim's identity, commonly using an accelerant in a disorganized manner
	Burglary: uses available materials to start the fire, characteristic of involvement multiple offenders
	Auto theft: strips and burns vehicle to eliminate fingerprints
	Destruction of records: sets fire in area where records usually kept
Common forensic findings	Determine if victim was alive at time of fire and why he/she did not escape
	Document injuries, particularly if concentrated around genitals
Investigative considerations	Common to find alcohol and recreational drug use
	Offender expected to have history of contacts of arrests by police and fire departments
	Offender is likely a young adult who lives in the surrounding community and is highly mobile
	Crime concealment suggests accompanied to scene by coconspirators
	Murder concealment is usually a one-time event
Search warrant suggestions	Refer to other category of primary motive
	Gasoline containers
	Clothing, shoes, glass fragments, burned paper documents

are fires set for the purpose of covering up a murder or burglary or to eliminate evidence left at a crime scene. The motive classification for crime concealment is referenced in *NFPA 921*, 2011 ed., pt. 22.4.9.3.5 (NFPA 2011).

Other examples include fires set to destroy business records to conceal cases of embezzlement, and fires set to destroy evidence of an auto theft. In these cases, the arsonist may set fire to the structure to destroy the evidence of the initial crime, to obliterate latent fingerprints and shoe prints, and sometimes to attempt to render useless DNA or serological evidence linking him or her to a victim left behind to die in the fire.

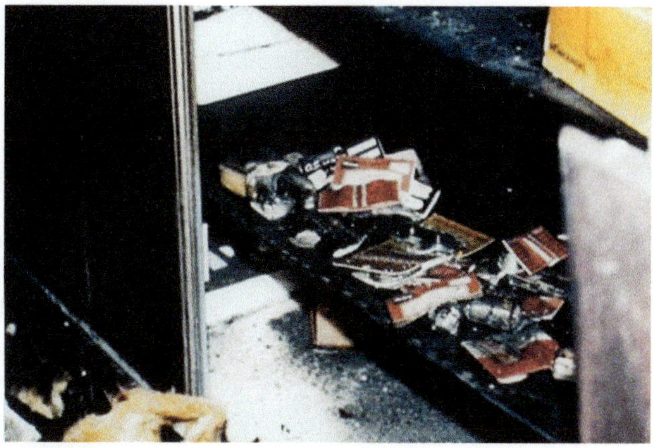

FIGURE 5-6 A crime concealment–motivated arson where valuable inventory was removed prior to the fire. *Courtesy of D. J. Icove.*

EXAMPLE 5-5 ▪ Case History of a Crime-Concealment Serial Arsonist

Three weeks after being released from prison, a 31-year-old unemployed laborer admitted responsibility for burglarizing and setting fires to 12 houses in a residential area over a period of 7 months. All but one of the fires occurred while the owners were out of town.

The offender stated, "I just drove through and looked for newspapers and other things where the people were gone for a while. I just took money and jewelry and stuff that would get lost or burn up in the fire. I would pour gasoline on everything and use a candle to light it off."

The fires were all identified as arson fires because of their rapid development and the detected presence of flammable liquids in the debris. The extensive damage made it difficult for the owners to determine whether valuables in the homes had been removed. The offender stated that he was identified when "one of the ladies spotted a ring in a pawn shop in another town. I would always take stuff somewhere else to sell it. What happened was she recognized the ring and called the cops."

His criminal history record consisted of a burglary conviction when stolen property was recovered and traced back to him. The offender's rationalization was, "I figured it out that if the fire burned everything nobody would know anything was gone. Last time, I got caught on a serial number, so this time I decided I would not leave any way for them to know what was gone" (Sapp et al. 1995).

PROFIT-MOTIVATED ARSON

Offenders in the category of *profit-motivated arson* expect to profit from their fire setting, either directly from monetary gain or more indirectly to profit from a goal other than money (see Table 5-5 and Figure 5-7). Examples of direct monetary gain include insurance fraud, liquidation of property, dissolution of businesses, destruction of inventory, parcel clearance, and gaining employment. The latter is exemplified by a construction worker's wanting to rebuild an apartment complex he destroyed or an unemployed laborer's seeking employment as a forest firefighter or as a logger to salvage burned timber. The motive classification arson-for-profit is referenced in *NFPA 921*, 2011 ed., pt. 22.4.9.3.6 (NFPA 2011) and a complete textbook for investigators is also available (Icove, Wherry, and Schroeder 1980).

Arson-for profit may have interesting twists when the offender benefits directly or indirectly. Arsonists have set fire to western forests to have their equipment rented to support part of the suppression effort. What may be the most disturbing of all are cases

TABLE 5-5	Profit-Motivated Arson
Characteristics	
Victimology: targeted property	Property targeted includes residential, business, and transportation (vehicles, boats, etc.)
Crime scene indicators frequently noted	Usually well-planned and methodical approach, with crime scene more organized because it contains less physical evidence
	With large businesses, multiple offenders may be involved
	Intent is complete destruction with excessive use of accelerants, incendiary devices, multiple points of fire origin, trailers
	Lack of forced entry
	Removal or substitution of items of value prior to fire
Common forensic findings	Use of sophisticated accelerants (water-soluble) or mixtures (gasoline and diesel fuel)
	Components of incendiary devices
Investigative considerations	Primary offender is an adult male with 10+ years of formal education
	Secondary offender is sometimes the "torch," who is usually a male, 25–40 years of age, and unemployed
	Offender generally lives more than 1 mile from the crime scene, may be accompanied to the scene, leaves, and usually does not return
	Indicators of financial difficulty
	Decreasing revenue with increasing and unprofitable production costs
	Technology outdates processes or equipment
	Costly lease or rental arrangements
	Personal expenses paid with corporate funds
	Hypothetical assets, overstated inventory levels
	Pending litigation, bankruptcy
	Prior fire losses and claims
	Frequent changes in property ownership, back taxes, multiple liens
Search warrant suggestions	Check financial records
	If evidence of fuel/air explosion at scene, check local emergency rooms for patients with burn injuries
	Determine condition of utilities as soon as possible

Sources: Updated from Icove and Estepp 1987; Icove et al. 1992; Sapp et al. 1995.

in which parents murder their own children for profit, with fire used to cover the intentional death of the child. Although this motive is uncommon, it is by no means rare (Huff 1997, 1999). Cases have been documented in which an insured child is murdered, but more commonly the parents wish to profit from getting rid of a perceived nuisance or hindrance—their own child. This is particularly true in a single-parent or divorce situation in which the child is viewed as an impediment to freedom or marriage. Filicide by fire is further discussed in a later section of this chapter.

Other nonmonetary reasons from which arsonists may profit range from setting brushfires to enhance the availability of animals for hunting, to burning adjacent properties to improve the view. Also, fires have been set to escape an undesirable environment, such as in the case of seamen who do not wish to set sail (Sapp et al. 1993, 1994).

FIGURE 5-7 Profit-motivated arsons to leverage overinflated insurance sometimes involve well-planned fires set in vacant dwellings fueled with excessive amounts of accelerants. *Courtesy of D. J. Icove.*

EXAMPLE 5-6 ■ Case History of a Profit-Motivated Serial Arsonist

Arnold (not his real name) was a professional arsonist (a "torch") and was serving 2 years in prison for one of his arsons. He admitted having burned down 35 to 40 vacant houses over his career. He became involved in setting fires for profit when talking with a real estate agent who was trying to find him an apartment. "He asked me if I would find somebody to burn a place. Did I know someone who was involved in demolition? I told him yeah and it started from there."

Arnold then partnered with the real estate agent and the demolition specialist. The real estate agent identified the targeted properties and made sure they were vacant. The demolition specialist taught Arnold how to set the fires, even accompanying him on his first job.

Arnold claimed that the fire-setting technique he used never resulted in a confirmed arson. "They could not tell. They would say it is under investigation. When they say it is under investigation, they are pretty sure it is arson but they cannot prove it." His method of operation was to pour 19 to 38 L (5 to 10 gal) of odorless white gas in the attic and leave a time-delayed chemical timer, literally burning the house from the top down. The ignited gasoline resulted in a fast-developing fire that collapsed the roof and overwhelmed the suspicions of the fire investigators, who found it hard to determine that an accelerant was used.

Arnold was concerned that no one would be injured in the fires he set. "I never burned any place where anybody was inside. I made sure I was not hurting anybody. That was important, it really was."

After being implicated with the real estate agent for an attempted arson, Arnold served 2 years in prison. "It was wrong, but at the time I was making good money and I was not hurting anybody. It was easy, but it is a dangerous job. I was afraid every minute while I was doing it."

As an afterthought, Arnold claimed that he was also a victim of his arsons. "I am the one that got hurt. I was underpaid. I was making $700 or $800 a fire, the real estate man was making about $10,000 on his share of the insurance policy, and I am the one that got sent away and everything" (Sapp et al. 1995).

EXTREMIST-MOTIVATED ARSON

Offenders involved in *extremist-motivated arson* may set fires to further social, political, or religious causes (see Table 5-6). Examples of extremist-motivated targets include abortion clinics, slaughterhouses, animal laboratories, fur farms, furrier outlets, and even, now, sport-utility vehicle (SUV) dealers. The targets of political terrorists reflect the focus of the terrorists' wrath. Random target selection also simply generates fear and confusion. Self-immolation has also been carried out as an extremist act. The motive classification for extremism is referenced in *NFPA 921*, 2011 ed., pt. 22.4.9.3.7 (NFPA 2011).

TABLE 5-6	Extremist-Motivated Arson
Characteristics	Analysis of targeted property essential in determining specific motive and represents the antithesis of the offender's belief
Victimology: targeted property	Targets include research laboratories, abortion clinics, businesses, religious institutions
Crime scene indicators frequently noted	Crime scene reflects organized and focused attack by the offender(s)
	Frequently employ incendiary devices, leaving a nonverbal warning or message
	Overkill when setting fire
Common forensic findings	Extremist arsonists are more sophisticated offenders and often use incendiary devices with remote or time-delay ignition
Investigative considerations	Offender is frequently identified with cause or group in question
	May have previous police contact or an arrest record with crimes such as trespassing, criminal mischief, or civil rights violations
	Postoffense claims should undergo threat assessment examination
Search warrant suggestions	Literature: writings, paraphernalia pertaining to a group or cause, manuals, diagrams
	Incendiary device components, travel records, sales receipts, credit card statements, bank records indicating purchases
	Flammable materials: materials, liquids

Sources: Updated from Icove and Estepp 1987; Icove et al. 1992; Sapp et al. 1995.

EXAMPLE 5-7 ■ Extremist-Motivated Arson

This is a fictionalized account of an actual extremist-motivated arson targeting a U.S. government agency (ADL 2003).

A federal jury convicted the head of a former antigovernment extremist group for setting fire to an Internal Revenue Service office. The jury found the 48-year-old leader guilty of destruction of government property and interference with IRS employees.

Prosecutors said he and two other accomplices used 19 L (5 gal) of gasoline and a timing device to start a fire that destroyed the IRS office. The fire caused $2.5 million in damage, and a firefighter was seriously injured while battling the blaze. The leader was also convicted of witness tampering and suborning perjury for asking a witness to lie when testifying before the federal grand jury. The leader reportedly also threatened a witness to prevent his cooperation with law enforcement officers. Prosecutors stated that the three men were motivated by antigovernment sentiments and set the fire as a protest against paying taxes.

Other Motive-Related Considerations

Other factors involving motives appear in the literature and are important to address. This information is provided to clarify inaccuracies and misconceptions.

PYROMANIA

Perhaps most conspicuous by its absence is any mention of *pyromania* in this discussion of motivations. For an authoritative definition of the term, refer to the American Psychological Association's (APA 2000) *Diagnostic and Statistical Manual of Mental*

Disorders, 4th edition, text revised (*DSM-IV-TR*), the standard for psychological and psychiatric diagnoses. The manual defines pyromania as a pattern of deliberate setting of fires for pleasure or satisfaction derived from the relief of tension experienced before the fire setting. The root derivation of the name of the disorder comes from two Greek words "pyro" meaning "fire" and "mania" meaning "loss of reason" or "madness."

A review reveals that each edition of the *DSM* has treated this topic differently. The current edition, *DSM-IV-TR,* does not list pyromania as a diagnosed personality disorder. The manual classifies pyromania and the other five disorders under the heading of impulse-control problems, meaning that a person diagnosed with pyromania fails to resist the impulsive desire to set fires. The definition of this disorder has cycled through the years owing to varying opinions and the lack of a solid definition.

DSM-IV-TR's diagnostic criteria (APA 2000) for pyromania include one or more of the following conditions.

- Deliberate and purposeful fire setting on more than one occasion.
- Tension or affective arousal before the act.
- Fascination with, interest in, curiosity about, or attraction to fire and its situational contexts (e.g., paraphernalia, uses, consequences).
- Pleasure, gratification, or relief when setting fires, or when witnessing or participating in their aftermath.
- The fire setting is not done for monetary gain, as an expression of sociopolitical ideology, to conceal criminal activity, to express anger or vengeance, to improve one's living circumstances, in response to a delusion or a hallucination, or as a result of impaired judgment (e.g., in Dementia, Mental Retardation, Substance Intoxication).
- The fire setting is not better accounted for by Conduct Disorder, a Manic Episode, or Antisocial Personality Disorder.

Researchers ask if the fire-setting impulses—characteristic of the various definitions of pyromania—could be a manifestation of some other disorder. Some report that the "irresistible impulse" to set fires may actually just be an impulse not resisted (Geller, Erlen, and Pinkas 1986). An evaluation of the term *pyromania* has shown that there is no consensus even among mental health professionals and behavioral scientists as to what constitutes pyromania, and, indeed, whether there really is such a disorder (Gardiner 1992). Even the FBI examined (Huff, Gary, and Icove 1997) the "myth of pyromania" at their National Center for the Analysis of Violent Crime in Quantico, Virginia. Doley (2003b) explores the common misperceptions that exist in the literature and attempts to clarify the true magnitude of "pyromania" as a motive by Australia's arsonists.

Sexual fantasies or desires have often been linked to pyromania in the popular literature, but this is not borne out by the interview research. Sex as a motive is highly overrated. In fact, experiments using penile response as an indicator of sexual arousal showed no correlation between sexual motivation and arson (Quinsey, Chaplin, and Upfold 1989). In the study, the penile responses of 26 fire setters and 15 non–fire setters were recorded and compared when the subjects were exposed to audiotaped narratives of neutral heterosexual activity and of fire-setting activities linked to several motives (sexual, excitement, insurance, revenge, heroism, and power). There were no significant differences in the responses of fire setters and non–fire setters to any of the narratives they heard.

The so-called motiveless arsonist in many cases knows his or her motive for setting fires, but it does not necessarily make sense to normal outsiders. The arsonist may also lack the verbal skills or capacity to express such concerns.

Fire investigators are cautioned not to label a subject a pyromaniac, since this is a diagnosis to be made by a mental health professional. Each field has a bias or different slant on its view of pyromania, so investigators, criminologists, psychologists, and psychiatrists are working from slightly different definitions.

MIXED MOTIVES

Interviews conducted with incarcerated arsonists underscore the complexity of human behavior, particularly when *mixed motives* arise. When questioned about the motives for their arsons, they gave responses indicating that there were often secondary and supplementary motives in addition to their primary motive (vandalism, excitement, revenge, crime concealment, profit, or extremism).

It should also be noted that arson can be used directly as a weapon, with the intent only of killing the target. Such homicides can be linked to a wide variety of motives, including self-defense. A thorough reconstruction of the entire incident, rather than just the fire-setting event, may reveal the offender's actual intent for the homicide.

Researchers, most notably Lewis and Yarnell (1951), assert that revenge is present as a motive in all arsons to a greater or lesser degree. Motives for arson, like other aspects of human behavior, often defy a structured, unbending definition. Another rationale is that arsonists may lack the social and communicative skills necessary to articulate their motives clearly. Embarrassment about the true motive may also lead to providing an alternative or false motive to investigators or mental health professionals.

Adding the elements of power and revenge reveals the problem with strict, unyielding classification. Fire investigators also should be aware that since arson is a criminal tool, motivations may change based on the target or situation.

The best examples of many of the factors that arise in a serial arson case are given in an actual account of an arson from an offender's point of view: the well-articulated account by a female serial arsonist who used the pen name Sarah Wheaton (2001). In her published article she included edited excerpts from her discharge summary, which noted a diagnosis of borderline personality disorder. Her treatment therapy included biofeedback, social skills training, and clomipramine, a psychotropic drug. At the time of writing the article, Ms. Wheaton stated that she had been fire-thinking-free for 8 months.

EXAMPLE 5-8 ■ Memoirs of a Serial Fire Setter

Sarah Wheaton (not her real name), a former self-admitted compulsive serial fire setter, is now working on a master's degree in psychology. At the end of her college freshman year, during the summer of 1993, she was involuntarily admitted to a psychiatric hospital for 2 weeks for treatment for fire setting.

The following excerpt is a direct quote from Ms. Wheaton's (2001) description of her treatment and perception of what she experienced as a serial arsonist.

Reason for admission: This 19-year-old single female was living in a dorm at the University of California at the time of admission. The patient had been involved in bizarre activity, including lighting five fires on campus that did not remain lit.

History of present illness: This young lady is a highly intelligent, highly active young woman who had been the president of her class in her high school for all four years. She had been going full-time to the university and working full-time at a pizzeria. She had called the police threatening suicide. When she was brought in, she was in absolute and complete denial. Patient claimed that everything was wonderful and she did not need to be here. She was admitted for 72-hour treatment and evaluation.

Hospital course: The hospital course initially was quite tempestuous. It was difficult to determine what the diagnosis was, as the patient was on the one hand very bright and very endearing to the staff and on the other hand was very unpredictable. She jumped over the wall on the patio (AWOL). The police were called and ultimately brought her back. She was subsequently certified to remain in custody for up to 14 days for intensive treatment because she attempted to cut herself with plastic and/or glass. She became more open after this, more tearful and at times more vulnerable. She tended to run from issues, to try to help everyone else, and not look at herself. Her father was seen in family therapy with her by a social worker.

Mother is reported by father to be both an alcoholic and have a history of bipolar illness. The patient herself reported sexual abuse by an older stepbrother when she was about age 9 to age 11.

Initially I intended to use antimanic medication with her, either carbamazepine or lithium, but opted not to as she was adamant against medication. The patient did seem to stabilize without medication. She showed dramatic improvement, although the staff and I still have concern. The stable environment she has been able to pull together may not exist after discharge. She is scheduled to go to Washington, DC, on July 1 to work as an intern in the office of one of the congressional representatives. She had done this two years ago working as a page.

Aftercare instructions: No follow-up appointment is scheduled because she is discharged today and will be leaving for Washington on the first.

Prognosis given by psychiatrist: Prognosis is very guarded given the severity of her condition.

Mental status: She firmly denies . . . destructive ideation, including fire setting at this time.

Discharge diagnosis [33 hospitalizations after the initial hospitalization]: Axis I. Major depressive disorder, recurrent, with psychosis. Axis II. Obsessive-compulsive personality disorder; history of pyromania; Borderline personality disorder. Axis III. Asthma. Axis V. GAF 45. Courtesy Dr. Jeffrey L. Geller, (Personal accounts: Memoirs of a compulsive firesetter. *Psychiatric Services* 52:1035–36. Copyright 2001, the American Psychiatric Association; http:// PS.psychiatryonline.org. Reprinted by permission.)

As a student of psychology Ms. Wheaton's article accurately traces many of the traits and characteristics of serial fire setters, particularly those reported in the literature and supported by interviews of law enforcement agents specializing in arson. Quoted next are her observations on how fire dominated her life from preschool to college.

Fire became a part of my vocabulary in my preschool days. During the summers our home would be evacuated because the local mountains were ablaze. I would watch in awe.

Below I have listed some of my thoughts and behaviors eight years after the onset of deviant behavior involving fire. I have also included suggestions for helping a fire setter.

Firesetting behaviors on a continuum: Each summer I look forward to the beginning of fire season as well as the fall—the dry and windy season. I set my fires alone. I am also very impulsive, which makes my behavior unpredictable. I exhibit paranoid characteristics when I am alone, always looking around me to see if someone is following me. I picture everything burnable around me on fire.

I watch the local news broadcasts for fires that have been set each day and read the local newspapers in search of articles dealing with suspicious fires. I read literature about fires, fire setters, pyromania, pyromaniacs, arson, and arsonists. I contact government agencies about fire information and keep up-to-date on the arson detection methods investigators use. I watch movies and listen to music about fires. My dreams are about fires that I have set, want to set, or wish I had set.

I like to investigate fires that are not my own, and I may call to confess to fires that I did not set. I love to drive back and forth in front of fire stations, and

I have the desire to pull every fire alarm I see. I am self-critical and defensive, I fear failure, and I sometimes behave suicidally.

Before a fire is set: I may feel abandoned, lonely, or bored, which triggers feelings of anxiety or emotional arousal before the fire. I sometimes experience severe headaches, a rapid heartbeat, uncontrollable motor movements in my hands, and tingling pain in my right arm. I never plan my fire, but typically drive back and forth or around the block or park and walk by the scene I am about to light on fire. I may do this to become familiar with the area and plan escape routes or to wait for the perfect moment to light the fire. This behavior may last anywhere from a few minutes to several hours.

At the time of lighting the fire: I never light a fire in the exact place other fires have occurred. I set fires at random, using material I have just bought or asked for at a gas station—matches, cigarettes, or small amounts of gasoline. I do not leave signatures to claim my fires. I set fires only in places that are secluded, such as roadsides, back canyons, cul-de-sacs, and parking lots. I usually set fires after nightfall because my chances of being caught are much lower then. I may set several small fires or one big fire, depending on my desires and needs at the time. It is at the time of lighting the fire that I experience an intense emotional response like tension release, excitement, or even panic.

Leaving the fire scene: I am well aware of the risks of being at the fire scene. When I leave a fire scene, I drive normally so that I do not look suspicious if another car or other people are nearby. Often I pass in the opposite direction of the fire truck called to the fire.

During the fire: Watching the fire from a perfect vantage point is important to me. I want to see the chaos as well as the destruction that I or others have caused. Talking to authorities on the phone or in person while the action is going on can be part of the thrill. I enjoy hearing about the fire on the radio or watching it on television, learning about all the possible motives and theories that officials have about why and how the fire started.

After the fire is out: At this time I feel sadness and anguish and a desire to set another fire. Overall it seems that the fire has created a temporary solution to a permanent problem.

Within 24 hours after the fire: I revisit the scene of the fire. I may also experience feelings of remorse as well as anger and rage at myself. Fortunately, no one has ever been physically harmed by the fires I have set.

Several days after the fire: I revel in the notoriety of the unknown fire setter, even if I did not set the fire. I also return again to see the damage and note areas of destruction on an area map.

Fire anniversaries: I always revisit the scene on anniversary days of fires that I or others set in the area.

Fires not my own: A fire not my own offers excitement and some tension relief. However, any fire set by someone else is one I wish I had set. The knowledge that there is another fire setter in the area may spark feelings of competition or envy in me and increase my desire to set bigger and better fires. I am just as interested in knowing the other fire setters' interests or motives for lighting their fires.

Suggestions for helping a firesetter. The likelihood of recidivism is high for a fire setter. The fire setter should be able to count on someone always being there to talk to about wanting to set fires. Fire setting may be such a big part of the person's life that he or she cannot imagine giving it up. This habit in all aspects fosters many emotions that become normal for the fire setter, including love, happiness, excitement, fear, rage, boredom, sadness, and pain.

A fire setter should be taught appropriate problem-solving skills and breathing and relaxation techniques. Exposure to burn units and disastrous fire scenes may be therapeutic and may enable the fire setter to talk openly about physical and emotional reactions. Doing so will not only help the fire setter but also give mental health professionals a deeper understanding of the fire setter's obsession.

The self-reported insight of this individual clearly confirms the findings of scientific, psychiatric, and law enforcement research. Of particular note are the pre- and postoffense behaviors at the fire scene, the significance of anniversaries, and the highly suggestive impact of the publicity of fires set by other arsonists. Astute fire investigators can use this information to assist them in identifying and solving cases set by compulsive serial arsonists.

EXAMPLE 5-9 ■ The Case of a Mother–Son Serial Arson Team

For more than 2 years a residential upper-middle-class suburb with a population of some 59,000 was plagued by a series of small fires and false alarms. Most of the fires involved roadside grass and median-strip ground cover and garbage cans, but several were in large commercial Dumpsters alongside wood-sided commercial and residential structures (with possibly serious consequences). Nearly all the false alarms (phone calls and pull stations) targeted one school. After a review of approximately 800 incidents queried from the district's NFIRS database (by "Incident Type" and by "Location"), investigators compiled a list of some 50 similar incidents. When these were plotted on a local street map (Figure 5-8), a pattern focused on the school and extending the length of the city along a major north–south corridor (Novato Blvd.) was discernible. The concentration around the school led investigators to suspect the arsonist lived nearby.

The actual ignitions were not observed, although a witness in one incident reported seeing a dark-colored SUV with two suspects leaving the scene. The scenes were sufficiently separated by enough distance to suggest the arsonist was not on foot but most probably in a vehicle. At one grass fire, a matchbook cover found near the roadside bore an identifiable latent fingerprint, but it did not have any hits in the database. (The book of matches itself was found burned up in the grass several feet away.) Sunflower seed hulls found near one fire suggested a lead, but no good suspects. Temporal profiling revealed that most of the similar fires occurred between 4:00 and 8:00 p.m. on weeknights, with a higher frequency during the summer (May–July) (see Figure 5-9).

When a fire seriously damaged a house (1320 Monte Maria) on August 15, 2004, investigators found several leads. Interviews with neighbors revealed that two of the residents—a 42-year-old mother (widow) and her 24-year-old developmentally disabled son—had been seen going into and out of the house just minutes prior to the fire. A drug-addicted daughter who lived at the house most of the time had confronted her brother during the fire suppression. In view of the neighbors and firefighters, she had slapped him and demanded, "Now what have you done?" The son, who did not drive, had a police–fire scanner in his possession. He had reportedly been out with his mother driving when he heard the fire call on the scanner and they returned to find the garage and an interior bedroom ablaze. Investigators determined that the fire had been ignited in furniture in both the garage and the adjacent bedroom. The fire patterns and duration (based on fire damage) of both fires indicated the fires were started separately at about the same time and could not have been the result of fire spread of a single fire through the connecting door (which was very badly burned from both sides).

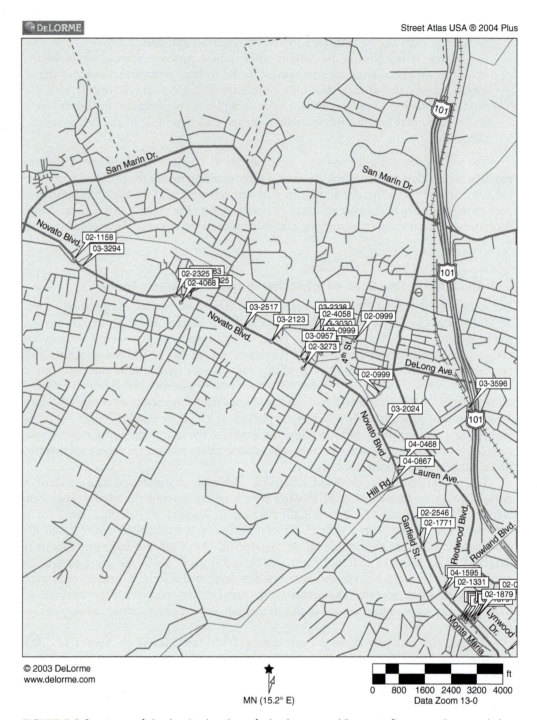

© 2003 DeLorme
www.delorme.com

MN (15.2° E)

Data Zoom 13-0

FIGURE 5-8 Street map of city showing locations of school, grass, and Dumpster fires over a 2-year period. Note the pattern along a major thoroughfare near the offenders' house (in cluster of fires in lower right corner). *Based on public records. Case courtesy of Novato Fire Protection District.*

TEMPORAL ANALYSIS OF SERIAL ARSON INCIDENTS, 2002

Variable	Jan.	Feb.	Mar.	Apr.	May	June	July	Aug.	Sept.	Oct.	Nov.	Dec.	Total
0800 – 1200											O		1
1200 – 1600					*	◈					O		3
1600 – 2000				◈			O**	□					5
2000 – 2400				O						×			2
2400 – 0400				◈							□		2
0400 – 0800	O		□										2
Monday				◈									1
Tuesday	O			O							O		3
Wednesday							O	□			□		3
Thursday							*						1
Friday		□			*		*			×			4
Saturday				◈							O		2
Sunday						◈							1
Total	1	0	1	3	1	1	3	1	0	1	3	0	15

Fire Type:

O Trash
□ Dumpster
◈ Pull Alarm
✳ Roadside
▲ Structure
♠ Wildland
× Railroad

TEMPORAL ANALYSIS OF SERIAL ARSON INCIDENTS, 2003

Variable	Jan.	Feb.	Mar.	Apr.	May	June	July	Aug.	Sept.	Oct.	Nov.	Dec.	Total
0800 – 1200											□		1
1200 – 1600			◈	◈◈◈		◈✳✳	□						8
1600 – 2000			□◈	O◈	◈	◈✳	✳✳♠	♠	✳		V		12
2000 – 2400				◈◈		□□	□♠♠			O	□		9
2400 – 0400													1
0400 – 0800			□		◈		O						3
Monday			□					♠			□		3
Tuesday				◈			O		✳				3
Wednesday						✳	✳✳♠♠						5
Thursday			□	O	◈								4
Friday				◈◈		□✳✳	□□						7
Saturday			◈◈	◈◈	◈	□◈	♠				□		9
Sunday				◈						O	V		2
Total	0	0	4	7	2	7	8	1	1	1	3	0	34

Fire Type:

O Trash
□ Dumpster
◈ Pull Alarm
✳ Roadside
♠ Wildland
× Railroad
V Vehicle

TEMPORAL ANALYSIS OF SERIAL ARSON INCIDENTS, 2004

Variable	Jan.	Feb.	Mar.	Apr.	May	June	July	Aug.	Sept.	Oct.	Nov.	Dec.	Total
0800 – 1200		*											*
1200 – 1600					*								1
1600 – 2000			☽					◈✳▲▲					5
2000 – 2400													
2400 – 0400		□											1
0400 – 0800	□												1
Monday													
Tuesday		□											1
Wednesday								*◈					2
Thursday		□	☽										2
Friday													
Saturday	□												1
Sunday		□			*			▲▲					4
Total	1	3	1	0	1	0	0	4	0	0	0	0	10

* Time not available
  for two of the
  dumpster fires.

Fire Type:

O Trash
□ Dumpster
◈ Pull Alarm
✳ Roadside
▲ Structure
☽ Porta-john

FIGURE 5-9 Target and temporal analysis (time, date, day) of fire events for 2002–2004. (Persons responsible were arrested August 2004.) *Based on public records. Case courtesy of Novato Fire Protection District.*

The next day, an accusatory letter was posted on the craigslist.com/ website accusing the mother of drug use, endangerment, and arson. The person who posted this notice was identified as an acquaintance of the family who admitted she had done it to draw attention to the problems in the family. During interviews with police investigators, the son admitted frequently burning materials in the barbecue in the backyard and listening on his scanner for reports of smoke. If the smoke was reported, he would extinguish the fire in the barbecue. Examination of the scene revealed burned twigs, leaves, and paper in the barbecue. The son admitted to being driven to various locations by his mother so he could start fires. While she drove down the main street, he would toss lighted matches from the windows of the vehicle. When the vehicle was examined, burned paper matches were found in the door assembly (accounting for a fire in the vehicle's door on November 9, 2003). He admitted repeatedly pulling the alarms on the school or calling in alarms and then listening on the scanner or watching the fire department response from his backyard, which overlooked the school. He did not ride a bicycle.

When the son was fingerprinted, the latent print on the matchbook cover from a wildland fire (July 23, 2003) was found to be his. He explained that when he got depressed, setting fires made him feel better. His mother said she helped him because she wanted him to feel better and be happy. She admitted to setting the fires in their home on August 15, 2004, and four other fires. (She was a recovering drug addict using methadone at the time of the fire.) Summertime fires coincided with the anniversary of his father's death.

Except for a pull alarm at the school just 4 days before the house fire, all other false alarms were generated in 2002. Dumpster and median fires also occurred during 2002, and they continued into 2003 and 2004. Wildland or grass fires became dominant in 2003, and a mixture of all types took place in 2004. After their arrests in August 2004 the mother and son negotiated a plea agreement, the mother pleading quickly to two counts, the son to four counts. Both were sent to jail. Dumpster and roadside fires in town reportedly dropped to zero after August 2004.

Case study courtesy of Novato (CA) Fire Protection District.

FAKED DEATHS BY FIRE

Insurance fraud by burning a substituted body in an attempt to obliterate the true identity is a crime unique to profit-motivated arson. This form of arson often involves detailed planning, particularly in acquiring the body and evading suspicion by the investigating authorities.

Case studies have shown that the warning signs for *faked deaths* include, but are not limited to, the following (Reardon 2002).

- The death occurs shortly after filing of the insurance application and/or during the contestability period.
- The death occurs abroad.
- The insured has financial problems.
- Large policies or multiple small policies that do not require medical examinations exist. The policies have misrepresentations or omissions.
- The insured uses aliases.
- Previous policies have been canceled.
- The levels of insurance are inappropriate for the actual earned income.
- There is no body, or the body is in an unidentifiable condition.

Reardon emphasizes that the fundamental component in the payment of any life insurance claim is proving that the insured is actually dead. He recommends thorough investigation of these types of claims and litigation when appropriate to reduce the likelihood of payment to the claimant.

EXAMPLE 5-10 ■ Out-of-Country Faked Death by Fire Scheme

In 1998, Donny Jones (not his real name) completed an application for a term life policy bearing a face value of $4 million. Jones already had a $3 million policy with another insurer. Six months after the $4 million policy was written, the insurance company received the news that Jones had died in a car accident outside the United States.

The insured prepared his plan carefully. While out of the country with a friend, Jones rented a large SUV, placed his bicycle in the car and, drove away at 10:00 p.m., supposedly to take a 3-hour drive through the desert to an adjoining town. Early the next morning, the SUV was found burning at the side of a desert highway. Local authorities found no traces of the mountain bike in the SUV, no signs of collision, and, initially, no body. Later, a body was found in the vehicle at an impound yard, but the remains were quickly claimed and cremated by the traveling companion of Mr. Jones.

Owing to the unique circumstances, the insurer initiated an immediate investigation into the claim by arranging for a forensic anthropologist to examine the skeletal bones and for a fire protection engineer to document and impartially determine the cause of the vehicle fire. The forensic anthropologist determined that the remains were those of an elderly man of Native American heritage, not of the insured, who was a 33-year-old Caucasian. The engineer also interpreted for the insurance company technical information on the incendiary nature of the fire, the staged accident, and forensic evidence collected during the investigation by the foreign authorities.

The insurance company denied the claim and filed suit in the U.S. District Court for declaratory judgment, seeking rescission of the policy. Initiating the litigation permitted the insurance company to begin to gather and preserve evidence from the authorities, police, and public and private sectors. The investigation outside the United States required the involvement of the U.S. consulate followed by litigation, for which the Hague Convention generally requires approval from a U.S. court and the appropriate foreign authority.

Eventually, the insured was discovered working in the United States through a routine background investigation by a firm for which the insured had been working, under an alias. The insurance policy was rescinded, and the insured later pled guilty in federal court to criminal charges of wire fraud and was ordered to pay full restitution (Reardon 2002).

FILICIDE

Filicide by fire is a crime in which the murderer is the parent of the victim and has used fire to mask the death, often making it appear to be an accident. Even though cases of filicide appear to be rare, these events are becoming more frequently studied by academics (Stanton and Simpson 2006).

Studies by the FBI, although based on limited case samples, still provide an insight into this crime. This agency speculates that owing to the thoroughness needed to conduct such investigations, this phenomenon may be widespread and underreported (Huff 1997, 1999). One of the present authors has been involved in the investigation of at least eight incidents in which a fire was set deliberately to kill one or more of the fire setter's children. One such case was even the subject of a best-selling book (Rule 1998).

Various motives are found for filicide, as for any crime, including the following:

■ *Unwanted child:* Falsely believing that she and her spouse or lover can then live unburdened, a mother commits child murder to remove the perceived obstacle or nuisance child(ren).

■ *Acute psychosis:* The parent is psychotic, such as the severely depressed single mother who stabbed her three small children to death and then set their apartment on fire.

■ *Spousal revenge:* An estranged husband cruelly decides to deprive his wife of her most cherished treasure, her child.

- *Murder-for-profit:* Parents take out large life insurance policies on their children shortly before killing them by fire. There may be cases in which only a few premiums were paid, and the policy lapsed for nonpayment.

In any of these cases, the authorities may not be suspicious initially, for various reasons, including the lack of a thorough investigation of the fire; the presumption that the fatal fire was accidental, caused by the child's playing with matches; masking by overwhelming compassion for the parents; and sadness for the children. These emotional reactions may override indications (red flags) of suspicion (Huff 1997).

Although there is no single indicator, several common factors appear in cases of filicide, including the following:

Victimology: The children are often young, of preschool age; have little training in how to behave in a fire; and are thus less likely to escape. The perception may be that younger children are more disposable and more easily "replaced" later.

Preoffense behavior: There is unusual behavior just before the fire, such as preparation of the children's favorite meal or a visit to a special place. For example, in one case a fast-food meal was served to two teenage sons to deliver a dose of common decongestant to make them sleepy when the fire was set (it was identified in the postmortem blood of the decedent and in dried blood stains on the shirt of the survivor). One of the boys died in the fire; the other survived to testify about the prefire events. A young daughter was not targeted and was kept out of the fire area of the home.

Temporal: The fires take place at night or early in the morning when the children are most likely to be in bed asleep. This gives the parents time to plan the event and, in some cases, lock the children in their rooms during the fire setting.

Crime scene characteristics: Children are asked to change their sleeping arrangements the night of the fire and sleep in a room not normally used. Sometimes, scenes are staged when the children have already been shot, stabbed, or strangled to death and a fire is set to cover the crime, frequently with a flammable liquid such as gasoline. In some cases escape routes are blocked and doors locked. Drugs may be administered to make the child sleepy or unconscious. A full forensic postmortem including X-ray and toxicology assays is essential on any child victim.

Offender characteristics: Halfhearted rescue attempts by adults are reported, with no appreciable element of danger encountered. Therefore, they do not display smoke-filled clothing, burns, or watery eyes, as would be expected. The parents most commonly live in manufactured housing and are no older than their mid-30s.

Postoffense behavior: After the incident, the adults exhibit inappropriate behavior, such as little or no grief; seldom speak about the victims, favoring the discussion of material losses (including insurance coverage); and are too quick to return to life as usual.

There will be one or a combination of two or more of these indicators. However, investigators are cautioned not to allow one factor alone necessarily to raise suspicion. They should move forward in a cautious yet professional manner and not yield to an emotional concern for the parent. As every person deals with trauma differently, investigators should not build a case around or place significant emphasis on "lack of expected grief."

EXAMPLE 5-11 ▪ Mother Involved in Filicide

Firefighters responding to a late-summer fire found a 22-month-old female hidden under a pile of clothes in a locked bedroom closet, dead from smoke inhalation. The child's mother told firefighters that her young daughter had been known to play with the cigarette lighter, which was found in the kitchen.

Firefighters doubted the story, but it became believable when the deceased child's brother, who was about 4 years old, demonstrated that he was able to operate the lighter. Investigators then ruled the fire accidental, initially concluding that the deceased girl had lit some paper napkins in the kitchen, taken them into the bedroom, and gone into the closet, with the door locking behind her.

All the facts surrounding this fire were consistent with a child's playing with a cigarette lighter and causing her own death. There was no reason to suspect foul play until 5 years later, when the son, now 9 years old, suffered third-degree burns in another fire. When firefighters arrived at the home, they found the child unconscious behind a locked bedroom door. The mother fled immediately after the fire.

One week later, fire investigators located and interviewed the mother, who allegedly confessed to setting both fires, telling them that she was angry with her husband. In each of these fires, individual factors indicated filicide by arson (Huff 1997).

Geography of Serial Arson

Offender-based classification of arson extends beyond merely identifying the motive. Investigators should also closely examine the *geographic locations* selected by the arsonist. An analysis of these sites may reveal much about the intended target, add insight into the arsonist's motive, and provide a potential surveillance schedule for attempting to identify and apprehend the arsonist.

The results of a joint research effort on the geography of serial violent crime illustrate the patterns found in serial arson cases. Results of the study indicate that criminal offenders who repeatedly set fires exhibit temporal, target-specific, and spatial patterns that are often associated with their modus operandi (method of operation; m.o.) (Icove, Escowitz, and Huff 1993; Fritzon 2001).

Many geographers, criminologists, and law enforcement professionals are concerned with the rising tide of serial violent crimes, including arson, that is plaguing the United States and other Free World countries. Many of these violent offenders purposely use jurisdictional boundaries to evade detection from law enforcement officials.

Serial arsonists often create a climate of fear in entire communities. Community leaders compound this problem by pressuring law enforcement agencies to identify and apprehend the fire setter quickly. Often, the arsonist evades apprehension for months, frustrating even the most experienced investigators. Unpredictable gaps often occur between incidents, leaving law enforcement authorities questioning whether the arsonist has stopped his or her fire setting, left the area, or been apprehended for another offense.

PREVIOUS RESEARCH FINDINGS

Few studies have been conducted on the geographic distribution of serial violent crimes, including arson. The studies that do exist in the literature focus on specific characteristics found in particular crimes (Icove 1979; Rossmo 2000).

A keynote analysis of violent crime measured it in terms of a spatial and ecological perspective (Georges 1967, 1978). The social-ecological view is that there is a direct relationship between crime and the environment. In this view, an absence of community controls, such as found in inner cities, correlates with high crime rates. Several historical approaches trace methods for interpreting the geography of crime, particularly as it affects arson and fire-related crimes.

Transition Zones The Chicago School of Sociology approach first proposed that crime was related to the environment (Park and Burgess 1921). This ecological approach assumed that high crime rates flourish in urban areas called *transition zones*. These zones have mixed land uses and high population fluxes.

In many cities, concentric rings form zones of transition, representing diffusion between the central business district and residential housing. These zones often contain rooming houses, ghettos, red-light districts, and diverse ethnic groups.

The transition zone theory was illustrated in a study of 15 fire bombings in Knoxville, Tennessee, during 1981 (Icove, Keith, and Shipley 1981). An analysis of the fire bombings investigated by the Knoxville Police Department's Arson Task Force showed that 66 percent of the incidents occurred within specific census tracts having significant population decline or growth.

The Knoxville study also examined a census tract in a low-income area where three fire bombings occurred during the reporting period. U.S. census data indicated that this tract had the highest minority population, the highest percentage of poverty, and the second-lowest mean income of all the tracts where bombings occurred.

Centrography   The technique of *centrography* uses descriptive statistics to measure the central tendency of crime on a two-dimensional spatial plane. In certain crimes, the offender's residence may be close to the site of the crime commission. Another hypothesis is that the older the offender, the greater the mean distance from the crime scenes to his or her residence. This mobility is due, in part, to the offender's access to bicycles and vehicles, as well as being over the curfew age.

A study for the U.S. Department of Justice explored the concept of centrography and described geographic *anchor points* for criminals (Rengert and Wasilchick 1990). Examining the geographic locations of buildings targeted by burglars, the study concluded that criminals select targets close to an anchor point that minimizes the travel and time required to commit their crimes. Often, the dominant anchor point is close to the offender's residence. Other anchor points away from the criminal's residence include bars, schools, and video arcades. Kocsis and Irwin (1997) also examined serial arson cases where the offenders tended to operate from their residence and moved out in various directions to set fires.

Other geographic analysis approaches take place on both the micro and the macro levels, as described in the case study earlier in this chapter. The microlevel approach examines the exact physical location of the crime, such as the type of building, vehicle, or street. The macrolevel approach tends to aggregate the data into zones. These zones can be census tracts, police beats, or other geographic areas. This approach tends to increase the scale of the analysis.

Centrography has been used to examine spatial–temporal relationships of other violent crimes (LeBeau 1987). Observations of the spatial distribution of crime use the x- and y-axes of a Cartesian coordinate system overlaid on a street-block map. The locations of numerous incidents are analyzed using the *mean center* approach to measure the central tendency by tracking the movement of these mean centers and to determine the standard deviational ellipse to describe the distribution of incidents.

Fritzon (2001) examined the relationship between the distance traveled by an arsonist and his or her motivation. The research used the concept of *smallest space analysis* (SSA) to show the relationships between the distances traveled to set fires, the crime-scene features, and the offender's background characteristics. The research findings demonstrated that expressive crimes occur closer to the arsonist's home, which could be considered an anchor point, than do instrumental crimes. In the case of revenge-motivated attacks, the arsonist will travel the greatest overall distance. Also, arsonists who recently separated from a partner were more likely to travel greater distances. The study certainly adds value to existing research in the field.

The *spatial and temporal trends* of serial arsonists and bombers also have been examined on a local level (Icove 1979; Rossmo 2000). Patterns not previously documented among serial incendiary crimes were found using cluster analysis techniques—individual and group serial arsonists and bombers tend to be geographically bounded by natural and artificial boundaries. (See Figure 5-8 for an example.) When the offender moves his or her residence, the cluster of activity normally follows.

Studies on the historical changes of locations of arsons in the city of Buffalo, New York, have also confirmed many of these observations using cluster analysis. The concept of spatial surveillance brought new meaning to the monitoring of the locations of fires over time. The researchers noted that changes in these spatial patterns may occur for many differing reasons, and conclusions should not be reached merely on visual interpretations (Rogerson and Sun 2001).

Temporal trends have also been examined by Icove (1979) using time-based cluster analysis techniques. This form of analysis detects subtle time-of-day and day-of-week trends correlated with geographic changes in the cluster centers. Other temporal trends such as increases in arsons with lunar activity have also been documented in New York City (Lieber and Agel 1978). A research thesis at the University of California, Berkeley, School of Criminology, explored the relationship between lunar phenomena (not necessarily full moon) and the crime of arson (Netherwood 1966).

EXAMPLE 5-12 ■ Incendiary Crime Detection

This is an example of the application of geographic analysis to detect local clusters of activity of serial arsonists. After plotting the locations of arsons in an eastern city from February through September 1974, researchers detected a major cluster center of activity. Three comparative techniques are shown in Figures 5-10 and 5-11, illustrating the use of two-dimensional gridded incident, shaded, and three-dimensional contour maps by Icove (1979).

A further analysis showed that a major cluster center was formed by two grids totaling 12 arson fires. Investigation revealed that during that time period, a gang of juvenile boys engaged in starting incendiary fires within a section of the large city. Properties targeted by these youths included garages and wooded areas.

When plotted on a map along with other arsons in the area, the incidents formed a major cluster. Data from these fires were collected and analyzed to provide a surveillance schedule for local law enforcement. A temporal analysis of the time of day and day of the week for the incidents contained in this major cluster was performed and revealed that the juveniles set these incendiary fires during breaks in the school day. In the summer months, fires occurred in the evening hours, but the trend reverted to the original pattern at the start of school in the fall. With few exceptions, the fires were on weekdays. During a temporal analysis, lunar patterns are often detected. In this example, 75 percent of the fires occurred within 2 days of either the new or the full moon phase.

Urban Morphology Georges (1978) examined the concept of *urban morphology*, which posits that crime tends to concentrate within certain zones or within areas with close access to transportation routes. For example, in Georges' earlier study on the geographic distribution of arsons during the 1967 Newark riots, fires were found to have been set along main commercial routes (Georges 1967). The study further documented earlier theories that as cities grow, crime tends to spread along main transportation routes (Hurd 1903). The mother–son case described previously illustrates this type of analysis, where fires were predominantly set along a major traffic route in the city.

Crime pattern analysis is a methodology for detecting recurring patterns, trends, and periodic events in the times, dates, and locations of incidents. Once patterns are detected in a serial arson case, much can be learned to help predict the next event and classify the

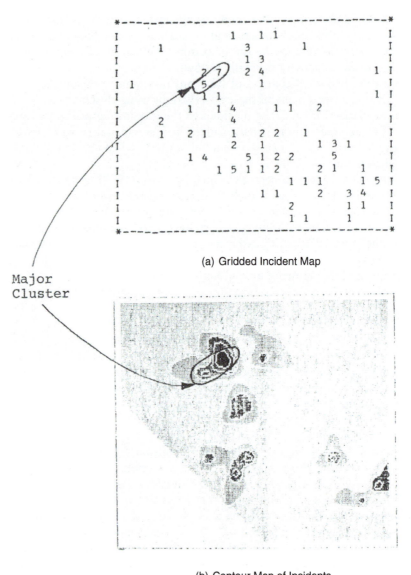

(a) Gridded Incident Map

Major
Cluster

(b) Contour Map of Incidents

FIGURE 5-10 Gridded incident and contour maps with major cluster centers of arson fires. *Source:* Icove 1979. *Courtesy of D. J. Icove.*

behavioral patterns exhibited by the offender. For a comprehensive discussion, see the National Institute of Justice Web page on crime mapping, which highlights crime reduction projects, geospatial tools, data sources, research, conferences, training, and publications, http://www.nij.gov/maps/.

Authoritative studies continue to reinforce that co-occurrence of offense characteristics also can be tied to both the geographic selection of targeted properties and the residence/workplace of the offender (Canter and Fritzon 1998). A systematic algorithm has been patented that performs crime site analysis in an area of criminal activity to determine a likely center of such activity (Rossmo 2000).

In Australia, the Australian Institute of Criminology (funded by the Bushfire Cooperative Research Centre) produced a handbook that is a crime analysis-driven approach to

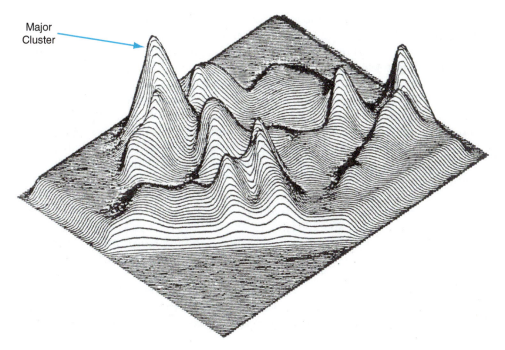

Major Cluster

FIGURE 5-11 The detection of geographic clusters of arsons in three dimensions with a 35° counterclockwise rotation. *Source:* Icove 1979. *Courtesy of D. J. Icove.*

detecting, predicting, and preventing arson (Willis 2004; Anderson 2010). The handbook serves as a resource tool for local, fire, and police agencies to develop prevention strategies. The project website contains downloadable manuals, worksheets, and an electronic spreadsheet to assist in capturing the information.

A *criminal investigative analysis* by the FBI (also referred to as a *profile*) is usually prepared by an expert, based on exposure to numerous case histories, personal experiences, educational background, and research performed. Suspects developed by crime analysis do not always fit every prediction, since the predictions are made based on similar historical cases, and the offender may often modify his or her m.o. in response to an active investigation.

The application of computer-assisted geographic profiling systems has been a long-established theme in the crime analysis field. The FBI's work in the 1980s in automated crime profiling (Icove 1986) used many of the concepts derived from incendiary crime analysis (Icove 1979).

Several key points are developed during a criminal investigative analysis involving the geography of serial incidents: temporal analysis, target selection, spatial analysis of the cluster centers, and standard distance.

1. *Temporal analysis.* The first type of pattern analysis in serial arson cases is temporal analysis, which determines if any time-of-day or day-of-week trends exist. (See Figure 5-9 for an example.) These trends may vary if the offender modifies his or her m.o. in response to an active investigation. This modification may be a conscious act by the suspect.
2. *Target selection.* Another form of pattern analysis deals with the type and location of the target selected by the offender. A commonly observed characteristic in cases involving serial arsonists is that the offender often escalates the targets

selected, and situations become more life-threatening with time. A typical scenario is that the arsonist starts with small grass and brush fires, then moves to outbuildings and vacant structures, and then to occupied structures. Distance traveled is also a factor in target selection (Fritzon 2001).

3. *Spatial analysis.* Using spatial analysis, an investigator can determine that a series of arson fires set repeatedly within the same geographic area forms a cluster of activity (as in Figure 5-8). Finding this center of activity often reveals information about future target selections or where the offender may live or work. As indicated previously, serial arsonists and bombers often tend to be geographically bounded by natural and artificial constraints. Therefore, the offenders either consciously or unknowingly maintain their activities within a bounded area, not often crossing major highways, rivers, or railroad tracks. Spatial monitoring of geographic patterns of arsons in Buffalo, New York, showed the value of searching for changes in spatial patterns (Rogerson and Sun 2001).

4. *Geographic profiling.* The use of geographic profiling has proved to be a trainable and heuristic talent. In a recent research study on geographic profiling, 215 individuals participated in an experiment to predict the residential locations of serial offenders based on information about where their crimes were committed. In the study, these individuals were pre- and posttested when given formal training on one actuarial profiling technique (Snook, Taylor, and Bennell 2004). Analysis of the study participants' performance showed that 50 percent of them used heuristics that led to accurate predictions before receiving the formal training. Almost 75 percent of the study participants improved in their predictive ability after receiving the formal training. Bennell and Corey (2007) propose the use of geographic profiling in the context of detecting and predicting terrorist attacks. See also work by (Canter and Fritzon 1998) and Rossmo (2000) in geographic profiling.

5. *Cluster centers.* The *mean center* is the primary measurement of a spatial distribution of a cluster of data points. Given that the columns on a map equate to *x*-values on a Cartesian coordinate system, and the rows to *y*-values, the equations for calculating the mean center (Ebdon 1983) are

$$\bar{x} = \frac{1}{n}\sum_{i=1}^{n} x_i,$$

$$\bar{y} = \frac{1}{n}\sum_{i=1}^{n} y_i.$$

For the general case of cluster center z, the formula is

$$z_i = \frac{1}{n}\sum_{i=1}^{n} x.$$

Long-term analysis of these *cluster centers* is useful for determining whether the offender has changed his or her residence or place of work. In cases spanning 3 years or more, a cluster center should be calculated for each year and plotted on a map (Icove 1979). If the cluster centers do not move significantly, or if they hover in a specific area, the offender has not likely changed residence or job. When a marked change is detected, this information is conveyed to the law enforcement agency, which compares this knowledge with the potential movement of the suspects.

Studies continue to suggest a correlation between motivation and distance traveled by arsonists (Fritzon 2001). Those offenders studied whose behavior had a strong emotional component, such as a revenge motivation, tended to travel greater distances.

A more suitable center of activity for crime pattern analysis is known as the *center of minimum travel* or what is often referred to as the *centroid*. This center is the point or coordinate at which the total sum of the squared Euclidean distances to all other points on the map is the lowest.

The centroid is more meaningful than a mean center when examining the geographic locations of crimes committed by a single offender. The reasoning behind this assumption is that an offender who walks to his or her crime scenes will choose the minimum travel distance. The mean center does not always coincide with the centroid.

The calculation of the centroid is not a simple formula but requires an iterative process of minimizing a performance index. The best clustering procedure using this technique is referred to as the *K*-means algorithm (MacQueen 1967). The application of this algorithm (Tou and Gonzalez 1974) is best broken down into logical steps.

6. *Standard distance.* Once a cluster center is calculated for the geographic locations of a series of crimes, law enforcement officials can use this knowledge to concentrate their investigative and surveillance efforts within that neighborhood. This search radius from the cluster center is useful in focusing a criminal investigation within a reasonable distance from the cluster center of activity. For example, in a detailed crime analysis of a serial arson case, profilers often recommend a surveillance schedule and area of concentration within the area bounded by a radius of one standard deviation from the cluster center (Icove 1979). This area is known as the *standard distance.*

Several case examples are presented to illustrate the phenomena associated with the geography of crime. The following cases were submitted for crime pattern analysis after an intensive investigation by a local law enforcement agency had exhausted all logical leads (Icove, Escowitz, and Huff 1993).).

EXAMPLE 5-13 ■ Shifting Cluster Centers

During the fall and winter months in a southwestern state, an unknown arsonist was suspected of setting 24 fires involving fields, vehicles, mobile homes, residences, and other structures. Some of the fires were set after the offender broke into the structures.

A temporal analysis of the 24 incidents revealed that the arsonist favored the late evening and early morning hours. The majority of the fires occurred on the weekends, as shown in Table 5-7. An analysis of the targets selected by the arsonist showed that the structures were either unoccupied, vacant, or closed for business. Over a period of time, the arsonist escalated the severity of his fire setting, first using available materials, and later turning to flammable liquids to accelerate the fires.

A geographic cluster analysis showed that the first three fires clustered tightly on the west side of the town. The remaining fires concentrated in a larger cluster of activity on the east side (Figure 5-12).

A crime analysis of this case was conducted, and the results were forwarded to the agency that requested the assistance. The investigators were told to concentrate their efforts on any suspects who first lived near the west cluster and then changed their residence to close to the center of the east-side cluster.

Armed with this information, the law enforcement agency later charged and convicted a 19-year-old single white male who matched the characteristics determined by the crime analysis. A key element in the case was that the offender lived near both cluster centers of the arsons. The geographic center of the activity shifted when the offender moved from the west to the east side of town (Icove, Escowitz, and Huff 1993).

TABLE 5-7	Temporal and Target Analysis of Geographically Shifting Cluster Centers

VARIABLE	AUG.	SEPT.	OCT.	NOV.	DEC.	TOTAL
8 a.m.–4 p.m.					1	1
4 p.m.–12 a.m.	1		4	3		8
12 a.m.–8 a.m.		1	4	6	3	14
Unknown			1			1
Monday					1	1
Tuesday	1					1
Wednesday					2	2
Thursday			1	2	1	4
Friday				2		2
Saturday		1	4	6		11
Sunday				2	1	3
Field				1	1	2
Vehicle			2			2
Mobile home				2	1	3
House, vacant	1	1	1			3
House, occupied					1	1
Structure		1	3	5	2	11
Total	**1**	**1**	**9**	**9**	**4**	**24**

Source: Icove, Escowitz, and Huff 1993.

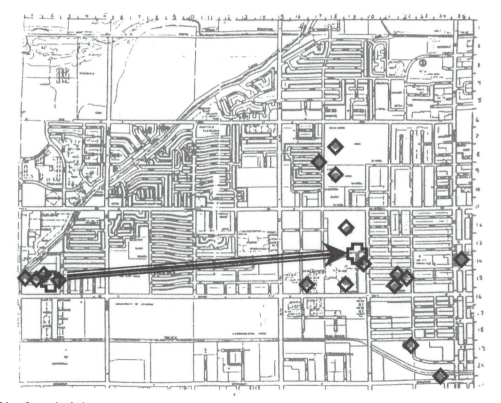

FIGURE 5-12
Geographic display of 24 fires showing the shift from west to east of centers of activity of two clusters. *Source:* Icove, Escowitz, and Huff 1993. *Courtesy of D. J. Icove.*

EXAMPLE 5-14 ▪ Temporal Clustering

A southeastern city was plagued with 52 arson fires in vacant buildings, vehicles, commercial businesses, residences, and garages over a 2-year period. The law enforcement agency investigating the case was stumped and turned to crime analysis after exhausting all traditional leads.

A crime analysis revealed temporal and geographic patterns in the time and locations of the arsons. The majority of the fires were set during the afternoon and evening weekday hours. There were two 6-month gaps in activity. An overwhelming majority of the arsons occurred within a 1-mile radius of the downtown area of the city, with two additional clusters adjacent to two lakes. The crime analysis returned to the agency stated that the offender lived close to the cluster center of activity of his fires. When later apprehended, the arsonist admitted in most cases walking to the scenes of his fires and using available materials and matches or a lighter carried to the scenes. Even though it is a high-risk scenario for the offender to set a fire in the afternoon and evening hours, his intimate knowledge of the geographic area diminished the possibility of his being detected and followed.

Four months after the department received the crime analysis, a 29-year-old white male was arrested after fleeing the scene of a fire he had just set. The suspect later told police that he first started setting fires when his relationship with a girlfriend failed. The different clusters of fires were the result of the subject's changing his residence. Periods of inactivity in his fire setting correlated directly with times when the offender had ongoing social relationships. Only when these relationships failed did the offender return to setting fires. His motive appeared to be revenge.

When arrested, the suspect had in his possession a map that marked the locations of the fires he had set. The offender's prior arrest history included charges for disorderly conduct, criminal mischief, harassment, and filing a false report to law enforcement (Icove, Escowitz, and Huff 1993).

CHAPTER REVIEW

Summary

Arson is defined as the willful and deliberate destruction of property by fire. Although establishing a motive for fire setting is not a legal requirement of the elements of the criminal offense, it can help focus investigative efforts and aid in the prosecution of the arsonist. Motives for fire setting usually include one or more of the following general categories: vandalism, excitement, revenge, crime concealment, profit, and political terrorism. Motiveless arson or pyromania is not considered an identifiable classification.

Arson crime scene analysis is still in its infancy. Problems associated with its use and applications include the broad degree of knowledge required to assess and interpret the crime scenes, particularly using a motive-based approach.

A combined background in the principles of fire protection engineering as well as in behavioral and forensic sciences can certainly enhance the analysis. Clearly, the concept that the "ashes can speak" is paramount in the application of this technique.

Future work in this area should combine the knowledge of fire pattern analysis with comprehensive photographic techniques to assist in the visualization of fire scenes long after they occur and deteriorate.

Problems

5.1. Research a local serial arson case that appears in the media. Plot the locations of these fires and conduct a temporal analysis. What can you learn from this analysis?

5.2. For the case from problem 5.1, determine what motive was developed. Is the assessment by the media correct? What was the motive stated by the prosecution? By the defense?

5.3. What published studies support the popular notion that sex is a major motive for arson? What is the reliability of such conclusions today?

Suggested Readings

Icove, D. J., V. B. Wherry., & J. D. Schroeder 1980. *Combating arson-for-profit: Advanced techniques for investigators*. Columbus, OH: Battelle Press.

Canter, David. 2003. *Mapping murder: The secrets of geographical profiling*. London: Virgin Books.

Nordskog, E. 2011. *"Torchered" minds: Case histories of notorious serial arsonists*. Xlibris.

References

ADL. 2003. Anti-government extremist convicted in Colorado IRS arson. Anti-Defamation League.

Allen, D. H. 1995. The multiple fire setter. Paper presented at the International Association of Arson Investigators, Los Angeles, CA.

Anderson, J. 2010. Bushfire arson prevention handbook. *Handbook No. 10*. Canberra: Australian Institute of Criminology.

APA. 2000. *Diagnostic and statistical manual of mental disorders (text revision) (DSM-IV-TR)*, 4th ed. Arlington, VA: American Psychiatric Association.

Bennell, C., & Corey, S. 2007. Geographic profiling of terrorist attacks. Chap. 9 in *Criminal profiling: International theory, research, and practice*, ed. R. N. Kocsis. Totowa, NJ: Humana Press.

Canter, D., & Fritzon, K. 1998. Differentiating arsonists: A model of firesetting actions and characteristics. *Legal and Criminological Psychology* 3 (1): 73–96, doi: 10.1111/j.2044-8333.1998.tb00352.x.

Decker, J. F., & Ottley, B. L. 2009. *Arson law and prosecution*. Durham, NC: Carolina Academic Press.

DeHaan, J. D., & Icove, D. J. 2012. *Kirk's fire investigation*, 7th ed. Upper Saddle River, NJ: Pearson-Prentice Hall.

Doley, R. 2003a. Making sense of arson through classification. *Psychiatry, Psychology and Law* 10 (2): 346–52, doi: 10.1375/pplt.2003.10.2.346.

———. 2003b. Pyromania: Fact or fiction? *British Journal of Criminology* 43 (4): 797–807, doi: 10.1093/bjc/43.4.797.

Douglas, J. E., Burgess, A.W., Burgess, A. G., & Ressler, R. K. 1992. *Crime classification manual: A standard system for investigating and classifying violent crimes.* San Francisco, CA: Jossey-Bass.

———. 2006. *Crime classification manual: A standard system for investigating and classifying violent crimes,* 2nd ed. New York: Wiley.

Douglas, J. E., Ressler, R. K., Burgess, A. W., & Hartman, C. R. 1986. Criminal profiling from crime scene analysis. *Behavioral Sciences & the Law* 4 (4): 401–21, doi: 10.1002/bsl.2370040405.

Ebdon, D. 1983. *Statistics in geography.* Oxford: Blackwell, 109.

FBI. 2010. *Crime in the United States, 2010: Arson.* Washington, DC: Federal Bureau of Investigation.

Fritzon, K. 2001. An examination of the relationship between distance travelled and motivational aspects of firesetting behaviour. *Journal of Environmental Psychology* 21 (1): 45–60, doi: 10.1006/jevp.2000.0197.

Fritzon, K., Canter, D., & Wilton, Z. 2001. The application of an action system model to destructive behaviour: The examples of arson and terrorism. *Behavioral Sciences & the Law* 19 (5–6): 657–90, doi: 10.1002/bsl.464.

Gannon, T. A., & Pina, A. 2010. Firesetting: Psychopathology, theory and treatment. *Aggression and Violent Behavior* 15 (3): 224–38, doi: 10.1016/j.avb.2010.01.001.

Gardiner, M. 1992. Arson and the arsonist: A need for further research. *Project Report.* London, UK: Polytechnic of Central London.

Geller, J. L. 1992. Arson in review: From profit to pathology. *Psychiatric Clinics of North America* 15:623–46.

———. 2008. Firesetting: A burning issue. In *Serial murder and the psychology of violent crimes,* ed. R. N. Kocsis, 141–77). Totowa, NJ: Humana Press.

Geller, J. L., Erlen, J., & Pinkus, R. L. 1986. A historical appraisal of America's experience with "pyromania": A diagnosis in search of a disorder. *International Journal of Law and Psychiatry* 9 (2): 201–29.

Georges, D. E. 1967. The ecology of urban unrest in the city of Newark, New Jersey, during the July 1967 riots. *Journal of Environmental Systems* 5 (3): 203–28.

———. 1978. The geography of crime and violence: A spatial and ecological perspective. Resource paper for college geography. No. 78-1. Washington, DC: Association of American Geographers.

Harmon, R. B., Rosner, R., & Wiederlight, M. 1985. Women and arson: A demographic study. *Journal of Forensic Science* 30 (2): 467–77.

Home Office. 1999. Safer communities: Towards effective arson control; The report of the arson scoping study. London, UK: Home Office.

Huff, T. G. 1994. Fire-setting fire fighters: Arsonists in the fire department; Identification and prevention. Quantico, VA: Federal Bureau of Investigation, National Center for the Analysis of Violent Crime.

———. 1997. Killing children by fire. Filicide: A preliminary analysis. Quantico, VA: Federal Bureau of Investigation, National Center for the Analysis of Violent Crime.

———. 1999. Filicide by fire. *Fire Chief* 43 (7): 66.

Huff, T. G., Gary, G. P., & Icove, D. J. 1997. The myth of pyromania. Quantico, VA: National Center for the Analysis of Violent Crime, FBI Academy.

Hurd, R. M. 1903. *Principles of city land values.* New York: Record and Guide.

Hurley, W., & Monahan, T. M. 1969. Arson: The criminal and the crime. *British Journal of Criminology* 9 (1): 4–21.

Icove, D. J. 1979. *Principles of incendiary crime analysis.* PhD diss., University of Tennessee, Knoxville, TN.

———. 1986. Automated crime profiling. *FBI Law Enforcement Bulletin* 55 (12): 27–30.

Icove, D. J., Douglas, J. E., Gary, G., Huff, T. G., & Smerick, P. A. 1992. Arson. Chap. 4 in *Crime classification manual,* ed. J. E. Douglas, A. W. Burgess, A. G. Burgess, & R. K. Ressler. San Francisco, CA: Jossey-Bass.

Icove, D. J., Escowitz, E. C., & Huff, T. G. 1993. The geography of violent crime: Serial arsonists. Paper presented at the 8th Annual Geographic Resources Analysis Support System (GRASS) Users Conference, March 14–19, Reston, VA.

Icove, D. J., & Estepp, M. H. 1987. Motive-based offender profiles of arson and fire related crimes. *FBI Law Enforcement Bulletin* 56 (4): 17–23.

Icove, D. J., and Horbert, P. R. 1990. Serial arsonists: An introduction. *Police Chief* (Arlington, VA) (December): 46–48.

Icove, D. J., Keith, P. E., and Shipley, H. L. 1981. An analysis of fire bombings in Knoxville, Tennessee. U.S. Fire Administration, Grant EMW-R-0599. Knoxville, TN: Knoxville Police Department, Arson Task Force.

Icove, D. J., Wherry, V. B., & Schroeder, J. D. 1980. *Combating arson-for-profit: Advanced techniques for investigators.* Columbus, OH: Battelle Press.

Inciardi, J. A. 1970. The adult firesetter: A typology. *Criminology* 8 (2): 141–55, doi: 10.1111/j.1745-9125.1970.tb00736.x.

Karter Jr, M. J. 2011. *Fire loss in the United States during 2010.* Quincy, MA: National Fire Protection Association.

Kocsis, R. N., & Cooksey, R. W. 2006. Criminal profiling of serial arson offenses. Chap. 9 in *Criminal profiling: Principles and practice,* ed. R. N. Kocsis, 153–74. Totowa, NJ: Humana Press.

Kocsis, R. N., & Irwin, H. J. 1997. An analysis of spatial patterns in serial rape, arson, and burglary: The utility of the circle theory of environmental range for psychological profiling. *Psychiatry, Psychology and Law* 4 (2): 195–206, doi: 10.1080/13218719709524910.

Kocsis, R. N., Irwin, H. J., & Hayes, A. F. 1998. Organised and disorganised criminal behaviour syndromes in arsonists: A validation study of a psychological profiling concept. *Psychiatry, Psychology and Law* 5 (1): 117–31, doi: 10.1080/13218719809524925.

LeBeau, J. L. 1987. The methods and measures of centrography and the spatial dynamics of rape. *Journal of Quantitative Criminology* 3 (2): 125–41.

Levin, B. 1976. Psychological characteristics of firesetters. *Fire Journal* (March): 36–41.

Lewis, N. D. C., & Yarnell, H. 1951. *Pathological firesetting: Pyromania.* New York: Nervous and Mental Disease Monographs.

Lieber, A. L., & Agel, J. 1978. *The lunar effect: Biological tides and human emotions.* Garden City, NY: Anchor Press.

MacQueen, J. B. 1967. Some methods for classification and analysis of multivariate observations. Paper presented at the Fifth Berkeley Symposium on Mathematical Statistics and Probability.

Miller, S., & Fritzon, K. 2007. Functional consistency across two behavioural modalities: Fire-setting and self-harm in female special hospital patients. *Criminal Behaviour & Mental Health* 17 (1): 31–44.

Netherwood, R. E. 1966. The relationship between lunar phenomena and the crime of arson. Master's thesis, University of California, Berkeley.

NFPA. 2011. *NFPA 921: Guide for fire and explosion investigations.* Quincy, MA: National Fire Protection Association.

NVFC. 2011. Report on the firefighter arson problem: Context, considerations, and best practices. Greenbelt, MD: National Volunteer Fire Council.

Park, R. E., & Burgess, E. W. 1921. *Introduction to the science of sociology.* Chicago, IL: The University of Chicago Press.

Quinsey, V. L., Chaplin, T. C., and Upfold, D. 1989. Arsonists and sexual arousal to fire setting: Correlation unsupported. *Journal of Behavioral and Experimental Psychiatry* 20 (no. 3): 203–8.

Reardon, J. J. 2002. The warning signs of a faked death: Life insurance beneficiaries can't recover without providing "due proof" of death. *Connecticut Law Tribune,* 5.

Rengert, G., & Wasilchick, J. 1990. *Space, time, and crime: Ethnographic insights into residential burglary.* Final Report to the U.S. Department of Justice. Philadelphia, PA: Temple University, Department of Criminal Justice.

Repo, E., Virkkunen, M., Rawlings, R., & Linnoila, M. 1997. Criminal and psychiatric histories of Finnish arsonists. *Acta Psychiatrica Scandinavica* 95 (4): 318–23, doi: 10.1111/j.1600-0447.1997.tb09638.x.

Rider, A. O. 1980. The firesetter: A psychological profile. *FBI Law Enforcement Bulletin* 49 (June, July, August): 7–23.

Robbins, E., & Robbins, L. 1967. Arson with special reference to pyromania. *New York State Journal of Medicine* 67:795–98.

Rogerson, P., & Sun, Y. 2001. Spatial monitoring of geographic patterns: An application to crime analysis. *Computers, Environment and Urban Systems* 25 (6): 539–56, doi: 10.1016/s0198-9715(00)00030-2.

Rossmo, D. K. 2000. *Geographic profiling.* Boca Raton, FL: CRC Press.

Rule, A. 1998. *Bitter harvest: A woman's fury, A mother's sacrifice.* Thorndike, ME: G.K. Hall.

Santtila, P., Fritzon, K., & Tamelander, A. 2004. Linking arson incidents on the basis of crime scene behavior. *Journal of Police and Criminal Psychology* 19 (1): 1–16, doi: 10.1007/bf02802570.

Sapp, A. D., Gary, G. P., Huff, T. G., Icove, D. J., & Horbett, P. R. 1994. *Motives of serial arsonists: Investigative implications.* Monograph. Quantico, VA: Federal Bureau of Investigation.

Sapp, A. D., Gary, G. P., Huff, T. G., & James, S. 1993. *Characteristics of arsons aboard naval ships.* Monograph. Quantico, VA: Federal Bureau of Investigation.

Sapp, A. D., Huff, T. G., Gary, G. P., D J Icove, D. J., & P Horbert, P. 1995. *Report of essential findings from a study of serial arsonists.* Monograph. Quantico, VA: Federal Bureau of Investigation.

Snook, B., Taylor, P. J., & Bennell, C. 2004. Geographic profiling: The fast, frugal, and accurate way. *Applied Cognitive Psychology* 18 (1): 105–21, doi: 10.1002/acp.956.

Stanton, J., & Simpson, A. I. F. 2006. The aftermath: Aspects of recovery described by perpetrators of maternal filicide committed in the context of severe mental illness. *Behavioral Sciences & the Law* 24 (1): 103–12, doi: 10.1002/bsl.688.

Steinmetz, R. C. 1966. Current arson problems. *Fire Journal* 60 (no. 5): 23–31..

Stewart, M. A., & Culver, K. W. 1982. Children who set fires: The clinical picture and a follow-up. *British Journal of Psychiatry* 140:357–63, doi: 10.1192/bjp.140.4.357.

Tou, J. T., & Gonzalez, R. C. 1974. *Pattern recognition principles.* Reading, MA: Addison-Wesley.

USFA. 2003. Special report: Firefighter arson. *Technical Report Series.* Emmitsburg, MD: U.S. Fire Administration.

Vandersall, T. A., and Wiener, J. M. 1970. Children who set fires. *Archives of General Psychiatry* 22 (January).

Virkkunen, M., Nuutila, A., Goodwin, F. K., & Linnoila, M. 1987. Cerebrospinal fluid monoamine metabolite levels in male arsonists. *Archives of General Psychiatry* 44 (March): 241–47.

Vreeland, R. G., & Waller, M. B. 1978. The psychology of firesetting: A review and appraisal. Chapel Hill, NC: University of North Carolina.

Wheaton, S. 2001. Personal accounts: Memoirs of a compulsive firesetter. *Psychiatric Services,* 52 (8), doi: 10.1176/appi.ps.52.8.1035.

Willis, M. 2004. Bushfire arson: A review of the literature. *Research and Public Policy Series No. 61.* Canberra, Australia: Australian Institute of Criminology.

Wolford, M. R. 1972. Some attitudinal, psychological and sociological characteristics of incarcerated arsonists. *Fire and Arson Investigator* 22.

Woodhams, J., Bull, R., & Hollin, C. R. 2007. Case linkage: Identifying crimes committed by the same offender. Chap 6 in *Criminal profiling: Principles and practice,* ed. R. N. Kocsis, 117–33. Totowa, NJ: Humana Press.

Yang, S., & Geller, J. L. 2009. Firesetting. In *Wiley encyclopedia of forensic science.* Chichester, UK: Wiley. (See online edition at www.wiley.com/.)

6

Fire Modeling

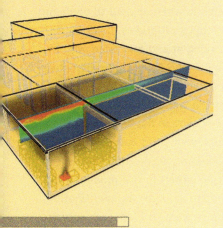

> ❝Data! Data! Data! I can make no bricks without clay. ❞
>
> —Sir Arthur Conan Doyle
> *The Adventure of the Copper Beeches*

Courtesy of D. J. Icove.

OBJECTIVES

After reading this chapter, the student should be able to:

- Identify proper application of fire modeling in fire scene reconstruction.
- Evaluate data necessary for fire model applications.
- Apply fire modeling to case studies.
- Interpret fire modeling results

Two general types of fire models are used in support of fire investigations, namely, physical models and computational models. This chapter addresses the latter—computational models. Physical models are addressed in Chapter 8, Fire Testing.

The concept of fire modeling as it applies to forensic fire investigation is unique to the last two decades, although models have existed since the 1960s. Previously, fire modeling was centered on explaining the physical phenomena of fires, particularly when applied to verifying existing experimental data. Now, computer modeling is commonplace (DeHaan 2005a, 2005b).

It was the effort of a few fire scientists and engineers working at and associated with the National Institute of Standards and Technology (NIST) and the Building Research Establishment (BRE), Fire Research Station (FRS) in the United Kingdom that pushed the acceptability and application of fire modeling out of laboratory conditions and into the world of forensic fire scene reconstruction. The application of these early successes in fire modeling to the field of fire litigation and reconstruction further underscored its usefulness (Bukowski 1991; Babrauskas 1996). Several of the keynote studies that contributed significantly to this effort are described in this text.

The purpose of this chapter is not to make the reader an expert in fire modeling but to allow him or her to gain a better appreciation for its added value to an investigation. Ample references are provided should more information be needed. This chapter answers the following questions that often arise when fire investigators are confronted with a fire of exceptional scale or impact: (1) What exactly is a fire model? (2) In which aspect of my investigation can fire modeling help? (3) What are the realistic and reliable results of a fire model? (4) Should more than one type of model be used to increase confidence in the results? (5) What is the future of fire modeling?

History of Fire Modeling

Research by Mitler (1991) at NIST best establishes the historical framework for the application of fire scene modeling. Work in 1927 at the National Bureau of Standards (NBS), the predecessor agency of NIST, was the first attempt to understand and explain in scientific terms the issues surrounding postflashover compartment fires by relating the room gas temperature to the available mass of fuel being consumed (Inberg 1927).

The first fire model was developed through work in Japan in 1958 to relate the ventilation factor to steady-state fire development (Kawagoe 1958). The second model was constructed in Sweden (Magnusson and Thelandersson 1970), followed by the work of Babrauskas at the University of California, Berkeley (1975).

Historical reasons for developing and applying mathematical fire modeling were also accurately articulated and predicted by NIST (Mitler 1991). A good mathematical fire model of a structure can be helpful by

- Avoiding repetitious full-scale testing,
- Helping designers and architects,
- Establishing flammability of materials,
- Increasing the flexibility and reliability of fire codes,
- Identifying needed fire research, and
- Helping in fire investigations and litigation.

The impact of environmental limitations and financial restrictions on fire testing has made fire modeling a valuable supplement to most fire testing programs. Factors such as the placement and combination of fuel items, changes in ventilation, and thickness of materials can be evaluated with fire models, reducing the need for repeated full-scale testing. Fire modeling can also supplement fire testing programs when used as a screening tool for full-scale testing scenarios.

Designers and architects have found modeling useful in assessing the impact of fire on buildings and occupants as they specify new construction methods and materials. New flexible fire performance codes are becoming more generally accepted as modeling introduces alternatives or new designs (such as atrium construction) that traditionally have not been addressed in historical prescriptive codes. Fire modeling can now also address the optimum designs for evacuation of people from buildings during emergencies.

As the science of fire dynamics advances, computational modeling identifies needed new areas of fire research into the phases of growth, flame spread, and flashover. Modeling can aid in gap assessments of certain areas that previously were not fully understood by designers, engineers, researchers, and even fire investigators.

Finally, fire modeling has had significant impact on forensic fire investigations and litigation (DeHaan and Icove 2012). Advancements in these areas are discussed in this chapter.

Fire Models

Computer **fire models** have a broad range of applications in the area of fire science and engineering. The fire model works to supplement the information gleaned from forensic evidence, witness interviews, media film footage, and preliminary fire scene examinations. Historically, these fire models, as shown in Table 6-1, have included eight major overlapping categories (Hunt 2000).

fire model ■ A method using mathematical or computer calculations that describes a system or process related to fire development, including fire dynamics, fire spread, occupant exposure, and the effects of fire. The results produced by fire models can be compared with physical and eyewitness evidence to test working hypotheses.

TABLE 6-1	Classes of Computer Fire Models and Commonly Cited Examples	
CLASS OF MODEL	**DESCRIPTION**	**EXAMPLE(S)**
Spreadsheet	Calculates mathematical solutions for interpretations of actual case data	FiREDSHEETS, NRC spreadsheets
Zone	Calculates fire environment through two homogeneous zones	FPETool, CFAST, ASET-B, BRANZFIRE, FireMD
Field	Calculates fire environment by solving conservation equations, usually with finite-element mathematics	FDS, JASMINE, FLOW3D, SMARTFIRE, PHOENICS, SOFIE
Postflashover	Calculates time–temperature history for energy, mass, and species and is useful in evaluating structural integrity in fire exposure	COMPF2, OZone, SFIRE-4
Fire protection performance	Calculates sprinkler and detector response times for specific fire exposures based on the response time index (RTI)	DETACT-QS, DETECT-T2, LAVENT
Thermal and structural response	Calculates structural fire endurance of a building using finite-element calculations	FIRES-T3, HEATING7, TASEF
Smoke movement	Calculates the dispersion of smoke and gaseous species	CONTAM96, Airnet, MFIRE
Egress	Calculates the evacuation times using stochastic modeling using smoke conditions, occupants, and egress variables	Allsafe, buildingEXODUS, EESCAPE, ELVAC, EVACNET, EXIT89, EXITT, EVACS, EXITT, Simplex, SIMULEX, WAYOUT

Sources: Updated from Bailey 2006; Friedman 1992; and Hunt 2000.

Fire models usually emulate the impact of fires, not the physical fire itself. There are two recognized approaches to modeling fires: probabilistic and deterministic. *Probabilistic* models usually center on the application of stochastic mathematics to estimate or predict an outcome (within certain likelihoods), such as human behavior (SFPE 2008, chaps. 3-11 and 3-12). In *deterministic* models, the investigator relies on mathematical relationships that form the underpinnings of the physics and chemistry of fire science. Deterministic models can range from a simple straight-line approximation to correlating test data with complex fire models solving hundreds of simultaneous equations.

Mathematical models are usually formulated from interpretations of actual test data. These types of models are usually calculated by hand using formulas and a scientific calculator, a spreadsheet, or a simple computer program. Smoke-filling rates, flame heights, virtual origin, and other approximations can usually be calculated by hand in several minutes. When based on sound mathematical representations and properly applied, fire models can often assure that the scientific method has been satisfied.

Because mathematical models are discussed throughout this text in the form of quick problem solutions, the zone and field models are emphasized here. The discussion is limited to a survey of the concepts and is by no means intended to serve as an instruction manual for their use. In fact, most guidance issued with fire models cautions that they are for use only by those having significant competency in the fire engineering field and that the results should supplement the user's professional judgment.

Spreadsheets Models

One of the better-known spreadsheets for solving simple fire protection engineering relationships emerged from a May 1992 study, "Methods of Quantitative Fire Hazard Analysis," prepared for the Electric Power Research Institute (EPRI) by the University of Maryland, Department of Fire Protection Engineering. The study was distributed through the Society of Fire Protection Engineers (SFPE) (Mowrer 1992).

The study describes fire hazard analysis models used in the Fire-Induced Vulnerability Evaluation (FIVE) methodology, which is used by the Nuclear Regulatory Commission (NRC). The FIVE analysis uses realistic industrial loss histories to evaluate hazards quantitatively, including those losses produced in fires resulting from ignitable liquid spills, cable trays, and electrical cabinets. Many of these examples predict the impact of the fire plumes and ceiling jet temperatures, hot gas layers, thermal radiation to targets, and critical heat fluxes. Many of these concepts are applicable to problems previously posed in fire scene reconstruction and analysis (Mowrer 1992).

From this study came FiREDSHEETS from the University of Maryland, Department of Fire Protection Engineering (Milke and Mowrer 2001). These spreadsheets concentrated on the specifics of compartment fire analysis and included extensive property tables for typical materials first ignited.

Mowrer has continued to develop spreadsheet templates for fire dynamics calculations, which are presently posted along with ample documentation on the Fire Risk Forum website, www.fireriskforum.com/ (Mowrer 2003). These templates incorporate a number of enclosure fire dynamics calculations used by FPETool (Nelson 1990) and other suites, as exemplified in Figure 6-1. Mowrer spreadsheets also include key calculations used in the fire and arson investigation training course for agents of the Bureau of Alcohol, Tobacco, Firearms and Explosives given by the University of Maryland's Department of Fire Protection Engineering. Listed in Table 6-2 are the calculations contained in the Mowrer spreadsheets.

A later and more industry-oriented set of spreadsheets known as the Fire Dynamics Tools (FDTs) was developed for the NRC fire protection inspection program (Iqbal and Salley 2004, 2004). The format of the spreadsheet output is more intuitive in its approach than are financial spreadsheets. Table 6-3 is a synopsis of the available FDTs spreadsheets.

PLUMETMP.XLS: Estimate of temperature rise in a fire plume

INPUT PARAMETERS		
FIRE HEAT RELEASE RATE (Q)	500	kW
CONVECTIVE FRACTION (Xc)	0.7	–
HEIGHT ABOVE FIRE (Z)	5	m
FIRE LOCATION FACTOR (kLF)	1	–
CALCULATED PARAMETERS		
CONVECTIVE HRR (Qc)	350	kW
FLAME HEIGHT (Zfl)	2.4	m
PLUME TEMPERATURE (dTpl)	95	C

FIRE LOCATION FACTORS	kLF
FIRE IN OPEN	1
FIRE ALONG WALL	2
FIRE IN CORNER	4

LAYERTMP.XLS: Estimate of upper layer temperature using modified MQH correlation

INPUT PARAMETERS		
ROOM LENGTH (L)	6.16	m
ROOM WIDTH (W)	6.7	m
ROOM HEIGHT (H)	2.8	m
OPENING WIDTH (Wo)	1.8	m
OPENING HEIGHT (Ho)	2.7	m
BOUNDARY CONDUCTIVITY (k)	0.00017	kW/m.K
BOUNDARY DENSITY (p)	960	kg/m3
BOUNDARY SPEC. HEAT (cp)	1.1	kJ/kg.K
BOUNDARY THICKNESS (d)	0.0125	m
FIRE HRR (Q)	2000	kW
FIRE LOCATION FACTOR (kLF)	4	–
CALCULATION TIME (t)	400	s
CALCULATED PARAMETERS		
HEAT TRANSFER COEFF (hk)	0.0211849	kW/m2.K
BOUNDARY AREA (At)	154.56	m2
VENTILATION FACTOR	7.99	m^5/2
TEMPERATURE RISE (dT)	582	C

FIGURE 6-1 Examples of two of the 21 fire dynamics calculations that can be performed using the Mowrer spreadsheets. *www.fireriskforum.com/.*

TABLE 6-2	Mowrer's Fire Risk Forum Spreadsheets

TEMPLATE	SPREADSHEET DESCRIPTION
ATRIATMP	Estimates the approximate average temperature rise in the hot gas layer that develops in a large open space such as an atrium
BUOYHEAD	Estimates the pressure differential, gas velocity, and unit mass flow rate caused by the buoyancy the of hot gases beneath a ceiling
BURNRATE	Estimates the burning rate history of a flammable liquid fire
CJTEMP	Estimates the temperature rise in an unconfined ceiling jet
DETACT	Estimates the response time of ceiling-mounted fire detectors
FLAMSPRED	Estimates the lateral flame spread rates on solid materials
FLASHOVR	Estimates the heat release rate needed to cause flashover in a compartment with a single rectangular wall vent
FUELDATA	Contains thermophysical and burning rate data for common fuels
GASCONQS	Estimates gas species concentrations
IGNTIME	Estimates the time to ignite a thermally thick solid exposed to a constant heat flux
LAYDSCNT	Estimates the smoke layer interface position in a closed room due to entrainment
LAYERTMP	Estimates the average hot gas layer temperature in an enclosure with a single rectangular wall opening
MASSBAL	Estimates the mass flow rate through an enclosure with a single wall opening
MECHVENT	Estimates the fire conditions in a mechanically ventilated space without natural ventilation

(continued)

TABLE 6-2	(continued)
TEMPLATE	**SPREADSHEET DESCRIPTION**
PLUMEFIL	Estimates the volumetric rate of smoke flow in a fire plume
PLUMETMP	Estimates the temperature rise in an axisymmetric fire plume
RADIGN	Estimates the potential for radiant ignition of a combustible target
STACK	Estimates the mass flow rate through an enclosure
TEMPRISE	Estimates the average temperature rise in a closed room
THERMPRP	Contains thermal property data for 15 types of boundary materials

Source: F. W. Mowrer, Spreadsheet Templates for Fire Dynamics Calculations, University of Maryland, Department of Fire Protection Engineering, September 2003. Spreadsheets and documentation are posted on the Fire Risk Forum website, fireriskforum.com/.

TABLE 6-3	NRC Fire Dynamics Spreadsheets
CHAPTER	**FDTs SPREADSHEET DESCRIPTION**
2	Predicting Hot Gas Layer Temperature and Smoke Layer Height in a Room Fire with Natural Ventilation
	Predicting Hot Gas Layer Temperature in a Room Fire with Forced Ventilation
	Predicting Hot Gas Layer Temperature in a Room Fire with Door Closed
3	Estimating Burning Characteristics of Liquid Pool Fire, Heat Release Rate, Burning Duration, and Flame Height
4	Estimating Wall Fire Flame Height
5	Estimating Radiant Heat Flux from Fire to a Target Fuel at Ground Level under Wind-Free Condition Point Source Radiation Model
	Estimating Radiant Heat Flux from Fire to a Target Fuel at Ground Level in Presence of Wind (Tilted Flame) Solid Flame Radiation Model
	Estimating Thermal Radiation from Hydrocarbon Fireballs
6	Estimating the Ignition Time of a Target Fuel Exposed to a Constant Radiative Heat Flux
7	Estimating the Full-Scale Heat Release Rate of a Cable Tray Fire
8	Estimating Burning Duration of Solid Combustibles
9	Estimating Centerline Temperature of a Buoyant Fire Plume
10	Estimating Sprinkler Response Time
13	Calculating Fire Severity
14	Estimating Pressure Rise Due to a Fire in a Closed Compartment
15	Estimating Pressure Increase and Explosive Energy Release Associated with Explosions
16	Calculating the Rate of Hydrogen Gas Generation in Battery Rooms
17	Estimating Thickness of Fire Protection Spray-Applied Coating for Structural Steel Beams (Substitution Correlation)
	Estimating Fire Resistance Time of Steel Beams Protected by Fire Protection Insulation (Quasi-Steady-State Approach)
	Estimating Fire Resistance Time of Unprotected Steel Beams (Quasi-Steady-State Approach)
18	Estimating Visibility Through Smoke

Source: Iqbal and Salley 2004.

Zone Models

Two-zone models are based on the concept that a fire in a room or enclosure can be described in terms of two unique zones, and the conditions within each zone can be predicted. The two zones refer to two individual control volumes known as the upper and lower zone or layer. The upper zone of heated gases and by-products of combustion is penetrated only by the fire plume. These hot gases and smoke fill the ceiling layer, which then slowly descends. A boundary layer marks the intersection of the upper and the lower areas. The only exchange between the upper and the lower layers is due to the action of the fire plume.

Zone models use a set of differential equations for each zone solving for pressure, temperature, carbon monoxide, oxygen, and soot production. The accuracy of these calculations varies with not only the model but also the reliability of the input data describing the circumstances and room dimensions and, very importantly, the assumptions as to the heat release rates of the materials burned in the fire. Complex two-zone models of up to 10 rooms typically take a few minutes to run using multi-core-class computer workstations or laptops.

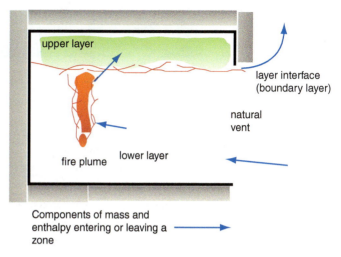

FIGURE 6-2 Representation of the two-zone fire model showing the location of the upper and lower layers, layer interface, fire plume, and compartment vent. *Courtesy of NIST, from Forney and Moss 1992.*

Figure 6-2 shows a schematic representation of the concepts behind a two-zone model. A survey published in 1992 by the SFPE showed ASET-B and DETACT-QS to be the most widely used zone models at that time (Friedman 1992).

Today, CFAST, the Consolidated Model of Fire Growth and Smoke Transport, has become the most widely used zone model. CFAST combines the engineering calculations from the FIREFORM model and the user interface from FASTlite in a unique comprehensive zone modeling and fire analysis tool (Peacock et al. 2008).

CFAST OVERVIEW

The CFAST two-zone model is a heat and mass-balance model based on not only the physics and chemistry but also the results of experimental and visual observations in actual fires. CFAST computes

- Production of enthalpy and mass by burning objects;
- Buoyancy and forced transport of enthalpy and mass through horizontal and vertical vents; and
- Temperatures, smoke optical densities, and species concentrations.

CFAST 3.1.7 has a graphical user interface (GUI) that allows the user to enter and modify characteristics about the physical form of the rooms, the fire signature, and graphics output. An enhancement in CFAST 5.0.1 and, most recently, CFAST 6.2 is that it is linked to Smokeview, the graphical interface of the NIST's Fire Dynamics Simulator (FDS) program. Table 6-4 summarizes the present capabilities of CFAST version 6.2.

CFAST has been in use since its first public release in June 1990, with many enhancements over the years. CFAST version 6.1 includes a newly designed user interface and incorporates improvements in calculating heat transfer, smoke flow through corridors, and more accurate combustion chemistry. NIST added the capability to define a general t^2 growth rate fire that allows the user to select growth rate, peak heat release rate, steady burning time, and decay time, including predefined constants for slow, medium, fast, and ultrafast t^2 fires. The latest CFAST guide for users and documentation follows the

zone model ■ A computer fire model based on the assumption that a fire in a room or enclosure can be described in terms of two unique zones known as the upper and the lower zone or layer, and the conditions within each zone can be predicted. Zone models use a set of differential equations for each zone solving for conditions such as, for example, pressure, temperature, carbon monoxide, oxygen, and soot production.

TABLE 6-4	Summary of Numerical Limits in the Software Implementation of the CFAST Model, Version 6.2

FEATURE	MAXIMUM
Simulation time(s)	86,400
Compartments	30
Object fires	31
Fire definitions in database file	30
Material thermal property definitions	125
Slabs in a single surface material	3
Fans in mechanical ventilation systems	5
Ducts in mechanical ventilation systems	60
Connections in compartments and mechanical ventilation systems	62
Independent mechanical ventilation systems	15
Targets	90
Data points in a history or spreadsheet file	900

Source: Peacock et al. 2005.

guidelines set forth in *ASTM E1355-11: Standard Guide for Evaluating the Predictive Capability of Deterministic Fire Models* (ASTM 2011b; Peacock et al. 2008).

Because CFAST does not have a pyrolysis model to predict fire growth, a fuel source is entered or described by its fire signature (heat release rate). The program converts this fuel information into two characteristics: enthalpy (heat) and mass. In an unconstrained fire, the burning of this fuel takes place within the fire plume. In constrained fires, the pyrolyzed fuel may burn in the fire plume where there is sufficient oxygen, but it may also burn in the upper or lower layer of the room of fire origin, in the plume in the doorway leading to an adjoining room, or even in the layers or plumes in adjacent rooms.

FIRE PLUMES AND LAYERS

As in previous mathematical models in which calculations of heat flux used the fire plume's virtual origin, the CFAST model instructs the fire plume to act as a "pump" to move enthalpy and mass into the upper layer from the lower layer. Mixed with this enthalpy and mass are inflows and outflows from horizontal or vertical vents (doors, windows, etc.), which are modeled as plumes.

Some assumptions of the model do not always hold in real fires. Some mixing of the upper and lower layers does take place at their interfaces. At cool wall surfaces, gases will flow downward as they lose heat and buoyancy. Also, heating and air-conditioning systems will cause mixing between the layers.

Horizontal flow of the fire from one room to the next occurs when the upper layer descends below the opening of an open vent. As the upper layer descends, pressure differentials between the connected rooms may cause air to flow into and out of a room or compartment in the opposite direction, resulting in the two flow conditions. Upward vertical flow may occur when the roof or ceiling of a room is opened.

HEAT TRANSFER

In the CFAST model, unique material properties of up to three layers can be defined for surfaces of the room (ceiling, walls, flooring). This design consideration is useful, since in

CFAST, heat transfers to the surface via convective heat transfer and through the surfaces via conductive heat transfer.

Radiative heat transfer takes place among fire plumes, gas layers, and surfaces. In radiative heat transfer, emissivity is dominated primarily by species contributions (smoke, carbon dioxide, and water) within the gas layers. CFAST applies a combustion chemistry scheme that balances carbon, hydrogen, and oxygen in the room of fire origin among the lower-layer portion of the burning fire plume, the upper layer, and air entrained in the lower layer that is absorbed into the upper layer of the next connecting room.

LIMITATIONS

Zone models have their limitations. Six specific aspects of physics and chemistry either are not included or have very limited implementations in zone models: flame spread, heat release rate, fire chemistry, smoke chemistry, realistic layer mixing, and suppression (Babrauskas 1996). Some errors in species concentrations can result in errors in the distribution of enthalpy among the layers, a phenomenon that affects the accuracy of temperature and flow calculations. CFAST has also been known to overpredict the upper-layer temperatures, partly because heat losses due to window radiation are not presently incorporated in the model.

Even with these known limitations, zone models have been extremely successful in forensic fire reconstruction and litigation (Bukowski 1991; DeWitt and Goff 2000). CFAST zone models have been successfully introduced in federal court litigation (see Chapter 1, Example 1-2). These successes are a result of their correct application by knowledgeable scientists and engineers, continued revalidation in actual fire testing, and applied research efforts, include verification and validation studies by the NRC (Salley and Kassawar 2007a, 2007b, 2007c, 2007d, 2007e, 2007f, 2007g).

Field Models

Field models, the newest and most sophisticated of all deterministic models, rely on *computational fluid dynamics* (CFD) technology. These models are attractive, especially as a tool in litigation support and fire scene reconstruction, owing to their ability to display the impact of information visually in three dimensions.

However, the downside of field models is that creating input data is typically very time consuming and often requires powerful computer workstations to compute and display the results. The computation time to run field models may be days or weeks depending on the complexity of the model and the relative computational power of the computer system available.

COMPUTATIONAL FLUID DYNAMICS MODELS

CFD models estimate the fire environment by dividing the compartment into uniform small cells, or control volumes, instead of two zones. The program then simultaneously solves the conservation equations for combustion, radiation, and mass transport to and from each cell surface. The present shortcomings include the level of background and training needed to codify a model, particularly for complex layouts.

There are many advantages to using CFD models over zone models such as CFAST. The higher geometric resolution of CFD models makes the solutions more refined. Higher-speed laptops and workstations allow CFD models to be run, whereas in the past, larger mainframe and minicomputers were required. Multiple workstations have been successfully linked together in parallel processing arrays to perform complex CFD calculations more quickly.

field model ■ A computer fire model that uses *computational fluid dynamics* (CFD) to estimate the fire environment by dividing the compartment into uniform small cells, or control volumes. The computer program then attempts to predict conditions in every volume such as, for example, pressure, temperature, carbon monoxide, oxygen, and soot production.

Because CFD models are used in companion areas such as fluid flow, combustion, and heat transfer, their underlying technology is generally more accepted over simpler and coarser models. Once these models become more broadly validated, they will gain further acceptance in situations where scientifically based testimony is sought, such as in forensic fire scene reconstructions.

CFD has a long history of use in forensic fire scene reconstructions and is gaining in popularity owing to commercially available and intuitive graphical interfaces (e.g., PyroSim [Thunderhead 2010, 2011]). CFD programs now outpace the capabilities of the zone models, which, for the most part, operate under maintenance modes only. Moreover, CFD programs such as the FDS make the technology economically feasible owing to its wide use, testing, and growing acceptance.

NIST CFD TECHNOLOGY–BASED MODELS

The recommended field model using CFD technology, owing to its availability at no cost, focus of research, and acceptance in the community is the FDS. This model is available from NIST's Building and Fire Research Laboratory (McGrattan et al. 2010a). The first version of FDS was publicly released in February 2000 and has become a mainstay of the fire research and forensic communities.

FDS is a fire-driven fluid flow model and numerically solves a form of the Navier-Stokes equations for thermally driven smoke and heat transport. The companion program to FDS is Smokeview, which is a graphical interface that produces a variety of visual records of FDS calculations. These include concentration of species, temperature, and heat flux in various animations. User guides for both FDS (McGrattan et al. 2010b) and Smokeview (Forney 2010) are available at NIST's website, fire.nist.gov/.

Traditional use of FDS to date has been divided among the evaluation of smoke-handling systems, sprinkler and detector activation, and fire reconstructions. It is also being used to study fundamental fire dynamics problems encountered in both academic and industrial settings. FDS uses a combination of models to handle the computational problems: a hydrodynamic model to solve the Navier-Stokes equations, a Smagorinsky form of large eddy simulation (LES) for smoke and heat transport, a numerical grid simulation to handle turbulence, a mixture fraction combustion model, and a finite volume method approach for radiation heat transfer.

Unfortunately, FDS uses a rectilinear grid to describe the computational cells. This makes it slightly difficult to model sloping roofs, rounded tunnels, and curved walls without some form of approximation. The boundary conditions for material surfaces consist of assigned thermal constants that include information on their burning behavior. FDS, like other CFD models, breaks up the room (or fire area) into cells and calculates the mass heat energy and species transport (fluxes) into and out of each of the six faces of each cell. Figure 6-3 shows the output grids from a typical FDS problem.

The latest release of FDS at the time of this publication is version 5, which is a significant enhancement over previous versions. A version 6 release is under development and testing. These enhancements include the ability to evaluate heat transfer through walls and the impact of water suppression and initial conditions, and to add image texturing to solid objects and surfaces. The latest companion release of Smokeview is version 5. Its enhanced features include controls over viewing the rectilinear mesh used in constructing the FDS model. Scene clipping allows the visualization of obstructed boundary surfaces in complex models having numerous walls. A more robust feature allows more control over the movement and orientation of the scene as rendered by Smokeview.

Other features include several visualization modes such as tracer particle flow, animated contour slices of computed variables such as vector heat flux plots, animated surface data, and multiple isocontours. New visualization tools allow the user to compare fire burn patterns with computed uniform surface contour planes. The option for

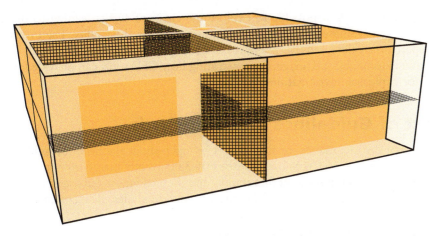

FIGURE 6-3 CFD (field) models break up the room (or fire area) into cells and calculate the mass heat energy and species transport (fluxes) into and out of each of the six faces of each cell.

animated flow vectors and particle animations allows observed heat flow vectors to be compared with computed vectors.

A horizontal time bar displays the duration of the run, and a vertical color-coded bar displays other variables, such as temperature and velocity. Graphics options include the ability to capture the output of the Smokeview program on a screen-by-screen basis. Screens may be exported and made into a movie, but this requires additional (non-NIST) software and can be tedious to produce.

FM GLOBAL CFD FIRE MODELING

FM Global has always served as a leader in the development of many of the fire protection engineering tools, fire dynamics relationships, fire growth and behavior models, and testing protocols cited in this textbook. FM Global maintains a unique center for engaging in property loss–prevention strategies, scientific research, and product testing at their Research Campus in West Glocester, Rhode Island.

FM Global is presently developing a number of key physical models relating to fluid mechanics, heat transfer, and combustion into a new open-source software package called FireFOAM. This program is a CFD toolbox for fire and explosion modeling applications. To promote outside cooperation, FM Global has released its modeling technologies.

FireFOAM utilizes the finite volume method on arbitrarily unstructured meshes, and is highly scalable on massively parallel computers. The capabilities of FireFOAM include many complex simulations, including fire growth and suppression. For more information about FireFOAM, please refer to FM Global's website, www.fmglobal.com/ (FireFOAM 2012).

FLACS CFD MODEL

The commercial FLACS CFD model, maintained by CMR GexCon, has historically been used for gas dispersion modeling and explosion calculations within many industries (GexCon 2012). GexCon emphasizes the need for validation of CFD models such as FLACS. The software has many applications, including quantitative risk assessments, investigations into explosion venting of coupled and nonstandard vessels, predicting the effect of explosions, and modeling toxic gas dispersion.

FLACS has been successfully used in hypothesis testing during large industrial explosion investigations. One of the most impressive uses of FLACS documented by GexCon US, Bethesda, Maryland, was in their analysis of the November 22, 2006, explosion and fire at an ink and paint manufacturing facility explosion in Danvers, Massachusetts (Davis et al. 2010).

Using FLACS made it possible to explore the chain of events leading to the explosion as well as to evaluate potential ignition sources within the facility. The investigators also compared the calculated overpressures of the exploding fuel/air cloud with observed internal and external blast damage, demonstrating how CFD tools can provide invaluable analyses for explosion investigations.

Impact of Guidelines and Standards

Within the last decade, the use of fire models brought about the need for guidelines and standards for their use. ASTM Subcommittee E05.33 currently maintains and continues to work on improving four guidelines that standardize the evaluation and use of computer fire models. The following is a synopsis of the current standard guides (Janssens and Birk 2000).

ASTM GUIDELINES AND STANDARDS

- *ASTM E1355-11:* Evaluates the predictive capacity of fire models by defining scenarios, validating assumptions, verifying the mathematical underpinnings of the model, and evaluating its accuracy (ASTM 2011b).
- *ASTM E1472-07:* Describes how a fire model is to be documented and includes a user's manual, a programmer's guide, mathematical routines, and installation and operation of the software (ASTM 2007a). *Note:* Withdrawn in 2010.
- *ASTM E1591-07:* Covers and documents available literature and data that are beneficial to modelers (ASTM 2007b).
- *ASTM E1895-07:* Examines the uses and limitations of fire models and addresses how to choose the most appropriate model for the situation (ASTM 2007c). *Note:* Withdrawn in 2011.

SFPE GUIDELINES AND STANDARDS

SFPE (2011) recently published *Guidelines for Substantiating a Fire Model for a Given Application*, Engineering Guide SFPE G.06 2011. This guide supplements the ASTM standards and assists the user of a fire model in defining the problem, selecting a candidate model, interpreting the verification and validation studies for various models, and understanding user effects. An appendix covers fire-related phenomena with guidance as to application to specific models and interpreting the underlying key physics.

VALIDATION

All fire models must be *validated* by comparing their predictions against the results of real fire tests if their results are to be relied on. This testing sometimes reveals flaws or blind spots—where some types of enclosures or conditions can produce predicted values that differ from observed fire behavior. If a model has not been shown to yield valid results for a particular type of fire, its predictions in an unknown scenario should not be accepted without question. For instance, FDS has been shown to be very accurate in predicting growth of fires in large compartments, but data on how FDS predictions relate to real fires in very small compartments are just now being gathered and published.

It should be noted that FDS was derived from earlier work on modeling the growth and movement of smoke plumes from large stationary fires such as oil pools. This work, based on large eddy simulation (LES), focused on atmospheric interaction with the smoke plumes and not on movement of the fire itself. More recently, FDS has been applied to studying wildland–urban interface (WUI) fires, considering the impact of vegetation fires on nearby structures. These developments depend on validation by comparison with actual incidents, since planned, real-world tests of such fires would be prohibitively expensive and dangerous.

Unfortunately, each year in the United States there are numerous opportunities to collect data from WUI fires after they occur. Because FDS predicts fire growth and spread based on the thermal characteristics of solid fuel surfaces, when fuel masses are porous (i.e., where fire can spread quickly *through* them as well as across them), the physics of heat transfer and ignition for that growth is too complex for FDS to accurately model such fires. The effects of wind-driven fire spread through such porous arrays complicate the process even further. Wildland fires in heavy brush or timber are almost always "crown fires" in their most destructive phase; that is, the fire spreads upward and through porous fuel arrays like leaves or needles, often driven by wind—either atmospheric (external) wind or fire-induced drafts. Although FDS has been validated against tests on flat grasslands (no vertical spread factors) with wind-aided spread only, it has not been shown to give accurate predictions for crown fires through heavier brush or timber.

THE DALMARNOCK TESTS

In 2006, the Dalmarnock fire tests were conducted in the United Kingdom, coordinated by the BRE Center for Fire Safety Engineering, University of Edinburgh. The tests consisted of a series of large-scale fire experiments conducted in a high-rise 23-story building located in Dalmarnock, Glasgow.

Prior to the tests, seven teams were given mutual access to the available information on the compartment geometry, fuel packages, ignition source, and ventilation conditions of fire. The teams attempted to build models that would predict the fire scenarios. The primary intent of the tests was to see how accurate the teams' predictions might be. Unfortunately, their results showed a wide range of predictions, emphasizing the difficulties of modeling complex fire scenarios.

Rein et al. (2009) concluded that there was an inherent difficulty in modeling fire dynamics in complex fire scenarios like Dalmarnock, and the modelers' ability to accurately predict fire growth was poor. Several studies and one textbook describe these results in detail (Rein et al. 2009; BRE Center 2012; Abecassis-Empis et al. 2008; Rein, Abecassis-Empis, and Carvel 2007).

However, it is interesting that previous work by Rein et al. (2006) reported that when a combination of a first-order, a zone, and a field model was used, the combined results of these three modeling approaches were in relatively good agreement, particularly in the early stages of the fire. The researchers' work also indicated that this approach to modeling can be used as a first step toward confirming the order of magnitude of the results from more complex models.

IMPACT OF VERIFICATION AND VALIDATION STUDIES

Recent work sponsored in part by the NRC examined the **verification and validation (V&V)** of fire models, including CFAST and FDS (Salley and Kassawar 2007a, 2007b, 2007c, 2007d, 2007e, 2007f, 2007g). Although the NRC's study centered on fire hazards specific to nuclear power plants, it addressed many of the concerns raised by those who think fire models are not accurate or appropriate for forensic fire scene reconstruction, such as in the previously mention Dalmarnock tests.

These concerns include the ability of the models to accurately predict common features of fires, such as upper-layer temperatures and heat fluxes. One of the features of this report was a comparison between actual fire test results and predictions of hand calculations, zone models, and field models. As shown in Figure 6-4 and in other studies, when the models are applied correctly, there is generally good agreement among them and the variability of real-world fires.

The V&V reports are presented in seven publications. Volume 1 (Salley and Kassawar 2007g), the main report, provides general background information, programmatic and technical overviews, and project insights and conclusions. Volumes 2 through 6 provide detailed discussions of the V&V of the Fire Dynamics Tools (FDTs) (Salley and Kassawar

verification and validation (V&V) ■ A formal process of establishing acceptable uses, suitability, and limitations of fire models. *Verification* determines that a model correctly represents the developer's conceptual description. *Validation* determines that a model is a suitable representation of the real world and is capable of reproducing phenomena of interest (Salley and Kassawar 2007g).

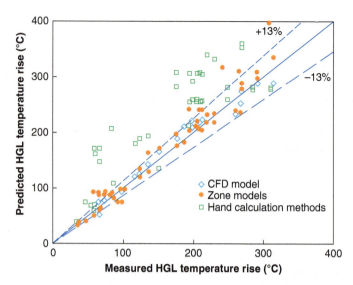

FIGURE 6-4 Comparison of hot gas layer (HGL) temperatures measured in full-scale tests compared with predictions of hand calculations (□), zone models (●), and FDS models (◊). A predicted ±13% variability range is included. Note that the hand calculations tended to overpredict layer temperatures, whereas both zone and field model predictions were generally within variability limits. *Courtesy of U.S. Nuclear Regulatory Commission.*

2007d); Fire-Induced Vulnerability Evaluation, Revision 1 (FIVE-Rev1) (Salley and Kassawar 2007e); Consolidated Model of Fire Growth and Smoke Transport (CFAST) (Salley and Kassawar 2007a); MAGIC (Salley and Kassawar 2007f); and the Fire Dynamics Simulator (FDS) (Salley and Kassawar 2007c). Volume 7 discusses in detail the uncertainty of the experiments used in the V&V study of these five fire models (Salley and Kassawar 2007b).

Fire Modeling Case Studies

Science and litigation are two major driving factors behind the evolution of fire modeling as an important tool in both the analysis of building codes and forensic fire scene reconstructions. Fire protection engineering principles developed from correlating tests with actual case histories to stretch the knowledge of fire science. Attorneys seeking answers regarding responsibility and liability for damages due to fires also have a reason to seek the answers to similar questions. Therefore, science and law meet in the courtroom.

Moreover, the public demands to know why and how incidents occur and whether fire codes perform as anticipated during actual fires. People not only are curious but they have a legitimate concern for answers to questions such as: Is it safe for my child to live in a nonsprinklered dormitory?

Traditionally, fire models have been used in two separate yet important areas:

- Building fire safety and codes analysis
- Forensic fire scene reconstruction and analysis

Specific fire modeling evaluations have looked at the impact or performance of smoke handling systems, sprinkler and detector activation, and fire/burn patterns in reconstructions. Fire modeling found its way into forensic reconstruction through a long history of applications to real-world events. Table 6-5 is a partial listing of cases in which NIST-produced fire models provided valuable insight into fire investigation. These and other reports can be obtained from the NIST fire publications website, fire.nist.gov/.

Several keynote case histories are worth mentioning to show the development of fire modeling in forensic evaluations of fire scenes. These historically significant cases are discussed along with the important concepts gleaned from the referenced and published studies.

UK FIRE RESEARCH STATION

The UK government operated the outstanding Fire Research Station (FRS) in Borehamwood, England, until the facility was privatized in 1997 and moved to Garston. Now a modest engineering consultancy, the facility has a history of more than 50 years of assisting with fire investigations for the UK government. FRS investigations and reconstructions have included fire incidents at the Stardust Disco in Dublin, at Windsor Castle, in the Channel Tunnel, and in the King's Cross Underground station. The FRS uses a combination of a statistical database of loss histories, scene investigations, a fire testing laboratory, and fire models to test hypotheses or investigate unusual phenomena (FRS 2012).

Early work at the FRS developed a compartment fire model to address smoke and heat venting, when its researchers examined the similarities of the Livonia, Michigan, and Jaguar auto plant fire losses (Nelson 2002). These fires at two separate parts

TABLE 6-5	Representative NIST Fire Investigations Using Fire Modeling

NIST REPORT NUMBER OR CITATION	CASE TITLE AND AUTHOR
NBSIR 87-3560 May 1987	*Engineering Analysis of the Early Stages of Fire Development: The Fire at the Dupont Plaza Hotel and Casino, December 31, 1986* H. E. Nelson
NISTIR 90-4268; August 1990, vol. 1	*Full Scale Simulation of a Fatal Fire and Comparison of Results with Two Multiroom Models* R. S. Levine and H.E. Nelson
NISTIR 4665 September 1991	*Engineering Analysis of the Fire Development in the Hillhaven Nursing Home Fire, October 5, 1989* H. E. Nelson and K. M. Tu
Journal of Fire Protection Engineering 4, no. 4 (1992): 117–31	*Analysis of the Happyland Social Club Fire With HAZARD I* R. W. Bukowski and R. C. Spetzler
NISTIR 4489 June 1994	*Fire Growth Analysis of the Fire of March 20, 1990, Pulaski Building, 20 Massachusetts Avenue, NW, Washington, DC* H. E. Nelson
Fire Engineers Journal 56, no. 185 (November 1996): 14–17	*Modeling a Backdraft Incident: The 62 Watts Street (New York) Fire.* R. W. Bukowski
NISTIR 6030 June 1997	*Fire Investigation: An Analysis of the Waldbaum Fire, Brooklyn, New York, August 3, 1978* J. G. Quintiere
NISTIR 6510 April 2000	*Simulation of the Dynamics of the Fire at 3146 Cherry Road, NE, Washington, DC, May 30, 1999* D. Madrzykowski and R. L. Vettori
NISTIR 7137 May 2004	*Simulation of the Dynamics of a Fire in the Basement of a Hardware Store, New York, June 17, 2001* N. P. Bryner and S. Kerber
NISTIR 6923 October 2002	*Simulation of the Dynamics of a Fire in a One-Story Restaurant, Texas, February 14, 2000* R. L. Vettori, D. Madrzykowski, and W. D. Walton
NISTIR 6854 January 2002	*Simulation of the Dynamics of a Fire in a Two-Story Duplex, Iowa, December 22, 1999* D. Madrzykowski, G. P. Forney, and W. D. Walton
NIST SP 995 March 2003	*Flame Heights and Heat Release Rates of 1991 Kuwait Oil Field Fires* D. Evans, D. Madrzykowski, and G. A. Haynes
NIST Special Pub. 1021 July 2004	*Cook County Administration Building Fire, 69 West Washington, Chicago, Illinois, October 17, 2003: Heat Release Rate Experiments and FDS Simulations* D. Madrzykowski and W. D. Walton
NISTIR 7137 2004	*Simulation of the Dynamics of a Fire in the Basement of a Hardware Store, New York, June 17, 2001* N. P. Bryner and S. Kerber
NIST NCSTAR September 2005	*Reconstruction of the Fires in the World Trade Center Towers. Federal Building and FireSafety Investigation of the World Trade Center Disaster* R. G. Gann, A. Hamins, K. B. McGrattan, G. W. Mulholland, H. E. Nelson, T. J. Ohlemiller, W. M. Pitts, and K. R. Prasad
Fire Technology 42, no. 4, (October 2006): 273–81	*Numerical Simulation of the Howard Street Tunnel Fire* K. B. McGrattan and A. Hamins
Fire Protection Engineering 31 (Summer 2006): 34–36	*NIST Station Nightclub Fire Investigation: Physical Simulation of the Fire* D. Madrzykowski, N. P. Bryner, and S. I. Kerber

Note: Many of these and other similar reports can be obtained on the NIST website, fire.nist.gov/.

manufacturing centers had devastating effects on the production of automobiles. What began as a search for a means to vent hot by-products of combustion through the ceiling formed the basis for the zone fire model, which was based on actual fire testing.

DUPONT PLAZA HOTEL AND CASINO FIRE

The December 31, 1986, fire at the Dupont Plaza Hotel and Casino, San Juan, Puerto Rico, killed 98 persons. Owing to the extreme and needless loss of life, this case has become a textbook example of the application of fire reconstruction of the initial development of what became a large, complex fire. The late NIST fire researcher Harold "Bud" Nelson worked in cooperation with ATF investigators to collect and analyze data during the scene investigation. This led to a successful prosecution of the individuals who started the fire simply to harass hotel management.

The analysis combined technical data and fire growth models to explain the dynamics exhibited by the fire (Nelson 1987). The analysis looked at mass burning rate, heat release rate, smoke temperatures, smoke layer, oxygen concentration, visibility, flame extension and spread, sprinkler response, smoke detector response, and fire duration. Fire models used in this analysis included FIRST, ROOMFIR, ASETB, and HOTVENT.

The analysis confirmed the area of origin to be the south ballroom. The fire spread next to a foyer, then to the lobby and casino area. Figure 6-5 documents the conditions at 60 seconds after the fire started.

KING'S CROSS UNDERGROUND STATION FIRE

One of the first important cases using CFD models for fire investigation was the analysis of the King's Cross Underground Station fire on November 18, 1987. The CFD model FLOW3D tracked the fire upward along a 30° incline of the wooden sides and treads of the escalators to the ticket concourse level. The theory of the "trench effect" was verified by the CFD model that described this unique phenomenon. Thirty-one people died in the fire, including a senior station officer (Moodie and Jagger 1992).

In determining the area of origin for this fire, investigators homed in on the area where the fire first started as observed by witnesses, some 21 m (68 ft) below the top of escalator No. 4. This area was at the lowest point of fire pattern damage to wooden handrails, treads, and risers. An analysis of the paint blistering and ceiling damage was also used to locate the fire origin. The physical indicators were corroborated by witnesses who saw the fire partway up the escalator from below.

The investigation concluded that the fire started beneath the wooden steps on the moving staircase from discarded smoking materials. The fire consumed much of the available combustible material in the escalator. Fire damage also extended to the other escalators and the upper-level ticket hall. Conditions in the latter were made untenable by the fact that a temporary plywood construction partition had been erected that was able to burn despite a coating of allegedly fire-retardant paint.

In addition to using a CFD model, government health and safety personnel obtained ignition characteristics of the samples taken from nearby the area of fire origin. Investigators were surprised to determine that escalator lubricating grease was able to be ignited only if combined with fibrous debris (paper, hair, lint) that acted as a wick.

To assess and validate the results of the CFD model, three types of tests were

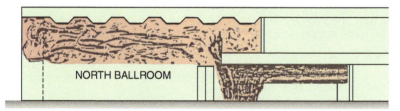

FIGURE 6-5 The analysis combined technical data and NIST fire growth models to explain the dynamics exhibited by the fire at the Dupont Plaza. The analysis looked at mass burning rate, heat release rate, smoke temperatures, smoke layer, oxygen concentration, visibility, flame extension and spread, sprinkler response, smoke detector response, and fire duration. *Courtesy of NIST, from Nelson 1987.*

run. Full-scale fire growth tests determined that the initial fire on the escalator was about 1 MW per meter of escalator burning. Small-scale models implied that above a certain fire size, the flame in the channel of the escalator did not rise vertically but was pushed down into the trench by entrained air, facilitating rapid surface flame spread. Scaled open-channel tests examined the 30° inclined plane and used plywood with various surface coverings to evaluate flame spread. The hot gases moved quickly up the tunnel to fill the ticket lobby. When the gases ignited in a flameover, the fire engulfed the lobby (Moodie and Jagger 1992).

Lessons learned from this fire included the influence of the trench effect on a fire and the speed at which such a fire can develop and spread. Witnesses underestimated the speed of the smoke spread, the growing intensity of the fire, and the need for rapid evacuation. Smoking on underground rail services has since been banned, and combustible escalators, enclosures, and signage have been eliminated.

FIRST INTERSTATE BANK BUILDING FIRE

Another NIST engineering analysis was conducted on the May 4, 1988, fire at the 62-story, steel-framed high-rise known as the First Interstate Bank Building, in Los Angeles, California (Nelson 1989). The fire involved an after-hours accidental ignition in an office cubicle on the 12th floor that spread across the floor of origin. The fire propagated to the 13th, 14th, and 15th floors and a section of the 16th floor as exterior windows failed, allowing large flame plumes to entrain against the sheer sides of the building, which caused windows above to fail. The fire burned for more than 2 hours before being suppressed by firefighters' hose streams inside the building. One of the maintenance personnel died after taking an elevator to the fire floor.

The NIST analysis of the fire used the Available Safe Egress Time (ASET) program to predict the smoke layer's temperatures, smoke level, and oxygen content after the calculated time for flashover (smoke temperature of 600°C [1150°F]). The DETACT-QS model was used to estimate the response of smoke detectors and sprinklers to the fire growth. These studies revealed how much damage could have been avoided if there had been sprinklers on the floor of origin.

This analysis was one of the first to evaluate burn patterns and flame spread on combustible furniture, glass breakage, and communication of the fire to upper floors by flame extension from broken windows, and compartment burning rates in reconstructing the fire. The study also advanced the theory of a transfire triangle, which demonstrates the important interdependent relationships among the mass burning (pyrolysis) rate, the heat release rate, and the available air (oxygen) (Nelson 1989).

HILLHAVEN NURSING HOME FIRE

A fire just after 10 p.m. on October 5, 1989, at the Hillhaven Rehabilitation and Convalescent Center, Norfolk, Virginia, claimed the lives of 13 persons. None of these individuals were burned, but all breathed sufficient carbon monoxide (CO) to produce fatally high carboxyhemoglobin (COHb) levels.

The late NIST fire researcher Harold E. "Bud" Nelson recognized the importance of evaluating the fire dynamics when reconstructing this incident. His analysis of the room of fire origin in the nursing home showed that the temperature rose to at least 1000°C (1800°F), with a CO concentration of 40,000 ppm (Nelson and Tu 1991).

Using the fire model FIRE SIMULATOR contained in its engineering package, FPE-Tool predicted the smoke temperatures, smoke layer, toxic gas concentrations, and velocity of the smoke front. An important aspect learned in the NIST analysis was the impact of the "what if" questions that are posed in fire reconstructions (Figure 6-6). These questions centered on the hypothetical impact of different fire activation devices on fire suppression as well as the predicted time of flashover.

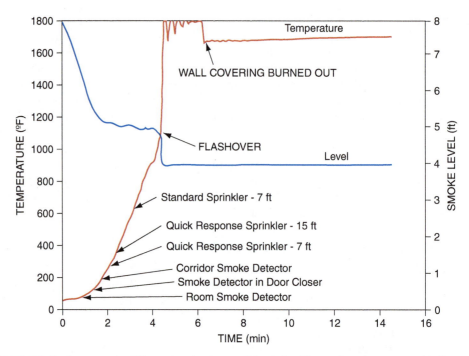

FIGURE 6-6 The impact at the Hillhaven Nursing Home of the "what if" questions that are posed in fire reconstructions, centering on the hypothetical impact of different fire activation devices on fire suppression as well as the predicted time of flashover. *Courtesy of NIST, from Nelson 1991.*

PULASKI BUILDING FIRE

A fire in Suite 5127 of the Pulaski Building, 20 Massachusetts Avenue, NW, Washington, DC, on March 23, 1990, damaged the offices of the U.S. Army Battlefield Monuments Commission (ABMC). A NIST engineering review of the fire was able to apply sound fire protection engineering principles to estimate the rate of fire growth (Nelson 1994).

At 11:24 a.m., an alert ABMC staff member noticed smoke coming from the open door of the unoccupied audio/visual conference room. The room contained an estimated 4530 kg (10,000 lb) of combustible materials consisting of publications and photos stored in corrugated board boxes and cardboard mailing tubes stacked about 1.5 m (5 ft) high. The fire was thought to have started when cartons of mailing tubes were exposed to a frayed energized electrical lamp cord that passed through a small hole from a podium into the projection room.

FPETool was used to analyze the fire and arrive at a scenario based on witness testimony and observations. FIRE SIMULATOR estimated the environmental conditions, temperature, depth of smoke, thickness of smoke, and energy vented from the room. Also included was the impact of the failure of the suspended ceiling.

The analysis showed that the fire reached flashover approximately 268 seconds into the fire. A unique aspect of this fire engineering analysis was the introduction of a timeline of events (Figure 6-7) that compared the approximate time with the observed fire conditions, behavior of the occupants, impact of existing fire protection systems, and "what if" scenarios of the impact of potential fire protection systems.

HAPPYLAND SOCIAL CLUB FIRE

On the morning of March 25, 1990, an arsonist set a fire in the entryway of a neighborhood club using 2.8 L (0.75 gal) of gasoline. The fire killed 87 occupants of the two-story club. A NIST fire engineer used the HAZARD I fire model to examine the fire

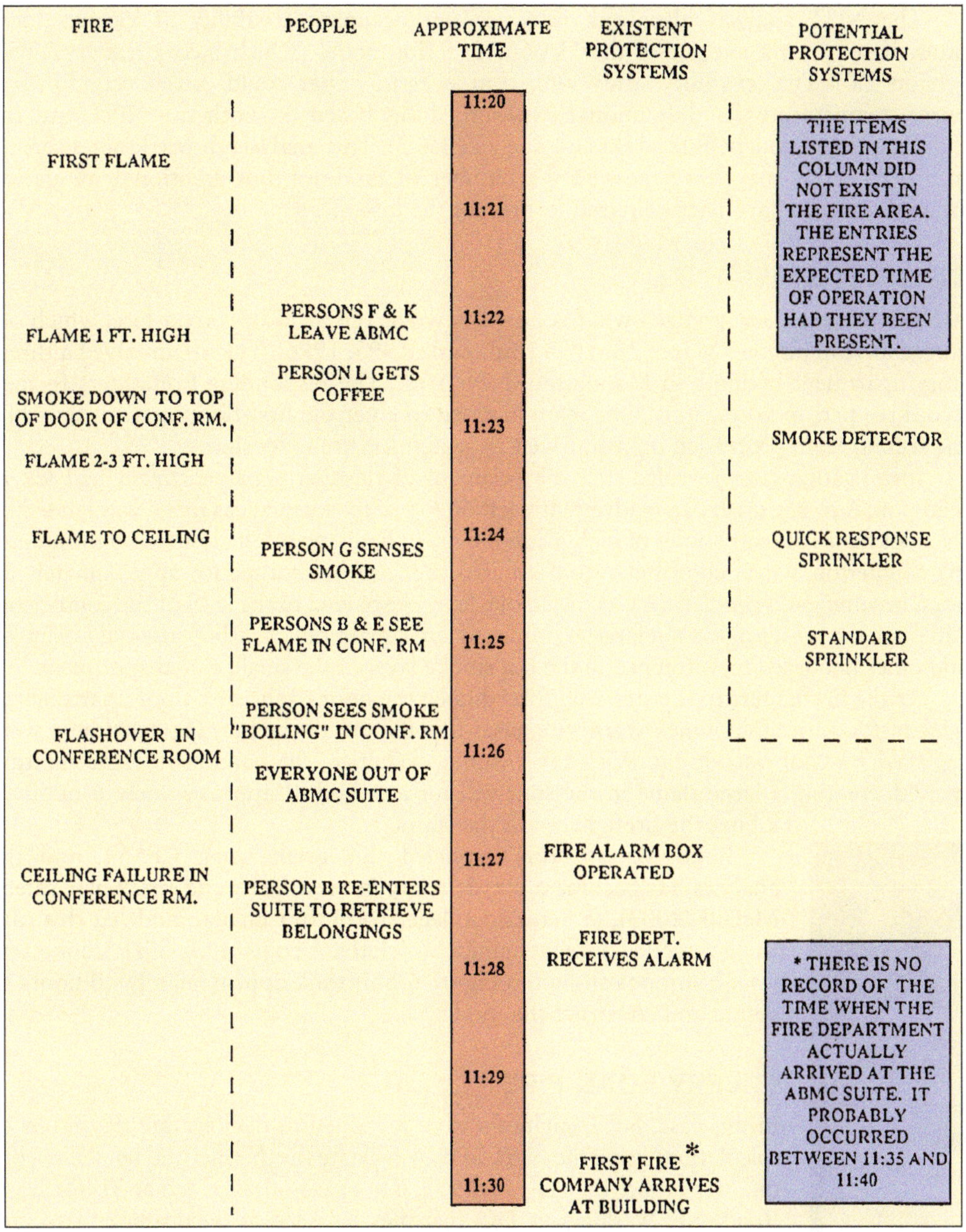

FIRE	PEOPLE	APPROXIMATE TIME	EXISTENT PROTECTION SYSTEMS	POTENTIAL PROTECTION SYSTEMS
		11:20		THE ITEMS LISTED IN THIS COLUMN DID NOT EXIST IN THE FIRE AREA. THE ENTRIES REPRESENT THE EXPECTED TIME OF OPERATION HAD THEY BEEN PRESENT.
FIRST FLAME		11:21		
FLAME 1 FT. HIGH	PERSONS F & K LEAVE ABMC	11:22		
SMOKE DOWN TO TOP OF DOOR OF CONF. RM.	PERSON L GETS COFFEE	11:23		SMOKE DETECTOR
FLAME 2-3 FT. HIGH				
FLAME TO CEILING	PERSON G SENSES SMOKE	11:24		QUICK RESPONSE SPRINKLER
	PERSONS B & E SEE FLAME IN CONF. RM	11:25		STANDARD SPRINKLER
	PERSON SEES SMOKE "BOILING" IN CONF. RM			
FLASHOVER IN CONFERENCE ROOM	EVERYONE OUT OF ABMC SUITE	11:26		
CEILING FAILURE IN CONFERENCE RM.	PERSON B RE-ENTERS SUITE TO RETRIEVE BELONGINGS	11:27	FIRE ALARM BOX OPERATED	
		11:28	FIRE DEPT. RECEIVES ALARM	* THERE IS NO RECORD OF THE TIME WHEN THE FIRE DEPARTMENT ACTUALLY ARRIVED AT THE ABMC SUITE. IT PROBABLY OCCURRED BETWEEN 11:35 AND 11:40
		11:29		
		11:30	FIRST FIRE * COMPANY ARRIVES AT BUILDING	

FIGURE 6-7 The timeline of events in the Pulaski Fire, which compares the approximate time with the observed fire conditions, behavior of the occupants, impact of existing fire protection systems, and "what if" scenarios of the impact of potential fire protection systems. *Courtesy of NIST, from Nelson 1989.*

development as well as potential mitigation strategies that could have influenced the outcome of the fire (Bukowski and Spetzler 1992).

These mitigation strategies included the use of automatic sprinkler protection, solid wooden doors at the base of the stairway, a fire escape and enclosed exit stairwell, and a noncombustible interior finish. The analysis also weighed the economic cost estimates to implement these strategies. The study concluded that the combustible wall coverings and the opening of an entry door, which allowed the entry of gasoline-fed flames, caused the fire to grow and combustion gases to spread so rapidly that everyone in the upstairs club was affected within 1–2 minutes, precluding their escape.

The NIST analysis also took into account occupant tenability at various locations, based on the output of the HAZARD I fire model (which incorporated CFAST version 2.0). The tenability study looked at heat flux that could cause second-degree burns, temperatures, and fractional exposure doses based on both the NIST and the Purser toxicity models (Bukowski and Spetzler 1992). This analysis showed that a second means of egress may have reduced the number of fatalities, but additional mitigation strategies would have been required to prevent the loss of life.

62 WATTS STREET FIRE

A NIST fire engineer modeled what is now known as the 62 Watts Street Fire, which occurred at 7:36 p.m. on March 28, 1994 (Bukowski 1995, 1996). The fire involved a three-story apartment building in Manhattan, New York. The responding firefighters formed two three-person hose teams. One team planned to enter the first-floor apartments while the second team proceeded up a stairwell to search for fire extension.

Investigation later revealed that the occupant of the first-floor apartment had left at 6:25 p.m. A plastic trash bag inadvertently left on top of the kitchen gas range was ignited by the pilot flame. Several bottles of high-alcohol-content liquid were left close by, on the adjoining countertop, which contributed to the initial fire. The fire burned for approximately an hour, becoming oxygen-deficient as the smoke layer descended to the level of the countertop. Although the fire itself was not large, it produced large quantities of carbon monoxide, smoke, and other unburned fuels trapped in the hot smoke layer in the small, closed apartment.

As the firefighters in the stairwell forced open the door to the first-floor apartment, a *backdraft* occurred, in which warm fire gases flowing out of the burning apartment were replaced by cool outside air. With the ensuing exchange, the combustible gas mixture ignited, creating a large flame in the stairwell for a period of approximately 6 minutes, killing the firefighters on the stairs.

Bukowski at NIST modeled this event using CFAST, and the analysis has become a classic study in backdrafts and firefighter safety. McGill (2003), at Seneca College, Toronto, Canada, modeled this case using the NIST Fire Dynamics Simulator. A copy of McGill's Smokeview visualization is shown in Figure 6-8. It took approximately 20 hours to define and construct the model.

CHERRY ROAD FIRE

In this case, NIST engineers were requested to demonstrate the results of calculations using the NIST FDS to evaluate the thermal and tenability conditions of a fire on May 30, 1999, in a townhouse located at 3146 Cherry Road, NE, Washington, DC, that may have led to the deaths of two firefighters and burned four others (Madrzykowski and Vettori 2000). The fire started in a ceiling electrical fixture on the lower level of the townhouse. Responding firefighters entered the first and basement floors of the structure. Both on-scene observations and subsequent modeling showed that the opening of the lower-level sliding glass doors increased the heat release rate of the fire, so that 820°C (1500°F) gases moved up the stairwell. The fire may have smoldered for several hours before flame ignition, producing fuel-rich gases that ignited while firefighters were inside the building searching for the fire and any victims in the heavy smoke.

FIGURE 6-8 Model of the 62 Watts Street fire using the NIST Fire Dynamics Simulator. Room of origin at lower left. The fire plume extending from the apartment door (center) into the hallway trapped and killed three firefighters on the stairs. *Courtesy of D. McGill (2003).*

The Cherry Road fire FDS model may become one of the best examples of application of the NIST FDS. Furthermore, the fire pattern damage observed correlated with the model's predictions. As the technology expands, so does the ability to address harder questions concerning transient heat and ventilation, rapid fire growth, and stratified or localized conditions.

Westchase Hilton Hotel Fire

A March 6, 1982, fire at the Westchase Hilton Hotel, Houston, Texas, resulted in the deaths of 12 people and serious injuries to three others. Hot gases and smoke progressed from a fire in a guest room on the fourth floor down the corridor and, to some degree, throughout the building, as illustrated in Figure 6-9. Records indicate that 137 guests were registered at the hotel the evening of the fire (Bryan 1983).

The fire started in room 404 on the fourth floor (Figure 6-10), and the first odor of smoke was reported at 2:00 a.m. on the tenth floor, followed by light smoke reported at 2:10 a.m. on the eighth floor; the fire was discovered at 2:20 a.m. by an occupant returning to his room.

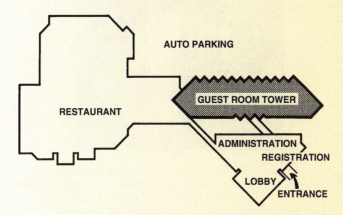

FIGURE 6-9 The March 6, 1982, fire at the Westchase Hilton Hotel resulted in the deaths of 12 people and serious injuries to three others. *Reprinted with permission from NFPA Westchase Hilton Fire Investigation. Copyright © 1982 and 1983, National Fire Protection Association, Quincy, MA 02269.*

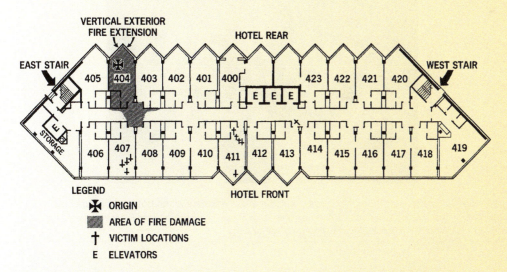

FIGURE 6-10 The Westchase Hilton Hotel fire started on the fourth floor, Room 404. *Reprinted with permission from NFPA Westchase Hilton Fire Investigation. Copyright © 1982 and 1983, National Fire Protection Association, Quincy, MA 02269.*

FIGURE 6-11 At 2:31 a.m., the district fire chief arrived and saw a fire plume projecting out the exterior window of the room of fire origin on the south side of the Westchase Hilton high-rise tower. Note the areas of external, horizontal, and vertical fire extension. *Reprinted with permission from NFPA Westchase Hilton Fire Investigation. Copyright © 1982 and 1983, National Fire Protection Association, Quincy, MA 02269.*

FIGURE 6-12 Exterior view, from the hallway, of Room 404, Westchase Hilton Hotel, whose combustible contents were almost totally consumed. *Reprinted with permission from NFPA Westchase Hilton Fire Investigation. Copyright © 1982 and 1983, National Fire Protection Association, Quincy, MA 02269.*

At 2:31 a.m., the district fire chief arrived and saw a fire plume projecting out the exterior window of the room of fire origin (Figures 6-11 and 6-12). Extinguishment of the property started at 2:38 a.m. and ended at 2:41 a.m. Fire investigating officials concluded from these times that flashover in the room occurred between 2:20 and 2:31 a.m.

TABLE 6-6	Numerical Constants in the Fire Investigation and Reconstruction Model
FEATURE	
Room floor area	24.51 m² (263.8 ft²)
Room ceiling height	2.44 m (8.00 ft)
Door (vent) soffit	1.99 m (6.56 ft)
Vent width (varied)	0.152–0.457 m (6–18 in.)
Fire height (estimated)	0.914 m (3.0 ft)
Heat loss fraction	0.66
Flashover	
0.152-m (0.50-ft) opening[a]	199 s
0.076-m (0.25-ft) opening[a]	299 s
Untenability, 1.2–1.5 m (4–5 ft)	183°C (361°F)

Source: Derived from Janssens and Birk 2000.
[a]Predicted value.

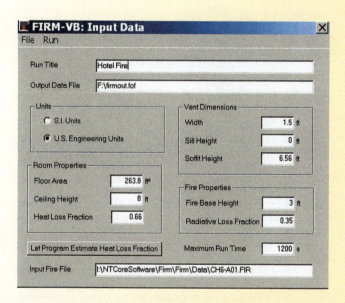

FIGURE 6-13 Input screen for the Fire Investigation and Reconstruction Model (FIRM).

An investigation revealed that a cigarette had smoldered in an upholstered chair in the room of origin. The U.S. Consumer Product Safety Commission (CPSC) later had fire tests performed on the furniture. Heat release rate data from these tests along with assumptions about how wide the door was ajar formed the basis for the input to the model (Table 6-6). Data from previous NIST tests of smoldering ignition of furniture were also used in evaluating the time factors between detection of smoke odors and observation of a flaming fire.

The zone model FIRM (Fire Investigation and Reconstruction Model) was later used to evaluate the single compartment of fire origin using data from the CPSC tests (Janssens and Birk 2000). The input screen using these data is shown in Figure 6-13. A sensitivity analysis using FIRM was conducted to examine the impact of the position of a door ajar on the room of fire origin.

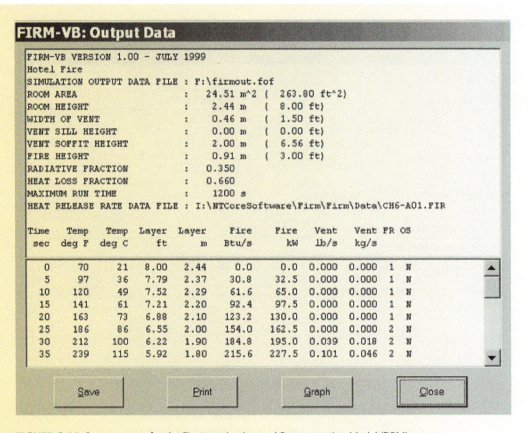

FIGURE 6-14 Output screen for the Fire Investigation and Reconstruction Model (FIRM).

The FIRM model also was used to predict the fire development within the room of origin, tenability conditions, and an evaluation of the time to flashover. Table 6-6 lists the observed and predicted data for the room of fire origin. The output file from the FIRM program is shown in Figure 6-14.

Later studies on the behavior of persons in this fire used a data collection interview form reproduced in Figure 6-15 (Bryan 1983). Approximately a month after the fire, this form was mailed to 130 guests who could be identified as registered at the time of the fire by the National Fire Protection Association (NFPA) in coordination with the Houston Fire Department. A total of 55 persons responded, which was more than 27 percent of the registered guests, representing 42 males and 13 females.

The study showed that some of the guests were aware of the fire for as long as almost 2 hours before the fire department was notified. The majority of the notification was by way of people yelling. The most frequent first actions after notification of the fire were dressing, calling the front desk, and attempting to exit. The study failed to reveal any nonproductive behavior by those persons escaping the fire, and some spent time knocking on doors to alert other occupants. Persons escaping the fire did not reenter the building. Movement through the smoke is always a question in research studies. More than 50 percent of the guests passed through the smoke for distances of from 4.67 to 182 m (15 to 600 ft).

Lessons learned from the analysis of this fire by several sets of experts included the need for better enforcement of the life safety code, education of the public, training of hotel staff, and emergency preparedness. Other lessons showed the utility of scientific review of fire scenes, development of a list of behaviorally related questions and answers, and application of a fire model.

FIRE EXPERIENCE SURVEY
Westchase Hilton Hotel, Houston, Texas
March 6, 1982

We are grateful for your willingness to share your fire experience with us. Your information will help us find ways to avoid recurrences of such tragedies.

Occupation _____ Sex ____ Age ____
Room No. ____

1) How did you first become aware that there was something unusual occurring in the Hotel? _____
2) What time was it? _____ How did you determine the time? _____
3) When did you realize that what was occurring was a fire? What time was it? _____ How did you determine the time? _____
4) How serious did you believe the fire to be at first?
  ( ) Not at all serious, ( ) Only slightly serious, ( ) Moderately serious, ( ) Extremely serious
5) Were you alone when you became aware of the fire?
  ( ) No ( ) Yes
6) How many persons were with you? ____ They were ( ) relatives ( ) others.
7) Were you injured? ( ) No ( ) Yes:
  If Yes, What was the nature and cause of the injury? _____
8) What did you do when you realized there was a fire? (state exact sequence of actions)
  First _____
  Second _____
  Third _____
  Fourth _____
  Fifth _____
9) Once aware of the fire, did you:
  (a) Call or attempt to call the hotel operator (switchboard) ( ) No ( ) Yes; If Yes, At what time? _____
  (b) Call or attempt to call the Fire Department directly ( ) No ( ) Yes;
   If Yes, at what time? _____
  (c) Operate a manual fire alarm pull station?
   ( ) No
   ( ) Yes; If Yes, At what time? _____
10) Did you voluntarily leave (i.e., without being requested by hotel or Fire Department personnel)?
  ( ) No ( ) Yes
  If No, Why not? _____
  If Yes, At what time? _____
  How? ( ) Stairway to ground ( ) Stairway to roof
   ( ) Elevator ( ) Window ( ) Exterior Door
   ( ) Other (Specify) _____
11) I left the building: ( ) Unassisted; Assisted by:
  ( ) Hotel staff ( ) Guest ( ) Fire Department
  ( ) Other (Specify) _____
12) After you realized there was a fire, how long did you wait before leaving the building? ____ minutes.
  What were you doing while waiting? _____
13) In your leaving, was there any visible smoke?
  ( ) No ( ) Yes
  Any smoke odor? ( ) No ( ) Yes
  Any flames? ( ) No ( ) Yes
  Did you try to move through smoke? ( ) No
  ( ) Yes: If Yes, How far did you move? ____ feet

(approx.). How far could you see at the time? ____ feet.
  Did you turn back? ( ) No ( ) Yes. If Yes, Why did you turn back? _____
14) Did you notice any lighted exit signs? ( ) No ( ) Yes
15) During your escape, was there any difficulty in following the direction marked by exit signs?
  ( ) No ( ) Yes:
  If Yes, What was difficulty? _____
16) Any obstructions to escape? _____
17) What aids helped you to escape? _____
18) Did smoke enter your room? ( ) No ( ) Yes:
  If Yes, How? _____
  ( ) Heating/cooling unit ( ) Bathroom vent
  ( ) Around door ( ) Window
  ( ) Don't know ( ) Other (Specify) _____
19) Beginning with your first action, number the sequence of events you took while in your room.
  ( ) Put materials over the heating/cooling unit vent
  ( ) Around door ( ) Around window
  ( ) Turned on TV ( ) Turned off radio
  ( ) Took no action ( ) Other _____
20) Did the smoke detector in your room sound an alarm? ( ) No ( ) Yes
21) Did you hear the building fire alarm?
  ( ) No ( ) Yes: If Yes, At what time? ____
  How long did it operate? _____
22) Did you receive any instructions from hotel staff during the fire emergency? ( ) No ( ) Yes: If Yes, What were they? _____
  Did you receive fire safety instructions from hotel employees prior to the fire? ( ) No ( ) Yes:
  If Yes, What were they? _____
  Did you observe fire safety information within your room? ( ) No ( ) Yes. If Yes, What was the information? _____
23) Did you have previous training on actions to take in a fire?
  ( ) No ( ) Yes: If Yes, Number of times ____
  Type _____
  Given by _____ Last time: Date ____
  Did the training help in this fire?
  ( ) No ( ) Yes
24) Did you receive previous fire safety information from: ( ) Radio ( ) TV ( ) Publication.
  What was the message? _____
  Did it help in this fire? ( ) No ( ) Yes
25) Were you ever involved in a fire before?
  ( ) No ( ) Yes Last time: Date _____
26) Please report any additional comments that you think might help others in a similar fire situation.
  _____
  _____
27) Please mark your escape route on the diagram provided. Please identify room number and floor. Thank you.

FIGURE 6-15 Later studies on the behavior of persons in this fire used this data collection interview form, mailed to 130 identified guests. A total of 55 persons responded, which was more than 27 percent of the registered guests, 42 males and 13 females. *Reprinted with permission from NFPA Westchase Hilton Fire Investigation. Copyright © 1982 and 1983, National Fire Protection Association, Quincy, MA 02269. This reprinted material is not the complete and official position of the National Fire Protection Association on the referenced subject, which is represented only by the standard in its entirety.*

The use of FDS in the forensic engineering analysis of fires, particularly ones where the origin and cause is disputed, has begun to gain further popularity and acceptance in the fire investigation community. The following case study consists of a series of fires for which FDS simulations provided compelling arguments to match fire burn pattern damage with witness accounts to prove or disprove hypotheses (Vasudevan 2004).

EXAMPLE 6-1 ■ Fire in a Minimart: Accident or Incendiary?

A fire was discovered soon after the owner-manager of a minimarket had locked up and left the building. A fire some time after the owner left raised questions as to origin and cause. Two disputed hypotheses were that (1) the fire was a result of an electrical malfunction in the attic area over the storage/display area, or (2) the fire was intentionally set in the floor-level liquor storage area.

FDS was used to model the hypothesis that the fire did not originate in the attic space but was an intentionally set fire in the liquor storage area. The fire patterns at the scene (Figure 6-16) were found to be consistent in extent and shape with the prediction of the FDS/Smokeview results for the incendiary fire hypothesis (Figure 6-17). (*Note*: FDS has not been validated for this specific application.)

FIGURE 6-16 Photo of actual fire burn pattern damage to a minimart. *Courtesy of R. Vasudevan, Sidhi Consultants, Inc.*

EXAMPLE 6-2 ■ Placement of Smoke Detector

A small fire in a living room of a one-bedroom apartment generated sufficient carbon monoxide to cause the death of one person and serious injuries to a second. It was disputed whether the lack or failure of a smoke detector affected the occupants' ability to respond appropriately to the fire. Although the placement of the smoke detector complied with the local applicable fire code, it did not provide adequate notification time for the occupants to respond to the toxic by-products of combustion (see Figures 6-18 and 6-19).

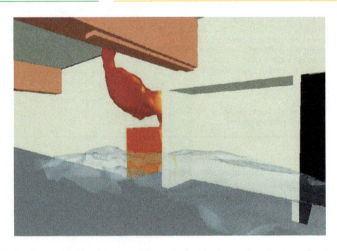

FIGURE 6-17 NIST Fire Dynamics Simulator model results for the incendiary fire hypothesis of the minimart. *Courtesy of R. Vasudevan, Sidhi Consultants, Inc.*

EXAMPLE 6-3 ■ Deadly Kitchen Cooking Fire

A kitchen cooking fire in a two-bedroom apartment resulted in the death of a child and carbon monoxide poisoning of two other persons. The issue concerned (1) the location of the smoke detector, (2) its ability to detect the fire, and (3) tenability levels of carbon monoxide within the children's rooms. The FDS analysis showed that the location of the smoke detector close to the incipient fire would damage the unit within seconds of its alarming. The analysis also showed that unsafe carbon monoxide levels would rapidly be reached. Finally, the model results closely matched the fire pattern damage (see Figures 6-20 and 6-21).

FIGURE 6-18 Photo of actual fire burn pattern damage to the living room of the one-bedroom apartment. *Courtesy of R. Vasudevan, Sidhi Consultants, Inc.*

EXAMPLE 6-4 ■ Wildland Case Study

In a recent case, a proponent in a large civil lawsuit attempted to use FDS to predict fire spread on a steeply sloped hillside that was covered in heavy brush (2–3 m [6–10 ft] tall). This hillside (some 100 m × 200 m [330 ft × 660 ft] in area) was known to be the general area of origin of a larger fire. Extensive calorimetry tests were carried out on arrays of the same type brush to collect data on the rates of heat release and mass loss for such fuels. The steep hillside was modeled in FDS as a series of rectilinear steps 1 m (3.3 ft) high and 1–2 m (3.3–6 ft) deep, with computational cells of the same proportions.

Because FDS currently does not adequately recognize porous fuel arrays, the fuel load on the hillside was encoded within the model as a series of thin horizontal and vertical panels (in steplike array) whose heat release curves approximated the calorimetry data from testing carried out on those woody shrubs. (The calorimetry testing, however, did not include any wind-driven growth factors, so it did not duplicate a real-world fire environment.) A mild, cross-slope wind was also encoded within the model. The resulting Smokeview animation showed horizontal (cross-slope) propagation and downslope propagation in a stop-start fashion roughly equal in spread velocity to the upslope propagation. It was proposed that this result supported an origin of the fire very different (well upslope) from the origin identified by the original scene investigators.

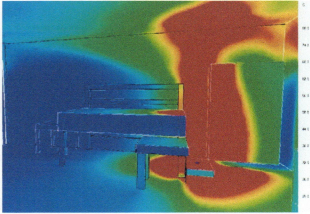

FIGURE 6-19 NIST Fire Dynamics Simulator model results for the living room of the one-bedroom apartment showing (a) fire and smoke obscuration and (b) surface boundary conditions. *Courtesy of R. Vasudevan, Sidhi Consultants, Inc.*

FIGURE 6-20 Photos of (a) rebuilt kitchen and (b) actual fire burn pattern damage. *Courtesy of R. Vasudevan, Sidhi Consultants, Inc.*

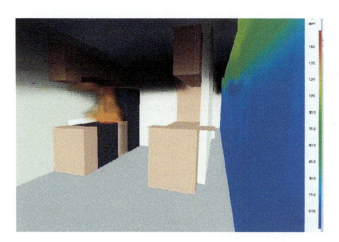

FIGURE 6-21 NIST Fire Dynamics Simulator model results for the kitchen in the smoke detector placement and carbon monoxide hypothesis tests. Compare these results with the burn patterns in Figure 6.19(b). *Courtesy of R. Vasudevan, Sidhi Consultants, Inc.*

When this model was offered to the court in a *Frye* admissibility hearing, the proponents argued that FDS had been developed and studied for exterior fires (smoke plumes and WUI fires) and therefore should yield valid results for exterior fires of all types. Although proponents could offer peer-reviewed papers discussing FDS modeling for flat grasslands (two-dimensional spread) and WUI (established fires in trees that spread to structures), they could not offer a single reference regarding FDS and fire spread in porous (three-dimensional) fuels on hillsides. Ultimately, the judge ruled that FDS modeling was not generally accepted for this situation and therefore its predictions could not be used as evidence.

Because the predictions of the model (downslope propagation in low-wind conditions) were contrary to logic and observations of countless wildland fires, the proponents should have acknowledged that FDS was not yielding reliable results. This is an example in which *ASTM E355* can be used to evaluate the predictive capacity of fire models by defining scenarios (see guidelines referenced in this chapter). That is, if the model results run contrary to normal behavior, there is something wrong with the model or with the data input. Models should be used only for the applications for which they were designed and developed and have been demonstrated to yield results in agreement with actual fires of that type.

Modeling carried out by an opposing consultant using FARSITE, the most widely used and tested computer model for wildland fires, predicted upslope growth whose features paralleled the observations of the first-responding fire crews. These crews had staged near the top of the slope and watched the fire progress rapidly upslope (with a mild cross-slope atmospheric wind) toward them from an area well down the slope. These crews were forced to abandon the house being used as an observation post when it was overrun by the growing fire. Growth of the fire was aided by the draft produced by the fire, which quickly developed into a wind-driven crown fire. This wind-aided growth was not duplicated in the calorimetry tests conducted.

Case study courtesy of John DeHaan.

Modeling Used to Confirm Forensic Findings of the Impact of Full and Partial Compartmentation and the Prediction of Conditions for Structural Collapse

Per *NFPA 921*, 2011 ed., pt. 3.3.59 (NFPA 2011), a fire analysis is "the process of determining the origin, cause, development, responsibility, and, when required, a failure analysis of a fire or explosion." The purpose of this example case analysis is to demonstrate the use of modeling in the evaluation of the initial development of a fire at an industrial facility, in February 2001. Using accepted principles of fire protection engineering, this fire analysis focused primarily on the fire's early development and growth, and evaluated the impact of the lack or presence of compartmentation between the room of fire origin and adjoining building areas.

Compartmentation is a primary fire protection engineering design concept that stresses the construction of structures so that fire is confined to its room of origin, and smoke movement to other areas of a building is minimized. This design concept consists of coordinating fire-resistive construction, fire-stopped pipe chases and utility openings in fire walls, and construction techniques that minimize smoke and flame movement. Structural designs that fail to utilize compartmentation are considered less fire safe and often enable smoke and fire to move outside the room of fire origin.

The commercial building was built in 1983, and had additions in 1987, 1994, and 1997. The approximate physical layout of the building at the time of the fire is best represented in Figure 6-22. Only the original 1983 building had fire sprinkler protection. The shaded area shows the location of the box room and approximate area of fire origin.

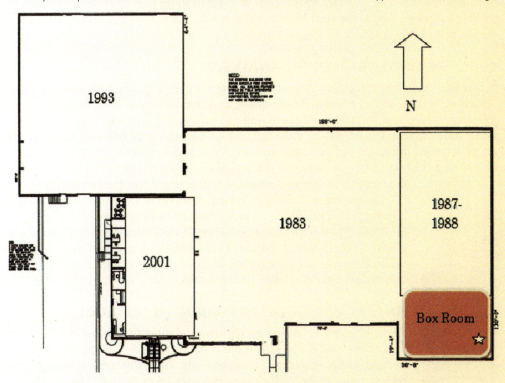

FIGURE 6-22 Dates of major expansion plans overlaid on the the building plans. Only the original 1983 building had fire sprinkler protection. Shaded area shows location of box room and approximate area of fire origin (not to scale). *Courtesy of D. J. Icove.*

FIRE SUPPRESSION

The fire department received an alarm call at 7:42 p.m. at the industrial facility. The first apparatus arrived at 7:48 pm and reported fire coming from the southeast corner of the building, which was known as the warehouse area (later identified as the storage box room). Additional responding units arrived approximately 15 minutes later, and they found the fire difficult to control owing to the fuel load of the corrugated boxes and polyethylene bags stored in the warehouse portion of the building. Approximately 30–45 minutes after the first apparatus arrived, the roof of the warehouse collapsed.

Weather conditions at 7:53 p.m., recorded at the local airport, were overcast condition, temperature −2.8°C (27°F), 53 percent humidity, barometric pressure 1022.1 hPa (30.19 in.), wind speed 7.4 km/hr (0.6 mi/hr, 2.1 m/s) from the east-southeast, visibility 16.1 km (10 mi) with no precipitation.

FIRE ORIGIN AND CAUSE INVESTIGATION

Scene investigation placed the area of origin within the box storage room. The point of fire origin was in or around two pallets of materials or in the debris around the base of these pallets along the interior eastern wall of the box storage room. The southeast corner of the roof above the box storage area was partially collapsed.

The freezer area located just north of the box storage room had heavy heat and smoke damage at the upper levels. Within the freezer, fire ventilation damage was seen in the southeastern corner of the room. This damage extended over the partition wall separating the freezer and box storage rooms. A gap between the partition wall and steel I-beam allowed smoke and heat to travel from the box room over the freezer and into the mechanical room. Above the steel I-beam, there was also a 6.35-cm (2.5-in.) gap between the beam and the underside of the roof deck.

FORENSIC FIRE ENGINEERING ANALYSIS

A forensic fire engineering analysis can assist in refining and testing scenarios for a fire's initiation, area of origin, growth, development, and impact on the structure. The specific purpose of the use of forensic fire engineering analysis in this case was to evaluate (1) the initial development and growth of a the fire and (2) the impact of a lack of compartmentation between the room of fire origin and adjoining building areas caused by a combination of a 0.61-m (2-ft) gap between the top of the brick partition wall along with a 6.35-cm (2.5-in.) gap between the steel beam and the ceiling.

In forensic fire scene reconstruction, there exist three accepted analytical approaches to modeling compartment fires (Mowrer 2002; Rein et al. 2006): (1) start with first-order analytical calculations; (2) use a zone model, and (3) should sufficient time remain, use a field model. Carried out in the order presented, each of these methods can refine both the accuracy and the complexity of the results.

ANALYTICAL MODELING: BURNING DURATION

Recording and inventorying the major combustible materials within the area of fire origin and remaining compartments is a necessary initial step in conducting a fire engineering analysis. From this information, the heat release rates of the various materials can be estimated. The information on heat release rate is often used in various analytical calculations to estimate the fire's duration, growth, flame height, and upper/lower layer temperature.

Forensic investigations seek to answer the following questions, among others: How long did the fire burn? How tall did the fire plume reach? What were the temperatures of the upper and lower zones? How much smoke escaped through the open doorway? Could fire sprinklers have extinguished or controlled the fire?

Investigation determined that a loose stack of approximately 50 corrugated paper boxes was the first item ignited in the box room. The fire subsequently spread to adjacent boxes, plastics, and wooden pallets. Figure 6-23 illustrates the typical configuration of these materials.

FIGURE 6-23 Typical exposed pallets of folded corrugated cardboard boxes (foreground) and wooden pallets (background) as seen in the inspection. *Courtesy of D. J. Icove.*

TABLE 6-7	A Preliminary Inventory of Exposed Fuels within the Box Storage Room			
			ESTIMATED TOTAL EXPOSED FLOOR AREA	
ITEM	TYPE OF MATERIAL	EXPOSED UNIT FLOOR AREAS	ft²	m²
Pallet of 500 folded boxes	Corrugated cardboard	22	352	32.7
Pallets	Wood	1	16	1.49
Plastic rolls	Polyethylene	6	96	8.9

TABLE 6-8	Pallet Dimensions of Two Types of Flat Corrugated Boxes Found during the Inspection										
	PALLET DIMENSIONS (IN.)			FLOOR AREA (W)(D)		VOLUME (W)(D)(H)		DENSITY* (D)	MASS (V)(D)	
BOX TYPE	W	D	H	ft²	m²	ft³	m³	kg/m³	lb	kg
Type #1	40.0	57.5	31.1	16.0	1.48	41.3	1.17	76	196	88.8
Type #2	46.0	50.0	39.5	16.0	1.48	41.3	1.17	76	250	113.2

Source: Derived from Ryu et al. 2007.

The inventory of the estimated fuels within the box room is listed in Table 6-7, which is based on information from the investigators along with an on-site inspection of the present facility. Each unit item represents coverage of approximately 16 ft² (1.49 m²) of floor area. Table 6-7 also provides the estimated total exposed floor area for these items within the box room.

The initial fire in the box room would likely have confined itself first to the exposed stacked pallets of corrugated cardboard boxes. The burning duration for these solid combustibles for a given compartment size and ventilation condition is driven mainly by the fuel load.

Fuel load is traditionally based on an estimate of the amount of combustible materials per unit floor area, expressed in kilograms per square meter (kg/m²) or pounds per square foot (lb/ft²). Given the mass of material burned per second and the amount of material available, the fire's total burning duration can be estimated. Other calculations can estimate the flame heights and upper/lower layer temperatures, particularly the ceiling temperature.

For purposes of the preliminary calculations for this report, only the pallets of corrugated boxes were taken into consideration. Table 6-8 summarizes the material properties of pallets of flat corrugated cardboard boxes along with estimates for the mass of each pallet.

The intensity and duration of a fully developed fire depends on the amount of combustible materials available, their burning rates, and the available air to support the combustion of these materials. The burning duration of a solid combustible fuel can be estimated (Buchanan 2001) as

$$t_{solid} = \frac{m_{fuel} \, \Delta H_c}{\dot{Q}'' A_{fuel}}, \qquad (6.1)$$

where

t_{solid} = burning duration (s),
m_{fuel} = mass of solid combustible fuel (kg),
ΔH_c = effective heat of combustion (kJ/kg),
$\dot{Q}''$ = heat release rate per unit floor area (kW/m²), and
A_{fuel} = exposed floor area (length × width) of fuel (m²).

	MASS OF COMBUSTIBLE FUEL (22 PALLETS) (kg) m_{fuel}	EFFECTIVE HEAT OF COMBUSTION (MJ/kg)[*] ΔH_c	HEAT RELEASE RATE PER UNIT FLOOR AREA (kW/m²)[**] $\dot{Q}''$	EXPOSED FLOOR AREA OF FUEL (m²) A_{fuel}	BURNING DURATION (s) t_{solid}
TABLE 6-9 — Two Preliminary Estimates for the Burning Duration of the Initial Fire in the Box Room					
BOX TYPE					
Type #1	88.8 × 22 = 1953	15.7	176	32.7	5327 (88.8 min)
Type #2	113.2 × 22 = 2490	15.7	176	32.7	6792 (113 min)

Sources: Derived from [*]Ryu et al. 2007 and [**]Mowrer 1992.

This calculation of burning duration is a first-order estimate, which is subject to several assumptions. The first is that the combustion is confined to the compartment, it is incomplete, and some of the unburned fuel is left behind. A cursory observation of the box room after extinguishment showed that some unburned fuel still remained, including corrugated boxes and plastic rolls, also an indication that flashover had not occurred.

Note that equation (6.1) is used by the NRC's Fire Dynamics Tools and gives a realistic range of somewhere between 88 and 113 minutes as the estimated burning duration for fuel from the initial fire in the box room considering only the boxes and no other contribution from other combustible materials present in the room. These calculations are summarized in Table 6-9.

Finally, the heat release rate of the fire can be calculated by multiplying the area of the exposed fuel surface times the heat release rate per unit floor area. In this case,

$$\dot{Q} = \left(176 \text{ kW/m}^2\right)\left(32.7 \text{ m}^2\right) = 5775 \text{ kW} = 5.78 \text{ MW}.$$

In this case if other combustible materials in the room were considered, it would be fair to state that the fire might easily have reached approximately 6000 kW, or 6 MW.

ANALYTICAL MODELING: HOT GAS LAYER TEMPERATURE

Another first-order calculation for a fire within an enclosure is the temperature of the upper hot gas layer or zone as it comes into contact with the underside of the exposed ceiling and other structural elements. Such a fire is typically confined to a single compartment, with ventilation through open windows and doors. For this problem, the initial fire was considered to have natural ventilation, with only one vent opening through the door from the adjoining loading dock.

The temperatures within the compartment of fire origin are affected by several factors but mainly include the amount of air supplied to the fire along with the location of where this air enters the compartment.

The approximation for the upper hot gas layer is by the method of McCaffrey, Quintiere, and Harkleroad (1981):

$$\Delta T_g = 6.85 \left[\frac{\dot{Q}^2}{\left(A_v \sqrt{H_v}\right)\left(A_T h_k\right)} \right]^{1/3}, \tag{6.2}$$

TABLE 6-10 **Preliminary Calculations Estimating the Temperature of the Upper Gas Layer in the Box Room Fire**

TIME AFTER IGNITION (t)		UPPER LAYER GAS TEMPERATURE (T_g)	
(min)	(s)	(°C)	(°F)
0	0	25	77
1	60	227.16	440.89
2	120	251.92	485.45
3	180	267.78	514.01
4	240	279.71	535.47
5	300	289.36	552.84
10	600	321.73	611.12

where

$\Delta T_g = \left(T_g - T_a\right)$ upper layer gas temperature rise above ambient (K);
$\dot{Q}$ = heat release rate of the fire (kW);
A_v = total area of ventilation openings (m^2);
H_v = height of ventilation opening (m);
h_k = heat transfer coefficient [kW/(m$^2 \cdot$ K)];
A_T = total area of the compartment enclosing surfaces, excluding area of vent openings (m^2).

Table 6-10 lists calculations as produced by the Fire Dynamics Tools spreadsheets (Iqbal and Salley 2004) for the initial 10 minutes of the fire. Note that this preliminary estimate for heat release rate took into account only the corrugated cardboard, and it was a conservative estimate.

Note that since the area of fire origin in the box room was near both a wall and a corner, the resultant temperatures for the upper gas layer should be higher than calculated in Table 6-10. Work by Mowrer and Williamson (1987) recommends that ΔT for a fire flush to a wall be multiplied by a factor of 1.3, and a fire in a corner multiplied by 1.7.

ANALYTICAL MODELING: FLASHOVER

Another first-order calculation for a fire within an enclosure is the potential for onset of flashover. A cursory observation of the box room by investigators after extinguishment showed that some unburned fuel still remained, including corrugated boxes and plastic rolls, a visual confirmation that flashover had not occurred within the room of fire origin. However, a forensic fire scene reconstruction should usually include calculations to authoritatively back up these observations.

The growth of a fire confined to a room is usually constrained by the ventilation-limited flow of air, smoke, and hot gases into and out of the enclosure. Confining variables may include the ceiling height, ventilation openings formed by windows and doors, room volume, and location of the fire in the room or compartment. While the fire gases are constrained, a flashover may occur as a direct consequence of the heat release rate of the combustible materials contained within the room.

However, in the absence of flashover, the fire may not become fully developed and instead may reach a maximum size because of limited available fuels, flame spread, or a ventilation limit. The following two recommended methods are first-order equations for predicting the minimum heat release rate needed to cause flashover in a compartment (Iqbal and Salley, 2004).

Method of Babrauskas (1980):

$$\dot{Q}_{fo} = 750 A_V \sqrt{H_v}\,; \tag{6.3}$$

Method of Thomas (1981):

$$\dot{Q}_{fo} = 7.84 A_T + 378 A_V \sqrt{H_v}, \tag{6.4}$$

where

$\dot{Q}_{fo}$ = heat release rate to cause flashover (kW);
A_T = total area of the compartment enclosing surfaces (m^2), excluding the area of the vent opening;
A_V = area of the ventilation openings (m^2); and
H_v = height of the ventilation openings (m).

In cases where there are multiple openings to the compartment, such as doors, windows, and penetrations, the weighted average height (H_V) and area (A_V) of the ventilation of n openings of base b and height H can be calculated as follows (Peterson et al. 1976; Karlsson and Quintiere 2000):

$$A_V = (A_1 + A_2 + \cdots + A_n) = (b_1 H_1 + b_2 H_2 + \cdots + b_n H_n) \tag{6.5}$$

$$H_V = \frac{(A_1 H_1 + A_2 H_2 + \cdots + A_n H_n)}{A_V}. \tag{6.6}$$

In the case of the box room, which was the area of origin as determined by all the investigators, two scenarios for openings to that room existed.

Case 1 is the architectural design with the 2.44 m × 2.44 m (8 ft × 8 ft) door opening along with a 0.0635 m × 18.3 m (2.5 in. × 60 ft) opening above the steel beam from the box room into the freezer. Assume a 6.71 m (22 ft) ceiling.

For Case 1, the calculations are as follows:

$$
\begin{aligned}
A_V &= (A_1 + A_2) = (b_1 H_1 + b_2 H_2) \\
&= (8 \times 8) + (60 \times 0.21) = (64 + 12.6) \\
&= 76.6 \text{ ft}^2 \ (7.12 \text{ m}^2).
\end{aligned}
$$

$$
\begin{aligned}
H_V &= \frac{(A_1 H_1 + A_2 H_2)}{A_V} \\
&= \frac{(64 \times 8) + (12.6 \times 0.21)}{76.6} \\
&= \frac{(512 + 2.65)}{76.6} = 6.72 \text{ ft } (2.05 \text{ m}).
\end{aligned}
$$

$$
\begin{aligned}
A_T &= 2\big[(D \times H) + (D \times W) + (H \times W)\big] - A_V \\
&= 2\big[(60 \times 22) + (60 \times 40) + (22 \times 40)\big] - 76.6 = 9123 \text{ ft}^2 \ (848 \text{ m}^2).
\end{aligned}
$$

Method of Babrauskas:

$$\dot{Q}_{fo} = 750 A_V \sqrt{H_V} = 750 \times 7.12 \sqrt{2.05} = 7645 \text{ kW} = 7.6 \text{ MW}.$$

Method of Thomas:

$$\dot{Q}_{fo} = 7.8 A_T + 378 A_V \sqrt{H_v} = (7.8 \times 848) + \left(378 \times 7.12 \sqrt{2.05}\right)$$
$$= 10467 \text{ kW} = 10.5 \text{ MW}.$$

Therefore, for Case 1, there was insufficient heat release rate to cause flashover, which was visually confirmed by the on-site investigators, who found remnants of unburned combustible materials in the box room.

Case 2 is the actual construction with the 2.44 m × 2.44 m (8 ft x 8 ft) door opening, along with a 0.0635 m × 18.3 m (2.5 in. × 60 ft) opening above the steel beam, and a 0.609 m × 18.3 m (2 ft × 60 ft) opening below the beam from the box room into the freezer. Assume a 6.71 m (22 ft) ceiling.

For Case 2, the calculations are as follows:

$$A_V = (A_1 + A_2 + A_3) = (b_1 H_1 + b_2 H_2 + b_3 H_3)$$
$$= (8 \times 8) + (60 \times 0.21) + (60 \times 2) = (64 + 12.6 + 120)$$
$$= 196.6 \text{ ft}^2 \left(18.3 \text{ m}^2\right).$$

$$H_V = \frac{(A_1 H_1 + A_2 H_2 + A_3 H_3)}{A_V}$$
$$= \frac{(64 \times 8) + (12.6 \times 0.21) + (120 \times 2)}{196.6}$$
$$= \frac{512 + 2.65 + 240}{196.6} = 3.84 \text{ ft} (1.17 \text{ m}).$$

$$A_T = 2\left[(D \times H) + (D \times W) + (H \times W)\right] - A_V$$
$$= 2\left[(60 \times 22) + (60 \times 40) + (22 \times 40)\right] - 196.6 = 9003 \text{ ft}^2 \left(836 \text{ m}^2\right).$$

Method of Babrauskas:

$$\dot{Q}_{fo} = 750 A_V \sqrt{H_V} = 750 \times 18.3 \sqrt{1.17} = 14,145 \text{ kW} = 14.1 \text{ MW}.$$

Method of Thomas:

$$\dot{Q}_{fo} = 7.8 A_T + 378 A_V \sqrt{H_v} = (7.8 \times 836) + \left(378 \times 18.3 \sqrt{1.17}\right)$$
$$= 14,003 \text{ kW} = 14 \text{ MW}.$$

Therefore, for Case 2, there was also insufficient heat release rate to cause flashover, which was visually confirmed by the on-site investigators, who found remnants of unburned combustible materials in the box room.

ANALYTICAL MODELING: STRUCTURAL STEEL MEMBERS

The standard method for assessing structural fire protection is to use *ASTM E119* (ASTM 2011a) or *NFPA 251: Standard Methods of Test for Fire Endurance of Building Construction and Materials* (NFPA 2006). Computer correlations based on actual testing have been developed for calculating the fire resistance of steel columns, beams, and trusses.

Of interest in this forensic fire reconstruction is the potential behavior of the steel beam supporting the roof directly between the box room and freezer. Roof assemblies held up by beams are known to absorb and conduct heat away from the beam.

According to U.S. standards for all structural steel elements, steel loses strength to support a load at a critical average temperature of 538°C (1000°F). The average temperature development of an unprotected steel member can be shown to be (Buchanan 2001, 179; Iqbal and Salley 2004, 17–19)

$$\Delta T_s = \frac{F}{V} \frac{1}{\rho_s c_s} \left[h_c \left(T_f - T_s \right) + \sigma \varepsilon \left(T_f^4 - T_s^4 \right) \right] \Delta t, \tag{6.7}$$

where

ΔT_s = temperature rise in the steel member (°F);
F/V = ratio of heated surface area to volume of beam per unit length (m^{-1});
ρ_s = density of the steel (kg/m^3);
c_s = specific heat of steel [J/(kg · K)];
h_c = convective heat transfer coefficient [W/(m^2 · K)];
T_f = fire exposure temperature (°F);
T_s = steel temperature (°F);
σ = Stefan-Boltzmann constant [kW/(m^2 · K^4)];
ε = flame emissivity; and
Δt = time step (s).

The FDTs package of spreadsheets incorporates several standard correlation methods for assessing structural fire protection. The only other variables needed by the program are the ambient air temperature, type of beam, the flame emissivity, and maximum allowable time step.

Emissivity (with values between 0 and 1) gives the fraction of energy emitted relative to a perfect radiator. The emissivity ranges from $\varepsilon = 0.7$ for a column exposed to fire on all sides, to $\varepsilon = 0.3$ for the column façade, to $\varepsilon = 0.5$ for a floor girder with floor slab of concrete, to $\varepsilon = 0.5$ for a floor girder with slab on the top flange of I-section for which the width/depth ratio is not less than 0.5, to $\varepsilon = 0.7$ for a girder of I-section where the width/depth ratio is less than 0.5.

Architectural plans showed that steel I-beams used between the box room and freezer were type W24 × 61, above the freezer was a type a W24 × 61, and between the freezer and the mechanical room was a type W18 × 55. Note that the designation W24 × 61 for a steel I-beam means that it has a wide flange (W), is approximately 24 in. in depth, and weighs approximately 61 lb/linear ft.

Results of the FDTs analysis using a 10-second step interval for the estimated time to failure for the full range of emissivities for the W24 × 61 steel beam are given in Table 6-11. Note that the closest value for this beam was a type W24 × 62.

Structural deformation and failure of the roof structure in this case was consistent with loss histories and structural fire protection engineering design studies (Buchanan 2001; Tinsley and Icove 2008).

ZONE MODELING

A second level of forensic analysis of the initial conditions of the fire used a zone model. Figure 6-24 shows the floor plan representing the layout of the plant at the time of the fire, and Figure 6-25 illustrates the three-dimensional view. This dimensional view was produced through an interface between CFAST and the NIST fire visualization program called Smokeview.

Two CFAST models were run representing Case 1 and Case 2, as described in the previous section. A majority of the closed internal and exterior doors had an appropriate "leakage" factor included to address air and smoke movement between compartments. This leakage factor included plastic strip curtains. The materials data for polyurethane was used in place of the polystyrene foam. A typical design curve with an approximate 10 MW limit, as shown in Figure 6-26, was used to run the model.

The results of the CFAST model (Figure 6-27) showed that in both cases, the temperature of the box room reached 400°C (752°F). With the exception of the ceiling space, all the other compartments remained well under 100°C (212°F).

The most significant result was that even with the architectural design, which still had a 2.5-in. gap above the steel I-beam, heat transfer took place from the box room into the ceiling space above the freezer. Visualization of both cases using Smokeview at 760 seconds showed the only difference to be a slightly increased temperature of the mechanical and engine rooms (Figure 6-28). The color sidebar to the right of each display gives the temperature range in degrees Fahrenheit.

FIELD MODELING

The purpose of using the FDS field model in the building case was to take advantage of the program's ability to model smoke movement throughout the building. The results of this analysis gave an approximation of the impact, if any, of compartmentation on the ability of smoke to migrate through the building.

TABLE 6-11	Estimated Fire Resistance Time of a W24 × 62 Unprotected Steel Beam Using the Quasi-Steady-State Approach with Fire Dynamics Tools	

TYPICAL CONSTRUCTION TYPE EXAMPLES	EMISSIVITY	TIME TO FAILURE OF BEAM (min)
Column/Beam outside column façade	0.3	10.8
Floor girder with floor slab of concrete (only the underside of the bottom flange being directly exposed to fire)		
Floor girder with floor slab on the top flange Girder of I-section for which the width-depth ration is not less than 0.5	0.5	8.8
Floor girder with floor slab on the top flange Girder of I-seciton for which the width-depth ratio is less than 0.5	0.7	7.5
Floor girder with floor slab on the top flange		
Box girder and lattice girder		
Column/Beam exposed to fire on all sides		

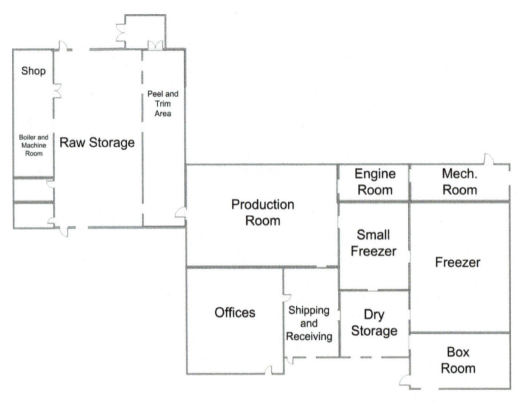

FIGURE 6-24 Floor plan representing the building at the time of the fire (not to scale). *Courtesy of D. J. Icove.*

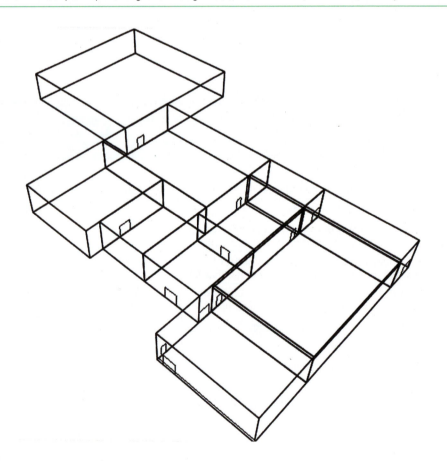

FIGURE 6-25 CFAST representation of the building using the NIST Smokeview fire visualization program. *Courtesy of D. J. Icove.*

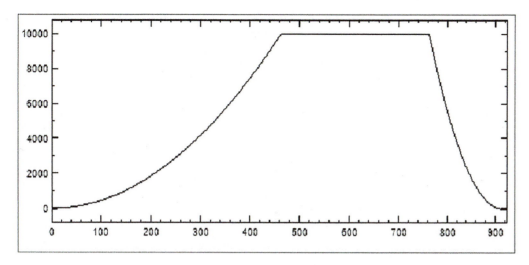

FIGURE 6-26 The design fire curve of time (s) versus heat release rate (kW) used in the CFAST model.
Courtesy of D. J. Icove.

Figure 6-29 shows the impact of smoke and heat distribution at 120 seconds into the simulation for a 600 kW fire. As would be expected, Figure 6-29(a) shows that more smoke and heat are retained within the box room as compared with the situation illustrated in Figure 6-29(b), where smoke and heat have migrated into the space above the freezer. However, further testing is needed to compare and contrast the results from FDS and CFAST computer runs for an identical design fire curve as previously shown in Figure 6-28.

Clearly, as demonstrated in this example, the use of a forensic fire engineering analysis can assist in refining and sometimes supporting the scenario for a fire's initiation, area of origin, growth, development, and impact on the structure. In this case, the forensic fire engineering analysis evaluated the initial development and growth of the fire and the impact of a lack of compartmentation between the room of fire origin and adjoining building areas.

If tenability of the occupants had been an issue, the computer models could have been modified to include calculation of the necessary information to conduct this analysis. Other variables that could be modeled include the activation of detectors or sprinklers and their resulting impact on notifying occupants and controlling or suppressing the incipient fire. Thus the versatility of the modeling programs may raise more questions than provide answers. The fire models may often give users the ability to address more complicated issues, such as the impact of fire detection and suppression systems, notification and response times of the fire service, or habitability of the occupancy.

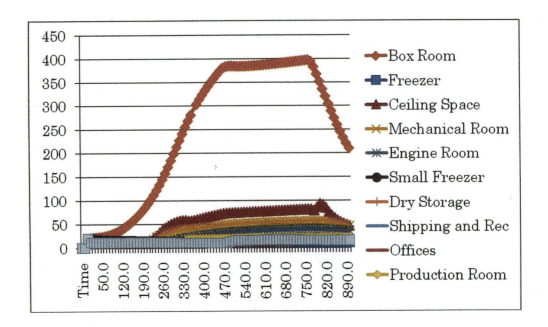

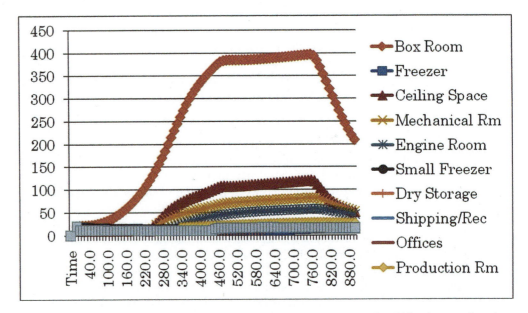

FIGURE 6-27 (a) Case 1 (architectural design): Plot of temperature (°C) versus time (s) for the upper layer in each compartment. (b) Case 2 (as built): Plot of temperature (°C) versus time (s) for the upper layer in each compartment. *Courtesy of D. J. Icove.*

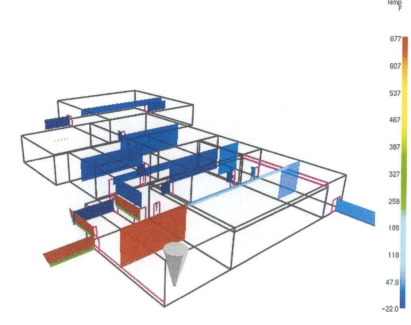

Smokeview 5.2.2 – Jul 18 2008

Zone
Temp
F

677
607
537
467
397
327
258
188
118
47.9
-22.0

Frame: 76
Time: 760.0

(b)

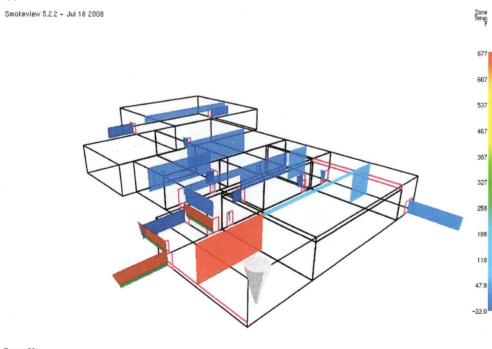

Smokeview 5.2.2 – Jul 18 2008

Zone
Temp
F

677
607
537
467
397
327
258
188
118
47.9
-22.0

Frame: 76
Time: 760.0

FIGURE 6-28 (a) Case 1 (architectural design): CFAST three-dimensional representation of temperatures and upper-layer height in each compartment at 760 seconds (12.6 minutes). (b) Case 2 (as built): CFAST three-dimensional representation of temperatures and upper-layer height in each compartment at 760 seconds (12.6 minutes). *Courtesy of D. J. Icove.*

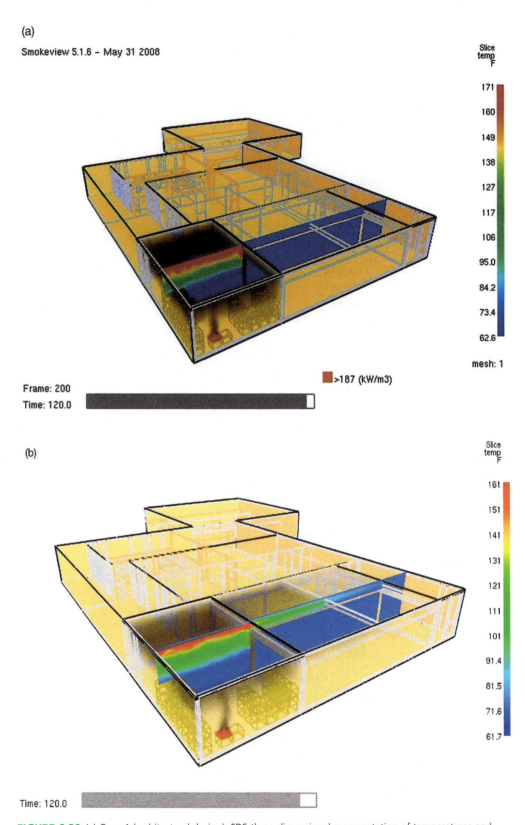

FIGURE 6-29 (a) Case 1 (architectural design): FDS three-dimensional representation of temperatures and upper-layer height. (b) Case 2 (as built): FDS three-dimensional representation of temperatures and upper-layer height. *Courtesy of D. J. Icove.*

Summary

This chapter introduced the concept and application of fire modeling to forensic fire investigation. Although models have existed since the 1960s, fire modeling centered on explaining the physical phenomena of fires, particularly when applied to verifying existing experimental data.

Though the effort of a few fire scientists and engineers working at NIST and FRS, the acceptability and application of fire modeling moved out of the laboratory and into the world of forensic fire scene reconstruction. As indicated, the purpose of this chapter was not to make the reader an expert in fire modeling but to allow him or her to gain a better appreciation of its value in an investigation.

The next chapter explores in more detail the issues of tenability as it applies to the modeling of toxicity of fires.

Problems

6.1. Choose one of the cases mentioned in this chapter and model it using the selected program. Obtain a similar fire modeling program and compare and contrast the results derived using it for the analysis.

6.2. Many computer fire models contain a database of materials and their burning properties. Examine the databases of two fire models. Compare and contrast the results obtained using each model.

6.3. Using the Fire Dynamics Tools (FDTs) spreadsheets, conduct a sensitivity analysis of several of its calculations by varying the input data (such as room dimensions or ventilation openings).

References

Abecassis-Empis, C., Reszka, P., Steinhaus, T., Cowlard, A., Biteau, H., Welch, S., Rein, G., & Torero, J. L. 2008. Characterisation of Dalmarnock fire test one. *Experimental Thermal and Fluid Science* 32 (7): 1334–43, doi: 10.1016/j.expthermflusci.2007.11.006.

ASTM. 2007a. *ASTM E1472-07: Standard guide for documenting computer software for fire models* (withdrawn 2011). West Conshohocken, PA: ASTM International.

———. 2007b. *ASTM E1591-07: Standard Guide for Obtaining Data for Deterministic Fire Models.* West Conshohocken, PA: ASTM International *ASTM International Subcommittee E05.33 on Fire Safety Engineering.*

———. 2007c. *ASTM E1895-07: Standard guide for determining uses and limitations of deterministic fire models* (withdrawn 2011). West Conshohocken, PA: ASTM International.

———. 2011a. *ASTM E119-11: Standard test methods for fire tests of building construction and materials.* West Conshohocken, PA: ASTM International.

———. 2011b. *ASTM E1355-11: Standard guide for evaluating the predictive capability of deterministic fire models.* West Conshohocken, PA: ASTM International.

Babrauskas, V. 1975. COMPF: A program for calculating post-flashover fire temperatures. Berkeley, CA: Fire Research Group, University of California, Berkeley.

———. 1980a. Estimating room flashover potential. *Fire Technology* 16 (2): 94–103, doi: 10.1007/bf02481843.

———. 1996. Fire modeling tools for FSE: Are they good enough? *Journal of Fire Protection Engineering* 8 (2): 87–96.

Bailey, C. 2006. One stop shop in structural fire engineering. http://www.mace.manchester.ac.uk/project/research/structures/strucfire/.

BRE-Center. 2012. The Dalmarnock fire tests, 2006, UK. Retrieved February 5, 2012, from http://www.see.ed.ac.uk/fire/dalmarnock.html/.

Bryan, J. L. 1983. An examination and analysis of the dynamics of the human behavior in the Westchase Hilton Hotel Fire, Houston, Texas, on March 6, 1982. Quincy, MA: National Fire Protection Association.

Buchanan, A. H. 2001. *Structural design for fire safety.* Chichester, UK: Wiley.

Bukowski, R. W. 1991. Fire models: The future is now! *Fire Journal* 85 (2): 60–69.

———. 1995. Modelling a backdraft incident: The 62 Watts Street (New York) fire. *NFPA Journal* (November/December): 85–89.

———. 1996. Modelling a backdraft incident: The 62 Watts Street (New York) fire. *Fire Engineers Journal* 56 (185): 14–17.

Bukowski, R. W., & Spetzler, R. C. 1992. Analysis of the Happyland Social Club fire with HAZARD I. *Fire and Arson Investigator* 42 (3): 37–47.

Davis, S. G., Engel, D., Gavelli, F., Hinze, P., & Hansen, O. R. 2010. Advanced methods for determining the origin of vapor cloud explosions case study: 2006 Danvers explosion investigation. Paper presented at the International Symposium on Fire Investigation Science and Technology, September 27–29,College Park, MD.

DeHaan, J. D. 2005a. Reliability of fire tests and computer modeling in fire scene reconstruction: Part 1. *Fire & Arson Investigator* (January): 40–45.

———. 2005b. Reliability of fire tests and computer modeling in fire scene reconstruction: Part 2. *Fire & Arson Investigator* (April): 35–45.

DeHaan, J. D., & Icove, D. J. 2012. *Kirk's fire investigation*, 7th ed. Upper Saddle River, NJ: Pearson-Prentice Hall.

DeWitt, W. E., & Goff, D. W. 2000. Forensic engineering assessment of FAST and FASTLite fire modeling software. *National Academy of Forensic Engineers*, 9–19.

FireFOAM. 2012. FM Research: Open Source Fire Modeling. Retrieved May 27, 2012, from http://www.fmglobal.com.

Forney, G. P. 2010. *Smokeview (version 5): A tool for visualizing fire dynamics simulation data*. Vol. I: *User's guide*. (NIST Special Publication 1017-1). Gaithersburg, MD: National Institute of Standards and Technology.

Friedman, R. 1992. An International Survey of Computer Models for Fire and Smoke. *SFPE Journal of Fire Protection Engineering* 4 (3): 81–92.

FRS. 2012. Fire Research Station, the research-based consultancy and testing company of BRE. Retrieved January 2012 from www.bre.co.uk/frs/.

GexCon. 2012. GexCon US. Retrieved February 5, 2012, from http://www.gexcon.com/.

Hunt, S. 2000. Computer fire models. *NFPA Section News* 1 (2): 7–9.

Inberg, S. H. 1927. Fire tests of office occupancies. *NFPA Quarterly* 20 (243).

Iqbal, N., & Salley, M. H. 2004. *Fire Dynamics Tools (FDTs): Quantitative fire hazard analysis methods for the U.S. Nuclear Regulatory Commission Fire Protection Inspection Program*. Washington, DC.

Janssens, M. L., & Birk, D. M. 2000. *An introduction to mathematical fire modeling*, 2nd ed. Lancaster, PA: Technomic.

Karlsson, B., & Quintiere, J. G. 2000. *Enclosure fire dynamics*. Boca Raton, FL: CRC Press.

Kawagoe, K. 1958. Fire behavior in rooms. Report No. 27. Tokyo: Building Research Institute.

Madrzykowski, D., & Vettori, R. L. 2000. Simulation of the dynamics of the fire at 3146 Cherry Road, NE, Washington, DC, May 30, 1999. *NISTIR 6510*. Gaithersburg, MD: National Institute of Standards and Technology, Center for Fire Research.

Magnusson, S. E., & Thelandersson, S. 1970. Temperature–time curves of the complete process of fire development. Theoretical study of wood fuel fires in enclosed spaces. *Civil and Building Construction Series No. 65*. Stockholm: Acta Polytechnica Scandinavia.

McCaffrey, B., Quintiere, J., & Harkleroad, M. 1981. Estimating room temperatures and the likelihood of flashover using fire test data correlations. *Fire Technology* 17 (2): 98–119, doi: 10.1007/bf02479583.

McGill, D. 2003. Fire Dynamics Simulator, FDS 683, participants' handbook. Toronto, Ontario, Canada: Seneca College, School of Fire Protection.

McGrattan, K., Baum, H., Rehm, R., Mell, W., McDermott, R., Hostikka, S., & Floyd, J. 2010. *Fire Dynamics Simulator (version 5) technical reference guide*. NIST Special Publication 1018-5. Gaithersburg, MD: National Institute of Standards and Technology.

McGrattan, K., McDermott, R., Hostikka, S., & Floyd, J. 2010. *Fire Dynamics Simulator (version 5) user's guide*. NIST Special Publication 1019-5. Gaithersburg, MD: National Institute of Standards and Technology.

Milke, J. A., & Mowrer, F. W. 2001. Application of fire behavior and compartment fire models seminar. Paper presented at the Tennessee Valley Society of Fire Protection Engineers (TVSFPE), September 27–28, Oak Ridge, TN.

Mitler, H. E. 1991. Mathematical modeling of enclosure fires. Gaithersburg, MD: National Institute of Standards and Technology.

Moodie, K., & Jagger, S. F. 1992. The King's Cross fire: Results and analysis from the scale model tests. *Fire Safety Journal* 18 (1): 83–103, doi: 10.1016/0379-7112(92)90049-i.

Mowrer, F. W. 1992. Methods of quantitative fire hazard analysis. Boston, MA: Society of Fire Protection Engineers, prepared for Electric Power Research Institute (EPRI).

———. 2002. The right tool for the job. *SFPE Fire Protection Engineering Magazine* 13 (Winter): 39–45.

———. 2003. Spreadsheet templates for fire dynamics calculations. College Park, MD: University of Maryland.

Mowrer, F. W., & Williamson, R. B. 1987. Estimating room temperatures from fires along walls and in corners. *Fire Technology* 23 (2): 133–45, doi: 10.1007/bf01040428.

Nelson, H. E. 1987. An engineering analysis of the early stages of fire development: The fire at the Dupont Plaza Hotel and Casino on December 31, 1986. Gaithersburg, MD: National Institute of Standards and Technology.

———. 1989. An engineering view of the fire of May 4, 1988, in the First Interstate Bank Building, Los Angeles, California. *NISTIR 89-4061*. Gaithersburg, MD:

National Institute of Standards and Technology, Center for Fire Research.

———. 1990. FPETool: Fire protection engineering tools for hazard estimation. Gaithersburg, MD: National Institute of Standards and Technology.

———. 1994. Fire growth analysis of the fire of March 20, 1990, Pulaski Building, 20 Massachusetts Avenue, NW, Washington, DC. *NISTIR 4489*. Gaithersburg, MD: National Institute of Standards and Technology, Center for Fire Research.

Nelson, H. E., & Tu, K. M. 1991. Engineering analysis of the fire development in the Hillhaven Nursing Home fire, October 5, 1989. *NISTIR 4665*. Gaithersburg, MD: National Institute of Standards and Technology, Center for Fire Research.

NFPA. 2006. *NFPA 251: Standard methods of tests of fire endurance of building construction and materials.* Quincy, MA: National Fire Protection Association.

———. 2011. *NFPA 921: Guide for fire and explosion investigations.* Quincy, MA: National Fire Protection Association.

Peacock, R. D., Jones, W. W., Reneke, P. A., & Forney, G. P. 2008. *CFAST: Consolidated Model of Fire Growth and Smoke Transport (version 6) user's guide.* NIST Special Publication 1041. Gaithersburg, Maryland: National Institute of Standards and Technology.

Peterson, O., Magnusson, S. E., & Thor, J. 1976. *Fire engineering design of steel structures* Publication No. 50. Stockholm: Swedish Institute of Steel Construction.

Rein, G., Bar-Ilan, A., Fernandez-Pello, A. C., & Alvares, N. 2006. A comparison of three models for the simulation of accidental fires *Journal of Fire Protection Engineering* 16 (3): 183–209.

Rein, G., Abecassis-Empis, C., & Carvel, R. 2007. *The Dalmarnock fire tests: Experiments and modelling.* Edinburgh, Scotland, UK: University of Edinburgh.

Rein, G., Torero, J. L., Jahn, W., Stern-Gottfried, J., Ryder, N. L., Desanghere, S., Lazaaro, M., et al. 2009. Round-robin study of a priori modelling predictions of the Dalmarnock fire test one. *Fire Safety Journal* 44 (4): 590–602, doi: 10.1016/j.firesaf.2008.12.008.

Ryu, C., Phan, A. N., Yang, Y.-b., Sharifi, V. N., & Swithenbank, J. 2007. Ignition and burning rates of segregated waste combustion in packed beds. *Waste Management* 27 (6): 802–10., doi: 10.1016/j.wasman.2006.04.013.

Salley, M. H., & Kassawar, R. P. 2007a. *Verification and validation of selected fire models for nuclear power plant applications: Consolidated Fire Growth and Smoke Transport Model (CFAST).* Washington, DC: U.S. Nuclear Regulatory Commission.

———. 2007b. *Verification and validation of selected fire models for nuclear power plant applications: Experimental uncertainty.* Washington, DC: U.S. Nuclear Regulatory Commission.

———. 2007c. *Verification and validation of selected fire models for nuclear power plant applications: Fire Dynamics Simulator (FDS)* Washington, DC: U.S. Nuclear Regulatory Commission.

———. 2007d. *Verification and validation of selected fire models for nuclear power plant applications: Fire Dynamics Tools (FDTs).* Washington, DC: U.S. Nuclear Regulatory Commission.

———. 2007e. *Verification and validation of selected fire models for nuclear power plant applications: Fire-Induced Vulnerability Evaluation (FIVE-Rev1).* Washington, DC: U.S. Nuclear Regulatory Commission.

———.2007f. *Verification and validation of selected fire models for nuclear power plant applications: MAGIC.* Washington, DC: U.S. Nuclear Regulatory Commission.

———. 2007g. *Verification and validation of selected fire models for nuclear power plant applications: Main report.* Washington, DC: U.S. Nuclear Regulatory Commission.

———. 2010. Methods for Applying Risk Analysis to Fire Scenarios (MARIAFIRES)-2008 NRC-RES/EPRI Fire PRA Workshop (vol. 1). Washington, DC: U.S. Nuclear Regulatory Commission.

SFPE. 2008. *SFPE handbook of fire protection engineering,* 4th ed. Quincy, MA: National Fire Protection Association, Society of Fire Protection Engineers.

———. 2011. *Guidelines for substantiating a fire model for a given application.* Bethesda, MD: Society of Fire Protection Engineers.

Thomas, P. H. 1981. Testing products and materials for their contribution to flashover in rooms. *Fire and Materials* 5 (3): 103–11, doi: 10.1002/fam.810050305.

Thunderhead. 2010. PyroSim example guide. Manhattan, KS: Thunderhead Engineering.

———. 2011. PyroSim user manual. Manhattan, KS: Thunderhead Engineering.

Tinsley, A. T., & Icove, D. J. 2008. An assessment of the use of structural deformation as a method for determining area of fire origin. Paper presented at the International Symposium on Fire Investigation Science and Technology, May 19–21, Cincinnati, OH.

Vasudevan, R. 2004. Forensic engineering analysis of fires using Fire Dynamics Simulator (FDS) modeling program. *Journal of the National Academy of Forensic Engineers* 21, no. 2 (December): 79–86.

> **""** There's a scarlet thread of murder running through the colorless skein of life, and our duty is to unravel it, and isolate it, and expose every inch of it. **""**
>
> —Sir Arthur Conan Doyle
> *A Study in Scarlet*

KEY TERMS

anoxia, *p. 320*

cause of death, *p. 329*

fractional effective concentration (FEC), *p. 313*

Haber's rule, *p. 314*

hypoxia, *p. 320*

tenability, *p. 310*

OBJECTIVES

After reading this chapter, the student should be able to:

- Recognize problems and pitfalls of death or injury scenes.
- Interpret the various contributions to human injuries and deaths at fire scenes.
- Compare case histories and their outcomes.
- Evaluate scenes and the cooperation of investigators.

I n every country, particularly in highly industrialized ones, fire kills a significant number of people. In the United States, it is one of the five leading causes of accidental death. Fire investigators should evaluate the common problems and pitfalls when conducting forensic reconstructions, particularly when death occurs. The purpose of this chapter is to review these human tenability factors, present some analytical approaches, and provide illustrative case histories.

The involvement of the fire investigator or forensic specialist in fatal fires can come in any form, from any sector, and challenge his or her talents and knowledge to come to just and accurate conclusions. These cases require the highest degree of cooperation among the investigators, all of whom have contributions to make toward a successful investigation. When deaths occur in a fire, the event becomes the focus of the press and the public, as well as police, fire, insurance, and forensic professionals. When problems occur, they can have far-reaching consequences.

Problems and Pitfalls

There are several problem areas that can complicate investigations of fires involving fatalities and compromise the accuracy and reliability of the conclusions reached.

- *Linkage between the fire and death investigations:* Prejudging the fire and its attendant death as an accident and automatically treating the scene investigation accordingly is a major problem. Fires can be intentional, natural, or accidental in their cause, and deaths can be accidental, homicidal, suicidal, or natural. The linkage between the two events can be direct, indirect, or simple coincidence. The responsibility of the fire investigation team in these cases is to establish the cause of the fire and assist the medical examiner in the death investigation and to determine the connection (if any) between the two.

- *Time interval to death:* Sudden violent deaths are assumed to be due to an instantaneous exposure to insult followed by immediate collapse and death of a victim (e.g., a shot is fired, and the victim collapses to die shortly afterward). Many forensic investigations are considered (and successfully concluded) in this light. Fires, however, occur over a period of time, creating dangerous environments that vary greatly with time and can kill by a variety of mechanisms. A person may be killed nearly instantaneously by exposure to a flash fire or only after hours of exposure to toxic gases. Investigators must have an appreciation for the nature of fire, its lethal products, and variables that must be considered and not treat the event as a single exposure to a single set of conditions at a precise moment in time that results in instant collapse.

- *Heat intensity and duration:* There is little accurate information available to detectives, pathologists, and medical examiners about the temperatures and intensities of heat exposure that occur in a fire as it develops. Misunderstandings and misapprehensions can lead investigators seriously astray in their interpretations when they try to assess injuries or postmortem damage resulting from fires.

- *Fire-related human behavior:* In most violent deaths, the victim responds to what is often referred to as a "fight or flight" response to a threat, then suffers an injury, collapses, and dies. In fires, the potential responses of victims often include going to investigate; simply observing; failing to notice or appreciate the danger; failing to respond owing to infirmity or incapacitation from drugs or alcohol; returning to the fire or delaying escape to rescue pets, family, or personal or valuable property; as well as fighting or attempting to fight the fire. This variability of responses can vastly complicate the process of solving critical problems of why the victim failed to escape the fire (and perhaps why other people escaped).

- *Time interval between fire and death:* Fires can kill in seconds, or death can occur minutes, hours, days, or even months after the victim is removed from the scene. The longer the time interval between the fire and the death, the harder it is to keep track of the actual cause (the fire) and the result (the death). Evidence is lost when a living victim is removed from a scene and dies later, away from the scene; it may be too late to recover or document that evidence. Fatalities that occur after a victim is hospitalized are inevitably omitted from the NFIRS statistics and may also be omitted from the national vital statistics.
- *Conflicts among investigating agencies:* Conflicts can arise regarding the perceived or mandated responsibilities of police, fire, medicolegal, and forensic personnel who are often involved in fire scene deaths.
- *Postmortem effects:* After death, severe postmortem fire effects on the body can vastly complicate the investigation by obliterating evidence. The body can bear fire patterns of heat effects and smoke deposits that can be masked by exposure to fire after death. The body can be incinerated by exposure to flames, so that evidence of prefire wounds or even clinical evidence such as blood samples is destroyed. Structural collapse and effects of firefighting hose streams and overhaul can cause additional damage to the scene and to the body.
- *Premature removal of the body:* A major problem is the premature removal of a deceased victim from the fire scene. The compulsion to rescue and remove every fire victim is a very strong one, particularly among dedicated firefighters. However, once the fire is under control and unable to inflict further damage to the body of a confirmed deceased, there is nothing to be gained and much to be lost in the way of burn pattern analysis, recovery of body fragments (especially dental evidence), projectiles, clothing and associated artifacts (e.g., keys, flashlight, dog leash), and even trace evidence, by the undocumented and hurried removal of the remains.

Tenability: What Kills People in Fires?

Structural fires can achieve their deadly result in a number of ways—heat, smoke, flames, soot, and others—but fire conditions change continually as a fire grows and evolves, and the conditions of victim exposure can vary from conditions involving little or no threat or injury to almost instant lethality. Lethal agents of fires can and usually do act in combination and include heat, smoke, and inhalation of smoke or toxic gases, hypoxia, flames, and blunt trauma. These lethal agents of fires are discussed in further detail later in this chapter.

The ability of humans to escape a fire, **tenability**, is measured by the time frame during which their environment remains survivable. Fires can produce incapacitating effects on humans when they are exposed to heat and smoke. These physiological effects are generally categorized into the following areas (Purser 2008):

tenability ■ An assessment of the potential for harm that may be imposed from a fire based on (1) an analysis of the source, heat release rates, and rates of production of the combustion products; (2) when occupants might become exposed to these harmful conditions; and (3) the effects on the occupants to this exposure.

- *Toxic gases and irritants:* Toxic gas inhalation can cause mental confusion, respiratory tract injuries, loss of consciousness, asphyxiation, and skin reactions depending on the chemical compositions in the toxic gases.
- *Heat transfer:* Excessive heat irritates exposed skin and the respiratory tract, causing severe pain as well as varying degrees of burn injuries, hyperthermia, or heat stroke.
- *Visibility:* Optical opacity of the smoke and irritants from a fire produces impaired vision as the distribution of thick smoke descends toward the floor through rooms, stairwells, and hallways.

Fire investigators should also consider the synergistic effects of two or more of these factors when conducting a forensic analysis. Of primary concern is assessing the point at which exposure to one or more of the preceding variables would cause injury or block the individual from successfully escaping the fire, resulting in death. The psychological behavior of people in fires when exposed and reacting to these variables affects their decisions and the time required to travel via safe escape routes.

Critical limits to human tenability include a limit of visibility to 5 m (16.4 ft), an accumulated exposure rate of carbon monoxide of 30,000 parts per million minute (ppm-min), and a critical temperature of approximately 120°C (240°F). The synergistic effects of two or more of these critical factors may override an individual's limits (Jensen 1998).

The major goal of the investigator when conducting a *tenability analysis* is to determine how an individual attempting to escape a burning structure becomes impaired and how the fire changes his or her environment and perceptions. These tenability analysis techniques are grounded in both experimental and forensic data, giving the investigator a balanced and practical approach.

The physiological and toxicological effects of heat transfer and toxic smoke on animals and humans are based on scientifically valid experiments. For example, studies correlating carbon monoxide exposure to carboxyhemoglobin levels in the bloodstream used individual subjects ranging from laboratory rats to volunteer medical students (Nelson 1998). Some of these data were extrapolated to model the results at higher levels of exposure to carbon monoxide. These models must take into account variations in age, health, and stature of the individual, because an individual's height affects exposure when he or she is standing upright in the stratified upper smoke layer, where nearly all the toxic gases normally reside. See Figure 7-1 for a graphic illustration of the exposure of humans walking upright through smoke layers (Bukowski 1995b).

Forensic evaluations of incapacitation of a fire victim also come from forensic data derived from actual case histories and investigations. The major studies of the behavior of people in fires come from subject interviews of people surviving large fires and explosions (Bryan and Icove 1977).

Toxic Gases

The previous section on tenability discussed visibility and irritant effects on humans as they try to navigate and survive fires. *Toxic gases* contained within smoke can also have a narcotic effect that asphyxiates victims. The dominant narcotic gases in smoke that affect the nervous and cardiovascular systems are carbon monoxide (CO) and hydrogen cyanide (HCN). Carbon dioxide and reduced oxygen levels (hypoxia) that are not toxic individually may have severe synergistic effects on tenability.

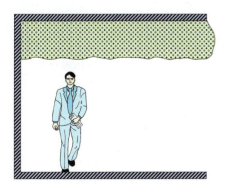

FIGURE 7-1 Illustration of human tenability in varying smoke layer levels. *R. W. Bukowski, "Predicting the Fire Performance of Buildings: Establishing Appropriate Calculation Methods of Regulatory Applications," National Institute of Standards and Technology, Gaithersburg, MD, 1995.*

Increased exposure to toxic gases may cause confusion, loss of consciousness, and eventually asphyxial death. The prediction of asphyxiation resulting in incapacitation and death in fires can be modeled (Purser 2008). Toxic products of combustion can include a wide variety of chemicals depending on what is burning and how efficiently it is burning (temperature, mixing, and oxygen concentration are all important variables in determining what chemical species are created).

Toxic gases can generally be classified into three basic categories:

- *Nonirritant gases:* (sometimes referred to as "narcotic gases"): CO, HCN, H_2S (hydrogen sulfide), and phosgene (CCl_2O).
- *Acidic irritants:* HCl (hydrogen chloride)—produced during the combustion of polyvinyl chloride (PVC) plastics; sulfur oxides (SO_x), which form H_2SO_3 (sulfurous acid) and H_2SO_4 (sulfuric acid)—produced by oxidation of sulfur-containing fuels; and nitrogen oxides (NO_x), which form HNO_2 (nitrous acid) and HNO_3 (nitric acid)—from nitrogen-containing fuels.
- *Organic irritants:* Formaldehyde (CH_2O) and acrolein (2-propenal, C_3H_4O) are produced by the combustion of cellulosic fuels. Isocyanates are produced by the combustion of polyurethanes.

Acidic irritant gases dissolve in the water of the mucous membranes and generate the corrosive acids listed. These acids disrupt the epithelial cell membranes and cause the cells to lyse, releasing fluids and producing edema, as evidenced by extreme watering of the eyes, coughing, and inability to breathe. Sulfur dioxide combines with water to form sulfurous acid, a strong irritant. These reactions can all result in incapacitation, preventing escape from lethal gases such as hydrogen cyanide and carbon monoxide. Exposure to hydrogen chloride (HCl) starting at concentrations of 50 ppm usually causes respiratory or visual impairment sufficient to affect walking movement. Total cessation of movement occurs at a concentration of hydrogen chloride gas approaching 300 ppm. The effects of concentrations of HCl above the 1000 ppm level are most likely severe enough to prevent escape (Purser 2001).

Table 7-1 shows the effects of various fire gases on both the escape and the incapacitation of humans. Most fires present occupants with a complex mixture of toxic and irritant gases and vapors combined with high carbon dioxide and *hypoxic* (low-oxygen) conditions. The fire investigator must assess what fuels were burning, under what conditions (smoldering, open flame, underventilated), and where the occupants would have been exposed to the toxic gases and vapors when assessing the tenability of the fire exposure.

TABLE 7-1	Irritant Concentrations of Fire Gases Predicted to Cause 50 Percent Impaired Escape or Incapacitation in the Human Population	
COMMON FIRE GASES	**IMPAIRED ESCAPE (ppm)**	**INCAPACITATION (ppm)**
Hydrogen chloride (HCl)	200	900
Hydrogen bromine (HBr)	200	900
Hydrogen fluoride (HF)	200	900
Sulfur dioxide (SO_2)	24	120
Nitrogen dioxide (NO_2)	70	350
Formaldehyde (CH_2O)	6	30
Acrolein (C_3H_4O)	4	20

Source: Derived from Purser 2001.

HCl is a major combustion/decomposition product of vinyl plastics, in both the flaming and the smoldering modes. Hydrogen bromide (HBr) or hydrogen fluoride (HF) is produced when some synthetic rubbers are burned. Acrolein is created when wood or cardboard or cellulosic materials are burned.

The term **fractional effective concentration (FEC)** was developed to assess the impact of toxic smoke and by-products of combustion on a subject (Purser 2001). The use of the concept of FECs by fire investigators will give them a greater appreciation and understanding of what tools in combustion toxicology are available to calculate the impact of lethal toxic products as hazardous to humans (Purser 2008).

The FEC is expressed in general terms as

$$\text{FEC} = \frac{\text{dose received at time } t(C_t)}{\text{effective } C_t \text{ dose to cause incapacitation or death}}. \tag{7.1}$$

The FEC, in special situations, is also referred to as the *fractional incapacitation dose* (FID) or the *fractional lethal dose* (FLD).

fractional effective concentration ■ A measurement that assesses the impact of toxic smoke and by-products of combustion on a subject. The FEC depends on the concentration of particular toxins within the fire gases, and the duration of exposure.

CARBON MONOXIDE

Carbon monoxide (CO) is produced in fires by the incomplete combustion of any carbon-containing fuel; however, it is not produced at the same rate in all fires. In free-burning (well-ventilated) fires, the CO concentration can be as little as 0.02 percent (200 ppm) of the total gaseous products of the fire. The CO concentrations in smoldering, postflashover, or underventilated fires range from 1 to 10 percent in the smoke stream.

CO must be inhaled into a living human to be transferred to the bloodstream. Thus, there is no measurable diffusion from an external atmosphere rich in CO into the blood or tissues of a dead body, owing to the lack of respiration. Similarly, CO content is stable in the blood of a dead body and is not lost until postmortem decomposition.

When inhaled and absorbed into the bloodstream, CO combines with the hemoglobin molecules and forms the complex carboxyhemoglobin (COHb) in the red blood cells. The binding affinity of CO for the heme portion of the hemoglobin is 200–300 times stronger than that of O_2. CO also binds with the heme group in myoglobin, which is the "red" in red muscle tissue. The affinity of CO for myoglobin is about 60 times that of O_2. Myoglobin stores and transports O_2 in muscle tissue, particularly in cardiac muscle. Under hypoxic conditions, CO can shift from the blood into the muscle and has a higher affinity for cardiac muscle than for striated muscle (Myers, Linberg, and Cowley 1979). This may explain why low concentrations of COHb are sometimes found in the blood of deceased victims with heart conditions.

Carbon monoxide also affects Cya_3 oxidase, an enzyme that catalyzes production of ATP in the cell (Feld 2002). The stability of the COHb complex reduces the O_2-carrying capacity of the blood. Without O_2 and water, ATP cannot be produced in the mitochondria of the cell, and the cell dies (Feld 2002). CO also impairs cellular tissue respiration by combining with cytochromes b and aa3 (Purser 2010, 129).

Goldbaum, Orellano, and Dergal (1976) reported that experimentally reducing the hematocrit (the blood-carrying capacity) of dogs by as much as 75 percent did not result in death. Even replacing blood with blood containing 60 percent COHb by transfusion or infusion of CO through the peritoneal cavity did not result in death. Only when these experimental animals inhaled the CO, did deaths occur. This suggests that respiration of CO and its interference with metabolism plays a critical role in causing death (Goldbaum, Orellano, and Dergal 1976).

The mere presence of CO in the blood is not a sign of breathing fire gases. The normal human body has COHb saturations of 0.5–1 percent as a result of degradation of heme in the blood. Higher concentrations (up to 3 percent) may be found in nonfire victims with anemia or other blood disorders (Penney 2000, 2008, 2010). Smokers can have levels of 4–10 percent, since tobacco smoke contains a high concentration of

CO. People in confined spaces with emergency generators, pumps, fuel-burning heaters, and compressors can have elevated—sometimes dangerous—blood COHb concentrations.

When a victim is removed from a CO-rich environment to fresh air, the CO is gradually eliminated from the blood. The higher the partial pressure of O_2 (such as administered by medical personnel), the faster CO is eliminated. In fresh air the initial concentration of COHb will be reduced by 50 percent in 250–320 minutes (approx. 4–5 hr). In 100 percent O_2 via mask, a 50 percent reduction can be achieved in 65–85 minutes (approx. 1 to 1.5 hr). In O_2 at hyperbaric pressures (3–4 atm), there is a 50 percent reduction of COHb in 20 minutes (Penney 2000, 2008, 2010). Treatment for high concentrations of CO or nitrogen gases in the blood may include time in a hyperbaric chamber to reduce the effects of CO or nitrogen gas poisoning.

The time at which a blood sample is drawn from a fire victim must be noted, as well as the nature of any medical treatment (such as the antemortem administration of O_2). The COHb saturation of blood in a dead body is very stable, even after decomposition has begun. CO poisoning kills many fire victims before they are ever exposed to fire. Because CO gas is colorless and odorless; it may go undetected by the victim. It can kill victims even some distance away from a fire when they are not exposed to any heat or flames, but it is not the only factor in many fire deaths.

Carbon dioxide is a product of nearly all fires. Levels of 4–5 percent CO_2 in air causes the adult respiratory rate to double (Purser 2010, 127). Levels of 10 percent cause it to quadruple and probably cause unconsciousness (Purser 2010). This increase in respiration increases the rate at which CO and other toxic gases are inhaled. High concentrations of CO_2 may also dilute the concentration of breathable oxygen to the point of inducing hypoxic collapse. Carbon dioxide concentration in the blood can be measured in living subjects, but blood chemistry begins to change after death. As a result, accurate CO_2 (and O_2) saturation measurements are difficult to obtain postmortem, since postmortem decomposition can also contribute to the CO_2 measured.

PREDICTING THE TIME TO INCAPACITATION BY CARBON MONOXIDE

The estimation of dosage levels for predicting times to incapacitation is an important concept. Under **Haber's rule** the dosage of toxic gases assimilated by an individual is assumed to be equivalent to the concentration times the duration of exposure. For example, a 1-hour exposure to a toxic gas at one concentration would be equivalent to a 2-hour exposure to half that concentration.

Haber's rule ■ The dosage of toxic gases assimilated by an individual is assumed to be equivalent to the concentration times the duration of exposure.

The Coburn-Forster-Kane (CFK) Equation In certain cases, Haber's rule does not hold exactly true for exposure to CO. The relationship between concentration and uptake is linear only for high CO concentrations and is not valid at extremely high concentrations. For lower concentrations, the time to incapacitation is an exponential relationship and is described by the CFK equation. The CFK equation (7.2) also predicts that the half-time elimination of CO is a hyperbolic function of the ventilation rate (Peterson and Stewart 1975).

$$\frac{A[COHb]_t - BV_{CO} - PI_{CO}}{A[COHb]_0 - BV_{CO} - PI_{CO}} = e^{-tAV_bB}, \tag{7.2}$$

where

$[COHB]_t$ = concentration of CO in blood at time t (mL/mL),
$[COHb]_0$ = concentration of CO in blood at beginning of exposure (mL/mL),
PI_{CO} = partial pressure of CO in inhaled air (mm Hg),
V_{CO} = rate of CO production (mL/min),
A = derived constant,
B = derived constant, and
V_b = derived constant.

TABLE 7-2	Standard Inhalation Values (RMV; L/min)				
ACTIVITY	MAN	WOMAN	CHILD	INFANT	NEWBORN
Resting	7.5	6.0	4.8	1.5	0.5
Light activity	20.0	19.0	13.0	4.2	1.5

Sources: Derived from Health Canada 1995; SFPE 2008, 2-102; Bide, Armour, and Yee 1997.

An obvious disadvantage of using the CFK equation is the number of variables needed. The CFK equation is appropriately used for exposure to CO concentrations of less than 2000 ppm (0.2 percent), exposure durations greater than 1 hour, or estimation of time to death where COHb is 50 percent (Purser in SFPE 2008, 2-117). The CFK equation is of limited value in reconstructing most fire death cases where the CO concentration is greater than 0.2 percent and the COHb saturations are often much greater than 50 percent.

The Stewart Equation When predictions of time to incapacitation deal with atmospheric CO concentrations higher than 2000 ppm (0.2 percent), and COHb in the blood is determined to be less than 50 percent saturation, a simpler equation known as the *Stewart equation* (7.3) applies:

$$\%COHb = \left(3.317 \times 10^{-5}\right)\left(ppm\ CO\right)^{1.036}(RMV)(t), \qquad (7.3)$$

where

CO $\quad$ = CO concentration (ppm),
RMV = respirations per minute of volume of air breathed (L/min), and
t $\qquad$ = exposure time (min).

Solving the Stewart equation for exposure time gives

$$t = \frac{\left(3.015 \times 10^4\right)(\%COHb)}{\left(ppm\ CO\right)^{1.036}(RMV)}. \qquad (7.4)$$

The standard inhalation values (RMV) are listed in Table 7-2, providing typical data for a man, woman, child, infant, and newborn. For additional information on RMVs, see Health Canada 1995; SFPE 2008, 2-102; and Bide, Armour, and Yee 1997.

According to Stewart, a few breaths of CO in concentrations of 1 to 10 percent (10,000 to 100,000 ppm) rapidly elevate the COHb saturation in the blood. For example, a 120-second exposure at 1 percent (10,000 ppm) CO results in a 30 percent COHb saturation, and a 30-second exposure at 10 percent (100,000 ppm) CO results in a 75 percent COHb saturation (Spitz and Spitz 2006).

The standard approach for evaluating incapacitation from CO is to calculate the fraction of the CO inhaled per minute in 1 hour. During moderate activity, human RMV value is approximately 25 L/min, and loss of consciousness occurs at 30 percent COHb in the blood (Purser in SFPE 2008, 2-117). The formula for the fractional incapacitating dose (FID) valid for up to 1 hour is

$$F_{ICO} = \frac{3.317 \times 10^{-5}[CO]1.036(V)(t)}{D}, \qquad (7.5)$$

where

$$F_{ICO} = \text{fractional incapacitating dose (FID)} = \frac{\begin{array}{c}\text{concentration of irritant to which}\\ \text{subject is exposed at time } t\end{array}}{\begin{array}{c}\text{concentration of irritant required to}\\ \text{cause impairment of escape efficiency}\end{array}}, \qquad (7.6)$$

[CO] = carbon monoxide concentration (ppm v/v 20°C)

t = exposure time (min), and

D = exposure dose (percent COHb) for incapacitation.

Resting or sleeping	$V = 8.3$ L/min and $D = 40\%$ COHb
Light work: walking to escape	$V = 25$ L/min and $D = 30\%$ COHb
Heavy work: slow running, walking up stairs	$V = 50$ L/min and $D = 20\%$ COHb

EXAMPLE 7-1 ■ CO Incapacitation

Problem. Rescuers found an adult female unconscious in her bed at the scene of a house fire. Assume that rough estimates suggest that she was exposed to a CO concentration of approximately 5000 ppm. Calculate the time to incapacitation and fractional incapacitating dose, assuming that the victim was at rest.

Suggested Solution. Use Table 7-3 for RMV data.

Volume of air breathed	RMV	$= 6.0$ L/min (resting)
Loss of consciousness	COHb	$= 40\%$ (resting)
CO concentration	CO	$= 5000$ ppm

Time to incapacitation (eq. 7.4) $t = \dfrac{(3.015 \times 10^4)(40)}{(5000)^{1.036}(6.0)} = 30\,\text{min}$

Fractional incapacitating dose (eq. 7.6) $F_{I_{CO}} = \dfrac{(8.2925 \times 10^{-4})(5000^{1.036})}{30} = 0.188$

EXAMPLE 7-2 ■ CO Incapacitation and Death of Firefighters

Problem. Computer fire modeling was used to reevaluate a U.S. Fire Administration investigation (Routley 1995) into a reported Pittsburgh fire that killed three firefighters (Christensen and Icove 2004). NIST's Fire Dynamics Simulator (FDS) was employed to model the fire and estimate the concentration of carbon monoxide present in the dwelling, which was the immediate cause of death of two of the firefighters, who were unable to escape the interior of a burning dwelling.

The fire occurred in a four-story townhouse when an arson fire was ignited using gasoline in a room on the ground floor. Firefighters entered on the street level and were attempting to locate the seat of the smoky fire. Details of the minutes prior to the deaths of the firefighters were unclear, but it appeared that at some point they realized they were running short of air in their self-contained breathing apparatus (SCBA), needed to leave, were unable to find an exit, and exhausted their air supplies. Two of the firefighters were believed to have removed or loosened their face pieces and made attempts to share the air that was available by "buddy breathing," involving the alternating the use of a single breathing apparatus. It was concluded that both had been rendered unconscious owing to toxic gas inhalation. They were found to have COHb saturations of 44 percent and 49 percent, respectively, at autopsy. A third firefighter was found with his face piece in place, and his COHb was 10 percent, indicating death from oxygen deficiency.

Solution. This estimate, along with an assumed respiration volume and known blood COHb levels, was used with the Stewart equation to estimate the time of exposure. The FDS model, as shown in Figure 7-2, indicated that 27 minutes into the fire, the CO concentration in the atmosphere at the location where the firefighters were found had already reached approximately 3600 ppm. At this concentration and a respiration rate of 70 L/min, an estimated 3 to 8 minutes of exposure would have been required to accumulate the average 47 percent COHb measured in two of the firefighters' blood at autopsy.

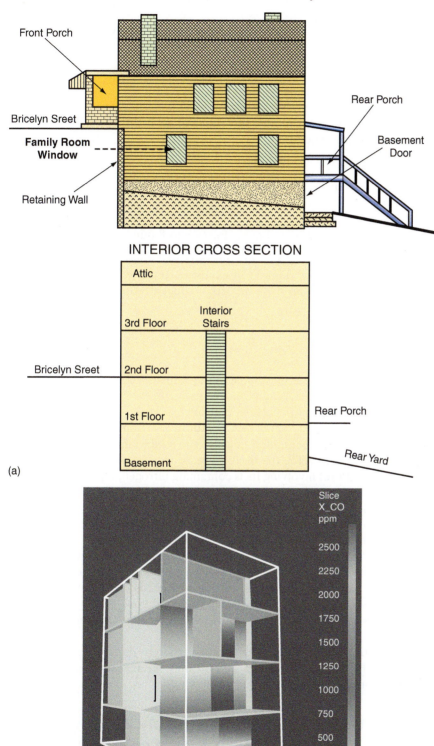

EAST SIDE OF DWELLING - 8361 Bricelyn Street

Front Porch

Bricelyn Sreet

Family Room Window

Retaining Wall

Rear Porch

Basement Door

INTERIOR CROSS SECTION

| Attic |
| 3rd Floor / Interior Stairs |
| 2nd Floor |
| 1st Floor |
| Basement |

Bricelyn Sreet

Rear Porch

Rear Yard

(a)

Slice X_CO ppm

2500
2250
2000
1750
1500
1250
1000
750
500
250
0.00

(b)

FIGURE 7-2 (a) Plan of town house where three firefighters were trapped. (b) FDS model of the distribution of carbon monoxide in town house at 27 minutes. *Courtesy of D. J. Icove.*

Solution of this problem with the Stewart equation using the CO value from the FDS model calculation gives

$$\% \text{COHb} = \left(3.317 \times 10^{-5}\right)(\text{ppm CO})^{1.036}(\text{RMV})(t),$$

where

CO   = CO concentration = 3600 ppm
%COHb = carboxyhemoglobin saturation = 47%
RMV   = respirations per minute of volume of air breathed = 70 L/min
t = exposure time (min).

Solving for time of exposure, we obtain

$$47\% = \left(3.317 \times 10^{-5}\right)(3600)^{1.036}(70)(t)$$

$$t = 4.2 \, \text{min}$$

Estimated range for $t=3$ to 8 min

A total exposure time of 4.2 minutes suggests that following removal of the firefighters' face pieces, only a few minutes of exposure to CO with no air from the SCBA would have been required to produce the lethal concentrations of COHb observed at autopsy.

HYDROGEN CYANIDE

Hydrogen cyanide (HCN) is readily soluble in the aqueous phase of blood plasma, cells, and organs, where it forms the CN radical. The main effect of HCN or CN in the cell is to inhibit the use of oxygen by the cell so that oxygen in the blood cannot be properly used by the cells (Purser 2010, 132). The CN radical also combines with Cya_3 oxidase, inhibiting its action in the mitochondria of cells. The inhibition of Cya_3 oxidase prevents the formation of water and ATP, the basic route of respiration in the cell (Feld 2002). HCN is produced in all fires involving fuels containing nitrogen, especially acrylic rubber, ABS plastics, and polyurethane.

As with CO, the production of HCN varies with the temperature and oxygen supply in the combustion zone. Unlike the effects of CO, those of HCN are immediate but complex and are often dependent on concentration and inhalation rate. Unlike with COHb, the concentration of HCN in the blood is unstable in the body, decreasing by 50 percent in 24 hours after death and also in stored blood samples (Purser 2010, 160). CN is suspected of contributing to many fire deaths but is often not measured immediately. Because it is produced under the same fire conditions as CO, it is likely to act quickly at low doses to incapacitate the victims, so that they are exposed to CO and other fire gases longer, causing their deaths.

Table 7-3 lists the tenability limits for incapacitation or death from exposure to CO, HCN, low O_2, and CO_2. The periods 5 and 30 minutes are common benchmarks for narcotic products of combustion.

PREDICTING THE TIME TO INCAPACITATION BY HYDROGEN CYANIDE

As discussed earlier, hydrogen cyanide (HCN) is another toxic gas in fires that incapacitates through biochemical asphyxiation. As with CO, the time to incapacitation depends on the uptake rate and dosage (Purser in SFPE 2008). Typical effects of HCN exposure are listed in Table 7-4.

Exposure to less than 80 ppm is expected to have minimal effect on a healthy adult (Purser 2010, 165). The effect of concentrations greater than 80 ppm can be calculated

TABLE 7-3	Tenability Limits for Incapacitation or Death from Exposures to Common Toxic Products of Combustion			
	5 MIN		**30 MIN**	
	INCAPACITATION	DEATH	INCAPACITATION	DEATH
CO (ppm)	6000–8000	12,000–16,000	1400–1700	2500–4000
HCN (ppm)	150–200	250–400	90–120	170–230
Low O_2 (%)	10–13	<5	<12	6–7
CO_2 (%)	7–8	>10	6–7	>9

Source: SFPE 2008, Table 2.6.B1, 2-185, Courtesy of the Society of Fire Protection Engineers, © 2008, reprinted with permission.

TABLE 7-4	Typical Effects of HCN Exposure

REPORTED EFFECTS ON HEALTHY ADULTS

Extremely toxic
  170—230 ppm = death in 30 minutes
  250—400 ppm = death in 5 minutes (SFPE)

Produced by any fuel that contains nitrogen (hair, wool, fur, leather, polyurethane, nylon)

Absorbed through inhalation and ingestion

Source: Lecture by J. D. DeHaan. "What Kills People in Fires?"

(Purser in SFPE 2008, 2-119). The formula for time to incapacitation for inhalation of 80–180 ppm HCN concentrations is

$$t_{ICN} (\text{min}) = \frac{185 - \text{ppm HCN}}{4.4}; \tag{7.7}$$

the formula for time to incapacitation for HCN exposure concentrations above 180 ppm is

$$t_{ICN} (\text{min}) = \exp\left[5.396 - (0.023)(\text{ppm HCN})\right]; \tag{7.8}$$

and the fractional incapacitating dose per minute (FID/min) is

$$F_{ICN} = \frac{1}{\exp\left[5.396 - (0.023)(\text{ppm HCN})\right]}. \tag{7.9}$$

Note that "exp" in equations (7.8) and (7.9) denotes the exponential form. As shown in Table 7-3, the HCN dosages required for incapacitation are much lower than the CO dosages. The lowest blood concentration associated with adult death by acute cyanide poisoning is 1–2 mg/ml. More typical levels are 2.4–2.5 at death (Purser 2010, 183). A case study using these equations is illustrated in Example 7-4.

EXAMPLE 7-3 ■ HCN Incapacitation

Problem. An adult male was found unconscious in the waiting room after being exposed to toxic by-products from the fire in Example 7-1. Fire modeling estimated that a burning polyurethane plastic mattress cushion had produced an HCN concentration of approximately 200 ppm in the air being breathed.

Solution. Calculate the time to incapacitation and the fractional incapacitating dose per minute (FID/min).

HCN concentration

$$HCN = 200 \text{ ppm}$$

Time to incapacitation

$$t = \exp\left[5.396 - (0.023)(200)\right] = 2.2 \text{ min}$$

Fractional incapacitating dose per minute

$$F_{I_{CN}} = \left(\frac{1}{\exp\left[5.396 - (0.023)(200)\right]}\right) = 0.45 \text{ FID/min}$$

INCAPACITATION BY LOW OXYGEN LEVELS

anoxia ■ Condition relating to an absence of oxygen.

hypoxia ■ Condition relating to low concentrations of oxygen.

Anoxia (absence of oxygen) or **hypoxia** (low concentration of oxygen) is the condition of inadequate oxygen to support life. This situation can occur when oxygen is displaced by another inert gas, such as nitrogen or carbon dioxide, by a fuel gas such as methane, or even by benign products of combustion such as CO_2 and water vapor. Normal air contains 20.9 percent O_2. At concentrations down to 15 percent O_2, there are no readily observable effects in humans owing to decreased oxygen concentrations. At concentrations between 10 and 15 percent, disorientation (similar to intoxication) occurs and judgment is affected. At levels below 10 percent, unconsciousness and death may occur. Hypoxia is aggravated by high levels of CO_2, which accelerate breathing rates. The effects of hypoxia include cerebral depression, causing lethargy, problems with memory and mental concentration, loss of consciousness, and death (Purser 2010, 184) (see Table 7-5).

The time to loss of consciousness for an adult exposed to a hypoxia equivalent is

$$\left(t_{lo}\right) = \exp\left[8.13 - 0.54(20.9 - \%O_2)\right], \tag{7.10}$$

where

t_{lo} = exposure time in minutes
$\%O_2$ = concentration at $20°C$ in the air being inhaled

PREDICTING THE TIME TO INCAPACITATION BY CARBON DIOXIDE

Exposure to carbon dioxide can produce a wide range of effects, ranging from respiratory distress to loss of consciousness (SFPE 2008, 2-119) (see Table 7-5).

Along with being an asphyxiant capable of displacing oxygen, carbon dioxide also increases the RMV, which in turn causes the individual to increase the uptake of the other toxic gases (Purser in SFPE 2008). This formula for the multiplication factor VCO_2 is

$$VCO_2 = \exp\left(\frac{CO_2}{5}\right), \tag{7.11}$$

TABLE 7-5	The Effects of Exposure to Low Oxygen Levels
PERCENT OXYGEN	**REPORTED EFFECTS ON HEALTHY ADULTS**
14.14–20.9	No significant effects, slight loss of exercise tolerance
11.18–14.14	Slight effects on memory and mental task performance, reduced exercise tolerance
9.6–11.8	Severe incapacitation, lethargy, euphoria, loss of consciousness
7.8–9.6	Loss of consciousness, death

Source: Derived from SFPE 2008.

| TABLE 7-6 | The Effects of Exposure to Carbon Dioxide |

PERCENT CARBON DIOXIDE	REPORTED EFFECT(S)
7–10	Loss of consciousness
6–7	Severe respiratory distress, dizziness, possible loss of consciousness
3–6	Respiratory distress increasing with concentration

Source: Derived from SFPE 2008.

$$VCO_2 = \left(\frac{\exp\left[(1.903)(\%CO_2) + 2.0004\right]}{7.1} \right). \tag{7.12}$$

The formula for the time to unconsciousness (incapacitation) from carbon dioxide is

$$t_{I_{CO2}} = \exp\left[6.1623 - (0.5189)(\%CO_2)\right], \tag{7.13}$$

and the fractional incapacitating dose per minute (FID/min) is

$$F_{I_{CO2}} = \left(\frac{1}{\exp\left[6.1623 - (0.5189)(\%CO_2)\right]} \right). \tag{7.14}$$

Heat

The human body is capable of surviving exposure to external heat as long as it can moderate its core temperature by radiant cooling of the blood through the skin and, more importantly, by evaporative cooling. This process occurs internally via evaporation of water from the mucosal linings of the mouth, nose, throat, and lungs and externally via evaporation of sweat from the skin. If the core body temperature exceeds 43°C (109°F), death is likely to occur.

Prolonged exposure to high external temperatures, 80–120°C (175–250°F), with low humidity can trigger fatal hyperthermia. Exposure to elevated temperatures accompanied by high humidity (which reduces the cooling evaporation rate of the water from the skin or mucosa) can also be lethal. Fire victims can die of exposure to heat alone even if they are protected from CO, smoke, and flames. These victims may have minimal postmortem changes, although skin blistering and sloughing can occur after death owing to heat denaturation of collagen and other proteins in the connective tissue and skin.

PREDICTING THE TIME TO INCAPACITATION BY HEAT

For exposure to convected heat in a fire environment, the fractional incapacitating dose per minute (FID/min) is calculated as follows:

$$F_{I_h} = \left(\frac{1}{\exp\left[5.1849 - (0.0273)\left(T\left[°C\right]\right)\right]} \right). \tag{7.15}$$

Assimilation of various data formed the basis for the *Toxic and Physical Hazard Assessment Model* (Purser in SFPE 2008, 2-179). Data such as toxic chemical and physical species concentration levels generated by fire models can serve as input to this hazard assessment model. According to this model, the normally accepted threshold for tolerance of radiant heat is 2.5 kW/m² for only a few minutes. Besides the burns to the skin (from both radiant and convected heat), thermal damage to the upper respiratory tract can also occur when dry gases at temperatures over 120°C (250°F) are inhaled.

Information needed to evaluate this hazard model is derived from two sets of information: concentration and time profiles of major toxic products, including time, concentration, and toxicity relationships. The estimated toxic products within the victim's breathing zone include concentrations of gases such as carbon monoxide, hydrogen cyanide, carbon dioxide, and factors such as radiant heat flux, and air temperature. Some of these values can be calculated by sophisticated computer models discussed previously.

INHALATION OF HOT GASES

Inhalation of very hot gases causes edema (swelling and inflammation) of mucosal tissues. This edema can be severe enough to cause blockage of the trachea and physical asphyxia. Inhalation of hot gases may also trigger *laryngospasm,* in which the larynx involuntarily closes up to prevent entry of foreign material, and *vagal inhibition,* in which the breathing stops and the heart rate drops.

Rapid cooling of the inhaled hot gases occurs on inhalation as the water evaporates from mucosal tissues, so thermal damage usually does not extend below the larynx if the inhaled gases are dry. If the hot gases include steam or are otherwise water saturated, evaporative cooling is minimized, and burns/edema can extend to the major bronchi and alveoli, which are tiny air sacs in the lungs. If inhaled gases are hot enough to damage the trachea and internal lung infrastructures, they will usually be hot enough to burn the facial skin and mouth as well as singe the facial or nasal hair.

EFFECTS OF HEAT AND FLAME

The human body is a complex target when being affected by heat from a fire. Skin consists of two basic layers. A thin upper outer layer of *epidermis* (dead, keratinized skin cells) overlies a thicker inner dermal layer of actively growing cells in which are embedded the nerve endings, hair follicles, and blood capillaries that supply nutrients to the growing skin. Beneath the dermal layer is a layer of tough elastic connective tissue, subcutaneous fat, and, finally, muscle and bone.

Each of these components is affected differently by heat and flames. Application of heat can cause the epidermis to separate from the underlying *dermis* and form blisters in much the way paint or wallpaper blisters away from the wood or plaster beneath when heated. Blistering of skin occurs when the tissue reaches temperatures in excess of 54°C (130°F). The raised epidermal layer is very thin and is more easily affected by continuing heat, which can cause the epidermal layer to char and turn black if the temperature is high enough. The epidermis can also separate in larger areas and form generalized skin slippage. Exposure of the denuded dermal layer to heat can cause extreme pain when its temperature exceeds 43°C–44°C (110°F–112°F) (Purser in SFPE 2008, 2-179).

More prolonged exposure can destroy the proteins of the dermal layer and cause further desiccation and discoloration. Higher heat fluxes can cause higher temperatures that cook and even char the tissues. As the skin desiccates, it shrinks, eliminating wrinkles and changing facial contours (making visual identifications of victims very risky). If the skin continues to shrink, it can split open, leaving jagged, irregular, torn surfaces (as opposed to the sharply defined surfaces of knife cuts), as shown in Figure 7-3 (Smith and Pope 2003).

FIGURE 7-3 Effects of heat and flame shrink skin, eliminating wrinkles and changing facial contours. If the skin continues to shrink, it can split, leaving jagged, irregular torn surfaces, as opposed to the sharply defined surfaces of knife cuts. This photo illustrates postfire recognition of knife cuts to the chest versus splitting of the skin on the arm. *Courtesy of Dr. Elayne J. Pope.*

Heat-split skin often exhibits subcutaneous bridging of underlying tissue, whereas cut skin does not.

If a victim survives for some time, shrinkage can constrict blood vessels, so incisions (called *escharotomies*) are made through the damaged dermal layer to relieve the pressure and maintain circulation. If a fire victim survived for some time after the fire and underwent medical interventions such as an escharotomy or skin graft harvesting, the investigator must recognize the effects of these and distinguish them from fire effects.

Owing to its small individual dimensions and low thermal mass, hair is affected very quickly by heat. Colors will change (typically changing to darker or redder colors or completely to a gray, ashen color). The hair shaft will bubble, shrink, and fracture as it singes from heat. This shrinkage causes the curling of hair shafts seen as singeing. The microscopic appearance of singed hair shafts is very distinctive (as compared with cut or broken hair shafts). If the hair is burned in large masses, it can form a black puffy entangled mass.

FIGURE 7-4 Shrinkage of muscle and tendons can cause the joints to flex, causing what is called *pugilistic posturing*. *Courtesy of Dr. Elayne J. Pope.*

If the body continues to be exposed to heat, the shrinkage can affect muscles. When the skin and muscles of the neck shrink, they can force the tongue out of the mouth. Shrinkage of muscle and tendons can cause the joints to flex, causing what is called *pugilistic posturing,* as shown in Figure 7-4. This flexing and posturing can cause bodies to move during fire exposure. If the body is on an irregular or unstable surface, this movement can cause the body to fall from a bed or chair and, possibly, change the direction of heat application, eliminating or obscuring previously protected areas. Pugilistic posturing has been observed during legal cremations after 10 minutes of exposure to flames at temperatures of 670°C–810°C (1240°F–1490°F) (Bohnert, Rost, and Pollak 1998; Pope 2007).

Direct flame impingement, with its high-temperature gases of 500°C–900°C (930°F–1650°F) and high heat fluxes (55 kW/m²), produces responses from the human body very quickly. Blisters will form in about 5 seconds, with charring following seconds later. Skin can be charred completely away in 5–10 minutes of direct flame contact, especially where stretched over the bone (joints, nose, forehead, skull) (Pope and Smith 2003, 2004; Pope 2007). Very short but intense (flash) fire exposure can cause blistering of the epidermal layer without the sensation of pain (because the pain sensors are in the dermal layer beneath), and the heat takes longer to penetrate deeply.

Even in the absence of fire, prolonged exposure of the body to high temperatures (over 50°C; 122°F) causes desiccation and shrinkage of muscle tissue, which causes pugilistic posturing. Exposure to flames can cause muscles to combust and major limb bones to fracture as they degrade where the bone is exposed to flames. Extreme heat causes bones to twist and fracture, and continued flame exposure (30 minutes or longer in observed cremations) causes calcination (in which the charred organic matter is burned away). As the bones become calcined, they become very fragile and may disintegrate of their own accord. The lower-density major bones from elderly victims of osteoporosis have been seen to disintegrate under fire exposure more quickly and completely than bones with normal densities (Christensen 2002). Tests have revealed that fire exposure can cause fluid that cushions the brain as well as the brain cell fluids to leak through fissures in the exposed skull but not cause the skull to explode. Internal organs that are exposed to the fire desiccate and char, thus requiring at least 30–40 minutes of cremation to burn away (Bohnert, Rost, and Pollak 1998). The adult torso will require longer exposure to a normal structure fire before the organs are affected. Bones exposed to fire can also shrink in length, thus giving rise to potential errors in the estimation of antemortem height.

The thin bones of the skull may delaminate, with the inner and outer layers failing separately. This delamination has given rise to thermally caused holes in the skull that have beveled edges similar to those produced by gunshots (Pope and Smith 2003). Recent experiments have demonstrated that failure of the skull can occur from fire damage whether or not there was preexisting mechanical or blunt force trauma. Some thermally induced failures can mimic bullet wounds, requiring extreme care in casework investigative interpretations (Pope and Smith 2004).

Heat exposure to the head can cause blood and fluids to accumulate and form a hematoma in the *extradural* or *epidural* space between the skull and the tough bag of tissue (the *dura mater*) that surrounds the brain. With further heating, these fluids boil, desiccate, and then char, producing a rigid, foamy, blackened mass. Physical trauma to the brain as a result of the impact from falling structures overhead also can cause *subdural* hematomas. However, fractures at the base of the skull have not been observed to be produced by fire exposure (Bohnert, Rost, and Pollak 1998; Pope 2007). Fire exposure chars and seals exposed tissues, so as a rule, bodies exposed to an enveloping fire do not bleed. When the body is moved, however, the fragile layer of char covering the tissues can be broken, allowing body fluids to seep out. Thus great care must be taken when moving a charred body with exposed charred tissues, and its condition prior to being moved must be carefully documented.

Within certain considerations, the amount of fire damage on a body can be related to the duration of exposure if the intensity of the fire can be estimated from the fuel, ventilation, and distribution factors described elsewhere in this text. Factors such as combustion rates, thermal inertia, and relative combustibility of body components have been explored in recent studies (DeHaan, Campbell, and Nurbakhsh 1999). Some aspects of postmortem fire damage to the human body are discussed later in this chapter.

FLAMES (INCINERATION)

When heat is applied to a surface, the rate at which it penetrates that surface is determined by the thermal inertia of the material (the numerical product of thermal capacity, density, and thermal conductivity). The thermal inertia of skin is not much different from that of a block of wood or polyethylene plastic (see Table 2-4). The pain sensors for human skin are in the dermis, about 2 mm (0.1 in.) below the surface. If heat is applied very briefly, there may not be any sensation of discomfort or pain. The longer the heat is applied, the deeper it will penetrate into the skin. The higher the intensity of the heat applied, the faster it will penetrate into the skin. Pain is triggered when skin cells reach a temperature of about 48°C (120°F), and cells are damaged if their temperature exceeds 54°C (130°F) (Stoll and Greene 1959).

Exposing skin to 2–4 kW/m^2 radiant heat for 30 seconds causes pain but no permanent cellular damage. Higher heat fluxes trigger damage such as blisters and skin slippage. Exposing skin to 4–6 kW/m^2 radiant heat for 8 seconds produces blisters (second-degree burns). Exposing skin to 10 kW/m^2 radiant heat for 5 seconds causes deeper, partial-thickness injuries. Exposing skin to 50–60 kW/m^2 radiant heat for 5 seconds produces third-degree burns with destruction of the dermis (Stoll and Greene 1959).

BURNS

Burns may appear in the absence of fire or flames as a result of prolonged exposure of the body or body parts to heat that raises their overall temperature above 54°C (130°F) and causes desiccation, sloughing, and blistering. These effects can also be caused by exposure of the body to caustic chemicals. It is often very difficult, if not impossible, to distinguish between these types of burns suffered near the time of death (*perimortem*) and those inflicted after death (postmortem). Fluid-filled blisters can result from fire exposure both antemortem and postmortem as well as from postmortem decomposition.

Most, but not all, medical personnel agree that *first-degree burns* involve only reddening of the skin. *Second-degree burns*, sometimes referred to as *partial-thickness burns*, involve damage to the epidermis with blistering and sloughing. Because the germinative layer of the dermis remains, the entire surface of such burns will usually heal, most often without grafts. *Third-degree burns* are called *full-thickness burns* because the dermis is damaged, and the wound heals from the edges only and requires grafting. *Fourth-degree burns,* sometimes also referred to as full-thickness burns, can include those in which the skin is destroyed, exposing muscle and the bones beneath the muscles.

When gasoline or a similar volatile liquid fuel with low viscosity and low surface tension is poured on bare skin, some is absorbed into the epidermis, but the bulk of the fuel runs off, leaving a very thin film of liquid, particularly on vertical surfaces. This thin film burns off very quickly (less than 10 seconds). The skin beneath may be spared completely or reddened (first-degree burns), except where folds of the skin or clothing retain enough fuel to sustain longer burning, producing severe blistering, and even charring of the epidermis in extreme cases. Deep penetrating burns exposing the subcutaneous fat or muscle below require flame exposures of several minutes, much longer than the typical thin-film gasoline fire. On horizontal skin surfaces, a deeper pool may be retained long enough to produce a halo or ring of blisters (second-degree burns) around the circumference of the pool (DeHaan and Icove 2012).

BLUNT FORCE TRAUMA

Blunt force trauma can also cause or contribute to the death of fire victims. Structural collapse or explosions can cause solid materials to strike victims. Falls or impacts with stationary surfaces (furniture or door frames) during escape attempts can induce blunt trauma that only a careful medical examination can distinguish from an assault. Wound patterns, bloodstains, or even trace evidence can be used to interpret blunt trauma injuries and establish whether they resulted from assault or from some fire-related event. The fire investigator should consider consulting with the pathologist and the criminalist to help evaluate blunt trauma injuries and possibly link them to features at the fire scene.

Visibility

The optical opacity of dense smoke and its irritants impairs the vision and respiration of normal-sighted people who may find their ability to travel impaired by the smoke. This optically dense smoke affects

- Exit choice and escape decisions,
- Speed of movement, and
- Wayfinding ability.

During structure fires, the occupants often depend on their ability to seek out exit signs, doors, and windows (Jin 1976, Jin and Yamada1985). *Visibility* of an object depends on several factors such as the smoke's ability to scatter or absorb the ambient light, the wavelength of the light, whether the item viewed (e.g., an exit sign) is light emitting or light reflecting, and the individual's visual acuity (Mulholland in SFPE 2008, 2–297).

An insight into the underlying principles on the first use of red exit signs can be found in historical studies by the U.S. Naval Medical Research Laboratory, Bureau of Medicine and Surgery. The studies examined the detectability of colored field targets at sea. Colors that were distinguishable at the greatest distances were the red fluorescents (yellow-red and orange-red) (USN 1955). Later research by the U.S. Air Force (Miller and Tredici 1992) on night vision expanded these findings when it was determined that the sensitivity of the eye changes from the red end of the visible spectrum toward the blue end when shifting the *photopic* vision (high illumination levels) to *scotopic* vision (reduced illumination levels, typically at night).

Note that the naval study on the ability of the eye to perceive the color red at greater distances gives credence to the logic of using the color red for fire exit signs, whether they be passive fluorescent or illuminated. OSHA regulations state that exit signs, when required, shall be lettered in legible red letters, not less than 6 inches high, on a white field [OSHA 29 CFR 126-200(d)].

OPTICAL DENSITY

Visibility can be estimated in terms of the *optical density* per meter. The calculation involves a collateral term known as the extinction coefficient, K, which is the product of an *extinction coefficient* per unit mass, K_m, and the mass concentration of the smoke aerosol, m:

$$K = K_m m; \tag{7.16}$$

$$D = \frac{K}{2.3}, \tag{7.17}$$

where

K = extinction coefficient (m^{-1})
K_m = specific extinction coefficient (m^2/g),
m = mass concentration of smoke (g/m^3), and
D = optical density per meter (m^{-1})

The values for K_m are typically 7.6 m^2/g for smoke produced during flaming combustion of wood or plastics, and 4.4 m^2/g for smoke produced during pyrolysis (SFPE 2008).

In terms of the extinction coefficient K, one problem at a fire is determining the visibility, S, of light-emitting and light-reflecting exit signs to occupants. The value S is a measure of how well an individual can see through the smoke. Light-emitting signs are two to four times more visible than light-reflecting signs (Mulholland in SFPE 2008, 2-297), as the values for KS reveal in equations (7.18) and (7.19),

$$KS = 8 \text{ for light-emitting signs}, \tag{7.18}$$

$$KS = 3 \text{ for light-reflecting signs}, \tag{7.19}$$

where

K = extinction coefficient (m^{-1}) and
S = visibility (m).

Methods for estimating visibility based on the mass optical density are considered realistic. In studies, the optical density at which people turned back from a smoke-filled area was a visibility distance of 3 m (9.84 ft) (Bryan 1983). The study also showed a tendency for women to be more likely to turn back than men. Other factors include the ability of the persons to see exit signs directing them to safe egress from the building. Height of exit signs as well as the height of the viewer may be critical to visibility.

Estimates of visibility are based on mass optical densities, D_m, derived from test data (Babrauskas 1981). Typical values of mass optical density produced by mattresses in flaming combustion are listed in Table 7-7.

The following equation is used to estimate the density, D, of the visible smoke:

$$D = \frac{D_m \Delta M}{V_C}, \tag{7.20}$$

TABLE 7-7	Mass Optical Density (D_m) for Flaming Mattresses	
TYPE OF MATERIAL	**MASS OPTICAL DENSITY (m²/g)**	
Polyurethane	0.22	
Cotton	0.12	
Latex	0.44	
Neoprene	0.20	

Source: Derived from Babrauskas 1981.

where

D = optical density per meter (m^{-1}),
D_m = mass optical density (m^2/g),
ΔM = total mass loss of sample (g), and
V_c = total volume of compartment or chamber (m^3).

EXAMPLE 7-4 ■ Visibility

Problem. A small, 300-g (0.66-lb), polyurethane mattress cushion on a waiting-room bench is set afire by a juvenile arsonist and is undergoing flaming combustion. The bench is located in a 6-m-square (20-ft-square) waiting room with a ceiling height of 2.5 m (8.2 ft). Determine the closest visibility of both light-emitting and light-reflecting signs leading to the exit door.

Solution. Assume that the smoke filling the waiting room is uniformly mixed.

Total mass loss of mattress	ΔM	= 300 g
Mass optical density	D_m	= (from Table 7-6) = 0.22 m²/g
Volume of compartment	V_c	= (6 m)(6 m)(2.5 m) = 90.0 m³
Optical density	D	= $(0.22^2/g)(300\,g)/(90.0\,m^3) = 0.733\,m^{-1}$
Extinction coefficient	K	= $2.3\,D = (2.3)(0.733m^{-1}) = 1.687m^{-1}$
Visibility (light emitting)	S	= $8/K = (8)/(1.687m^{-1}) = 4.74\,m$
Visibility (light reflecting)	S	= $3/K = (3)/(1.687m^{-1}) = 1.77\,m$

The calculations indicate that a light-emitting sign can be seen in this fire at a distance of up to 4.74 m (15.5 ft), compared with 1.7 m (5.58 ft) for an unlit sign. In real rooms the buoyancy of the hot smoke layer makes it harder to see signs in the upper half of the room than this example indicates. Exit signs are being placed at knee level or lower today to make them more visible longer.

FRACTIONAL EQUIVALENT CONCENTRATIONS OF SMOKE

The following formulas relate to enclosed spaces. As the value of FEC_{smoke} approaches 1, the level of visual obscuration increases, and the chance of escape decreases significantly.

$$FEC_{smoke} = \frac{D}{0.2} \text{ for small enclosures;} \qquad (7.21)$$

$$FEC_{smoke} = \frac{D}{0.1} \text{ for large enclosures,} \qquad (7.22)$$

where

D = optical density per meter of the smoke being encountered.

WALKING SPEED

As previously summarized, the optical density of the smoke affects a person's decision to choose the closest exit and ability to make proper escape decisions, as well as wayfinding ability and speed of movement. Normal adult walking speed is about 2 m/s in clear areas and with normal visibility. During smoky conditions furniture and other occupants impeding the escape can reduce this rate. The fire investigator should take this factor into consideration when assessing witness statements as to egress times from structures during fires.

Experiments with human subjects navigating through nonirritant smoke–filled corridors indicated that speed of movement decreases with increased smoke density (Jin 1976; Jin and Yamada1985).

Based on Jin's (1976) research, equation (7.23) provides an expression of this relationship:

$$\text{FWS} = (-1.738)(D) + 1.236 \text{ for the range } 0.13\,\text{m}^{-1} \le D \le 0.30\,\text{m}^{-1}, \qquad (7.23)$$

where

FWS = fractional walking speed (m/s), and
D = optical density per meter (m^{-1}).

For this equation, the smoke optical density ranges between 0.13/m (below normal walking speed) and 0.56/m (above walking speed of 0.3 m/s in darkness). The limits on the equation do not allow for delays such as erratic walking and sensory irritation. Jin used wood smoke in his experiments.

WAYFINDING

Jin's expression does not correlate midcourse corrections in wayfinding, reduced visibility, and irritability of the smoke. Studies have shown that the average density in smoke at which persons turn back is a visibility distance of 3 m (9.84 ft) ($D = 0.33\,\text{m}^{-1}$ and $K = 0.76$). Poor visibility and irritation of the eyes are the leading factors in reduced wayfinding, followed by irritation of the respiratory system (Jensen 1998).

SMOKE

Smoke contains water vapor, CO, CO_2, inorganic ash, toxic gases, and chemicals in aerosol form as well as soot. Soot is agglomerations of carbon from incomplete combustion large enough to produce visible particles. These particles may be very hot and are not cooled readily as they are inhaled, so they may induce edema and burns where they lodge in the mucosal tissue of the respiratory system. Soot particles are active adsorbents, so they may carry toxic chemicals and permit their ingestion or inhalation (with direct absorption by the mucosal tissues). Soot can be inhaled in quantities sufficient to physically block airways and cause mechanical asphyxiation. Soot, water vapor, ash, and aerosols in smoke can also obscure the vision of victims and prevent their escape.

Time Intervals

INTERVAL BETWEEN FIRE AND DEATH

One of the problems outlined earlier is the time interval between exposure to a fire and its fatal aftermath. Death can occur nearly instantaneously or within minutes or hours later. Under these conditions it is not difficult to connect the death to its actual cause. When a person dies weeks or even months after a fire, the cause can still be the fire, but the linkage can be obscured by extensive medical interventions.

There is a range of effects that can determine the time interval between the fire and death. Death can occur within seconds to minutes from *hyperthermia*, exposure to very hot gases or steam, or from anoxia, the lack of oxygen. With *instantaneous death,* vagal inhibition or laryngospasm occurs on inhalation of flames and hot gases, causing cessation of breathing followed very quickly by death. Explosion trauma and incineration from exposure to a fully developed fire (often resulting in structural collapses) results in nearly instantaneous death.

Inhalation of toxic gases such as hydrogen cyanide, carbon monoxide, or other pyrolysis products; blockage of airways by soot, exposure to flames; physical trauma (with loss of blood); internal injuries; and brain injuries can cause death within minutes.

Death can occur within hours from carbon monoxide exposure, edema from inhalation of hot gases, burns, and brain or other internal injuries. Dehydration and shock from burns cause death days after the fire. Even weeks or months after the fire, deaths can occur owing to infections or organ failure triggered by fire injuries.

Cause of death is defined as the injury or disease that triggers the sequence of events leading to death. In a fire, the cause of death may include inhalation of hot gases, CO, or other toxic gases, heat, burns, anoxia, asphyxia, structural collapse, or blunt trauma. The *mechanism of death* is the biological or biochemical derangement incompatible with life. Mechanisms of death can be respiratory failure, exsanguination, infection, organ failure, and cardiac arrest. The *manner of death* is a medicolegal assessment and classification of the circumstances in which the cause of death was brought about. In the United States, these classifications are most often *homicide, suicide, accident, natural,* or *undetermined.*

The longer the interval between the cause (the fire) and the onset of the mechanism of death (organ failure, septicemia, etc.), the more likely it is that the connection will be lost. This is especially true when victims are moved from trauma care hospitals to long-term care facilities, sometimes in other geographic areas. The investigator must be diligent to ensure that the cause of death is not listed on the final death certificate as some generic mechanism term such as respiratory failure, cardiac arrest, or septicemia.

cause of death ■ The injury or disease that triggers the sequence of events leading to death. In a fire, the cause of death *may* include inhalation of hot gases, carbon monoxide, or other toxic gases; heat; burns; anoxia; asphyxia; structural collapse; or blunt force trauma.

SCENE INVESTIGATION

The reconstruction of the activities of a fire victim may depend on finding and documenting bloodstains (from impact with a wall or door jamb) or handprints on a wall, mechanical (blunt trauma) injuries, or artifacts found with the body. The nature of dress (street clothes, robe, nightgown) or objects (dog leash, jewelry, flashlight, fire extinguisher, house keys, phone, keepsakes, etc.) can provide clues as to what the person was doing prior to collapse.

The position of the victim (face up or face down) is not usually significant, as people known to have died during a fire have been found in all positions. Owing to pugilistic posturing, the attitude of the body bears little reliable relation to antemortem actions. As mentioned earlier, pugilistic posturing occurs as a physical reaction of the body to the heat and can occur whether the person was dead before the fire or died during the fire. It can cause bodies to shift position, sometimes to the point where they roll off unstable surfaces like chairs or mattresses. Young children may hide under beds or in closets, but finding them in other locations is not proof that they did not have the time or physical capacity to seek safety.

POSTMORTEM DESTRUCTION

A body exposed to fire can support combustion, the rate and thoroughness of which depend on the nature and condition of exposure of the body to the flames. The skin, fat, muscles, and connective tissues will shrink as they dehydrate and char. If exposed to enough flame, they will burn and yield some heat of combustion. The relatively waterlogged tissues of the internal organs must be dried by heat exposure before they can combust, and that dehydration step increases their fire resistance and delays their consumption.

Bones have moisture and a high fat content, especially in the marrow, so they will shrink, crack, and split and contribute fuel to an external fire. The subcutaneous fat of the human body provides the best fuel, having an effective heat of combustion on the order of 32–36 kJ/g (DeHaan, Campbell, and Nurbakhsh 1999). Like candle wax, however, it will not self-ignite or smolder and will not normally support flaming combustion unless the rendered fat is absorbed into a suitable wick. The material for the wick can be charred clothing, bedding, carpet, upholstery, or wood in the vicinity (as long as it forms a porous, rigid mass). The size of the fire that can be supported by such a process is controlled by the size (surface area) of the wick. Depending on the position of the body and its available wick area, fires supported by the combustion of a body will be of the order of 20–120 kW, similar to a small wastebasket fire. Such a fire will affect only items close to it, and fire damage will often be very confined.

Flame temperatures produced by combustion of body fat range from 800°C to 900°C (1475°F–1650°F), and if flames impinge on the body surface, they can aid the destruction of the body. The process, if unaided by an external fire, is quite slow, with a fuel consumption rate on the order of 3.6–10.8 kg/hr (7–25 lb/hr). It is possible, given a long enough time (5–10 hr), that a great deal of the body can be reduced to bone fragments (DeHaan 2001; DeHaan, Campbell, and Nurbakhsh 1999; DeHaan and Pope 2007).

If a body is exposed to a fully developed room or vehicle fire, however, the rate of destruction is much closer to that observed in a commercial crematorium. In those cases, flames of 700°C–900°C (1300°F–1650°F) and a heat intensity of 100 kW/m^2 envelop the body and can reduce it to ash and fragments of the larger bones in 1.5–3 hours (Bohnert, Rost, and Pollak 1998; DeHaan and Fisher 2003; DeHaan 2012).

Summary of Postmortem Tests Desirable in Fire Death Cases

The complexity of many fire death cases may necessitate finding the answers to questions that were not apparent earlier. Having comprehensive forensic samples and data is the best route to a successful and accurate investigation. Although not all the following forensic tests and data may be needed in all fire death cases, once the body is released for burial or cremation, it will be too late for forensic examination (DeHaan and Icove 2012, chap. 15).

- *Blood* (taken from a major blood vessel or chamber of the heart, not from the body cavity): tested for COHb saturation, HCN, drugs (therapeutic and abuse), alcohol, and volatile hydrocarbons.
- *Tissue (brain, kidney, liver, lung):* Tested for drugs, poisons, volatile hydrocarbons, combustion by-products, CO (as a backup for insufficient blood).
- *Tissue (skin near burns):* Tested for vital chemical or cellular response to burns.
- *Stomach contents:* Tested for establishment of activities before death and possible time of death.
- *Eye fluid:* An uncontaminated source of drugs and metabolites.
- *Airways:* Full longitudinal transection of airways from mouth to lungs to examine and document the extent and distribution of edema, scorching/dehydration, and soot.
- *Internal body temperature:* Should be measured at the scene. It may be elevated owing to fire exposure, postmortem decomposition, or antemortem hyperthermia, but unexpectedly *low* body temperatures after a recent fire should indicate that the victim was dead well before the fire.
- *X-rays:* Full-body (including associated debris in body bag) and details of teeth and any unusual features discovered (fractures, implants).
- *Clothing:* All clothing remnants and associated artifacts removed and preserved.

- *Photographs:* General (overall), with close-ups of any burns or wounds, in color, with scale.
- *Postmortem Multidetector Computed Tomography (MDCT):* Experimentally has been found to be a valuable tool, should this resource be available (see Levy and Harcke 2011).
- *Postmortem weight:* Postmortem weight of the body with organs determined, *not* including the fire debris or body bag.

It is very useful for the fire investigator to be present when the postmortem examination and autopsy are conducted, not only to ensure that all appropriate observations are made but also to be on hand to answer any questions that arise during the examination. Few pathologists have extensive knowledge of fire chemistry or fire dynamics, or can interpret the effects of fire on the body, so the investigator is in a good position to advise the pathologist as to the fire conditions in the vicinity of the body.

Death Followed by Fire—Cause and Duration

CASE EXAMPLE 1

At about 9 a.m., Mr. John Doe (obviously not his real name) discovered the severely burned bodies of his sister-in-law and his 3-year-old nephew in the dining room of their home. He had been called minutes earlier by his brother, Mr. Jim Doe, a salesman, who reported that he had left home that morning and had not been able to reach his wife (who was 8 months pregnant) by telephone. Concerned, he called his brother and asked that he stop by and check on her.

On arrival, John Doe found the house full of smoke but only small flames showing in the vicinity of the bodies that were sprawled amid an area of burned carpet in the dining room. He reportedly batted out the small flames with a hand towel and then withdrew to call emergency 9-1-1. The local fire department arrived by 9:10 a.m. and, forewarned of the conditions within, limited its response to one firefighter with a pressurized water tank who entered the building, sprayed a very small amount of water on or near the body to extinguish the small flames still present, and then withdrew along the same path. Except for opening windows to vent the accumulated smoke, that was the extent of fire suppression.

Jim Doe returned to the house while the investigation was beginning and claimed that he had left the house at about 6:00 a.m. and had found everything to be in order. He had stopped to put gasoline in his car and then at a restaurant while making his sales calls for the morning, establishing his presence elsewhere after 6:00 a.m.

The fire damage was limited to an oval area of carpet about 2 m² (21.5 ft²) in size, centered on the two bodies. The mother was sprawled roughly facedown, and the child was spread-eagled on his back alongside her, as shown in Figure 7-5. There were areas of scorching on the walls and baseboards of the dining room near

FIGURE 7-5 The fire damage was limited to an oval area of carpet about 2 m² (22 ft²) in size, centered on the two bodies. The mother was sprawled roughly facedown, and the child was spread-eagled on his back alongside her. *Courtesy of Cosumnes Fire Department.*

FIGURE 7-6 Fire investigators quickly determined through laboratory analysis that automotive gasoline had been poured on the carpet around the bodies and in nearby areas of the living room, kitchen, and hallway. *Courtesy of Cosumnes Fire Department.*

the entry to the kitchen and two areas of damage to the vinyl floor of the kitchen immediately adjacent to the mother's lower right leg. There were areas of scorching on her legs adjacent to these floor patterns. The fire investigators quickly determined through laboratory analysis that automotive gasoline had been poured on the carpet around the bodies and in nearby areas of the living room dining room, kitchen, and hallway (Figure 7-6).

Both victims had zero COHb levels, and no soot was detected in the airways. Both bodies bore signs of physical (blunt force) trauma, and the child's cause of death was ascribed to severe multiple head wounds. Bloodstain patterns indicated that the child had been beaten or kicked in the vicinity of where the bodies were found, but fire damage to the mother's face, neck, and upper torso precluded the establishment of an exact cause of death (Moran 2001).

The major fire-related issues pertained to the duration of the fire: What had fueled the fire? How long would a body burn in such circumstances? What is the mass loss rate for a body when it represents the main fuel package for a fire (as opposed to being involved in a massive enveloping fire fueled by furnishings and other external fuel load)? A test protocol was designed to collect data by which some hypotheses could be tested.

Carpet and pad from the scene were recovered in large quantities to serve as comparison samples and test targets for fire testing. The carpet was 95 percent polypropylene and 5 percent nylon face yarns bonded to a polypropylene mesh backing. The pad underneath was low-density polyurethane foam. Testing revealed that this carpet/pad combination could be ignited to support a self-sustaining combustion by the ignition of a small quantity of gasoline. The burning carpet was observed in tests to produce small flames, 5–8 cm (2–3 in.) high that advanced to consume approximately 0.5 m² of carpet per hour. Based on the limited amount of carpet burned very deeply (i.e., in direct contact with the gasoline pool fire), it was estimated that less than 2 L of gasoline had been used. Such a quantity would be expected to produce a fire of approximately 500 kW if ignited on carpet, with a duration of 2 to 3 minutes. Testing confirmed these estimates.

As can be seen in the floor plan in Figure 7-7, the living room, dining room, kitchen, and family room formed one large room, with a high vaulted ceiling extending throughout those areas, and a partial-height wall separated the living/dining portion from the family room/kitchen. The vaulted ceiling over the location of the bodies was an estimated 3.6 m (12 ft) from the floor and bore no heat damage. A flame plume from a 500 kW fire approximately 1 m (3.3 ft) in diameter would be (per Heskestad) about 1.8 m (6 ft) high and would not be expected to produce any ceiling damage.

The wall nearest the main plume would have been the south dining room wall and would have been more than a meter away. All walls were painted gypsum wallboard, and none bore any heat damage. There was no heat damage to any of the wood furniture in the dining room. There was a narrow, arc-shaped burn mark on the carpet identified as a gasoline trail between the entry door and the bodies, but there was no damage to any furnishings except to the carpet.

Several areas of the carpet bore scorched and melted areas whose appearance could be reproduced by the burning of a splattering of small droplets of gasoline on the surface of such synthetic carpet pile. Smoke had penetrated throughout the house, and soot had lightly coated all horizontal surfaces and condensed out on vertical surfaces (showing the outlines of wall studs, ceiling joists, and other concealed structural elements).

Blood spatters on surfaces and coatings of soot on floors showed that objects such as vases, boxes, and cutlery that were strewn about to simulate a burglary had been placed in those positions after the blood was splattered but before the fire. The areas of scorching and sooting on the vinyl flooring of the kitchen were

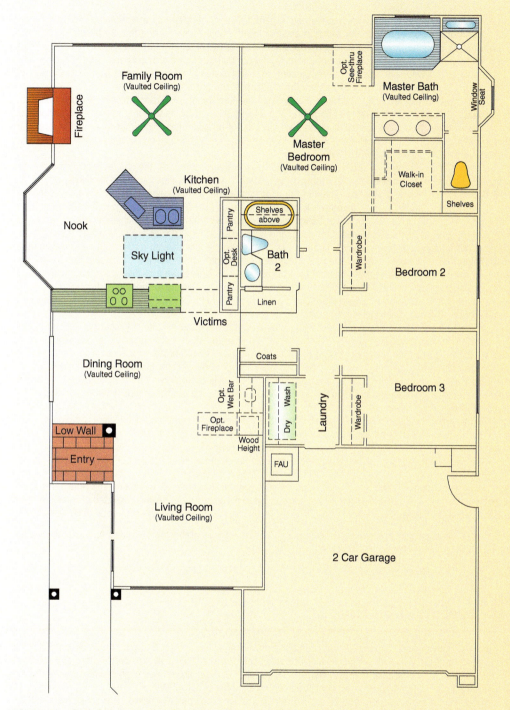

FIGURE 7-7 Floor plan showing the living room, dining room, kitchen, and family room, which formed one large area with a high vaulted ceiling. *Courtesy of Bruce Moran, Sacramento County District Attorney's Laboratory of Forensic Services, Sacramento, CA.*

identified as being produced by gasoline or a similar ignitable liquid of relatively low boiling point by comparison with tests of such fluids on similar floor coverings. The most significant fire damage was to the bodies of the two victims. No clothing could be identified on the child's body, and the remains of a knit top and panties on Mrs. Doe would not have constituted a significant fuel load.

A reasonably small quantity of gasoline had apparently been used, and so attention was focused on the combustion of the bodies as a major factor in establishing the time frame for the fire. Because Mrs. Doe had been receiving prenatal care, there was good documentation of her antemortem body weight. Compared with the postmortem weight, it was estimated that some 12 kg (25 lb) of body weight had been lost. No such data of comparable quality were available for the child victim, so it was decided that the mass lost from Mrs. Doe's body would be the most reliable measure of establishing duration of the fire.

Testing was conducted at the California Bureau of Home Furnishings (BHF) room calorimeter to establish the heat release rates and mass loss rates of body fat. Pork fat was selected for its similarity to human subcutaneous fat and availability. Tests of various small (1- to 2-kg, 2.2- to 4.4-lb) quantities of pork fat (with skin) wrapped in cotton cloth established that small fires, 20 to 70 kW, could be maintained by combustion of the rendered fat (DeHaan, Campbell, and Nurbakhsh 1999).

The size of the fire was controlled largely by the surface area of the wick (the charred cloth), and the duration of the fire was controlled by the supply of fuel. Mass loss rates appeared to be of the order of 1–2 g/s (3.6–7.2 kg/hr). Testing was also conducted in the cone calorimeter at BHF using a 10 cm × 10 cm (3.9 in. × 3.9 in.) tray, with paraffin wax as a control. It was found that at 35–50 kW/m^2 incident radiant heating, pork fat would melt and smoke but could not be ignited to support a continuous flame unless a cotton cloth wick was placed across its surface (DeHaan, Campbell, and Nurbakhsh 1999). The radiant heat would then char the cotton cloth, melting the fat beneath. The melted fat wicking up via the charred cloth could be readily ignited. Cone calorimeter testing established that the effective heat of combustion for both pork and human fat was approximately 34 kJ/g and that their rates of heat release (per unit surface area of fuel) were roughly the same, 200–250 kW/m^2. The observed properties of melting behavior and ignitability for human and pork fat were also indistinguishable, confirming that test results using pork fat or pig carcasses would yield results applicable to human cadavers (DeHaan, Campbell, and Nurbakhsh 1999).

A final series of tests in which pig carcasses clothed in knit garments and ignited by means of a 1-L (0.22-gal) pour of gasoline confirmed that the charred garments and charred carpet and pad served as a wick and would support combustion of the fat being rendered out of the carcass. This combustion produced a smoky fire of approximately 50 kW, with a mass loss rate of approximately 1.5 g/s (5.4 kg/hr) for times exceeding 1 hour. Further, the gasoline pour would burn off in approximately 3 minutes while producing a fire of 250–400 kW. This brief fire would initiate fire in the surrounding carpet that would continue to burn for various times, consuming carpet at a rate of about 0.5 m^2/hr.

Of interest in this case was the observation that the gasoline-fed fire was not of sufficient duration to cause the skin to shrink and split and allow the rendered fat to escape to support external flames. Only after the clothing and carpet had burned for 10 minutes or longer could the rendering process be seen to be supporting the fire.

The fire behavior supported by tests and data were considered when reaching the conclusion that the fire had been burning for 2 to 4 hours at the time of its extinguishment. Since the fire was still burning at approximately 9 a.m., it was concluded that it was ignited using gasoline sometime between 5 and 7 a.m. This time frame, combined with conflicting statements made by the suspect, led to his prosecution. He was convicted of three counts of second-degree murder and one count of arson. He is serving consecutive sentences for a total of 60 years to life.

This case is courtesy of Bruce Moran, Sacramento County District Attorney's Laboratory, Sacramento, California, and Jeff Campbell, Sacramento County Fire Department (retired).

CASE EXAMPLE 2   Execution by Fire

The burned body of a young woman was found in a remote agricultural area. She bore burns over most of her body and was dressed in the burned remains of cotton denim pants, a long-sleeved cotton shirt, and a bra, all identified by brand name tags that survived the fire (Figure 7-8). Her ankles had been bound together with a heavy leather belt wrapped over her cotton socks, prior to the fire (Figure 7-9).

The victim was found lying on her back with her arms and legs partially flexed. There was a burned pile of clothing some 2.6 m (8 ft 8 in.) away from her head (probably in a cloth duffel bag). The clothing and other contents indicated that this bag belonged to the victim.

There were a number of small areas of scorching or charring visible on the dry soil nearby, isolated from both the clothing and the body. Although these could have been the result of isolated burning splashes of flammable liquid, they were more consistent in appearance with having been the result of contact between the burning clothing and the soil.

Examination of the scene revealed that there was a faint pattern of wrinkled fabric (as from garments like shirt and pants) impressed into the dry sandy soil between the body and the pile/bag of clothing. There was an additional pattern of wrinkled fabric in the soil to the right of the body adjacent to the torso and upper right leg.

FIGURE 7-8 Victim of murder by burning as found at scene. Repeated scuff marks with limited scorching in soil adjacent to torso and limbs show movement during the fire. Impressions of creased fabric of shirt on soil next to torso were produced as victim rolled on ground. Charred residue of denim jeans remain on left leg and waist. Charred cloth fragments away from body also show movement after and while burning. Shoe print in lower left corner is partially overlain by "leg" fabric impression, showing sequence.

FIGURE 7-9 Scuff marks in soil near toes and feet show repeated movement. Flexion of toes is due to heat shrinkage of tendons. Feet of victim bear only first- and second-degree burns from burning of cotton socks. Sock material trapped beneath leather belt around ankles burned longer (supported by gasoline wicking out) and induced charring of skin beneath.

The soil at the fire scene revealed additional information. There were deep scuff marks in the soil, of the type made by bare or stocking feet, under and beside the toes of the victim. Scorched areas of soil to the right side of the body were displaced and disturbed by patterns of movement of the soil, probably by movement of the right arm and right leg of the victim. There were a number of shoe prints and tire prints in the vicinity of the body, some of which overlaid the scuff marks, but in most cases the scuff marks obscured the shoe impressions, confirming sequencing and timing of movements.

Laboratory tests revealed the presence of gasoline in the clothing of the victim, in the burned clothing pile, and in the soil sampled in several areas. There was no gasoline container found in the vicinity. No estimate could be made of the quantity of gasoline that had been used. No other fuels were present except for the clothing.

There was extensive scorching and blistering of the skin across nearly all exposed areas of the victim's body. The back was more burned than the front of the torso, and most of the clothing was burned away, particularly on the back and sides. There were signs of vital reaction to many burned areas of skin. The burns were only partial thickness (second degree) over nearly all the body, with the only exception being deep charring of the ankles and lower legs immediately adjacent to the cotton socks held in place by the leather belt. Nearly all the scalp hair was burned away, reduced to black masses. Charred remains of cloth were clutched in the fingers of the right hand.

Postmortem examination revealed internal heat/flame damage to the mucosa of the throat and larynx as a result of inhalation of extremely hot gases. Toxicological analysis reported a COHb level of approximately 14 percent. The fire issues here were whether gasoline had been poured and ignited on the victim while she was conscious and what the result of such an event would be.

Re-creations in the laboratory using a dressmaker's mannequin approximately the same height as the victim demonstrated that two full revolutions of a body that size as it rolled on the ground resulted in the same displacement as seen at the scene between the body and the clothing pile. Owing to the taper of the body from the shoulders to the feet, the path taken by a body rolling in this manner is in the form of an arc rather than a straight line.

A gasoline fire is known to produce heat transfer sufficient that direct flame contact induces blistering (partial-thickness burns) in about 5 seconds at 50 kW/m². Based on the distribution of burn damage to the body and the clothing, it was concluded that the fire damage could not have been the result of gasoline having been poured over an inert, unconscious body. The absence of protected areas and the extensive damage to the back and buttocks could have resulted only from the victim's consciously moving. The charred cloth clutched in the hand away from other fire damage indicated a conscious attempt to grasp the burning clothing during the fire.

Although flexion of the major joints as a result of fire exposure can result in some movement of an unconscious or even a deceased body, the pattern of cloth/fabric impressions in the soil was the result of a rolling movement of the body. The scuff marks in the soil to the right side of the body were most likely the result of movement of the right arm and right leg of the victim while she was lying on her right side. The body apparently fell or rolled onto its back, where it assumed its final resting position. Movement of the feet and toes would have produced the deep scuffs found there. Fire damage to the clothing and the isolated scorch marks on the soil were also consistent with conscious movement of the body.

The duration of a gasoline-fueled fire would be of the order of 2 minutes or less (assuming that less than a gallon was used). The duration would be shorter if the gasoline was spread on bare skin or thin cotton fabric. Such a short-duration fire would produce the limited-depth burns found here, with the only exception being the gasoline-soaked cotton socks, which acted as a wick to sustain flames long enough to induce deep burns to the adjacent tissue. The presence of shoe prints overlying some of the scuff marks in the soil was an indication that the victim was set afire in the vicinity of the clothing bag and that her assailants were present during (at least) the initial stages of the fire as she rolled on the ground in an attempt to extinguish the flames. It was concluded that she was conscious for many seconds during this fire.

The three people who carried out this execution were acquaintances of the victim who decided that it would be interesting to dispose of her in this manner and transported her in a truck while they purchased the can and the gasoline. All three pled guilty to murder charges.

CASE EXAMPLE 3   Fatal Van Fire

At about 5:30 on a cold November morning a farmer heard shouts of alarm coming from across the river. He looked out to see a camping-style van fully engulfed in flames. He phoned the fire department, but owing to the remote location of the site, it was 15 minutes or more before fire crews arrived and another 10 minutes before the fire was extinguished.

Inside the van were the remains of an adult female victim who had reportedly been asleep in the back of the van with her boyfriend at the time the fire was discovered. The boyfriend, after awakening and reportedly discovering the smoke and limited flames within the van, had allegedly tried to rouse the victim, who was non-responsive, and then escaped the van by butting his head against a rear window. He attempted to rescue his friend from outside but discovered that all but one of the doors to the van were locked, and he could not reach her through the one unlocked door because a blanket hung between the cargo compartment and the driver's compartment. He claimed that they had been camping on the river just for a day. Both were recreational drug users, and he had a record for breaking-and-entering and minor drug charges.

Investigators called to the scene discovered that all the windows to the "Sportsman-style" van had been broken but could not tell if any had been broken by mechanical force. The side and rear windows were tempered glass, which shatters into similar small pieces whether broken by thermal shock or mechanical impact, but all appeared to have been soot covered. The van had been gutted, with most damage to the cargo and driver's compartments (Figure 7-10). There were signs that the fire had penetrated into the engine from inside the van. The sheet metal of the roof had been badly distorted by the fire. There were no signs that the vehicle had been in operation at the time of the fire. The vehicle's fuel tank was still half full, and the fuel system was normal.

Fire damage outside the vehicle was limited to the grass immediately around and under it; there were no indications of an external grass fire growing to involve the vehicle. The van was heavily loaded with bags and boxes of clothing, food, camping supplies, tools, and blankets. There were a propane camping lantern and camp stove with two 1-lb propane bottles, but they were not in use. All exposed fuels in the van were fire damaged.

A friend who had visited the man and woman the night before the fire said that they were eating snacks (with no cooking) and were using small votive candles for light. Luckily the fire chief in this jurisdiction always kept a point-and-shoot camera in his vehicle and was able to get several photos of the still-smoldering interior of the vehicle before the contents were removed to permit overhaul. The upholstery of the seats and dash and paneling of the cargo compartment were completely consumed (Figure 7-11). The fire damage clearly demonstrated that

FIGURE 7-10 Exterior damage to the gutted van, showing most damage to the cargo and driver's compartments. *Courtesy of Solano County Sheriff's Dept., Fairfield, CA.*

FIGURE 7-11 Interior of the vehicle immediately after extinguishment and before overhaul. The victim is sitting facing the camera, with her back resting against the back of the passenger seat. *Courtesy of Solano County Sheriff's Dept., Fairfield, CA.*

FIGURE 7-12 Fire victim as found in the vehicle. Note the extensive destruction by flames, with exposed ribs, spinal column, and internal organs. The skull has calcination and fracture damage resulting from prolonged fire exposure. *Courtesy of Solano County Sheriff's Dept., Fairfield, CA.*

the fire within this compartment had gone to flashover, with ventilation supplied by the windows (closed at the start of the fire but failing as the fire grew).

The decedent was found in a partial sitting position on the right side of the cargo compartment (Figure 7-12) facing the rear of the van. The body had suffered massive fire damage; nearly the entire right half of the body had burned away (exposing the spinal column), with major bones reduced to fragments. The victim had a 41 percent COHb level, with no alcohol and very low levels of amphetamine metabolites. She was young (23 years old), healthy, and physically unimpaired. Extensive fire exposure had destroyed the soft tissues of the head and neck and had caused the exposed skull to calcine and fracture into a number of pieces. There were no underlying hemorrhages indicating antemortem trauma. The fire had compromised the lungs and airways such that the presence of combustion products could not be determined.

The COHb level left no doubt that the victim had been alive and breathing for some time during the fire and had died as a result of exposure to smoke and flames. The extensive damage to the body was ascribed by the pathologist to incineration by a flammable liquid fire, despite later laboratory findings that revealed no petroleum distillates in the small segment of carpet recovered from beneath her.

No flammable liquid residues were detected on the exterior of the van or beneath it. The extensive damage to the van, the absence of accidental ignition sources, and the nonoperational status of the van led investigators to conclude that the fire had been deliberately ignited with a flammable liquid on or around the decedent, which prevented her escape. The boyfriend was charged with murder.

The surviving victim (boyfriend) was found outside the van dressed in T-shirt and undershorts (he was the person shouting for help whom the farmer heard). He had a coating of soot on his face, arms, and hands but no burns. His facial hair (including mustache) was unburned, but hair at the top of his head was singed. He had fresh abrasions on his knuckles, one on the top of his head, and scratches and abrasions on his lower legs. Tiny fragments of glass were found in his scalp hair. His blood was not sampled immediately but was taken later and was found to bear low CO concentrations, low levels of amphetamine metabolites, and no alcohol. He claimed that he and the decedent had gone to sleep side by side wrapped in blankets on the floor of the cargo compartment and awakened to discover the fire. He denied any conflict with the decedent.

A review of the evidence (by one of the authors) at the request of the local district attorney's investigator resulted in the conclusion that the physical, medical, pathological, and toxicological evidence was more consistent with an accidental fire occurring as described by the accused than with a murder with flammable liquid. A fire-related human behavior analysis revealed that if the survivor had been sleeping on his back in a smoke-filled van, his face and arms would be expected to be coated with soot but not exposed to the hot gas layer until he sat up and attempted escape. At that time, his scalp hair would be exposed to the highest temperatures. Beating at the window glass or fumbling in the dark for the door locks could have produced the abrasions on his right hand. Butting his head against the rear window until it broke could have produced the abrasion on the top of his head and the glass fragments in his hair. Sliding out the high rear window would be expected to induce cuts and scratches to the lower legs.

The 41 percent COHb level in the decedent is similar to the levels encountered in victims of accidental house fires. In the opposing hypothesis, throwing gasoline on someone in sufficient quantities to prevent his/her escape would be likely to leave unburned residues after the fire in protected areas under the body and would also be likely to produce a flash fire of sufficient size to induce singed facial hair on someone close by (i.e., the thrower). The ignition of a quantity of gasoline vapor in the vicinity of a victim's face can induce rapid cessation of breathing so that CO levels in the blood would be very low. The first few seconds of a gasoline vapor/air fire are also very low in CO, so the gases that are inhaled are unlikely to produce an elevated COHb level. Such a scenario would not produce the glass fragments, singed scalp hair, or leg injuries. A gasoline-fueled fire would also be unlikely to induce such extreme damage to the decedent's body unless it involved massive quantities of fuel.

The fire issues involved were the following.

- What were the contributions of the fuel load in the van?
- Did the size of the van (compartment) and the breakage of windows play a role?
- Could a fire in a closed van produce a fire that would be escapable and produce the physical and clinical evidence seen on the survivor?
- Could such a fire progress to flashover?
- If the fire did progress to flashover, could postflashover fire produce the damage to the body and to the van without the presence of any ignitable liquids?

A long-wheelbase Sportsman van similar to the one in the incident was obtained by the district attorney's office. Based on calculations of the window area and height, it was determined that when all the windows failed, enough air would be admitted to the van to support a fire of 3.2 MW. The internal volume of the van was calculated to be 9.8 m^3, so there was considerably more air inside even a closed van than there would be in the average passenger vehicle.

The van obtained (Figure 7-13) had nearly the same arrangement of windows but had one solid side panel instead of a large window, so a vent panel was cut through the metal and left attached by a hinge so it could be opened during the fire test to simulate failure of the window by thermal shock. The interior finish was duplicated, with paneling, carpeting, and seats similar to the original.

Wooden tool boxes and a wool blanket curtain were fitted to simulate the original, and some 120 kg (265 lb) of clothing, blankets, plastic crates, cardboard, and miscellaneous combustibles was added, estimated to be about half the added fuel mass in the original van, as shown in Figure 7-14.

Temperatures were monitored via three thermocouples mounted on a tree in the center of the van near the position of the decedent and one thermocouple near the rear door where the survivor allegedly made his escape. Ignition was produced with an open flame in ordinary combustibles and the van was then closed up. A misplaced votive candle was probably the ignition source for the actual fire. Given the nature of the fire and the fuels available, it was very unlikely to have been a smoldering source.

In the test, within 3 minutes heavy smoke had formed throughout the van, and temperatures in the smoke layer ranged from 80°C to 100°C (176°F to 212°F), as shown in Figure 7-15. Soot was observed condensing on all windows. The rear window was broken from outside by mechanical impact at 3 minutes 45 seconds. Gas layer temperatures exceeded 600°C (1112°F) by 7 minutes, and flashover occurred as windows began to fail from thermal shock. The side vent was opened as all the windows failed between 6 and 8 minutes into the fire, and by 8 minutes, temperatures in the hot gas layer exceeded 1000°C (1832°F). Temperatures of over 600°C (1112°F) were maintained for over 12 minutes as the fire consumed the fuel load and went into decay. Total burn time

FIGURE 7-13 The test van obtained had nearly the same arrangement of windows but had one solid side panel instead of a large window, so a vent panel was cut through the metal and left attached by a hinge so it could be opened during the fire test to simulate failure of the window by thermal shock. *Courtesy of Solano County Sheriff's Dept., Fairfield, CA.*

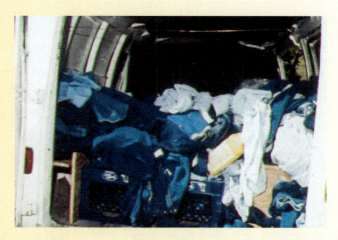

FIGURE 7-14 Wooden toolboxes and a wool blanket curtain were fitted to simulate the original, and some 120 kg of clothing, blankets, plastic crates, cardboard, and miscellaneous combustibles was added, estimated to be about half of the added fuel mass in the original van. *Courtesy of Solano County Sheriff's Dept., Fairfield, CA.*

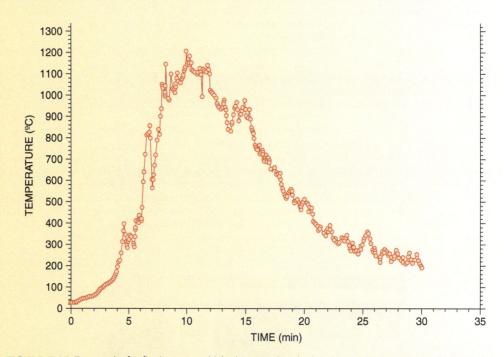

FIGURE 7-15 Test results for fire in a test vehicle documenting the interior temperature versus time at a thermocouple located near the victim's location. *Courtesy of Fred Fisher, Fisher Research & Development, Inc., Vacaville, CA.*

was 30 minutes. Damage to the exterior and interior of the van duplicated that of the original fire scene van, as shown in Figures 7-16 and 7-17.

Temperatures and heat fluxes in excess of those found in commercial crematoria were measured in the center of the van. The victim's body would have been exposed to these intense fire exposures from her right side, near the center of the compartment, where the postflashover burning would have been most intense. The fuel load available in this test limited the duration of the fire, but not its intensity. The intensity was determined by the surface area of fuel involved in the postflashover burning and the ventilation available to support the combustion.

FIGURE 7-16 Thermal damage to exterior of test van. *Courtesy of Solano County Sheriff's Dept., Fairfield, CA.*

FIGURE 7-17 Thermal damage to interior of test van. *Courtesy of Solano County Sheriff's Dept., Fairfield, CA.*

Although no heat release data were captured during this test, the test team, with their extensive forensic fire experience, estimated that the heat release rate was in excess of 3 MW during the peak of the fire (at 10 to 12 min). Because the ventilation inside the test van was the same as in the van involved in the fatality, the size and intensity of the fires would have been equivalent. With the more substantial fuel load in the scene van, it was potentially capable of supporting a full postflashover fire for much longer than the test van could.

Witness observation by the farmer confirmed that the van was fully engulfed when he first saw it and it burned that way until extinguished some 25–30 minutes later. Based on the results of this test, the district attorney agreed that an accidental fire could not be ruled out and was, in fact, more consistent with the physical evidence. All charges against the male suspect were dropped.

The test data in this case are courtesy of Fred Fisher, P.E., of Vacaville, California. The authors thank Professor R. Brady Williamson of the University of California, Berkeley; Dr. Charles Fleischman of the University of Canterbury, Christchurch, New Zealand; and members of the Solano County Fire Investigation Team for their contributions and for making this test possible.

At approximately 4 p.m. on a weekday, fire and ambulance services were alerted to attend to a female who had been burned and to a fire in a kitchen inside a suburban apartment block. On arrival, ambulance and fire personnel found a deceased adult female, with obvious burn injuries, at the street frontage gate. A male indicating he was the partner of the deceased provided the location of the suspected fire in an apartment kitchen. Firefighters located and inspected the apartment, only to find that a small fire had occurred in the kitchen, causing slight fire damage. They extinguished the fire and requested that an investigator attend owing to the fatality.

The initial theory (hypothesis) provided by firefighters at the scene was that the male was working on a model car engine, the fuel being used had ignited, and the female was involved by the ensuing fire. Another theory (hypothesis), developed by police at the scene, was that the male and female had been quarrelling, and fuel had accidentally been spilled on the kitchen floor. Subsequently, an accidental ignition had occurred that involved the clothing of the female. She ran out of the apartment and downstairs into the yard, where she was found on fire. She was found approximately 50 m away, at the front gate, where she succumbed to her injuries from the fire, and where she was found by attending emergency services.

The body of the deceased was inspected in situ, and it was noted that she was wearing a cotton-type material tracksuit, with long legs and long sleeves; underwear; and what appeared to be nylon-type socks on her feet. The victim's clothing was severely fire damaged, principally from the hips/thighs up and involved the upper body and head (Figure 7-18). The victim's upper body was severely burned, and her head and chin area had suffered extreme burn injuries. Her hands and forearms had suffered severe burn injuries as well. The lower legs of the pants, however, were intact and not consumed by the fire (Figure 7-19).

Burned remnants of clothing and skin, from the body, were found in a trail from the front gate along a path leading to the front door and up the internal stairs to the second level. Several pieces of burned skin adhered to the wall along the stairs outside the fire-affected apartment.

Inside the apartment, investigators found that a small fire had damaged sections of the kitchen. There were some small scorch marks on the tiled floor and some slight scorch marks on the bottom of the timber cupboards, but more important, several areas of the top surface of the cupboards had been fire damaged (Figure 7-20). A paper towel roll had a V-shaped pattern from the benchtop upward on its front surface; a microwave oven had a V-shaped pattern of scorching on its front window surface; and one of the plastic feet on a kitchen appliance, located on the benchtop, was slightly melted. An odor of methylated spirits (denatured alcohol) was evident within the kitchen and was emitted from beneath the microwave oven (Figure 7-21). A partially melted plastic container was located on the floor adjacent to the bottom of the kitchen cupboards (approximately 4-L capacity, or approximately 1 gal).

A one-fifth scale model car engine was located within an adjoining room, and details were recorded of the model and serial numbers. Subsequent research through a model car supplier revealed that this engine was fueled by methylated spirits (denatured alcohol) and not petrol (gasoline), as suggested by the male occupant as the fuel for the fire.

After inspection of the injuries suffered by the deceased and of the fire damage within the kitchen, the police and investigators held a discussion. The notion (hypothesis) that the deceased had been accidentally caught in

FIGURE 7-18 Upper body of fire victim, found outside residence. Note fire damage to clothing, upper body, and face. *Courtesy of Ross Brogan (ret.), NSW Fire Brigades, Greenacre, NSW, Australia.*

FIGURE 7-19 Legs of fire victim. Note pants legs are intact. Nylon socks are melted. *Courtesy of Ross Brogan (ret.), NSW Fire Brigades, Greenacre, NSW, Australia.*

FIGURE 7-20 The kitchen where the incident occurred. Note small fire damage at center bottom of cupboards and on paper towel roll at countertop near microwave. *Courtesy of Ross Brogan (ret.), NSW Fire Brigades, Greenacre, NSW, Australia.*

a fire, which had been ignited by spilled fuel on the kitchen floor, was considered. The reasoning used to test this theory was that the deceased was still wearing nylon socks and that the lower section of her cotton clothing was intact and not fire damaged. In fact, if the fire had actually been on the floor of the kitchen—based on normal fire behavior—the socks and lower body clothing would have been the first to ignite. The fire theory advanced was that the deceased had had fuel splashed/poured/thrown over the upper half of her body and ignited, causing the injuries and damage found. Subsequent scientific testing showed that the clothing and areas of the kitchen were indeed contaminated with methylated spirits.

Police interviewed the suspect male who subsequently provided information that he had quarreled with his female partner. The suspect had been in possession of a 4-L plastic container of methylated spirits (approximately half full) and had been so enraged that he had thrown some of the liquid onto her, had thrown the container onto the floor, and had kicked it around in his rage. He had then held a disposable lighter up and away from her and lit it in an effort to scare her. The liquid on the floor had accidentally ignited, and her clothing had caught on fire, so she had run down the stairs and out of the apartment. He had followed, smothered the fire on her clothing with a bed sheet, and carried her to the front gate.

A further search of the apartment revealed some of the male's clothing, burned and wet on the bathroom floor. This clothing evidence, along with burns that the suspect had suffered showed his clothing had caught fire as well during the incident and that he had gone to the shower and extinguished himself prior to following the female downstairs. The police subsequently charged the male suspect with murder.

FIGURE 7-21 Fire damage on kitchen countertop at scene of fire. Note V pattern on paper towel roll and burn mark on microwave front window. *Courtesy of Ross Brogan (ret.), NSW Fire Brigades, Greenacre, NSW, Australia.*

FIGURE 7-22 Testing of theory with clothing and flammable liquid in a fatal fire death case. *Courtesy of Ross Brogan (ret.), NSW Fire Brigades, Greenacre, NSW, Australia.*

In an effort to test the investigator's theory, the clothing that had been worn by the deceased was scientifically analyzed, and similar sets of clothing were purchased. Two store mannequins were also obtained, and a test was conducted to fire test the clothing. The Fire Service Training College Fire Training Tower was used, and all tests were videotaped and photographed, to be used as evidence (Figure 7-22). The mannequins were clothed in the identical clothing, and two tests were conducted using methylated spirits, one with liquid poured only on the floor around the base of the mannequin and another with liquid poured over the upper body of the mannequin and onto the floor.

The liquid was ignited in both cases and the resulting fire recorded. Even though the concrete room was bare of furnishings, fittings, and cupboards it was considered a basic test to determine whether the resulting fire would produce injuries similar to those of the deceased. In both cases with the liquid on the floor, the lower-body clothing and socks were consumed early in the fire (Figures 7-23, 7-24, 7-25, and 7-26).

With the liquid applied to the upper body of the test mannequin, the resulting burn damage to the face and upper body of the mannequin proved to be identical with the actual injuries suffered by the deceased; both sets of photographs could be placed side by side and the injuries were seen to be identical (Figures 7-27 and 7-28). The injuries suffered by the deceased showed the fire had risen from beneath the face and affected the underchin area, lower portion of the upper lip, the nostrils and underside of the eyelids, and the lower lobes of the ears.

With these data and evidence the case was prepared for trial on the murder charges. After a conference with the Crown Prosecutor it was decided that further tests would be conducted in a kitchen-type configuration to test the validity of the actual scene configuration to validate the veracity of the testing and the theories

FIGURE 7-23 Test underway with flammable liquid at floor level only. *Courtesy of Ross Brogan, NSW Fire Brigades, Greenacre, NSW, Australia.*

FIGURE 7-24 Fire test with fire being extinguished. *Courtesy of Ross Brogan (ret.), NSW Fire Brigades, Greenacre, NSW, Australia.*

FIGURE 7-25 Fire damage to mannequin after fire test from floor-level pour. *Courtesy of Ross Brogan (ret.), NSW Fire Brigades, Greenacre, NSW, Australia.*

FIGURE 7-26 Fire test underway with flammable liquid on floor and on upper body and clothing. Note that dark coloring on clothing is liquid on upper body. *Courtesy of Ross Brogan (ret.), NSW Fire Brigades, Greenacre, NSW, Australia.*

FIGURE 7-27 Mannequin fully involved in test with liquid at floor and on upper body. The judge considered that this graphic photo might have upset jurors. *Courtesy of Ross Brogan (ret.), NSW Fire Brigades, Greenacre, NSW, Australia.*

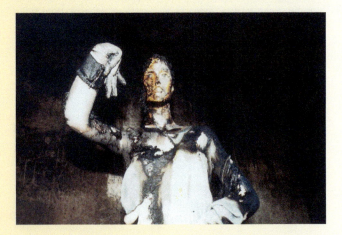

FIGURE 7-28 Results of test showing fire damage/injury to mannequin that was compared with photographs of the actual victim. *Courtesy of Ross Brogan (ret.), NSW Fire Brigades, Greenacre, NSW, Australia.*

involved. The public housing authority was approached, and a vacant townhouse was obtained with a kitchen configuration similar to that of the actual incident scene. This configuration, rather than the open concrete room used in the initial tests, was thought to lend authenticity to the test results.

First, a 4-L (approximately 1-gal) container similar to the one found in the apartment was obtained and marked with graduations on the outside to show liquid levels, in approximately 250-mL increments. The container was filled with colored water and flung to try to give an indication of how much liquid was expelled with each throw when tested at differing liquid levels. The results of these tests were used to estimate the volume of liquid thrown onto the victim during the actual incident. These tests were videotaped and photographed, again for evidentiary purposes, and were later used in the court trial.

The fire tests were conducted using clothing identical with what the victim had been wearing, this time in the kitchen similar in configuration to the fire scene. Videotaped and photographed, these tests were again available for the trial as evidence. Again, the tests in which the liquid was applied to the upper body of the mannequin resulted in burn damage and injuries identical with those of the victim. The investigator and the prosecutor agreed that the theory proposed was valid and practical, having been tested by scientific methodology.

At the trial, after viewing the video evidence of the tests, the trial judge refused to allow the videos to be shown to the jury, stating that they appeared to be prejudicial to the defendant and a possible appeal point after the trial. However, the judge allowed the theory testing and results to be discussed during presentation of evidence. In all, the investigator spent five and a half days in the witness box presenting evidence in chief and under cross examination, and presenting evidence to establish his own credibility, the credibility of the hypothesis, and the validity of the tests that proved the story of the defendant to be false. The male was subsequently convicted of manslaughter.

It was interesting that during the cross-examination the defense attorney proposed a theory aligned with the fact that it was (the investigator had explained the scientific basis of ignition of flammable liquid vapor and not the actual liquid), in his words, "a spiderweb of vapor hanging in the air" that ignited and ignited the upper clothing of the deceased. This spiderweb theory was vigorously debated, with the scientific basis of the theory explained during the cross-examination, and ultimately dismissed, partly because a question from the jury indicated that a member of the jury had some scientific training and complained that the defense was using "junk science theories" to deceive the jury on something that was quite plainly a valid scientific theory proposed by the prosecution. The other factor was that if the "spiderweb of vapor" was hanging in the air, then the lower clothing was just as susceptible to ignition as the upper clothing. The jury found the defendant guilty.

This case provided courtesy of Inspector Ross Brogan (ret) AFSM, CFI—NSW Fire Brigades, Sydney, Australia.

CHAPTER REVIEW

Summary

In a fatal fire, the cause of death and the cause of the fire are independent but linked by circumstances. Each must be established, and only then can the link between them be determined.

Accidental fires can accompany deaths by accident, suicide, homicide, or even natural causes. Incendiary fires can be associated with homicide (as a direct cause of death or simply as part of the crime event) but also with accidental or natural-cause deaths. For a fire death investigation to be successful (i.e., accurate and defensible), the following guidelines must be observed.

- *Treat the scene as a crime scene.* Every fire with a death or major injury should be treated as a potential crime scene and not prejudged as accidental. The scene should be secured, preserved, documented, and searched by qualified personnel acting as a cooperative team.
- *Document all critical features (as discussed in Chapter 4).* Documentation includes accurate floor plans, with dimensions and major fuel packages included, and comprehensive photographic coverage. Photos must include presearch survey photos, photos during search and layering, and photos of all views of the body prior to removal, during removal, and during postmortem examination by a medical examiner. This documentation is essential to proper forensic fire scene reconstruction (as described in Chapter 4).
- *Avoid moving the body.* The body must not be moved until it has been properly examined by the fire investigator and the pathologist or coroner's representative and thoroughly documented by photos and diagrams. The debris under and within 0.9 m (3 ft) of the body should be carefully layered and sifted. All clothing or fragments should be preserved. Artifacts (jewelry, weapons, etc.) must be documented and collected.
- *Assess the fuels.* The fire investigator has to assess the fuels already at the scene (structure as well as furnishings), and the role fuels may or may not have played in ignition, flame spread, heat release rates, time of development, and creation of flashover conditions.
- *Perform full forensic exams.* Every fire death deserves a full forensic postmortem examination including toxicology and X-rays. Toxicology samples should be tested for alcohol and drugs, as well as COHb and HCN, and should include both blood and tissue. The clothing should remain with the body and be documented and evaluated in situ before removal, if possible, then properly preserved. The internal (liver) body temperature should be taken as soon as possible (preferably at the scene).
- *Examine pets.* Deceased pets should be X-rayed and necropsied. Injuries to living pets should be noted and documented. Blood from deceased pets should also be tested for COHb saturation and drugs.
- *Examine living victims.* Fire victims who are survivors, whether burned or not, should be visually examined during interviews and, when possible, be photographed. Blood samples should be taken from them for analysis later if needed. External clothing (pants, shoes, shirt) should be saved and properly preserved.
- *Fully appreciate the fire environment.* Pathologists and homicide detectives must appreciate the fire environment—temperatures, heat and its transfer, flames, and smoke—and the distribution of fire products and the variables in human response to those conditions. In best practice, the medical examiner and/or pathologist visits the scene and sees the body in situ to appreciate its conditions of exposure (to flame, heat, and smoke), the nature of debris, and its location and the position.
- *Carry out forensic reconstructions.* A full reconstruction may involve criminalistics evidence such as blood spatter, blood transfers, fingerprints, tool marks, shoe prints, and trace evidence. The criminalist should be part of the scene investigation team along with the homicide detective, fire investigator, and pathologist.

As we have seen, a death involving fire is not just a simple exposure to a static set of conditions at a single moment in time. Fire is a very complex event, and a fire death investigation is even more complex and challenging. A coalition of talents and knowledge working together as a team is the only way to get the right answers to the three big questions: What killed the victim? Was the fire accidental or deliberate? and How did those two events interact?

Problems

7.1. Discuss the common problems and pitfalls associated with fatal fire investigations. Compare them with a recent ongoing case in the national media. What parallels can you find?

7.2. Referring to Example 7-1, determine the visibility of the same fire in an enclosed hallway measuring 3 m × 15 m (9.84 ft × 49.2 ft) with a ceiling height of 2.5 m (8.2 ft).

7.3. Referring to Example 7-2, calculate the fractional incapacitating dose for a male, a child, and an infant in the same scenario.

7.4. Review the newspaper coverage of a recent fatal fire in your community. Determine who investigated it, what the conclusions were, and whether criminal charges resulted.

References

Babrauskas, V. 1981. Applications of predictive smoke measurements. *Journal of Fire and Flammability* 12:51–66.

Bide, R. W., Armour, S. J., & Yee, E. 1997. Estimation of human toxicity from animal inhalation toxicity data: 1. Minute volume–body weight relationships between animals and man. Ralston, Alberta, Canada: Defence Research Establishment Suffield (DRES).

Bohnert, M., Rost, T., & Pollak, S. 1998. The degree of destruction of human bodies in relation to the duration of the fire. *Forensic Science International* 95 (1): 11–21, doi: 10.1016/s0379-0738(98)00076-0.

Bryan, J. L. 1983. An examination and analysis of the dynamics of the human behavior in the Westchase Hilton Hotel Fire, Houston, Texas, on March 6, 1982. Quincy, MA: National Fire Protection Association.

Bryan, J. L., & Icove, D. J. 1997. Recent advances in computer-assisted arson investigation. *Fire Journal* 71 (1): 20–23.

Bukowski, R. W. 1995a. How to evaluate alternative designs based on fire modeling. *Fire Journal* 89 (2): 68–70.

———. 1995b. Predicting the fire performance of buildings: Establishing appropriate calculation methods for regulatory applications. In Proceedings of ASIAFLAM95 International Conference on Fire Science and Engineering, 9–18, Kowloon, Hong Kong, March 15–16.

Christensen, A. M. 2002. Experiments in the combustibility of the human body. *Journal of Forensic Sciences* 47 (3): 466–70.

Christensen, A. M., & Icove, D. J. 2004. The application of NIST's fire dynamics simulator to the investigation of carbon monoxide exposure in the deaths of three Pittsburgh fire fighters. *Journal of Forensic Sciences* 49 (1): 104–7.

DeHaan, J. D. 2001. Full-scale compartment fire tests. *CAC News* (Second Quarter): 14–21.

———. (2012), Sustained combustion of bodies: Some observations. *Journal of Forensic Sciences,* doi: 10.1111/j.1556–4029.2012.02190.x.

DeHaan, J. D., Campbell, S. J., & Nurbakhsh, S. 1999. Combustion of animal fat and its implications for the consumption of human bodies in fires.

Science & Justice 39 (1): 27–38, doi: 10.1016/s1355-0306(99)72011-3.

DeHaan, J. D., & Fisher, F. L. 2003. Reconstruction of a fatal fire in a parked motor vehicle. *Fire & Arson Investigator* 53 (2): 42–46.

DeHaan, J. D., & Icove, D. J. 2012. *Kirk's fire investigation,* 7th ed. Upper Saddle River, NJ: Pearson-Prentice Hall.

DeHaan, J. D., & Pope, E. J. 2007. Combustion properties of human and large animal remains. Paper presented at Interflam, July 3–5, London, UK.

Feld, J. M. 2002. The physiology and biochemistry of combustion toxicology. Paper presented at Fire Risk and Hazard Assessment Research Applications Symposium, July 24-26, Baltimore, Md.

Goldbaum, L. R., Orellano, T., & Dergal, E. 1976. Mechanism of the toxic action of carbon monoxide. *Annals of Clinical and Laboratory Science* 6 (4): 372–76.

Health Canada. 1995. Investigating human exposure to contaminants in the environment: A handbook for exposure calculations. Ottawa, Ontario, Canada: Ministry of National Health and Welfare, H. C. Health Protection Branch.

Jensen, G. 1998. Wayfinding in heavy smoke: Decisive factors and safety products; Findings related to full-scale tests. IGPAS: InterConsult Group.

Jin, T. 1976. Visibility through fire smoke: Part 5, Allowable smoke density for escape from fire *Report No. 42.* Fire Research Institute of Japan.

Jin, T., & Yamada, T. 1985. Irritating effects of fire smoke on visibility. *Fire Science and Technology* 5 (1): 79–90.

Levy, A. D., & Harcke Jr, H. T. 2011. *Essentials of forensic imaging.* Baton Rouge, LA: CRC Press.

Miller II, R. E., & Tredici, T. J. 1992. Night vision manual for the flight surgeon. *Special Report AL-SR-1992-0002.* Brooks Air Force Base, TX: Armstrong Laboratory.

Moran, Bruce. 2001. Personal communication. March 29.

Mulholland, G. W. 2008. Smoke production and properties. In *SFPE Handbook of Fire Protection Engineering,* 4th ed., ed. P. J. DiNenno, pt. 2, chap. 13, 12–297. Quincy, MA: National Fire Protection Association.

Myers, R.A.M., Linberg, S. E., & Cowley, R. A. 1979. Carbon monoxide poisoning: The injury and its treatment. *Journal of the American College of Emergency Physicians* 8 (11): 479–84.

Nelson, G. L. 1998. Carbon monoxide and fire toxicity: A review and analysis of recent work. *Fire Technology* 34 (1): 39-58, doi: 10.1023/a:1015308915032.

Penney, D. G. 2000. *Carbon monoxide toxicity.* Boca Raton: CRC Press.

———. 2008. *Carbon monoxide poisoning.* Boca Raton, FL: CRC Press.

———. 2010. Hazards from smoke and irritants. In *Fire toxicity,* ed. A. A. Stec & T. R. Hull. Boca Raton, FL: CRC Press.

Peterson, J. E., & Stewart, R. D. 1975. Predicting the carboxyhemoglobin levels resulting from carbon monoxide exposure. *Journal of Applied Physiology* 39:633–38.

Pope, E. J. 2007. The effects of fire on human remains: Characteristics of taphonomy and trauma. PhD diss., University of Arkansas, Fayetteville, Arkansas.

Pope, E. J., & Smith, O. C. 2003. Features of preexisting trauma and burned cranial bone. Presentation lecture, American Academy of Forensic Sciences (AAFS) 55th Annual Meeting, Chicago, February 17–22.

———. 2004. Identification of traumatic injury in burned cranial bone: An experimental approach. *Journal of Forensic Sciences* 49 (3): 431–40.

Purser, D. A. 2001. Human tenability. The technical basis for performance-based fire regulations. Paper presented at the United Engineering Foundation Conference, January 7–11, San Diego, CA.

———. 2008. Assessment of hazards to occupants from smoke, toxic gases and heat. In *SFPE Handbook of Fire Protection Engineering,* 4th ed., ed. P. J. DiNenno, pt. 2, chap. 6, 96–193. Quincy, MA: National Fire Protection Association.

———. 2010. Asphyxiant components of fire effluents. In *Fire Toxicity.* Boca Raton, FL: CRC Press, 118–98.

Routley, J. G. 1995. Three firefighters die in Pittsburgh House Fire, Pittsburgh, Pennsylvania. In *Major fires investigation project.* Emmitsburg, PA: U.S. Fire Administration.

SFPE. 2008. *SFPE Handbook of Fire Protection Engineering,* 4th ed. Quincy, MA: National Fire Protection Association, Society of Fire Protection Engineers.

Smith, O.B.C., & Pope, E. J. 2003. Burning extremities: Patterns of arms, legs, and preexisting trauma. Paper presented at the 55th Annual Meeting of the American Academy of Forensic Sciences, Chicago, IL.

Spitz, W. U., & Spitz, D. J., eds. 2006. *Spitz and Fisher's medicolegal investigation of death: Guidelines for the application of pathology to crime investigation,* 4th ed. Springfield, IL: Charles C. Thomas.

Stoll, A. M., & Greene, L. C. 1959. Relationship between pain and tissue damage due to thermal radiation. *Journal of Applied Physiology* 14 (3): 373–82.

USN. 1955. Field study of detectability of colored targets at sea. *Medical Research Laboratory Report No. 265* (vol. 14, no. 5). New London, CT: U.S. Naval Medical Research Laboratory.

8

Fire Testing

Mark Campbell, Fire Forensics Research Center, Denver, CO

> ❝I have devised seven separate explanations, each of which would cover the facts as far as we know them. But which of these is correct can only be determined by the fresh information which we shall no doubt find waiting for us.❞
>
> –Sir Arthur Conan Doyle
> *The Adventure of the Copper Beeches*

KEY TERMS

autoignition temperature, *p. 355*
calorimetry, *p. 358*

fire point, *p. 358*
flash point, *p. 357*

spontaneous ignition temperature, *p. 355*

OBJECTIVES

After reading this chapter, the student should be able to:

- Recognize the value of fire testing.
- Describe the basic types of fire tests and give examples of each.
- Learn how to interpolate scaled test data to real-world scenarios.
- Appreciate the use of fire test data in the analysis and evaluation of a working hypothesis.

ire testing is a form of physical modeling in which testing or experimentation is conducted using materials related to a case or study. Testing conducted in support of fire reconstruction can span the range from simple field tests requiring no equipment to benchtop flame tests to extensively instrumented, full-scale test burns in real buildings (as shown in Figure 8-1).

NFPA 921, 2011 ed., pt. 20.5.1, states, "Fire testing is a tool that can provide data that complements data collected at the fire scene (see *NFPA 921*, 2011 ed., pt. 4.3.3), or can be used to test hypotheses (see *NFPA 921*, 2011 ed., pt. 4.3.6). Such fire testing can range in scope from bench scale testing to full-scale re-creation of the entire event" (NFPA 2011).

Fire tests can help confirm or reject hypotheses about the fire's ignition or spread, test and validate the predictions of computational models or those of experienced investigators, or demonstrate the roles of various factors in the fire and its effects on occupants. This chapter explores, in summary form, many of the tests useful in fire investigation.

ASTM and CFR Flammability Tests

A number of test standards and methods applicable to fire investigation were listed in Table 1-3 (see Chapter 1). Several of the ASTM tests can be used to assess the ignitability of materials or the nature and speed of flame propagation, and they are discussed briefly here. However, as you will see, there are limitations to the application of results of these tests to an actual case. Many limitations are due mostly to the geometry or size of the sample.

A flame will spread much more quickly upward than it will downward or outward in the same fuel. A fire started at the top edge of a sofa back will spread much more slowly

FIGURE 8-1 Testing for fire reconstruction purposes can span the range from simple field tests requiring no equipment to extensively instrumented full-scale test burns on real buildings. *Courtesy of Michael Dalton, Knox County Sheriff's Office.*

than if the same fabric is ignited at the base of the sofa. A fire ignited at the top of hanging draperies will be much more likely to cause the hooks and drapery rod to fail and collapse with the draperies, possibly resulting in dropdown ignition of materials beneath, whereas draperies ignited at the bottom are often nearly completely consumed before they cause failure of their supports.

Fire resistance properties today are sometimes built into a combination of layers of fabrics and backings that result in acceptable ignition resistance. Separating and testing the layers individually may yield very misleading results. The residual moisture in test samples also can affect their ignitability, so specimens are often conditioned for 24–48 hours under specific criteria prior to testing.

Most ASTM standard methods include an advisory comment regarding use of the method to compare materials or their performance under controlled conditions to assess the possible contributions a material *might* make under a specific set of fire conditions and not to predict what a product *will* do when exposed to different fire conditions. Some of the tests reference other entities such as the National Fire Protection Association (NFPA) and the U.S. Consumer Product Safety Commission (CPSC).

- *Flammability of Clothing Textiles (16 CFR 1610-U.S.):* This test requires that a sample of fabric 50 mm × 150 mm (2 in. × 6 in.) placed in a holder at a 45° angle and exposed to a small flame touching its surface for 1 second not ignite and spread flame up the length of the sample in less than 3.5 seconds for smooth fabrics or 4.0 seconds for napped fabrics. Test sample must be oven dried and cooled prior to the test.
- *ASTM D 1230: Standard Test Method for Flammability of Apparel Textiles (not identical with 16 CFR 1610):* CPSC requires fabrics introduced into commerce to meet requirements of *16 CFR 1610*. This test is suitable for textile fabrics as they reach the consumer or for apparel other than children's sleepwear or special protective garments. A sample of fabric 50 mm × 150 mm (2 in. × 6 in.) is held at a 45° angle in a metal holder and a controlled flame is applied to the bottom end for 1 second. The time required for the flame to proceed up the fabric a distance of 127 mm (5 in.) is recorded.
- *Flammability of Vinyl Plastic Film (16 CFR 1611-U.S.):* This test requires that vinyl plastic film (for wearing apparel) placed in a holder at a 45° angle and ignited not burn at a rate exceeding 3.0 cm (1.2 in.) per second.
- *Flammability of Carpets and Rugs: 16 CFR 1630 (Large Carpets) and 16 CFR 1631 (Small Carpets):* A specimen of carpet is placed under a steel plate with a 20.32-cm (8-in.)-diameter circular hole. A methenamine tablet is placed in the center of the hole and ignited. The duration of burning and heat release rate of the tablet duplicate those of a typical dropped match. If the specimen chars more than 3 in., in any direction, it is considered a "fail." The ambient radiant heat flux is minimal, because the sample is tested at room temperature. The ignition source is a modest 50 to 80 W flame of brief duration. Testing has shown that some carpets that pass this test will be readily ignited and spread flames if the radiant heat flux is larger as a result of a larger and more prolonged ignition source. The carpet is tested in the flat, horizontal position, so the same carpet mounted vertically *may behave very differently*.
- *ASTM D2859: Standard Test Method for Ignition Characteristics of Finished Textile Floor Covering Materials:* This method uses a steel plate 22.9 cm (9 in.) square and 0.64 cm (0.25 in.) thick with a 20.32-cm (8-in.)-diameter circular hole. A methenamine tablet is placed in the center and ignited in a draft-free enclosure. Samples are thoroughly oven-dried and cooled in a desiccator before being tested. The product fails if the charred portion reaches within 25 mm (1 in.) of the edge of the hole in the steel plate. Eight specimens are tested. The Flammable Fabrics Act (FFA) regulations require that at least seven of the eight specimens pass this test.

- *Flammability of Mattresses and Pads (16 CFR 1632-U.S.):* A minimum of nine regular tobacco cigarettes are burned at various locations on the bare mattress—quilted and smooth portions, tape edge, tufted pockets, and so forth. The char length of the mattress surface must not be more than 50 mm (2 in.) from any cigarette in any direction. The test is repeated with the cigarettes placed between two sheets covering the mattress.
- *Standard for the Flammability (Open Flame) of Mattress Sets (16 CFR 1633-U.S):* This test method is a full-scale test based on NIST research. The mattress or mattress and foundation set is exposed to a pair of T-shaped propane burners for a short time and allowed to burn freely for 30 minutes. The burners are designed to represent burning bedclothes. Measurements are taken to determine the heat release rate from the specimen and total heat energy generated from the fire. The standard establishes two test criteria, both of which the mattress set must meet to comply with the standard. (1) The peak heat release rate for the specimen must not exceed 200 kW at any time during the 30-minute test. (2) The total heat release must not exceed 15 MJ for the first 10 minutes of the test (*Federal Register* 71 (no. 50), March 15, 2006).
- *Flammability of Children's Sleepwear (16 CFR 1615 and 1616-U.S.):* Each of five 88.9 cm × 25.4 cm (35 in. × 10 in.) specimens is suspended vertically in a holder in a cabinet and exposed to a small gas flame along its bottom edge for 3 seconds. The specimens cannot have an average char length of more than 18 cm (7 in.), no single specimen can have a char length of 25.4 cm (10 in.) (full burn), and no single sample can have flaming material on the bottom of the cabinet 10 seconds after the ignition source is removed. These requirements are for finished items (as produced or after one washing and drying) and after the items have been washed and dried 50 times.
- *ASTM E1352: Standard Test Method for Cigarette Ignition Resistance of Mock-Up Upholstered Furniture Assemblies:* This test uses reduced-scale plywood mock-ups, 47 cm × 56 cm (18 in.× 22 in.), upholstered with simulations of seat, backrest, and armrests to test ignitability of furniture to dropped smoldering cigarettes. The test is used for furniture to be used in public and private occupancies such as nursing homes and hospitals. The cigarettes are positioned on each of the various surfaces and along crevices between seats, armrest, and back. The distance of char extension from each cigarette or ignition by open flame is recorded for comparison.
- *ASTM E1353: Standard Test Method for Cigarette Ignition Resistance of Components of Upholstered Furniture:* This test uses mock-ups to test individual components—cover fabrics, welt cords, interior fabrics, and filling or batting materials in the form, geometry, and combination in which they are used in real furniture. This test uses single cigarettes, but each is covered with a single layer of cotton sheeting that retains more heat and makes it a more severe test than one in open air.
- *ASTM E648: Standard Test Method for Critical Radiant Flux of Floor-Covering Systems Using a Radiant Heat Energy Source:* This test uses a horizontally mounted floor covering specimen 20 cm × 99 cm (8 in. × 39 in.) with a gas-fired radiant heater (with a gas pilot burner) mounted at a 30° angle above it that produces a radiant heat flux of 1–10 kW/m^2. The distance the flame propagates along the sample indicates the minimum radiant heat flux for ignition and propagation.

ASTM Test Methods for Other Materials

The ASTM International Committee E05 on Fire Standards identifies methods for testing other materials.

- *ASTM D1929: Standard Test Method for Determining Ignition Temperature of Plastics (ISO 871):* This test uses a cylindrical hot-air furnace to heat the test sample in a pan. A thermocouple monitors the temperature of the sample while the temperature is adjusted so that pyrolysis gases venting through the top of the chamber can be ignited by a small pilot flame held near it, which establishes the *flash ignition temperature* (FIT). The same apparatus can be used to establish the **spontaneous ignition temperature (SIT)** by eliminating the pilot flame and observing the sample visually for flaming or glowing combustion (or a rapid rise in the sample temperature). SITs for common plastics are 20°C–50°C (68°F–122°F) higher than FITs by this technique. The test requires 3 g of material for each test in the form of pellets, powder, or cut-up solids or films.

■ *ASTM E119: Standard Test Methods for Fire Tests of Building Construction and Materials:*   A specimen (wall assembly, floor, door, etc.) is exposed to a standard fire exposure to evaluate the duration for which those assemblies will contain a fire or retain their integrity. A large furnace (horizontal or vertical) is used to create a standard time–temperature environment. The assembly is exposed to the prescribed conditions, and its surface temperature is measured, its structural integrity is monitored, and the time for penetration of flame through it is determined. The standard time–temperature exposure does not accurately reflect all real-world fire exposures, so the performance rating of tested building components is for comparison only.

■ *ASTM E659: Standard Test Method for Autoignition Temperature of Liquid Chemicals:*   A small sample (100 µL) of liquid is injected into the mouth of a glass flask heated to a predetermined temperature, and the flask is observed for the presence of a flash flame inside the flask and a sudden rise in internal temperature (as monitored by an internal thermocouple). If no ignition is observed, the temperature is increased and the test repeated until the material ignites reliably. This method establishes the *hot-flame* **autoignition temperature** (AIT) in air. Ignition delay times may also be recorded. The temperature at which small sharp rises occur in the internal temperature alone is the *cool-flame autoignition temperature.*

   The method can also be used for solid fuels that melt and vaporize or sublime completely at the test temperatures, leaving no solid residues. The test conditions are controlled by the heat transfer between the glass flask and the fuel introduced and the confined geometry of the spherical flask. In real-world ignitions, the nature of the surface and the manner of contact will determine the convective heat transfer coefficient and may modify times or temperatures. Any geometry (tube or enclosure) that keeps the fuel in contact with the hot surface is going to produce ignitions at lower temperatures than an open, flat surface, where buoyancy of the heated vapors can transport the fuel away from the heated surface.

■ *ASTM D3675: Standard Test Method for Surface Flammability of Flexible Cellular Materials Using a Radiant Heat Energy Source:*   This method uses a gas-fired radiant heat panel 30 cm × 46 cm (12 in. × 18 in.) in front of an inclined specimen of material 15 cm × 46 cm (6 in. × 18 in.), oriented so that ignition with a pilot flame first occurs at the upper edge of the sample and the flame front moves downward. The rate at which the flame moves downward is observed, and a flame spread index is calculated. The test is suitable for any material that may be exposed to fire. At least four specimens must be tested.

■ *ASTM D3659: Standard Test Method for Flammability of Apparel Fabrics by Semi-restraint Method:*   This test simulates the burning characteristics of a garment hanging vertically from the shoulders of a wearer. Test specimens are 15 cm × 38 cm (6 in. × 15 in.) (five specimens required) and are oven-dried and weighed prior to testing. The sample is hung vertically from a crossbar, and the flame of a small gas burner is positioned against the bottom edge of the fabric for 3 seconds and then removed. The weight (percentage area) destroyed by the flame and the time required are the criteria for comparison. (Cross-reference to *FF 3-71: Flammability of Children's Sleepwear, Sizes 0–6X.*)

spontaneous ignition temperature (SIT) ■ See **autoignition temperature**.

autoignition temperature ■ The temperature at which a material will ignite in the absence of any external pilot source of heat; also referred to as the **spontaneous ignition temperature**.

- *ASTM E84: Standard Test Method for Surface Burning Characteristics of Building Materials:* This is the "Steiner Tunnel" test, which mounts specimens on the underside of the top of an insulated tunnel 7.6 m (25 ft) long, 30 cm (12 in.) high, and 45 cm (17.75 in.) wide. A gas burner at one end ignites the sample, and the rate of flame spread along the length of the specimen is observed and recorded. The test specimen must measure at least 51 cm × 7.32 m (20 in. × 24 ft). An optical density measurement system (light source and photocell) is mounted in the vent pipe of the apparatus. Samples of products of combustion can be taken downstream of the photometer. The upside-down configuration of this test apparatus is not often found in real-world situations except with combustible ceiling coverings.

- *ASTM E1321: Standard Test Method for Determining Material Ignition and Flame Spread Properties:* This method tests for the ignition and flame spread properties of a vertically oriented fuel surface when exposed to an external radiant heat flux. The results can be used to calculate the minimum radiant heat flux and temperature needed for ignition and flame spread. The test configuration is a large, vertically mounted specimen and a gas-fired radiant panel heater mounted at an angle to it. A gas pilot flame is used, and the rate and distance of flame spread along the length of the specimen are recorded (see Figure 8-2).

- *ASTM E800: Standard Guide for Measurement of Gases Present or Generated during Fires:* This guide describes methods for properly collecting and preserving combustion gas samples during fire tests and analytical methods for O_2, CO, CO_2, N_2, HCl, HCN, NO_x, and SO_x, using gas chromatography, infrared, or wet chemical methods. As we have seen, some fuels produce high concentrations of toxic or irritant gases under various fire conditions. Testing materials known to be present in a fatal fire may yield clues as to what role their combustion gases played in the fatality.

- *California TB 603 Standard for Mattresses, Box Springs, and Futons:* After January 1, 2005, the state of California required that bedding materials sold for residential and commercial use comply with new testing procedures. The test is designed to simulate real-world flaming ignition sources and measure the energy released from burning bedding materials. The top and side of the mattress are exposed to T-shaped gas burners for a period of time (as in Figure 8-3). Failure of the test can be either (1) a peak heat release rate of 200 kW or greater within 30 minutes after ignition or (2) a total heat release of 25 MJ or greater within 10 minutes of ignition. As of July 1, 2007, all mattresses, mattress and foundation in sets, and futons sold for residential use in the United States must pass the *16 CFR 1633* test, which is similar, except the total heat released must not exceed 15 MJ in the first 10 minutes.

FIGURE 8-2 Surface rate of flame spread test (*ASTM E 1321*) uses a radiant heat panel at an angle to the surface being tested. *Courtesy of NIST.*

Flash and Fire Points of Liquids

Tests for flash and fire (flame) points of ignitable liquids can be as simple as placing a few drops of the fuel in a petri dish or watch glass at room temperature and waving a lighted match near the surface of the liquid. Ignition will confirm that the material is a Class I A or B flammable liquid with a flash point below room temperature (23°C [73°F]). More reliable testing requires more controlled evaporation and confinement of the vapors with a more reproducible ignition source. Flash point testing in general involves taking a quantity of liquid fuel and placing it in the cup of a water bath whose temperature is gradually increased. The cup may be open to the atmosphere or closed with a small shutter. A very small flame is introduced periodically, and the cup is observed for the presence of a brief flash of flame. The fire point of many liquids may be established by the open-cup method by increasing the temperature until a self-sustaining flame is established rather than just a flash of flame.

Closed-cup testers retain some of the vapor produced and make it easier to ignite. Closed-cup flash points are typically a few degrees lower than open-cup determinations of the same fuel.

Tests for flash and fire points of liquids include the following:

- *ASTM D56: Standard Test Method for Flash Point by Tag Closed Cup Tester:* Applicable to low-viscosity liquids with flash points below 93°C (200°F). Requires 50 mL of liquid for each test. (Sometimes abbreviated TCC.)
- *ASTM D92: Standard Test Method for Flash and Fire Points by Cleveland Open Cup Tester:* Applicable to all petroleum products with flash points above 79°C (175°F) and below 400°C (752°F) except fuel oils. Requires at least 70 mL for each test. (Sometimes abbreviated COC.)
- *ASTM D93: Standard Test Methods for Flash Point by Pensky-Martens Closed-Cup Tester:* Applicable for petroleum products with a flash point in the range of 40°C–360°C (104°F–680°F) including fuel oils, lubricating oils, suspensions, and higher-viscosity liquids. Requires at least 75 mL of fuel for each test.
- *ASTM D1310: Standard Test Method for Flash Point and Fire Point of Liquids by Tag Open-Cup Apparatus:* Applicable for liquids with **flash points** between −18°C and 165°C (0°F and 325°F) and **fire points** up to 165°C (325°F) will work at subambient temperatures. Fire point criterion: When fuel ignites and burns for at least 5 seconds. Requires 75 mL of sample for each test.
- *ASTM D3278: Standard Test Methods for Flash Point of Liquids by Small Scale Closed-Cup Apparatus:* Suitable for flash point determination of fuels with a **flash point** between 0°C and 110°C (32°F to 230°F) in small quantities (2 mL required for each test). Employs a microscale apparatus sold as the Setaflash Tester. Test methods similar to those for ISO 3679 and 3680 will work on subambient determinations.
- *ASTM D3828: Standard Test Methods for Flash Point by Small Scale Closed Tester:* Similar to *ASTM D3278*. Used to establish whether a product will flash at a given temperature, using 2–4 mL of sample for each test.

Owing to the multiplicity of flash point test methods, the *ASTM E502-07e1* offers a *Standard Test Method for Selection and Use of ASTM Standards for the Determination of Flash Points of Chemicals by Closed-Cup Methods* (ASTM 2007a).

FIGURE 8-3 The TB 603 test in use on a mattress. All mattresses sold in the United States for residential use after July 1, 2007, must pass the *16 CFR 1633* test, which has a total heat release limit of 15 MJ, compared with the 25 MJ limit of this test. *Courtesy of Bureau of Home Furnishings.*

flash point ■ The minimum temperature at which (under defined test conditions) the vapor being produced by a material can be ignited to produce a brief flash (not sustained) of flame.

Calorimetry

Classical bomb **calorimetry** is used to measure the total heat of combustion of fuels (as in *ASTM D4809*). The term *bomb* refers to a sealed vessel that can withstand internal pressures. The procedure involves burning a weighed specimen in a sealed container with an excess of oxygen until the specimen is completely oxidized. The heat generated by the combustion is determined by measuring the increase in temperature of the sealed vessel and using its specific heat and mass to calculate the heat released in the combustion. This test is unsuitable for fuels comprising multiple substances, as it yields the *total* heat of combustion, not the *effective* heat of combustion (which accounts for losses due to incomplete combustion).

Because almost all common fuels yield nearly the same amount of heat per mass of oxygen consumed (13 kJ/g), one could easily calculate the heat generated by measuring the amount of oxygen consumed during the combustion process. This procedure is called *oxygen consumption* (or oxygen depletion) *calorimetry*. If a specimen is burned in air and the waste products are all drawn into a system of ducts, the O_2, CO, and CO_2 concentrations can be measured. If the ventilation rate into the test chamber and through the exhaust duct is measured, the amount of oxygen consumption can be determined. Optical sensors in the ductwork can also monitor the optical density and obscuration of smoke products generated. Note that both the oxygen consumption and the heat release rate are calculated from collected data.

This approach has been refined into four general categories of size and application: cone, furniture, room, and industrial calorimeters. In the *cone calorimeter*, developed by Babrauskas at NIST and described in *ASTM E1354-11b*, a 10 cm × 10 cm (4 in. × 4 in.) specimen in a metal tray is exposed to a uniform incident radiant heat flux from a cone-shaped electric heater mounted above it (as in Figure 8-4) (ASTM 2011).

The radiant flux striking the sample can be controlled by adjusting the temperature of the heater element. A small electric arc source is introduced to ignite the plume of smoke generated by the heating and then removed. The flame and smoke vent upward through the center of the cone heater, and the gases are ducted to where the flow rate; O_2, CO, and CO_2 concentrations; and smoke obscuration are measured. The entire specimen tray is mounted on a sensitive load cell, so the mass loss is continuously measured. The analysis calculates the heat release rate (expressed per unit surface area of fuel), mass loss rate, and effective heat of combustion. Even though the samples tested are reduced scale, testing has shown that the results have a high degree of correlation with real-world fire performance (Babrauskas 1997). The system works with liquid or solid fuels (even low-melting-point thermoplastic materials) in the horizontal configuration. Wood and other rigid fuels can also be tested in the vertical configuration.

Furniture calorimeters were developed to test the heat release rate of single items of full-scale real furniture by the same method. Here, the furniture is mounted on a load cell and burned in the open, with all combustion products drawn into a vent hood. The temperature; pressure; and O_2, CO, and CO_2 concentrations are measured in the ductwork. The calculated heat release rate can then be combined with the mass loss data from the load cell to calculate the effective heat of combustion.

When multiple pieces of furniture, along with carpet, wall linings, and other fuels, are to be evaluated as a fuel package, a *room calorimeter* is used. In this case a room is built (often to a standard size of 2.4 m × 3.6 m × 2.4 m (8 ft × 12 ft × 8 ft), and the entire room becomes the "collector" for the exhaust vent. The combustion products vent through the door opening and are drawn up into the exhaust hood and measured. This arrangement allows heat release rate to be calculated constantly as fire spreads from one item to another, even culminating in near-flashover conditions.

The largest-scale devices are often called *industrial calorimeters*. These are basically a large exhaust hood with appropriate fans, ducting, and instrumentation. They are generally located in high-bay buildings. Typical measuring capacities are 12 kW

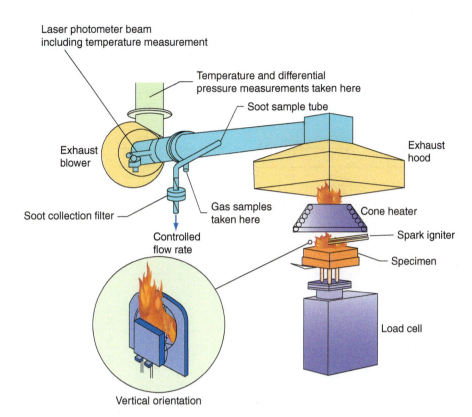

Laser photometer beam including temperature measurement

Temperature and differential pressure measurements taken here

Soot sample tube

Exhaust blower

Exhaust hood

Cone heater

Spark igniter

Specimen

Soot collection filter

Gas samples taken here

Controlled flow rate

Load cell

Vertical orientation

FIGURE 8-4 In the cone calorimeter, developed by Babrauskas at NIST and described in *ASTM E1354-11b*, a 10 cm × 10 cm (4 in. × 4 in.) specimen in a metal tray is exposed to a uniform incident radiant heat flux from a cone-shaped electric heater mounted above it. *Courtesy of Dr. Vytenis Babrauskas, Fire Science & Technology, Issaquah, Washington.*

(cone calorimeter), 1.5–5 MW (furniture calorimeter), 3–6 MW (room calorimeter), and 10–50 MW (industrial calorimeter). The latter would be most useful for performing fire scene reconstructions; unfortunately, most of these devices are owned by organizations (e.g., UL, ATF) that do not offer commercial testing services for fire reconstruction. Southwest Research Institute (San Antonio, Texas) has 4–10 MW calorimeters available for commercial fire reconstruction projects.

Furnishings

Tests for ignitability of upholstered furniture vary widely, from cigarette ignition to small flames to larger gas-fired burners of known heat output (typically 17–40 kW). They may be applicable to standardized small-scale mock-ups of furnishings as well as to the actual items taken from the production line. Pass/fail criteria may be based on observable flame spread, penetration, percentage mass loss, peak heat release rate, quantity of smoke produced, or toxicity of fire gases.

TEST METHODS

Various test methods in use today include those described under the U.S. Code of Federal Regulations (CFR), British Standards (BS), International Organization for Standardization (ISO), ASTM, and California Bureau of Home Furnishings Technical Bulletins (TB). Krasny, Parker, and Babrauskas (2001) have assembled a comprehensive description and discussion of the current techniques, which are summarized in Table 8-1. See also *Kirk's Fire Investigation*, 7th ed., for additional information (DeHaan and Icove 2012).

TABLE 8-1	Summary of Furniture Item Flammability Tests				
TEST	**CIGARETTE IGNITION**	**FLAME IGNITION**	**HRR**	**SMOKE**	**TOXICITY**
ISO 8191	X	X			
BS 5852	X	X			
ATSM E1352	X				
ASTM E1353	X				
NFPA 261	X				
NFPA 260	X				
California TB116, 117	X				
Upholstered Furniture Act Council	X				
BIFMA	X	X			
CFR 1632	X				
BS 6807	X	X			
Calif. TB 133		X	X	X	X
Calif. TB 129		X	X	X	X
Calif. TB 121		X		X	X
ASTM E1537		X		X	X
ASTM E1590		X		X	X
ASTM E162		X			
ASTM D3675		X			
ISO 5660			X	X	X
ASTM E1474			X	X	X
ASTM E1354			X	X	X
NFPA 264A			X	X	X
ASTM F1550M				X	
ISO TR 5924				X	
ASTM E662				X	
ASTM E906			X	X	
ISO TR 9122					X
ISO TR 6543					X
ASTM E1678					X

Source: Derived from Krasny, Parker, and Babrauskas 2001.

As with the bench-scale tests described earlier, there are limitations and cautions regarding the application of these test results to the reconstruction of real-world fires. For example, the mere passing of a particular test by a target material does not imply that it will be fire resistant or even fire safe in a real-world environment.

Other cautions to be observed include requirements for conditioning samples for 24–48 hours at a particular temperature and humidity. The investigator should be aware of the potential influence of such variables on the test result. The geometry of the test compartment is also important. The same piece of furniture may yield different heat outputs or fire patterns if tested in a corner or against a wall versus in the center of a large room that minimizes radiant feedback or ventilation effects.

The location, manner, means, and duration of ignition can play a significant role in the fire performance of furnishings. A vinyl-covered chair seat, for instance, can provide substantial resistance to small ignition sources if an ignition source is placed on the seat (PVC upholstery tends to char, intumesce, and shield the substrate). But even a small flaming source placed under the seat where the flame can contact the polyurethane padding will readily ignite the chair.

CIGARETTE IGNITION OF UPHOLSTERED FURNITURE

Cigarette ignition of upholstered furniture was a major fire cause for decades owing to the predominant use of cotton or linen upholstery fabrics over cellulosic padding of cotton, coir, kapok, or similar materials, all of which were easily ignited by a glowing cigarette.

Holleyhead (1999) published an extensive review of the literature regarding cigarette ignition of furniture. However, a great deal of misinformation once existed about the time required for such ignition to result in flaming combustion. Investigators erroneously concluded that such ignitions always required 1–2 hours, inferring that any flaming ignition in a shorter time must have been the result of deliberate ignition by open flame.

Extensive testing by the California Bureau of Home Furnishings and NIST (then NBS), first reported in the early 1980s, demonstrated that ignition time was variable. Transition to flaming combustion was reported in as little as 22 minutes from the placement of a glowing cigarette between the cushions, or between the cushions and an arm or backrest. Other tests required 1 to 3 hours, and some never transitioned to open flames and smoldered for hours before self-extinguishing. Even identically built mock-ups demonstrated a wide variety of behaviors. A statistical analysis of these results was published by Babrauskas and Krasny (1997).

Tests by the authors have demonstrated similar time and ignition results. If the fabric is torn and the cigarette drops into direct contact with cotton padding that has not been treated to be flame retardant, transition to flame has been observed in as little as 18 minutes in still air. The presence of drafts can influence the progress and increase the heat release rate of smoldering cellulosics. One informal test by DeHaan with a cigarette wrapped loosely in a cotton towel and left outside in a light, variable wind resulted in flames about 15 minutes after placement. The mechanism for transition from smoldering to flame has so far defied modeling or accurate predictions. Hand-rolled cigarettes present a lower ignition risk because they tend to self-extinguish more often than commercially made tobacco cigarettes. For references and testing of cannabis resin cigarettes on petrol-soaked clothing, see the work by Jewell, Thomas, and Dodds (2011).

According to the NFPA-sponsored website firesafecigarrettes.org/, all 50 states have passed legislation mandating what is termed as "fire-safe" or self-extinguishing cigarettes. The European Union (EU) established standards for its member countries and required as of November 17, 2011, that all cigarettes sold in the EU be "reduced ignition propensity" (RIP) cigarettes. The term RIP is a more technically accurate term than "fire-safe" for these cigarettes, because it means that they have been designed to be less likely than a conventional cigarette to ignite soft furnishings such as a couch or mattress. These redesigned cigarettes will only reduce chances for cigarette ignition of upholstered materials in the future but will not prevent all ignitions. Fire-safe cigarettes typically use a banded paper with alternating high and low porosity in the bands, which causes smoldering to cease.

Latex (natural) rubber foam is also easily ignited by contact with a glowing cigarette. It should be noted that a cigarette will be consumed in a maximum of 20–25 minutes after being lit (somewhat longer if tucked between seat cushions). Ignition times longer than that result from a self-sustaining smolder in the surrounding fuel initiated by the cigarette, as confirmed by the significant quantity of white smoke that is produced for some minutes prior to flame (far more than can arise from a single cigarette).

Additional Physical Tests

SCALE MODELS

A great deal of useful information can be obtained from building scaled-down replicas of rooms or furnishings for testing, at reduced cost and complexity compared with full-scale re-creations. These models are particularly valuable for testing ignition and incipient fire hypotheses (in which the interaction of room surfaces or conditions does not play a major role).

When radiant heat or ventilation factors are involved, the test must take into account that all factors in a fire do not scale down in the same linear fashion as the room dimensions. For instance, ventilation through an opening is controlled by the area of the opening and the square root of its height. It is possible to calculate corrections for such factors, but it is not as simple as building a dollhouse, with all doors and windows in the same proportion as in full scale. Radiant heat flux falls off with the inverse square ($1/r^2$) of the separation distance, so scale-model builders must keep those relationships in mind as well.

As with all fire investigation tools available to the investigator, a *reduced-scale enclosure* (RSE) can provide a lot of information that may be very helpful and is relatively simple to build and burn. The hot smoke is dynamically a fluid, so its movement through a *full-scale enclosure* (FSE) tends to be similar to that through a RSE. In an early paper, Quintiere evaluated various scaling considerations regarding heat transfer and smoke movement (Quintiere 1989).

Current research in ventilation openings (doors) in RSEs is investigating scaling the vertical dimension but adjusting the width to maintain the entrained air mass flow rate by modifying the ventilation factor $A_0\sqrt{h_0}$, where A_o is the area of the door opening, and h_o is the height of the door opening. For example, if the scale is one-fourth (0.25), then the width of the RSE door is $\sqrt{0.25}$ (or 0.5) times the width of the FSE door. Thus, a 30-in. door opening at one-fourth (0.25) scale would be 7-1/2 in. wide, but with the adjustment of the square-root factor, the opening would be 15 in. wide. Note that only the width of the ventilation opening is adjusted (Bryner, Johnsson, and Pitts 1995).

In addition, current research appears to suggest that the furniture construction does not require precise and intricate construction. The use of standard 1/2-in. oriented-strand board (OSB) as the framework, polyurethane as the padding, and furniture upholstery as the covering appears to provide adequate fuel packages. Overall dimensions are scaled, but the thickness of the furniture is not. One style of scaled furniture cuts the center out of the OSB to form a frame from which the chairs and couches are made. They are glued or stapled to form the structure. Furniture upholstering is used to wrap the frame and glued in place. The polyurethane is covered with furniture upholstering and placed on the frame. Beds have been made of two layers of polyurethane covered with bed sheets. Simple wood structures are placed under the bed to form the bed frame. See Figure 8-5 for innovative work by Mark Campbell on scale-model demonstrations.

Current testing has shown similarities between FSE, RSE, and FDS models. Some of these similarities include calcination plots, clean-burn areas, ventilation burn patterns, flashover, backdrafts, and thermocouple data. Thermocouple plots in RSEs have shown the growth phase of the fire and the transition to flashover, turbulent mixing that occurs in the enclosure, and decay of the fire. This type of information may be used to help the investigator examine patterns, locations, and intensities.

It should be recognized that a RSE will help provide information related to fire effects but should not be used to calculate heat release rate, fire spread time, mass loss rate, gas species, or gas concentrations, as these are more difficult to scale. Heat release rate, speed of smoke movement, and time to flashover will also not be linearly correlated. This valuable information is used to help the investigator refine the working hypothesis and should never be the stand-alone research or testing.

(a)

FIGURE 8-5 (a) Quarter-scale room built by Mark Campbell and Dan Hrouda. A front view of the Reduced Scale Enclosure. The drywall front will be added and will face the door opening. Two cameras will be added: one on side A, facing the back of the room, and one on side D, facing the couch. The propane sand burner is in corner B/C. Fifteen thermocouples are located throughout the cell. *Mark Campbell, Fire Forensics Research Center, Denver, CO*

(b)

FIGURE 8-5 (b) This is a side view from side D looking toward side B just after the propane sand burner was shot off after 90 seconds. The propane sand burner is located just to the right of the couch. The heat release rate was scaled to be approximately 3 kW so it would scale to the $l^{5/2}$, as required by the scaling laws. A 100 kW fire is used in the Full Scale Enclosure. *Mark Campbell, Fire Forensics Research Center, Denver, CO*

(c)

FIGURE 8-5 (c) The fire has spread across the couch, and now the hot ceiling gases have stared to pyrolyze the chair and caused it to ignite. The thermocouples in the middle of the room help determine when the room starts the transition through flashover. *Mark Campbell, Fire Forensics Research Center, Denver, CO*

(d)

FIGURE 8-5 (d) The fire exits the Reduced Scale Enclosure, as it is now in the postflashover burning stage. The enclosure is burned for 2 minutes postflashover. The Reduced Scale Enclosure provides the investigator with another tool to help assess the various fire effects and burn patterns. *Mark Campbell, Fire Forensics Research Center, Denver, CO*

Floor or wall materials proportionally reduced in thickness may respond to thermal insult as *thermally thin* targets if they are less than a few millimeters thick and may respond differently than thicker sections of the same material. Special materials may be selected for models of walls and ceiling so their thermal behavior will duplicate, in scale, that of gypsum or plaster walls.

FLUID TANKS

A great deal of the information used to create and test the models of smoke and hot gas movement discussed in previous chapters was developed in *fluid tanks* with no fire or smoke at all. Because the buoyancy of the hot gases relative to normal room air drives the movement of smoke in a building, the same process can be simulated by using two liquids of different densities (such as fresh- and saltwater). If one fluid is dyed, its movement and mixing can easily be observed and recorded. See Figure 8-6 for innovative work on fluid tank model demonstrations.

Often, a scale (one-eighth to one-fourth) model of the room(s) is built of clear Plexiglas and immersed and inverted in a large tank of the lighter (less dense) liquid. The heavier liquid is then introduced into the model in a manner to simulate the generation of gases from a fire in the room. Scaling factors are important, and the relative densities and viscosities of the liquids have to be calibrated to simulate the mixing of gases in a buoyant flow, but the technique can provide answers to some investigative problems and confirm or reject computational models and hypotheses (Fleischmann, Pagni, and Williamson 1994).

FIELD TESTS

Identification of the first fuel ignited is critical to the reconstruction process. If the first fuel is more easily susceptible to open-flame ignition than to glowing- or hot-surface

(a)

FIGURE 8-6 (a) Side view of quarter-scale fluid tank demonstrating cool denser fluid (representing cool air) entering the chamber through the lower half of the door opening after the door (extreme right) is opened. The buoyant fluid (blue) is venting from the top of the opening. These tests were conducted to evaluate flows and mixing in possible backdraft events in compartment fires. *Courtesy of Dr. Charles Fleischmann, University of Canterbury, Christchurch, New Zealand.*

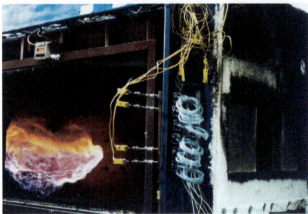

(b)

FIGURE 8-6 (b) Quarter view of half-scale fire test chamber. An underventilated burner in the closed chamber has produced a hot, fuel-rich, buoyant smoke layer. When the door (on the right) is opened, the buoyant smoke escapes, and cool fresh air flows in the bottom of the opening to mix with the hot smoke. When ignited by a pilot flame at the far end, the flame progresses quickly along the premixed interface as a backdraft event. This behavior was predicted by the fluid tank tests shown in (a). *Courtesy of Dr. Charles Fleischmann, University of Canterbury, Christchurch, New Zealand.*

ignition (for example, such fuels as natural gas, gasoline vapors, or polyethylene plastic), the fire scene search can focus on the appropriate type of potential ignition sources. Cellulosic fuels or other natural materials such as wool are much more easily ignited by hot-surface sources like discarded cigarettes or glowing electrical connections. Most such materials can be reliably identified by visual inspection, but information as to their behavior when ignited is much more within the realm of expertise of the fire investigator.

A simple *flame* or *ignition susceptibility test* using a match or lighter applied to one corner while the small test sample is held vertically in still air will reveal how readily it will ignite and whether it will support flaming combustion when the ignition source is removed, as described in *NFPA 705* (NFPA 2009). However, *NFPA 701* (NFPA 2010) is now considered the preferred practice.

Under the test conditions, cellulosic materials support a yellow flame with a gray smoke. When the flame is blown out, these materials tend to support glowing combustion. The ash left behind is gray to black in color and powdery or crumbly in texture. There is no droplet of hard-melted residue. Thermoplastic synthetic materials melt as they burn, so melted, burning droplets result. They also tend to shrink and shrivel as they melt and often burn with a blue-based flame, with smoke ranging from negligible to white (polyethylene) to heavy and inky black (polystyrene). These materials do not support smoldering combustion.

Thermosetting resins are usually more difficult to ignite than other fuels. They tend to smolder when flame is removed, produce an aggressive smoke, and leave a hard, semiporous residue. Elastomers (rubbers) can be either natural (latex) or synthetic and can behave like either. Some will burn very readily; others, like silicone rubber, burn much more reluctantly. Silicone rubber leaves a brilliant white, powdery ash, whereas other elastomers leave a hard, dark porous mass. Samples of any unknown material suspected of being the first fuel ignited should be collected for laboratory analysis if there is doubt concerning their contribution to the ignition or growth of the fire.

Caution should be exercised in both conducting and interpreting such informal ignition tests. The *NFPA 705* test has been criticized as being subjective and variable owing to the procedures and personnel involved. There is also a risk of injury to personnel and a reported cause of serious accidental fires. A strong suggestion is to use *NFPA 701: Methods of Fire Tests for Flame Propagation of Textiles and Films* when conducting flame or ignition susceptibility tests (as in Figure 8-7) (NFPA 2010).

NFPA 701 is basically a flammability test that applies to typical straight-hanging fabrics such as curtains, draperies, and window treatments. The test determines a fabric's rate of flammability subjected to certain ignition sources. In *NFPA 701* tests, the material is hung vertically and subjected to flame for a prescribed time period, is observed to see whether it self-extinguishes, is measured for char length, and is observed to see whether it does not continue to burn after reaching the floor of the test chamber.

NFPA 701 Test Method No. 1 applies to single-layer fabrics and to multilayer curtain and drapery assemblies. Vinyl-coated fabric blackout linings are tested according to *NFPA 701* Test Method No. 2. *NFPA 701* cautions that materials applied to surfaces of buildings or interior finishes should be tested in accordance with *NFPA 255: Standard Method of Test of Surface Burning Characteristics of Building Materials*, or *NFPA 265: Standard Methods of Fire Tests for Evaluating Room Fire Growth Contribution of Textile Coverings on Full Height Panels and Walls.*

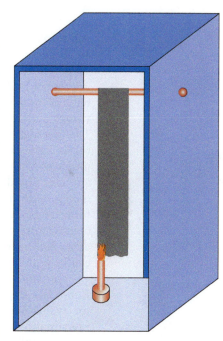

FIGURE 8-7 The *NFPA 701* test is used for small-scale testing of the flammability of fabrics. The *NFPA 701* apparatus uses a free-hanging vertical strip [150 mm × 400 mm/ (6 in. × 16 in.)] of the fabric to be tested in an enclosed test chamber. The flame from a laboratory burner is applied at the bottom of the strip for 45 seconds and then removed. The flame spread behavior of the fabric is observed, and after extinguishment, the residue is weighed.

The presence of fire retardants may significantly affect ignition and flame spread. Materials like carpet are much more easily ignited from a corner or edge than if the same ignition source is applied to the center of the specimen. Ignition and flame spread are enhanced by vertical sample orientation and may be retarded by draft or moisture in the sample, so the results should be noted but not taken as proof of identity or of actual fire behavior (NFPA 2009).

FULL-SCALE FIRE TESTS

Considerable knowledge has been gathered from fire tests conducted in real buildings slated for demolition. It is rare that the dimensions, ventilation, and construction materials of a test building will match those conditions of a specific fire that is being studied, but a great deal of reliable knowledge about fire behavior has been gleaned from these tests. Aside from physical suitability, there are often issues of fire exposure to surrounding properties, environmental concerns, logistics of access, and limitation of fire protection services. If such buildings are used, provisions should be made for multiple video camera viewports and the inclusion of thermocouple arrays and radiometers, as described next.

DeHaan found a house nearly identical to one in which six fire fatalities occurred. The test house had been scheduled for demolition. It was repaired and refurnished to duplicate the initial fire scene. Thermocouples, gas analyzers, and video cameras were used to document fire tests with accelerated and nonaccelerated ignitions. The test data revealed that a nonaccelerated fire could have been responsible for the structural damage and trapping the victims very quickly on the upper floor (DeHaan 1992).

Because of a number of training-fire deaths in the United States, fire departments have been considerably more reluctant to expose their personnel to the risks of large fires in original buildings. The majority of fire personnel recognize the value of training firefighters in such structures and plan live training burns in accordance with the safety provisions outlined in *NFPA 1403: Standard on Live Fire Training Evolutions* (NFPA 2012). The investigator should be aware that those provisions include covering floor openings, removing features such as glass windows and doors as well as debris contributing to an unsafe condition, identifying and evaluating all exits, and using only limited fuel loads (no ignitable liquids or gases). The resulting fire behavior and, most important, fire patterns therefore do not generally conform to those features of fires in buildings with normal furnishings, doors, glass windows, and intact ceilings and roofs. It is suggested that fire investigators not attempt to use such firefighting exercises to gather behavior/pattern data. Full-scale rooms that duplicate particular conditions and control other variables can be built for fire tests.

Guidance for room fire experiments is given in *ASTME 603: Standard Guide for Room Fire Experiments* (ASTM 2007b). This guide addresses assembling tests to be used for evaluating the fire response of materials, assemblies, or room contents in specific fire situations that cannot be evaluated in small-scale tests. This guide can assist those planning full-scale compartment-fire experiments and points out issues that should be resolved before testing commences. For example, *ASTM E603* suggests that a typical compartment size should be 2.4 m × 3.7 m (8 ft × 12 ft), with allowances for a 2.4-m (8-ft) high ceiling. The standard-size doorway (0.80 m ×2.0 m high) should be located in one wall, and the top of the doorway should be at least 0.4 m (16 in.) down from the ceiling to partially contain smoke and hot gases. The guide also recommends instrumentation having provisions for measuring the optical density of smoke, temperatures, and heat fluxes in the compartment. The documentation and controls necessary are also described.

Effective room mock-up cubicles can be assembled at low cost for full-scale tests. Based on a design by Mark Wallace, these are basically four 2.43 m × 2.43 m (8 ft × 8 ft) wood-framed panels (typically wood studs on 61-cm (24-in.) centers with gypsum wallboard, with a similar panel resting on top to form a ceiling (see Figure 8-8). A larger unit can easily be constructed, and designs provide for a knockdown wall that can easily be removed after the test to facilitate documentation.

Doors and windows can be cut as needed. A header must be left above each door 25–45 cm (10–18 in. deep). Cubicles are easily erected on four wood pallets covered with 13-mm (0.5-in.) sheet plywood or on (2 in. × 4 in. or 4 in. × 4 in.) joists. Since these are intended for short-duration (<30 min) tests, lack of insulation will not usually affect the results of interior fires. Electrical outlets can easily be added.

Viewports are easily added by cutting holes approximately 25 cm × 30 cm (10 in. × 12 in.) in size in one or more walls near floor level. An unframed piece of ordinary window glass is then glued to the *interior* surface of the wall using silicone sealant (and allowed to harden for 24 hours before the fire). The absence of a frame means that the entire sheet of glass is exposed to the same heat flux and will not develop the same stresses that cause failure before flashover. Experience reveals that such windows will usually last until flashover. Heat-resistant glass such as that used in fireplace screens or oven doors can ensure integrity postflashover if that is desired.

A *thermocouple* is a sensor for measuring temperature and consists of two different metal alloys joined together (twisted or welded) at one end. When that junction is exposed to heat, it generates a small voltage somewhat proportional to the temperature. The most common is Type K (Chromel/Alumel) and must be properly connected to extension cables with respect to their polarity to function correctly. Its most useful temperature measurement range is –200°C to 1250°C (–328°F to 2282°F). For a good source of commercial documentation and materials, consult the Omega Engineering Corporation website (www.omega.com/).

Thermocouples can easily be added through small holes drilled through the gypsum where desired. A minimum of three thermocouples on one wall is suggested: one near the ceiling, one midlevel, and one approximately

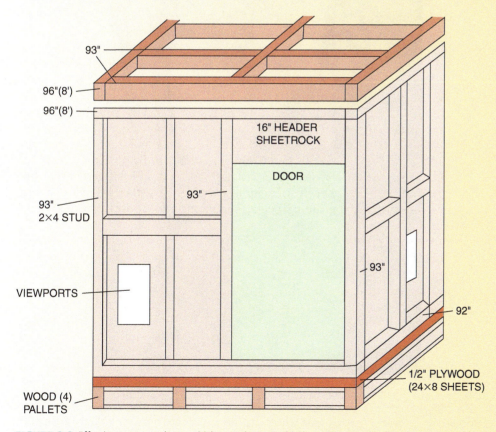

FIGURE 8-8 Effective room mock-up cubicles can be assembled at a low cost for full-scale tests. *Courtesy of J. D. DeHaan.*

15 cm (6 in.) above the floor at a point well away from the door or other vents. A second set can be installed on the opposite side of the room, or additional thermocouples can be placed inside the door header (to record flameover) or above target fuel packages. They can be shielded with a small-diameter metal, ceramic, or glass tube leaving only the tip exposed. The smallest-diameter thermocouples that are practicable should be used, since larger wire gauges lead to systematic depression in the measured temperatures. A wire size of 24 AWG is often suitable. Larger AWG numbers indicate a smaller wire diameter; however, thinner thermocouple wires can be more difficult to work with and more prone to breakage under field fire testing conditions. Data are best captured on a digital data logger such as Picolog, although they can also be captured on a video camera monitoring digital outputs of several individual meters at once for later manual transcription.

Such cubicles allow creation of environments that closely simulate real rooms (additional 1 m × 2.5 m [4 ft × 8 ft] segments can be added for larger rooms) while keeping safety risks to a minimum and reducing hazards for personnel (e.g., toxic material, asbestos exposure, or structural collapse). See Figure 8-9 for an example of a postflashover effect created in such a test cubicle. One entry wall can be framed separately so it can be easily opened and laid flat on the ground after the fire for ease of examination and photography. Video recordings should be made of all tests using multiple cameras and synchronized time clocks wherever possible. A visual and audio cue should accompany ignition so that it marks $t = 0$ on all recordings simultaneously. The small size (2.5 m × 2.5 m [8 ft × 8 ft]) will produce flashover in less time (approximately 30 percent less) than a typical 3 m × 4 m (10 ft × 13 ft) room given the same fuel load.

Scientifically valid small-scale tests can be conducted to test flame spread ignitability without special equipment as long as possible roles of the test variables are taken into account. For instance, carpet or furniture can be tested in the open with the realization that it may burn very differently if tested in a compartment because of the radiant heat reflected back from walls and ceiling and changes to ventilation conditions.

FIGURE 8-9 Results of postflashover fire in a furnished 2.4 m × 2.4 m (8 ft × 8 ft) office cubicle. Note extensive destruction of synthetic carpet and combustion of drywall paper on right. Glass viewport in right-rear corner failed prior to flashover. Note clean-burn ventilation pattern extending up the rear wall, and extensive damage to carpet beneath the opening. There was only carpet and pad in the right-rear corner.
Courtesy of J. D. DeHaan. Tests courtesy of California Association of Criminalists and Huntington Beach Fire Department.

Carpets need to be secured to the floor in the same manner as in the actual structure. Synthetic carpets shrink and curl as they burn. If the edges lift, the combustion of the carpet can be significantly enhanced and distort the heat release rate and fire patterns. Carpets must be backed with the same pad as during the fire being duplicated, since interaction between the pad and the carpet can cause self-sustaining combustion although neither material may sustain flame spread when exposed separately.

The ignition source used should duplicate the one known to be involved, or various kinds should be tested. For instance, polypropylene carpet will resist ignition by a single dropped match at room temperature, as it will pass the required methenamine tablet test; however, the carpet will self-sustain combustion if exposed to a slightly larger ignition source such as a burning crumpled sheet of newspaper or any additional external heat flux. Once ignited, such carpet will sustain combustion at rates up to approximately 0.5–1 m^2/hr (5–11 ft^2/hr) with small flames about 5–7 cm (2–3 in.) tall.

If a fire test is conducted in a compartment, care must be taken that the major fuel packages are located in the same position in each test (or in the same position as in the scene being replicated). Packages in corners will burn differently than packages against the wall or in the middle of the room. A large fire located away from an open door will burn differently than the same fire set near the door (Figure 8-10). Changes in the compartment vent opening size and location will alter air flow during testing and will influence the resulting fire patterns. Ventilation-effect fire patterns are to be expected if the fire test is allowed to proceed to postflashover (ventilation-limited) burning.

(a)

(b)

FIGURE 8-10 (a) Interior of 2.4 m × 2.4 m (8 ft × 8 ft) cubicle after a 20-minute fire. Bed on right was ignited by open flame to clothing on top and required nearly 16 minutes to be completely involved. The room went to flashover less than 1 minute later, igniting carpet and chair in left-rear corner (with pig haunch and legs). Note the extensive combustion of the corner of the bed nearest the door (ventilation effect). Glass viewport in rear corner failed prior to flashover. *Courtesy of J. D. DeHaan.*

FIGURE 8-10 (b) View from doorway of same compartment after removal of bed, chair, and remaining carpet from left side of compartment. Note exposed areas of carpet and 13-mm (0.5-in.) plywood floor consumed in less than 4 minutes of postflashover fire and top-down burning of 5 cm × 10-cm (2 in. × 4 in.) wood floor joists. No accelerants were used. *Courtesy of J. D. DeHaan. Tests courtesy of the Iowa Chapter of the IAAI.*

Examples of the valuable data obtained from cubicle tests include tests conducted September 25, 2002, in Waterloo, Iowa. In these tests, the Iowa Chapter of the International Association of Arson Investigators (IAAI) and the Bureau of Alcohol, Tobacco, Firearms and Explosives (ATF) built two cubicles of nearly identical fuel loads.

Each cubicle contained a synthetic upholstered chair, a wood dresser, curtains, a kitchen chair, and synthetic carpet with polyurethane padding. The first cubicle was ignited by an open flame to the skirting of one chair. This fire required 10.5 minutes to grow to flashover. The second cubicle was ignited with gasoline across the middle of the floor, with flashover occurring in approximately 70 seconds. Both cubicles experienced the same maximum temperatures and same damage to the carpet and wood floor despite the difference in time to reach flashover. The tests served as a good example of physical reconstruction and data gathering.

Figures 8-11 through 8-21 detail the Iowa cubicle tests.

FIGURE 8-11 Prefire view of the furnished 2.4 m × 2.4 m (8 ft × 8 ft) cubicle with dry-wall walls and ceilings. The fire was ignited on the skirt of the chair on the left. *Photo courtesy of Special Agent Mike Marquardt, CFI, ATF, Grand Rapids, MI. Test courtesy of the Iowa Chapter of the IAAI.*

FIGURE 8-12 Postfire view after 13 minutes, which was postflashover for 5 minutes. *Photo courtesy of Special Agent Mike Marquardt, CFI, ATF, Grand Rapids, MI. Test courtesy of the Iowa Chapter of the IAAI.*

FIGURE 8-13 Pattern on floor showing intense charring of plywood and irregular destruction of carpet and pad, which is most extensive at the door. *Photo courtesy of Special Agent Mike Marquardt, CFI, ATF, Grand Rapids, MI. Test courtesy of the Iowa Chapter of the IAAI.*

FIGURE 8-14 Same cubicle with furniture replaced in its prefire positions. Note the floor-to-ceiling combustion direction patterns on the second chair and dresser. *Photo courtesy of Special Agent Mike Marquardt, CFI, ATF, Grand Rapids, MI. Test courtesy of the Iowa Chapter of the IAAI.*

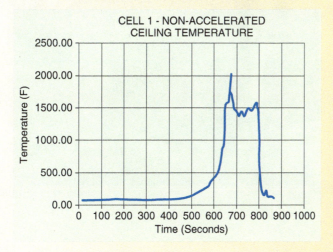

FIGURE 8-15 Temperature plot for the thermocouple tree in the center of the compartment at 0.3-m (1-ft) intervals. Note postflashover temperatures of approximately 816°C (1500°F). *Photo courtesy of Special Agent Mike Marquardt, CFI, ATF, Grand Rapids, MI. Test courtesy of the Iowa Chapter of the IAAI. Data courtesy of David Sheppard, Fire Protection Engineer, ATF Research Laboratory, Ammendale, MD.*

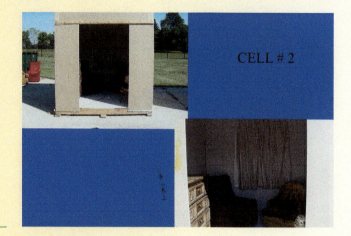

FIGURE 8-16 Prefire views of cubicle furnished in the same fashion as previously. The thermocouple tree is in the center of the cubicle. *Photo courtesy of Special Agent Mike Marquardt, CFI, ATF, Grand Rapids, MI. Test courtesy of the Iowa Chapter of the IAAI.*

CELL # 2

FIGURE 8-17 Accelerated cubicle fire approaches flashover temperature at 60 seconds. *Photo courtesy of Special Agent Mike Marquardt, CFI, ATF, Grand Rapids, MI. Test courtesy of the Iowa Chapter of the IAAI.*

FIGURE 8-18 Postfire view after 4 minutes of gasoline-accelerated fire. The ceiling and front drywall collapsed from firefighting action. Note protected areas on walls behind both large chairs. *Photo courtesy of Special Agent Mike Marquardt, CFI, ATF, Grand Rapids, MI. Test courtesy of the Iowa Chapter of the IAAI.*

FIGURE 8-19 Floor patterns indistinguishable from that of nonaccelerated fire. Note destruction of carpet and padding and charring of floor in areas toward door, where no accelerant was poured. *Photo courtesy of Special Agent Mike Marquardt, CFI, ATF, Grand Rapids, MI. Test courtesy of the Iowa Chapter of the IAAI.*

FIGURE 8-20 Reconstruction of furniture placement. Note fire patterns on chairs outward from the rear center of the floor, where the accelerant was poured. *Photo courtesy of Special Agent Mike Marquardt, CFI, ATF, Grand Rapids, MI. Test courtesy of the Iowa Chapter of the IAAI.*

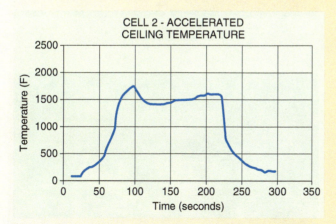

FIGURE 8-21 Temperature plot for thermocouple tree. Note ultrafast fire growth to flashover at 70–90 seconds. Postflashover period was approximately 130 seconds. *Photo courtesy of Special Agent Mike Marquardt, CFI, ATF, Grand Rapids, MI. Test courtesy of the Iowa Chapter of the IAAI. Data courtesy of David Sheppard, Fire Protection Engineer, ATF Fire Research Laboratory, Ammendale, MD.*

The most sophisticated (and most expensive) form of full-scale testing is the re-creation of an entire compartment in a laboratory with calorimetry, radiometry, continuous gas sampling, and video monitoring. An excellent example of this type of large-scale fire study was the re-creation of the entire seating bay of the Stardust Disco, conducted at the Fire Research Station, Garston, UK (FRS 1982).

The single-component or furniture mock-up tests described previously are useful for gauging ignitability and fuel properties, but only a full-scale reconstruction (such as the Stardust test) can reveal the complex interactions between the growing fire and the various fuels that, in that case, resulted in the rapid growth to flashover of a small initial fire.

There are very few facilities in the world that can provide the resources necessary. NIST, Factory Mutual Engineering, Underwriters Laboratories, California Bureau of Home Furnishings (Department of Consumer Affairs), Aberdeen Proving Grounds, Building Research Establishment (Garston, UK), and Southwest Research Institute all have large burn laboratory and testing facilities. These facilities have provided invaluable testing for many fire investigators over the years, often at no cost to public agencies.

The ATF Fire Research Laboratory (FRL), Ammendale, Maryland, is a scientific fire research laboratory dedicated to supporting many research needs of the fire investigation community. It also houses ATF's National Forensic Laboratory. The FRL was designed by a team of fire scientists, engineers, and fire investigation specialists from NIST, Factory Mutual Engineering, Underwriters Laboratories, Hughes Associates, and the University of Maryland.

An example of a large-scale test, simplified for the purposes of a single demonstration, is shown in Figures 8-22 through 8-25, corresponding to 3, 8, 11, and 14 minutes after initiation of a room fire, respectively. The BRE Fire

FIGURE 8-22 Full-scale living room fire test at 3 minutes after ignition, with only wastebasket and newspapers alight (incipient fire stage). *Courtesy of J. D. DeHaan. Test courtesy of FRS, Building Research Establishment, Garston, UK.*

FIGURE 8-23 Full-scale living room fire test at 8 minutes after ignition. Draperies have burned almost completely; dropdown has ignited chair and table in far-left corner (growth phase). *Courtesy of J. D. DeHaan. Test courtesy of FRS, Building Research Establishment, Garston, UK.*

FIGURE 8-24 Full-scale living room fire test at 11 minutes after ignition. Fire is postflashover, with maximum heat release rate of 5.2 MW observed at 10.7 minutes. Carpet is fully involved. *Courtesy of J. D. DeHaan. Test courtesy of FRS, Building Research Establishment, Garston, UK.*

FIGURE 8-25 Full-scale living room fire test at 14 minutes after ignition. Wood cabinet at left is alight. Filled with the remaining draperies, it will sustain a large fire that grows to 2 MW, then decays. Sofa and chairs are almost completely consumed (decay phase). *Courtesy of J. D. DeHaan. Test courtesy of FRS, Building Research Establishment, Garston, U.K.*

Research Station staff assembled a 2.5 m × 3.75 m × 2.4 m (8.2 ft × 12.4 ft × 8 ft) re-creation of a living room (lounge) under the 9 m × 9 m (30 ft × 30 ft) calorimeter in the Large Burn Hall at the FRS.

The structure was wood frame with ceramic fire insulation board liner (to permit reuse of the basic structure). The interior was lined with gypsum plasterboard, and the floor was covered with vinyl-backed carpet tiles (over a layer of sand to protect the concrete floor of the burn hall). There was a shallow 25-cm (10-in.) header across one side, which was otherwise left entirely open to simulate a "patio room" enclosure. This very large opening would provide enough air to ensure that the fire would not be ventilation limited even if it went to flashover. From the relationship for maximum mass flow of air through an opening of area A_o and height H_o,

$$\dot{m}_{air} = 0.5A_0\sqrt{H_O} = (0.5)(8.06)(1.46) = 5.88\,\text{kg / s}, \tag{8.1}$$

and since 1 kg of air will support 3 MJ of heat release,

$$\dot{Q}_{max} = (3000\,\text{kJ / kg}^3)(\dot{m}) = 17{,}640\,\text{kW}. \tag{8.2}$$

The ventilation limit for this room would be of the order of 15–17 MW. This calculation assumes 100 percent efficiency. At 50 percent, $\dot{Q}_{max}$ would be 8825 kW (8.8 MW).

The furnishings included non-flame-retardant draperies (no window); wood tables, chairs, and cabinets; a polyvinyl chloride beanbag chair; miscellaneous newspapers and books; and a three-seat sofa of recent production. This sofa was made using flame-retardant fabrics that delayed the growth of the fire substantially.

Three Chromel/Alumel thermocouples were placed in the room to monitor gas temperatures at ceiling and breathing level. Data from those thermocouples were logged every second during the test. The fire gases were collected in the calorimeter hood above (not visible in Figure 8-23).

Heat release rate and CO, CO_2, and O_2 gas concentrations were monitored continuously in the calorimeter. Continuous video recordings were made and still photos were taken by observers at 30-second to 1-minute intervals.

The newspapers in the wastebasket at the far end of the sofa (near the beanbag chair) were ignited by open flame. The heat release rate of this test is shown in Figure 8-26, and the temperatures are shown in Figure 8-27. They can be compared with the development stages of the fire shown in Figures 8-22–8-25.

This fire test was allowed to continue to burn for 25 minutes, and the remaining flames were then extinguished by water spray. Owing to the sustained postflashover burning in the room, there were no reliable indicators of duration or direction of propagation remaining on visual inspection. The absence of such indicators in postflashover rooms is one of the basic problems for fire investigators in their fire reconstruction analysis for such extended fires.

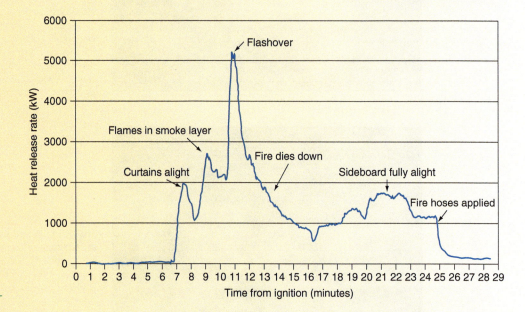

FIGURE 8-26 Plot of heat release rate versus time derived from data recorded in full-scale fire test. *Data courtesy of FRS, Building Research Establishment, Garston, UK.*

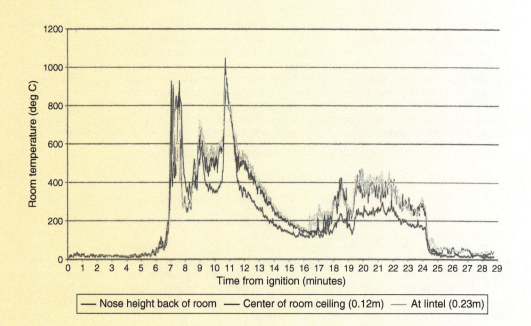

FIGURE 8-27 Plot of recorded temperatures versus time derived from data recorded in full-scale room fire test. *Data courtesy of FRS, Building Research Establishment, Garston, UK.*

This demonstration of fire development showed that when a single major fuel package such as a living room sofa is flame retardant, the growth to flashover can be delayed considerably (a 1990-vintage sofa would have promoted flashover in a similar room in less than 3 minutes if ignited by open flame in a similar fashion). Data from such tests can be used to confirm the accuracy of time and condition predictions of models, corroborate witness statements, demonstrate what stage of a fire might have incapacitated or killed a victim trapped in it, or demonstrate the role furnishings or fire protection systems may have in preventing future fires.

A series of full-scale fire tests were carried out under the auspices of the U.S. Fire Administration, National Institute of Justice (NIJ), National Association of Fire Investigators (NAFI), and Eastern Kentucky University by Ron Hopkins, Patrick Kennedy, and their co-investigators. The most recent series concerned the survivability of fire patterns in postflashover tests (Hopkins, Gorbett, and Kennedy 2009). Some of their results are shown in Figure 8-28.

These full-scale test burns provided a considerable amount of data concerning fire pattern development and evolution during fire growth and spread. The test burns demonstrated that fire pattern predictability is discernible during pre– and post–full room involvement fires and patterns can persist through brief postflashover fire exposure. The full-scale tests demonstrated that the fire patterns described in current literature are correct and when used properly can assist in the determination of the origin of a fire. Researchers found that if properly conducted, postfire testing utilizing full-scale burns and computer fire modeling may assist in the understanding of fire pattern development and fire growth.

Tests on postflashover fires include those by Carman (2008, 2010), who reported the results of tests at a 2005 fire training conference on fire dynamics led by ATF. The seminar included tests using two identical one-room cells with standard-size doorways, each burned for 7 minutes. Hours later, 53 fire investigator–students, who had not observed the fires, were asked to briefly examine the cells and identify the quadrant of each cell where they thought the fires had originated. The results showed that only 5.7 percent of the students correctly identified the quadrant of origin in each cell. Those who identified an incorrect origin typically reported they were misled during their analyses by extensive postflashover-generated burn patterns.

The authors note that these examinations of the scenes were only cursory inspections during which the investigators were unable to use accepted techniques, which would have included char depth measurements, burn pattern analysis, vector analysis, fire modeling, and laboratory examinations.

(a)

FIGURE 8-28 Room fire tests conducted at Eastern Kentucky University revealed that postflashover burning of short duration (2–4 min) does not obliterate V patterns and other useful fire patterns. (a) Fire pattern on wall near origin in postflashover bedroom test shows V pattern extending out from origin. Fire started on bed, flashover occurred at 790 seconds, and extinguishment at 1005 seconds. Postflashover fire duration: 215 seconds (3.6 min). *Courtesy of Ron Hopkins, TRACE Fire Protection and Safety, Richmond, KY.*

(b)

FIGURE 8-28 (b) Fire pattern on wall near origin in postflashover living room test shows V pattern extending out from origin, and effect of major thermal plume from sofa. Fire started on sofa, flashover occurred at 640 seconds, and extinguishment at 836 seconds. Postflashover fire duration: 196 seconds (3.2 min). *Courtesy of Ron Hopkins, TRACE Fire Protection and Safety, Richmond, KY.*

CASE EXAMPLE 4 — Investigation of a Multiple-Fatality Dormitory Fire

At approximately 4:30 a.m. on January 19, 2000, a fire occurred on the third floor of the north wing of Boland Hall, a dormitory on the South Orange, New Jersey, campus of Seton Hall University. The fire resulted in the deaths of three students and injury to more than 50 other individuals. The fire origin was determined to have been on a sofa along the west wall of the third-floor common area student lounge, and the fire cause was classified as incendiary.

BUILDING DESCRIPTION

Boland Hall is a freshman dormitory located in the northwest corner of the Seton Hall University grounds. This facility comprises two sections: South Boland, a five-story building constructed in the 1950s; and North Boland, a six-story building constructed in the 1960s. The two building sections are of similar construction and are connected with a corridor at each floor level. The building construction is Type II (noncombustible) with concrete floor slabs and masonry walls (NFPA 2008). The building had a fire alarm system with manual pull stations and smoke detectors. It also had a wet standpipe system but was not equipped with automatic sprinkler protection when this incident occurred. Figures 8-29 and 8-30 show the general layout of this building.

The North Boland building floor plan is a T-shape with a long corridor oriented east–west (north corridor) that is intersected at its midpoint by a shorter corridor (center corridor) that connects to South Boland. The north corridor is approximately 81 m (266 ft) long and varies in width between 2.13 m (7 ft) and 1.52 m (5 ft). The center corridor is 1.52 m (5 ft) wide and measures 14.45 m (47 ft 5 in.) from its intersection with the north corridor to the connection with South Boland. The ceiling height in the corridors is approximately 2.41 m (7 ft 11 in.) measured slab to slab, and 2.24 m (7 ft 4 in.) floor to suspended ceiling. There are three stairways for this building accessible by the north corridor. One is located at each end of north corridor, and one is at the midpoint adjacent to the intersection with the center corridor. There are 40 dormitory rooms, accommodating approximately 84 residents on the third floor of North Boland. The fire occurred in the third-floor lounge area of North Boland. This lounge area is adjacent to the T-intersection of the north and center corridors and is open to the corridors, as shown in Figures 8-30 and 8-31. The east side of the lounge area serves as the elevator lobby for this section of the building. The lounge measures 18.13 m (26 ft 8 in.) by 7.11 m (23 ft 4 in.) with the same ceiling height as the corridors. Two smoke detectors were located on the ceiling of the lounge. One was located near the center of the lounge area, and the other was adjacent to the elevator lobby area.

FIGURE 8-29 Aerial view of Boland Hall, from Haynes and Morris 2007. *Reproduced Courtesy of Interscience Communications, Ltd.*

The lounge area contained three sofas that were located against the south wall, the west wall, and adjacent to the north corridor. These sofas were crate-style furniture pieces with heavy wood frame construction and fabric-covered polyurethane foam cushions. A wood-framed bulletin board was mounted on the west wall over one of the sofas. The dimensions of the bulletin board were approximately 2.44 m (8 ft) wide by 1.22 m (4 ft) high. The bulletin board material was a medium-density fiberboard covered with decorative paper.

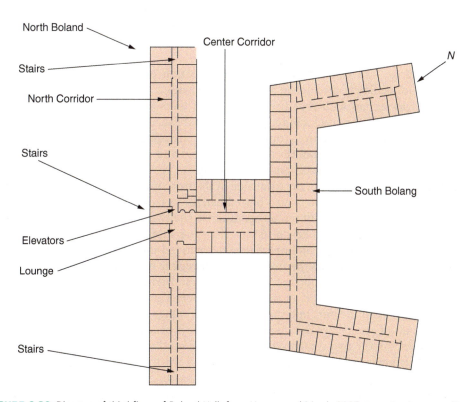

FIGURE 8-30 Diagram of third floor of Boland Hall, from Haynes and Morris 2007. *Reproduced courtesy of Interscience Communications, Ltd.*

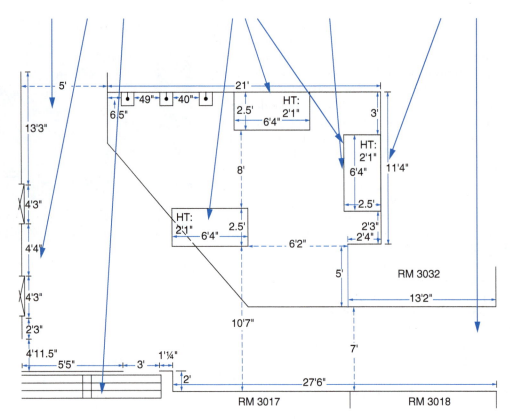

FIGURE 8-31 Diagram of third-floor lounge of North Boland Hall, from Haynes and Morris 2007. *Reproduced courtesy of Interscience Communications, Ltd.*

FIRE DEPARTMENT RESPONSE

The South Orange Fire Department received notification of the fire at approximately 4:32 a.m. and immediately dispatched an engine and truck company. Each of these pieces of apparatus was staffed with four firefighters. Both units arrived on scene at approximately 4:37 a.m., and the truck company officer established incident command and reported a working fire with multiple trapped victims. Firefighters reported that on arrival they observed smoke coming from third-floor windows and several students hanging out of windows calling for help (Frucci and Irwin 2001, 5).

The truck company set up on the front side of the structure and initiated rescue efforts using the aerial ladder. Firefighters also placed extension ladders against the building to third-floor windows to assist with rescue operations. The engine company was positioned behind the aerial apparatus and began to set up for fire suppression operations.

The truck company officer and one firefighter entered the structure with a high-rise hose pack and were directed to the fire area by a resident of the building. On reaching the third-floor landing of the north center stairs these firefighters could see fire in the lounge area through a glass panel in the stairway door. They connected the hose line to the stand pipe, charged the line, and then entered the lounge area from the stairway. The lounge area and adjacent corridors were charged with heat and thick smoke. The fire was concentrated on the two sofas located against the south and west walls of the lounge area. These firefighters stated that when they entered the lounge area, neither sofa had flames extending to ceiling height and that the fire was extinguished within one minute of their opening the hose line nozzle (Frucci and Irwin 2001, 5). After this initial fire attack was complete, the truck company officer returned to the first floor, and the firefighter remained in the third-floor lounge cooling the area with the water spray. The fire was marked under control by 4:44 a.m., but fire department search and rescue operations continued for some time owing to the heavy smoke conditions and the number of occupants remaining in many areas of the building (Sullivan 2000, 1).

Emergency medical services personnel set up on-scene triage, treatment, and patient transfer areas. These personnel provided emergency care and transportation for approximately 55 civilian casualties and four firefighters. Many of these victims sustained smoke inhalation and thermal burn injuries; at least five suffered serious and permanent injury. Three student residents died during this fire; all were from third-floor dormitory rooms. Two of the victims were found in the third-floor lounge area, and one was found in his dormitory room located at the west end of the main corridor (Frucci and Irwin 2001, 6).

SCENE PROCESSING

Examination and Excavation   This fire was investigated by the Essex County Prosecutor's Arson Task Force, with assistance from the South Orange Police Department, and ATF. The first investigator on location received a briefing about the incident from fire department command and then began processing the scene. Initial steps in the scene processing included a walk-around of the building perimeter and a walk-through of the fire-affected building sections. During these preliminary scene examination steps, the investigator made observations and notes regarding fire progression indicators, fire suppression and rescue operations, and fire victim locations.

Investigators continued processing by photographing and sketching the fire area to document the scene. The bodies of the three fire victims were removed from the scene and transported to the medical examiner's office for autopsy. A canine and handler team conducted an examination of the scene for ignitable liquids recorded no alerts. Electrical wiring and fixtures in the fire area were examined and eliminated as potential causes of the fire. No evidence of extension cords or electrical appliances was found during a thorough search and excavation of the lounge area. No ashtrays, lighter, or other evidence of cigarette smoking was found in the lounge area.

The excavation of the fire scene continued with examination and removal of the debris down to the floor of the lounge. The entire lounge area and the remains of the three sofas were carefully excavated and examined. On completion of the scene processing, investigators concluded that the origin of the fire was on the sofa located against the west wall of the third-floor lounge, as shown in Figure 8-32 (Frucci and Irwin 2001, 9–14).

Witness Information   Shortly after the fire scene excavation and examination were completed, investigators began interviewing residents of Boland Hall. More than 200 witness interviews were conducted, and 130 formal statements were taken. Witnesses were able to provide descriptions of the lounge before the fire. A few witnesses provided descriptions of the early fire development. Others described conditions in the lounge area just minutes before the fire was discovered. Several witnesses described an incident that occurred earlier in the morning before the fire. A student pulled paper from the bulletin board over the sofa against the west wall of the lounge. This paper was draped across this sofa, covering most of its seating area.

ENGINEERING AND SCIENTIFIC SUPPORT

The original request for engineering support and assistance was received from scene investigators within days of the fire's occurrence. The initial response to this request by an ATF fire protection engineer (FPE) was to provide some general heat release data and flame height calculations for common upholstered furniture items. A basic methodology

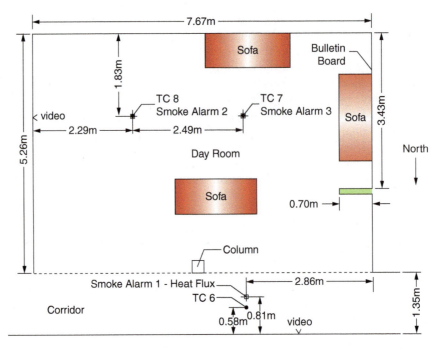

FIGURE 8-32 Diagram of day room showing sofa arrangement and instrumentation locations, from Haynes and Morris 2007. *Reproduced courtesy of Interscience Communications, Ltd.*

for incorporating this information into a simple timeline to describe possible fire development scenarios was also provided.

At the time of this investigation ATF and the NIST Building and Fire Research Laboratory, Fire Research Division, had in place a Memorandum of Understanding (MOU) and Reimbursable Agreement, which allowed the agencies to combine resources, share technical expertise, instrumentation, and facilities when conducting experiments and tests in fire research and measurement in support of arson investigation.

NIST and ATF fire protection engineers met with ATF investigators and members of the Essex County, New Jersey, Prosecutor's Office Arson Task Force on May 11, 2000. At the conclusion of this meeting a site visit of Boland Hall was conducted to collect information required for experimental set up and analysis. Data collected during this site visit included basic floor plans, sketches, limited photographs of fire-damaged areas, and measurements of significant building elements on the third floor. Samples of sofa cushions and ceiling tiles were provided for small-scale experiments and cone calorimeter testing to be conducted by NIST.

In May 2000 an ATF Special Agent assigned to the Charleston, South Carolina, ATF office notified an ATF FPE that a number of dormitory buildings were available to ATF and NIST for fire research and training. The buildings were slated for demolition as part of the Myrtle Beach Air Force Base Redevelopment Plan. In August 2000 ATF and NIST engineers visited the Myrtle Beach Air Force Base site to determine if these buildings were suitable for use to conduct fire experiments to support this investigation and various ATF and NIST fire research projects. The dormitory buildings were in reasonable condition and offered a wide range of fire experiment opportunities. The cooperation of the Myrtle Beach Fire Department and the Myrtle Beach Air Force Base Redevelopment Authority made these structures a viable location for conducting the necessary fire experiments to support this investigation.

SMALL-SCALE EXPERIMENTS

Cone calorimeter testing of the sofa cushion and ceiling tile materials was conducted by NIST in June 2000. The initial purpose of this testing was to provide material data for a planned computer modeling designed to support the large-scale dormitory experiments.

A series of small-scale ignition experiments on sofa cushion materials and decorative craft paper were conducted by NIST in August 2000 and January 2001. These experiments were designed to characterize the ignition propensity of these materials when subjected to open flame and smoldering cigarette ignition sources. The results of these small-scale ignition experiments are reported in final format in an NIST Letter Report of Test dated January 23, 2002 (Madrzykowski 2002).

LARGE-SCALE EXPERIMENTS

On September 20, 2000, NIST and ATF conducted a series of full-scale experiments in support of both this investigation and a U.S. Fire Administration (USFA)–funded research project. This joint project was part of a USFA initiative to improve fire safety in college housing. The objective of the study was to compare the levels of hazard created by room fires in a dormitory building with and without automatic fire sprinklers in the room of fire origin. Three fire experiments were conducted: (1) sprinklered, (2) unsprinklered with limited ventilation, and (3) unsprinklered with increased ventilation, all initiated in a day room area open to the corridor of the dormitory. The experiments were conducted on the first floor of a three-story building formerly used as a military dormitory. The building construction consisted of poured concrete floor and ceiling deck with concrete block walls. The vertical distance between the floor and the concrete ceiling was 2.60 m (8 ft 6 in.). No floor covering was installed. A drop ceiling, composed of fire-resistant aspen wood fiber tiles, was installed in the day room and corridor areas. Each ceiling tile was approximately 0.61 m (2 ft) by 1.22 m (4 ft) and 25 mm (1 in.) thick.

In the first of these experiments fire suppression began automatically with the activation of a sprinkler system installed for these experiments. This system consisted of four standard response sprinklers, with activation temperatures of 74°C (165°F) located under the drop ceiling of the day room. This sprinkler system was part of the NIST research project to examine the fire development, the spread of hot gases down the corridors, and the effectiveness of an automatic sprinkler system in suppressing the fire. In tests 2 and 3, manual fire suppression by Myrtle Beach Fire Department personnel was initiated approximately 15 minutes after ignition.

Each experiment utilized three sofas. Each of the sofas used for the site of ignition were similar in construction. Each was manufactured with an exposed wood frame and fabric-covered polyurethane foam. The ends of the sofa were composed of solid wood measuring 0.76 m (30 in.) wide, 0.58 m (23 in.) high, and 44 mm (1.75 in.) thick. The ends of the sofa were attached together with front and back solid wood supports. Each sofa had three back cushions and three seat cushions. The polyurethane was covered with a thin layer of polyester batting, which was covered with a textile material. The back cushions were approximately 0.61 m (24 in.) wide, 0.38 m (15 in.) high, and 0.18 m (7 in.) thick. The seat cushions were approximately 0.61 m (24 in.) wide, 0.53 m (21 in.) deep, and 0.20 m (8 in.) thick.

In experiment 1, the ignition sofa was similar in construction to the sofas used in experiments 2 and 3; however, two different types of upholstered sofas were used as "target" fuels in experiment 1. The target sofas were built with a wood frame and had seat cushions filled with polyurethane foam and back cushions filled with polyester batting. In experiments 2 and 3, all three sofas were similar.

The sofa used for the ignition site was located on the west wall of the day room, 0.91 m (3 ft) from the south wall. The second sofa was located on the south wall, 1.83 m (6 ft) from the west wall. The front face of the third sofa was located 3.2 m (10 ft 6 in.) north of the south wall, and the west side of the sofa was positioned 2.6 m (8 ft 6 in.) from the west wall, as shown in Figure 4.

A bulletin board was located above the sofa on the west wall. The dimensions of the bulletin board were 2.44 m (8 ft) wide by 1.22 m (4 ft) high. The bulletin board material was a medium-density fiberboard. It was 12 mm (0.5 in.) thick and was attached directly to the wall. The bulletin board material was framed with wood molding approximately 63.5 mm (2.5 in.) wide by 12 mm (0.5 in.) thick. Two sheets of craft paper were partially pulled down from the bulletin board and draped across the sofa. Each piece of paper was approximately 2.33 m (7 ft 7.75 in.) wide by 0.91 m (3 ft) high. In test 1 the paper was attached directly to the gypsum board wall; no bulletin board was in place.

Temperatures were measured with 0.51-mm (0.02-in.) nominal diameter bare-bead Type K thermocouples. Ten arrays of thermocouples were installed over the length of the corridor, and two thermocouple arrays were installed in the day room area. Each thermocouple array had a thermocouple located 25 mm (1 in.), 0.305 m (1 ft), 0.610 m (2 ft), 0.910 m (3 ft), 1.22 m (4 ft), 1.52 m (5 ft), and 1.83 m (6 ft) below the ceiling.

Thermocouple arrays in the corridor were located along the centerline of the corridor. The arrays were spaced on 7.62- m (25- ft) intervals, with the exception of the arrays near the east and west ends of the corridor.

Three pairs of Gardon-type heat flux gauges were installed near thermocouple arrays 4, 5, and 6. The heat flux gauges that were positioned closest to the day room, adjacent to thermocouple array 6, had a design heat flux level of 227 kW/m^2 [20 Btu/(ft$^2 \cdot$ s)]. The next pair of heat flux gauges positioned to the west, adjacent to thermocouple array 5, had a design heat flux level of 114 kW/m^2 [10 Btu/(ft$^2 \cdot$ s)]. The last pair of heat flux gauges, installed adjacent to thermocouple array 4, had a design heat flux level of 57 kW/m^2 ([5 Btu/(ft$^2 \cdot$ s)]. Each pair of gauges consisted of one gauge facing the ceiling and the other gauge facing the day room. The height of the gauges facing the ceiling was approximately 0.91 m (3 ft) above the corridor floor or 1.17 m (3 ft 10 in.) below the suspended ceiling. The height of the gauges positioned horizontally toward the fire was approximately 0.86 m (2 ft 10 in.) above the corridor floor or 1.22 m (4 ft) below the suspended ceiling.

Commercially available ionization smoke alarms were used. The alarms were mounted under the suspended ceiling in the day room and along the corridor. Each alarm was separately connected to the data acquisition system. The voltage change, as measured across the battery terminals at its alarm point, served as the data marker for the alarm time. New smoke alarms were used for each experiment.

Each experiment was documented using thermally protected video cameras. Two video cameras captured an east to west and a north to south view of the day room. Both cameras were installed close to the floor. A third camera was installed on the floor in the vent at the west end of the corridor looking east.

The Essex County, Prosecutor's Office Arson Task Force supplied the ignition sofa in experiment 1 and all three sofas used in each of the other experiments (2 and 3). The ceiling tile material used in these experiments was similar to that identified in Boland Hall. The building geometry and furniture arrangement for these experiments was established based on the furniture arrangement and building geometry of the third-floor student lounge area of Boland Hall. The location of the smoke alarms in the day room and corridors was selected based on the smoke detector arrangement in Boland Hall. The duration of experiments 2 and 3 (ignition to manual fire suppression) was established based on the timeline of events of the fire at Boland Hall. In experiments 2 and 3, an ATF FPE ignited the bulletin board paper with a single paper match at approximately the same location along the front edge of the sofa center seat cushion. This ignition scenario was chosen for consistency between experiments and to aid in testing the hypothesis of the original fire scene investigators as to the initiation of the fire at Boland Hall.

These experiments were documented in the draft report Day Room/Corridor Fire Experiments: Draft Report of Test with cover letter dated March 12, 2001, and in NISTIR

7120, Impact of Sprinklers on Fire Hazard in Dormitories: Day Room Fire Experiments, published in June 2004 (Madrzykowski, Stoup, and Walton 2004). These reports provide detailed descriptions of the experimental configuration, furnishings, instrumentation, and experimental procedure. They also provide the results, including experiment timelines, smoke alarm activation times, temperature data, and heat flux data.

ANALYSIS

The series of small-scale ignition experiments on the sofa cushion materials and decorative craft paper conducted by NIST were designed to characterize the ignition propensity of these materials when subjected to open flame and smoldering cigarette ignition sources. The results of these ignition experiments showed that the smoldering cigarette did not cause ignition of the craft paper alone, the sofa cushion foam alone, or the craft paper on the sofa cushion sample. These experiments did show that the open flame of a single burning match was sufficient to cause ignition of each initial fuel type considered—the craft paper, the sofa cushion foam, and the sofa cushion assembly. The qualitative results from these experiments indicated that a burning match was more likely to ignite the given samples than a smoldering cigarette.

Two full-scale fire experiments were conducted at the Myrtle Beach Air Force Base, both were unsprinklered and were initiated in a day room area open to the corridor of a dormitory. These experiments were set up to have a building geometry and fuel configuration similar to that of the third-floor student lounge area of Boland Hall and are referred in the NIST report as Experiment 2 and Experiment 3. Both experiments were initiated by ignition of the bulletin board paper with a single paper match at approximately the same location along the front edge of the sofa center seat cushion.

In Experiment 2 the fire grew rapidly from ignition of the bulletin board paper and quickly spread to the entire seat and back cushions of the ignition sofa. The fire grew to sufficient intensity to cause ignition of the sofa positioned along the south wall of the day room and began to cause thermal damage to the sofa adjacent to the corridor, positioned as shown in Figure 8-32. The peak temperatures near the ceiling ranged from 780°C (1436°F) at thermocouple array 7, to 120°C (248°F), at thermocouple array 1 near the east end of the corridor. The maximum temperature at the west end of the corridor was 170°C (338°F). Analysis of the temperature and heat flux data for this experiment showed that the fire grew for approximately 7 minutes and then began to decrease in intensity. The temperatures throughout the structure peaked at approximately 400 seconds (6.7 minutes) and then steadily declined. These temperature profiles along with the video record indicated that the fire became ventilation limited and as a result decreased in intensity until extinguished by the Myrtle Beach Fire Department. The postfire analysis of the day room area also supported this observation—specifically, the damage to the sofas, ceiling, and walls, and the amount of combustible material that remained of all three sofas.

This ventilation-limit cause for the decreasing fire was considered during the setup for Experiment 3, and changes were made in the ventilation openings. In Experiment 3, five sleeping room doors were left open on the east end of the corridor. Each of these sleeping rooms had windows that opened to the outside. This change resulted in increased temperatures relative to Experiment 2. The peak temperatures near the ceiling ranged from 900°C (1652°F) at thermocouple array 7 to 240°C (464°F) at thermocouple array 1 near the east end of the corridor. The maximum temperature at the west end of the corridor was 230°C (446°F).

In Experiment 3 the fire grew rapidly from ignition of the bulletin board paper and quickly spread to the entire ignition sofa seat and back cushions, similar to Experiment 2. The fire continued to grow in intensity until it became fully developed and involved all three sofas and caused damage to the suspended ceiling. The ceiling tiles began to burn, and the ceiling suspension grid failed in areas, dropping ceiling tiles to the floor.

The fire was suppressed by Myrtle Beach Fire Department after approximately 15 minutes. When the fire suppression crew entered the corridor and day room area, the smoke level was close to the floor, and visibility was limited. The main body of fire was quickly extinguished with minimal water application from a single hand line. There were small spot fires at several locations in the day room and corridor, some of which were attributed to burning ceiling tiles. This account of the fire and the suppression sequence was similar to the accounts of the firefighters who made the initial attack on the fire in Boland Hall.

The postfire analysis of the day room area showed that the fire consumed most of the sofa cushions. The fallen and fire-damaged ceiling tiles could also be seen as debris piled on the floor. Comparison of the postfire photographs of the sofas and corridor after Experiment 3 with the fire scene photographs of the sofas and corridor of Boland Hall showed striking similarity.

CONCLUSION

The small-scale ignition experiments on the sofa cushion materials and decorative craft paper conducted by NIST showed that the smoldering cigarette did not cause ignition of the craft paper alone, the sofa cushion foam alone, or the craft paper on the sofa cushion sample. These results supported the hypothesis of the scene investigators that the fire on the sofa was not likely to have been ignited by a carelessly discarded cigarette. An open flame, however, could have constituted a competent ignition source for any of these first fuels.

The large-scale fire experiments conducted at the Myrtle Beach Air Force Base both had similar early fire development. This fire growth was similar to that described in witness accounts of residents in Boland Hall during the early stages of that fire. The fire in Experiment 2 became ventilation limited before suppression.

The growth and development of the fire in Experiment 3 was consistent with that of the fire in Boland Hall during both the early stage and the later stage right up to suppression. The accounts of the firefighters and the postfire analysis of the damage in both the Boland Hall fire and Experiment 3 supported this comparison. These similarities between the fire in the third-floor student lounge of Boland Hall and the large-scale fire experiments conducted at Myrtle Beach Air Force Base supported the hypothesis and the ultimate conclusions of the scene investigators about the origin and cause for this fire.

Following the ATF and NIST testing, investigators concluded that the fire had been intentionally set through the introduction of an open flame to construction paper from a bulletin board over the sofa along the west wall of the third-floor lounge in Boland Hall.

The results of this extensive investigation were presented to a grand jury. In June 2003 the grand jury returned indictments against two Seton Hall students for arson, aggravated assault, reckless manslaughter, and felony murder. The fire investigators and their determination of arson were tested over the next 3-1/2 years in pretrial litigation.

A defense motion was filed challenging the techniques, methodology, and procedures used by investigators in the course of this investigation. After extensive review and examination the trial judge for this case ruled that the investigation into the Seton Hall fire was sound and that it would be admitted into evidence at trial.

On November 15, 2006, Joseph LePore and Sean Ryan, during their allocution at their guilty plea, confessed that they had introduced an open flame to the construction paper that covered the west wall sofa in the third floor lounge. These defendants were sentenced to five years in state prison for arson on January 26, 2007.

Case study courtesy of Gerald Haynes, P.E., Forensic Fire Analysis, LLC, and Michael Morris, Assistant Prosecutor Office of the Essex County Prosecutor. This study was previously published at InterFlam 2007, and reproduced by permission.

Summary

Fire testing includes a very large range of tests, from simple field flammability tests such as *NFPA 705,* to bench-scale fabric tests such as ASTM and *16 CFR* tests, to full-scale tests in buildings. Useful data from such tests can range from simple observations to temperatures and radiant heat fluxes, and owing to oxygen depletion calorimetry, even heat release rates of large fuel packages. Fire tests are often designed, conducted, and analyzed to collect more information about basic fire processes. This information is then published in peer-reviewed publications and authoritative treatises, or on websites. (See the Pearson-Brady Resource Central box.) The data are then available for use in predicting ignition and fire events. Data from such tests can fill a critical role in testing hypotheses about the ignition, spread, and effects of fire.

The criteria for using fire test information in forensic fire scene reconstructions are whether the test was correctly performed, whether the data were accurately collected and reported, and whether the test was appropriate and applicable to the situation under consideration. Fire tests are designed to collect certain data in a reproducible and valid manner. The investigator must ask whether the fuels, conditions, and ignition mechanism reproduced the fire in question. Was the test designed to follow a published test protocol such as one from ASTM or NFPA? If so, did it actually follow that protocol? If not, what factors were considered in its design? What variables were considered and how were they controlled? Fuel moisture, fuel mass and quantity, physical state, ambient temperature and humidity, heat flux, and oxygen content all play important roles in ignition, flame spread, and heat release. If it was a custom-designed test, what data could be fairly and accurately collected and analyzed? Was there a planned series of tests to examine the sensitivity and reproducibility of the data? Because of the importance of fire test data in both criminal and civil fire investigations, it is imperative that tests be conducted and data be interpreted without misrepresentation in a balanced, impartial, and reproducible manner.

Problems

8.1. If the test was intended to replicate an actual event, how closely did the materials, dimensions, and the ventilation match the original conditions?

8.2. Find an example of a commercial, academic, or governmental fire testing facility in your state. Visit or call this organization and determine what type of testing it performs.

8.3. Obtain a collection of at least 10 tests involving the use of a calorimeter from published research.

8.4. Research the methods of conducting field fire testing.

8.5. Obtain small samples of different clothing or upholstery fabrics from a fabric store (with identified content) and conduct *NFPA 705* ignition tests on each (in a safe location). Collect data and compare observations with descriptions included in the references.

References

ASTM. 2007a. *ASTM E502-07e1: Standard test method for selection and use of ASTM standards for the determination of flash point of chemicals by closed cup methods.* West Conshohocken, PA: ASTM International.

———. 2007b. *ASTM E603-07: Standard guide for room fire experiments.* West Conshohocken, PA: ASTM International.

———. 2011. *ASTM E1354-11b: Standard test method for heat and visible smoke release rates for materials and products using an oxygen consumption calorimeter.* West Conshohocken, PA: ASTM International.

Babrauskas, V. 1997. The role of heat release rate in describing fires. *Fire & Arson Investigator* 47 (June): 54–57.

Babrauskas, V., & Krasny, J. 1985. *Fire behavior of upholstered furniture*. Gaithersburg, MD: U.S. Dept. of Commerce.

———. 1997. Upholstered furniture transition from smoldering to flaming. *Journal of Forensic Sciences* 42:1029–31.

Bryner, N. P., Johnsson, E. L., & Pitts, W. M. 1995. Scaling compartment fires: Reduced- and full-scale enclosure burns. Paper presented at the International Conference on Fire Research and Engineering, September 10–15, Orlando, FL.

Carman, S. W. 2008. Improving the understanding of post-flashover fire behavior. Paper presented at the International Symposium on Fire Investigation Science and Technology, May 19–21, Cincinnati, OH.

———. 2010. Clean burn fire patterns: A new perspective for interpretation. Paper presented at Interflam, July 5–7, Nottingham, UK.

DeHaan, J. D. 1992. Fire: Fatal intensity; A third view of the Lime Street fire. *Fire and Arson Investigator* 43 (1): 5.

———. 2001. Full-scale compartment fire tests. *CAC News* (Second Quarter): 14–21.

DeHaan, J. D., & Icove, D. J. 2012. *Kirk's fire investigation*, 7th ed. Upper Saddle River, NJ: Pearson-Prentice Hall.

Fleischmann, C. M., Pagni, P. J., & Williamson, R. B. 1994. Salt water modeling of fire compartment gravity currents. Paper presented at Fire Safety Science: Fourth International Symposium, July 13–17, Ottawa, Ontario, Canada.

FRS. 1982. Anatomy of a fire (video). Garston, Watford, UK.

Frucci, J., & Irwin, R. 2001. Investigation of fire in Boland Hall on January 19, 2000. Report to Essex County Prosecutors Office Arson Task Force, Newark, NJ.

Haynes, G. A., & Morris, M. 2007. Investigation of a multiple fatality dormitory fire at Seton Hall University. Paper presented at InterFlam, July 3–5, London, UK.

Holleyhead, R. 1999. Ignition of solid materials and furniture by lighted cigarettes. A review. *Science & Justice* 39 (2): 75–102, doi: 10.1016/s1355-0306(99)72027-7.

Hopkins, R. L., Gorbett, G. E., & Kennedy, P. M. 2009. Fire pattern persistence and predictability during full scale comparison fire tests and the use for comparison of post fire analysis. Paper presented at Fire and Materials 2009, 11th International Conference, January 26–28, San Francisco, CA.

Jewell, R. S., Thomas, J. D., & Dodds, R. A. 2011. Attempted ignition of petrol vapour by lit cigarettes and lit cannabis resin joints. *Science & Justice* 51 (2): 72–76, doi: 10.1016/j.scijus.2010.10.002.

Krasny, J., Parker, W. J., & Babrauskas, V. 2001. *Fire behavior of upholstered furniture and mattresses*. Norwich, NY: William Andrew.

Madrzykowski, D. 2002. NIST letter report of test. Gaithersburg, MD: National Institute of Standards and Technology.

Madrzykowski, D., Stoup, D. W., & Walton, W. D. 2004. Impact of sprinklers on fire hazard in dormitories: Day room fire experiments. Gaithersburg, MD: National Institute of Standards and Technology.

NFPA. 2008. *Fire Protection Handbook* (20th ed.). Quincy, MA: National Fire Protection Association.

———. 2009. *NFPA 705: Recommended practice for a field flame test for textiles and films*. Quincy, MA: National Fire Protection Association.

———. 2010. *NFPA 701: Recommended practice for a field flame test for textiles and films*. Quincy, MA: National Fire Protection Association.

———. 2011. *NFPA 921: Guide for fire and explosion investigations*. Quincy, MA: National Fire Protection Association.

———. 2012. *NFPA 1403: Standard on live fire training evolutions*. Quincy, MA: National Fire Protection Association.

Quintiere, J. G. 1989. Scaling applications in fire research. *Fire Safety Journal* 15 (1): 3–29, doi: 10.1016/0379-7112(89)90045-3.

Sullivan, D. 2000. South Orange Fire Department incident report 2000015. South Orange, NJ.

Epilogue

" I had come to an entirely erroneous conclusion, which shows, my dear Watson, how dangerous it always is to reason from insufficient data. "

—Sir Arthur Conan Doyle
Adventures of the Speckled Band

The forensic sciences generally, and forensic fire scene reconstruction particularly, have entered a period of light-speed transformation owing to the convergence of scientific, investigative, forensic, and legal forces. The consequence of this contemporary paradigm shift has been the reshaping of fire investigative methodologies globally. As a result, current practitioners and aspiring neophytes are tasked with deciphering and implementing an array of near-term technological solutions as well as crafting viable strategies to address long-term occupational challenges.

Presently, fire investigative methodologies, including but not limited to fire pattern analysis, suffer from reliability deficiencies owing to an abundance of undereducated practitioners, an underdeveloped scientific foundation, and obsolete capture and documentation techniques. Untrustworthy fire-related testimony and evidence routinely circumnavigates an ill-informed and overwhelmed judiciary tasked with forensic gatekeeping responsibilities. Skepticism is mounting, and change is inevitable.

Because *forensic fire scene documentation* is emerging as an indispensable cornerstone in nearly all judicially contested controversies, modest reconstructive efforts such as repositioning furniture and other postfire artifacts into prefire positions no longer suffice. Contemporary fire investigators must fastidiously evaluate all fire scenes, fully process and record every piece of evidence, and meticulously document findings in a diversity of investigations ranging from a simple toaster malfunction to a multiple-fatality arson fire. After all, it is virtually inevitable that scientific validity and reliability will be contested by peer review, *Daubert* challenges, blistering depositions, and/or pointed cross-examinations.

During scene documentation, today's fire investigator must possess skill in scientifically based *fire pattern analyses* to employ analytical methodologies based on expert interpretation of discernible patterns surviving a fire event. An astute fire investigator must not only describe the patterns but must also explain foundational fire science theories and engineering concepts to articulate how the resulting fire patterns were in fact generated. *Fire dynamics* calculations are essential in evaluating various working hypotheses, such as the quantity of fuel consumed, the fire's duration, and the effect of ventilation in the room of fire origin.

Accurate assessments also require the use of a comprehensive strategy to systematically document the investigation with the objective of supporting and verifying investigative opinions and conclusions. Compulsory protocol includes the recordation of witness observations, evidence location, and fire scene delayering, along with still *photography* and video recordings supplemented by comprehensive *sketching* techniques to ensure process integrity. Photographs and sketches capture information to make certain that a fire scene's maximum evidentiary value is preserved.

Painstaking measurements are imperative to precisely documenting a fire scene, particularly if physical or fire modeling is later utilized, and accurate dimensions of doors, windows, and ceiling heights will be needed. For example, fire investigators can leverage *photogrammetric measurement* systems to compare the distances between competent ignition sources and fuel packages. The estimated spill areas of ignitable liquid patterns also can be calculated. Fire patterns can later be compared with atlases of existing patterns. Measurements inputted into fire models will add spatial and temporal evidentiary information concerning fire ignition, growth, travel, and tenability. Together, these innovative methods and technologies will soon revolutionize fire scene investigations.

Two-dimensional fire scene assessments will soon become antiquated propositions, since this methodology lends itself to a static rather than fluid interpretation of involved fire dynamics. Instead, 3D views of the scene are requisite for confirming witness observations or for demonstrating evidentiary endeavors. Forensic technologies continue to evolve, and such methods as *forensic 3D digital imaging and scanning* are becoming commonplace and will soon supplant 2D-based approaches. As 3D scanning equipment becomes more affordable, accessibility will increase.

Fire investigative practices on the horizon will entail the use of modern *laser-assisted imaging* methods to extrapolate visual and numerical information from images to compare and differentiate features and interpret image configuration. These actions will in turn allow a vigilant investigator to test proposed hypotheses against known circumstances and then answer reconstructive questions concerning dynamic fire pattern development to calculate the possibility of their linkage to assumed fuel packages.

The *educational and training deficiencies* of fire investigative personnel have become the subject of increased judicial scrutiny. Fire investigators will no longer be able to accumulate minimal educational and training credentials to obtain fire investigation and law enforcement certifications; instead, investigators will be required to function as uniquely skilled forensic detectives. In addition to the existing educational requirements of *NFPA 1033*, added prerequisites for fire investigators may include theoretical and working knowledge in areas such as burglary, narcotics, gambling, computer forensics, accident investigation, accounting and tax calculations, forensic evidence recognition and collection, and even participation in autopsies and medical examinations to evaluate physical injuries. Related formal educational requirements will be incorporated in fire-related science and engineering undergraduate and graduate degree programs.

Individuals who are successful in attaining these more stringent academic and training qualifications will have arrived only at the analytical threshold of any given fire-related inquiry. A proficient fire investigator still must be able to critically analyze a fire scenario by developing a comprehensive set of *working hypotheses*, a concept central to and within the framework and implementation of the *scientific method*. A working fire scene reconstruction hypothesis is based on how an investigator describes or explains a fire's origin, cause, and subsequent development. A working hypothesis is not the final conclusion but is instead subject to continuous review and modification. These viable propositions mirror the mosaic of those basic areas identified in this textbook: professional guidelines and standards, an understanding of fire dynamics and modeling, fire pattern analysis, comprehensive witness statements, correct application of forensic science, an analysis of similar fire and explosion loss histories, interpretation of published historical fire testing, and human behavior.

Reliance on *standard accepted practices* such as those outlined in this and other texts is becoming the norm. Experienced fire investigators now carry COM port–based devices capable of accumulating a canvas of initial fire scene patterns, assessing loss histories of comparable occupancies, querying meteorological data, capturing video of on-scene interviews, and constructing 3D high dynamic resolution (HDR) images.

Fire investigators increasingly must become knowledgeable in the discipline of *fire testing* as applied to field applications, in response to a case, or to elucidate historical tests. Investigators must comprehend that traditional fire testing conclusions are frequently, albeit inaccurately, applied by fire investigators to assist in fire pattern analysis, even though most testing was never conducted or designed to support such undertakings. Bench-scale testing, when correctly applied and buttressed by competent fire modeling, will allow investigators to evaluate the conditions, ignition scenarios, implications, and hypotheses for a fire event.

Fire investigative efforts are becoming increasingly dependent on *fire testing by public as well as private research laboratories*, such as the public-supported National Institute of Standards and Technology (NIST), the Bureau of Alcohol, Tobacco, Firearms and Explosives (ATF) Fire Research Laboratory, and progressively developing liaisons with private labs such as FM Global, Underwriters Laboratories, and Southwest Research Institute.

Because present-day full-scale room *fire pattern studies* offer only rudimentary explanations for critical questions requiring more discriminating interpretation, emerging 3D technologies will be aggressively developed and employed to navigate these impediments. Future fire pattern studies will rely on a robust database or atlas of pertinent fire patterns, and related fuel packages will allow for quantitative comparison of fire artifacts with known exemplars. Matching scans will be compared in virtual space by graphic overlap to calculate the degree of shared similarity.

The interface of fire-related forensic science and law dictates that contemporary fire investigators must acquire an in-depth knowledge of the fundamentals of basic legal rules and principles and then comprehend how the application of these edicts affects the admission and interpretation of expert evidence and testimony. Proffered evidence must persevere through a pivotal primary appraisal focused on assurances of legal and logical relevancy, reliability, validity, probative value, and presumed prejudicial effect. In fact, these imposing yet unavoidable juristic fortifications subject scientific, technical, and other specialized evidence to an incalculable litany of nonexclusive "factors" designed to purge from the courthouse a burgeoning barrage of fantastic imaginings and junk expert testimony that impersonates scientific evidence. Fact-specific inquiries include, but certainly are not limited to, the following:

- Is the opinion testable?
- Is the opinion grounded in sufficient facts and/or data?
- Are the facts being utilized supported by the evidence?
- Does the investigator possess the requisite qualifications to offer such an opinion?
- Are his or her qualifications specifically congruent with the proposed opinion?
- Is there general acceptance of the methodology utilized within the fire investigative community?
- Is the testing being properly applied and interpreted?
- Is the theory or technique relied on too abstract and therefore not helpful?
- How many experiments have been conducted?
- What is the error rate?
- What are the distinctions between the methodology utilized and the conclusions stated?
- Does the application of the scientific technique "fit" on this particular occasion?
- Is the underlying scientific theory empirically valid?
- Have professional peers weighed in, and if so, what is their opinion?
- Have the proposed theories been published?
- Is the expert's opinion certain, probable, or merely possible?

Indeed, twenty-first-century fire investigators must now "think like lawyers" to effectively synthesize scientifically grounded fire-related deductions with complex, uncertain, and ambiguous legal principles.

Lastly, even analytically sophisticated fire investigators must master the craft of effective written communication to become confident, polished composers of expert witness reports and other documents. Comprehensive, effective, and concise writing pertaining to complicated and slippery subject matter requires a logical, thorough, and precise methodology to satisfy a target audience demanding absolute clarity. While typical expert witness reports routinely include expressed opinions along with accompanying rationale, superior reports are organized, formatted, stylistic, and detailed. Documents relied on are disclosed and referenced in textual footnotes, and voluminous citations to authorities and learned treatises provide additional substantive validation. The author painstakingly assures compliance with the applicable civil and evidentiary rules of the forum jurisdiction. Meticulous drafters create a legal instrument that exudes absoluteness and simultaneously chills any contemplated *Daubert* challenges.

The high-tech future of fire investigation is on the horizon. The aforementioned techniques, in combination with other visual and laboratory examination, will improve quantitative and qualitative analyses and in turn facilitate more robust and accurate fire investigations.

Thomas R. May, MS, JD
Fire Litigation Strategies, LLC

Thomas May is the author of the recent landmark law review article "Fire Pattern Analysis, Junk Science, Old Wives Tales, and Ipse Dixit: Emerging Forensic 3D Imaging Technologies to the Rescue?" T.C. Williams School of Law, University of Richmond. 16 *Richmond Journal of Law & Technology* 13 (2010), http://jolt.richmond.edu/v16i4/article13.pdf.

Mathematics Refresher

A.1 Fractional Powers

When a number (or value) is followed by a superscript number or fraction, such as X^n, it means that the number (X) is raised to a power (n). If the power is 2, the number is squared (multiplied by itself) ($3^2 = 3 \times 3 = 9$). If the power is 3, the number is cubed (multiplied by itself and then by itself again) ($3^3 = 3 \times 3 \times 3 = 27$).

The number n can be a fraction, such as ½.

$$X^{1/2} \quad \text{or} \quad X^{0.5} = \sqrt{X} \quad \text{(the square root of } X\text{).} \tag{A.1}$$

The number n can be any value: $n = 2/5, 5/2$, and $3/2$ are common calculations in fire dynamics. Such values must be calculated using a "scientific notation" pocket calculator with a y^x or y^n function (where $y = X$, and x [or n] is the power to which y is raised). If n is a negative number, then X^{-n} denotes $1/X^n$ (X^n is put in the denominator of a fraction). The following are general examples of mathematical equations using powers:

$$Q^x Q^y = Q^{x+y} \tag{A.2}$$

$$\frac{Q^x}{Q^y} = Q^{x-y} \tag{A.3}$$

$$\left(Q^x\right)^y = Q^{xy} \tag{A.4}$$

A.2 Logarithms

Logarithms are numerical equivalents calculated by raising 10 to a power, where $\log_{10}$ = the value to which the number 10 must be raised to be equal to the original number. Thus,

$$\log_{10} \text{ of } 100 = 2, \tag{A.5}$$

because

$$10^2 = 100. \tag{A.6}$$

Also,

$$\log_{10} \text{ of } 10 = 1, \tag{A.7}$$

because

$$10^1 = 10. \tag{A.8}$$

This value can be found using a pocket calculator or mathematical tables. For instance,

$$\log_{10} 3.5 = 0.54, \tag{A.9}$$

and

$$\log_{10} 2330 = 3.3673. \tag{A.10}$$

The inverse function is sometimes called the *antilog:*

$$\text{antilog}_{10}(2) = 10^2 = 100. \tag{A.11}$$

When the base unit e is used ($e = 2.71$), the value is called the *natural log* (ln). For example,

$$\log_e 5 = \ln 5 = 1.6094, \tag{A.12}$$

and

$$\ln 10 = 2.3025. \tag{A.13}$$

The antilog e^x is the value calculated by raising e to that power:

$$e^5 = 148.413, \tag{A.14}$$

and

$$e^{10} = 22,026.5. \tag{A.15}$$

A modified "power of 10" notation is used as shorthand for very large or very small numbers. For example,

$$3.40E - 05 = 3.4 \times 10^{-5} = 0.000034,$$
$$2.88E + 00 = 2.88 \times 10^0 = 2.88,$$
$$5.00E + 02 = 5.0 \times 10^2 = 500.$$

A.3 Dimensional Analysis

A useful concept for checking calculations is called *dimensional analysis*. If one keeps track of the units used for variables and constants and applies the rules of canceling units when they appear in both the numerator and the denominator of a function, for instance, one can verify that the correct relationship was used.

For example

$$\dot{Q} = \dot{m}'' A \, \Delta H_c, \tag{A.16}$$

where

$\dot{m}''$ = mass flux [kg/(m$^2 \cdot$ s)],
A = area of burning surface (m^2), and
ΔH_c = heat of combustion (kJ/kg).

Multiplication gives

$$\left(\frac{\text{kg}}{\text{m}^2 \cdot \text{s}} \right) \left(\text{m}^2 \right) \left(\frac{\text{kJ}}{\text{kg}} \right) = \frac{\text{kJ}}{\text{s}} = \text{kW}, \tag{A.17}$$

which is the correct unit for $\dot{Q}$.

For conduction calculations,

$$\dot{Q} = k \frac{T_2 - T_1}{l} A, \tag{A.18}$$

where

$$k = \frac{\text{W}}{\text{m} \cdot \text{K}},$$
$l = \text{length (m)},$
$T_2 - T_1 = \Delta T(\text{K}), \text{ and}$
$A = \text{area (m}^2).$

Carrying out the calculation, we obtain

$$\dot{Q} = \frac{\left(\dfrac{\text{W}}{\text{m} \cdot \text{K}}\right)(\text{K})\left(\text{m}^2\right)}{\text{m}} = \text{W}, \tag{A.19}$$

which is the correct unit for $\dot{Q}$.

GLOSSARY

Included in this glossary are key terms used in this textbook as well as other terms often found useful in maintaining a broad understanding of the field of fire investigation. Note that in some cases multiple definitions exist for the same term, and the appropriate references are cited as to their source.

A

Abductive reasoning   The process of reasoning to the best explanations for a phenomenon.

Accelerant   A fuel (usually a flammable liquid) that is used to initiate or increase the intensity or speed of spread of fire (*Kirk's* 7th ed.).

Adiabatic   Conditions of equilibrium of temperature and pressure. Also describing a reaction occurring without loss or gain of heat.

Adsorption   Trapping of gaseous materials on the surface of a solid substrate.

Air entrainment   The process of air being drawn into a fire, plume, or jet (*NFPA 921*, 2011 ed.).

Aliphatic   Hydrocarbons with their carbon atoms in a straight chain; *normal* hydrocarbon.

Alligatoring   Rectangular patterns of char formed on burned wood.

Ambient   Surrounding conditions.

Ampacity   Current-carrying capacity of electric conductors (expressed as amperes).

Ampere   Quantity of electrical charge passing a point in an electrical circuit per unit time (1 coulomb per second). A unit of measurement of electrical current.

Annealing   Loss of temper in metal caused by heating.

Anoxia   Condition relating to an absence of oxygen.

Appliance   Equipment, usually nonindustrial, that is installed or connected as a unit to perform one or more functions such as clothes washing, air conditioning, food mixing, or cooking. Normally built in standardized sizes and types.

Arc   Flow of current across a gap between two conductors, generally producing high temperatures and luminous gases.

Arc-fault mapping   Locating and analyzing the pattern of arc faults in electrical circuits as a means of locating a possible area of origin.

Arc-tracking   Passage of current across a contaminated nonconductor or through a poor conductor causing localized heating, which can degrade the material and lead to current flow.

Arcing through char   Unintended passage of current through a semiconductive degradation product.

Area of confusion   Mixture of fire directional indicators in a wildlands fire.

Area of origin   The general locale in which a fire was ignited (*Kirk's* 7th ed.). A structure, part of a structure, or general geographic location within a fire scene, in which the "point of origin" of a fire or explosion is reasonably believed to be located. (See also **point of origin.**) (*NFPA 921*, 2011 ed., pt. 3.3.9).

Aromatic   Hydrocarbon compound whose structure is based on a benzene ring.

Arson   The intentional setting of a fire with intent to damage or defraud.

Atom   The smallest unit of an element that still retains its properties.

Auger spectroscopy   A means of identifying elements trapped within a substrate by analyzing the electrons emitted during electron microscopy.

Autoignition   Ignition due to sufficient surrounding temperature in the absence of an external source of ignition; nonpiloted ignition.

Autoignition temperature   The temperature at which a material will ignite in the absence of any external pilot source of heat; also referred to as the **spontaneous ignition temperature.**

B

Backdraft   A deflagrative explosion or rapid combustion of gases and smoke from an established fire that has depleted the oxygen content of a structure or compartment, most often initiated by the introduction of oxygen through ventilation or structural failure.

Boiling liquid expanding vapor explosion (BLEVE)   A mechanical explosion caused by the heating of a liquid in a sealed vessel to a temperature far above its boiling point (*Kirk's* 7th ed.). Failure of a confining vessel due to pressure increase facilitated by heating of liquids within the container. Failure often is associated with heat weakening the metal container and with a sudden fireball when contents are combustible (*NFPA 921*, 2011 ed., pt. 21.2.2).

Boiling point   The (pressure-dependent) temperature at which a liquid changes to its gas phase and that transition reaches equilibrium.

Branch circuit   The circuit conductors between the outlet(s) and the final overcurrent device protecting that circuit.

Brisance   The amount of shattering effect that can be produced by a high explosive.

British thermal unit (BTU)   The quantity of heat required to raise the temperature of one pound of water 1°F at a pressure of 1 atmosphere; a British thermal unit is equal to 1055 joules, 1.055 kilojoules, or 252.15 calories (*NFPA 921*, 2011 ed., pt. 3.3.19).

Buoyancy   Tendency or ability to rise or float in air or liquid as a result of a difference in density (*Kirk's* 7th ed.). The upward force exerted on a body or volume of fluid by the ambient fluid surrounding it. If the volume of fluid has positive buoyancy, then it is lighter than the surrounding fluid and tends to rise. Conversely, if it has negative buoyancy, it is heavier and tends to sink. Buoyancy of a fluid depends on both its molecular weight and its temperature (Drysdale in *NFPA Handbook* 2008, sec. 2, chap. 1).

Burn patterns   Patterns created when applied heat fluxes are above the critical thresholds to scorch, melt, char or ignite a surface.

Burning rate   The mass of fuel consumed per unit time in a fire.

C

Calcination   Loss of water of crystallization caused by heating.

Calorie   The amount of heat necessary to raise the temperature of 1 gram of water by 1 degree Celsius.

Calorimetry   An analytical method used to measure the total heat of combustion of fuels.

Cause   The circumstances, conditions, or agencies that brought about or resulted in the fire or explosion incident, damage to property resulting from the fire or explosion incident, or bodily injury or loss of life resulting from the fire or explosion incident (*NFPA 921*, 2011 ed., pt. 3.3.22).

Cause of death   The injury or disease that triggers the sequence of events leading to death. In a fire, the cause of death *may* include inhalation of hot gases, carbon monoxide, or other toxic gases, heat, burns, anoxia, asphyxia, structural collapse, or blunt trauma.

Ceiling jet   A relatively thin layer of flowing hot gases that develops under a horizontal surface (e.g., ceiling) as a result of plume impingement and the flowing gas being forced to move horizontally (*NFPA 921*, 2011 ed., pt. 3.3.23).

Ceiling layer   A buoyant layer of hot gases and smoke produced by a fire in a compartment (*NFPA 921*, 2011 ed., pt. 3.3.24).

Celsius   A temperature measurement system that designates the freezing point of water at 0° and the boiling point of water at 100°. Absolute zero is −273.16°C.

Chain of evidence (chain of custody)   Written documentation of possession of items of physical evidence from their recovery to their submission in court.

Char   Carbonaceous material resulting from pyrolysis, often appearing to be blackened blisters on the surface of cellulose and other solid organic fuels.

Char depth   The measurement of pyrolytic or combustion damage to a wood surface compared with its original surface height.

Chromatography   Chemical procedure that allows the separation of compounds based on differences in their

chemical affinities for two materials in different physical states, such as gas/liquid and liquid/solid.

Circuit breaker   A device designed to open a circuit automatically at a predetermined overcurrent without injury to itself when properly applied within its ratings.

Clean burn   An area of wall or ceiling or other surface where the charred organic residues have been burned away by direct flame contact or other source of high temperature.

Combustible   A material that will ignite and burn when sufficient heat is applied and when an appropriate oxidizer is present.

Combustible liquid   Any liquid that has a closed-cup flash point at or above 100°F (37.8°C), as determined by the test procedures and apparatus (*NFPA 30*, 2012).

Combustion   Oxidation that generates detectable heat and light (*Kirk's* 7th ed.). A chemical process of oxidation that occurs at a rate fast enough to produce heat and usually light in the form of either a glow or flame (*NFPA 921*, 2011 ed., pt. 3.3.31).

Competent ignition source   An ignition source that has sufficient energy and is capable of transferring that energy to the fuel long enough to raise the fuel to its ignition temperature (*NFPA 921*, 2011 ed., pt. 3.3.33).

Concealed wiring   Wiring rendered inaccessible by the structure or finish of the building. Wiring in covered raceways is considered concealed.

Conduction   Process of transferring heat through a material or between materials by direct physical contact (*Kirk's* 7th ed.). The transfer of energy in the form of heat by direct contact through the excitation of molecules and/or particles driven by a temperature difference (*NFPA 921*, 2011 ed.).

Configuration factor   The fraction of the radiation leaving one surface that is intercepted by a second surface. Also commonly referred to as the *view factor*.

Convection   Process of transferring heat by movement of a fluid (typically gas). In convective flow, a warm fluid is less dense than the surrounding fluid and rises, inducing a circulation.

Corpus delicti   Literally, the body of the crime. The fundamental facts necessary to prove the commission of a crime.

Crazing   Stress cracks in glass as the result of rapid cooling.

Criminalistics   The application of the methods and knowledge of the natural sciences (physics, chemistry, biology, botany, etc.) to legal inquiries.

D

Data   Facts or information to be used as a basis for a discussion or decision (*ASTM E1138-89*, 1989).

Dead load   The weight of a structure and any equipment and appliances permanently attached.

Deductive reasoning The process by which conclusions are drawn by logical inference from given premises (*NFPA 921,* 2011 ed., pt. 3.3.38).

Deep-seated Fire that has gained headway and built up sufficient heat in a structure to require great cooling for extinguishment; fire that has burrowed deep into combustible fuels (as opposed to a surface fire); deep charring of structural members.

Deflagration A very rapid oxidation (typically of fuel gas, vapor, or dust) with the evolution of heat and light and the generation of a low-energy pressure wave that can cause damage. The reaction proceeds between fuel elements at subsonic speeds in the fuel (<1000 m/s; 3300 ft/s).

Detonation An extremely rapid reaction that generates very high temperatures and an intense pressure/shock wave that produces violently disruptive effects. It propagates through the material at supersonic speeds (>1000 m/s; 3300 ft/s).

Device Any chemical or mechanical contrivance or means used to start a fire or explosion.

Diatomic Molecules consisting of two atoms of an element.

Diffusion flame Flame resulting from fire where oxygen mixes with the fuel at the combustion zone (*NFPA 921,* 2011 ed., pt. 3.3.44).

Direct attack Application of hose streams or other extinguishing agents directly on a fire, rather than attempting extinguishment by generating steam within a structure.

Dropdown The collapse of burning material in a room that induces separate, low-level ignition; fall-down.

Duty Conditions of use in electrical service: (1) *Continuous duty.* Operation at substantially constant load for an indefinitely long time. (2) *Intermittent duty.* Operation for alternating intervals of (a) load/no load, (b) load/rest, or (c) load/no load/rest. (3) *Periodic duty.* Intermittent operation in which the load conditions are regularly recurrent. (4) *Short-time duty.* Operation at substantially constant load for a short and definitely specified time. (5) *Varying duty.* Operation at loads and for intervals of time that are both subject to wide variation.

E

Endothermic Absorbing heat during a chemical reaction.

Entrainment The mixing of two or more fluids (gases) as a result of laminar flow or movement. In air entrainment, the process of air or gases being drawn into a fire, plume, or jet (*NFPA 921,* 2011 ed., p. 3.3.48).

Eutectic An alloy of two materials having special physical or chemical properties, typically having the lowest melting point of any combination of the two.

Exothermic Generating or giving off heat during a chemical reaction.

Explosion The sudden conversion of potential energy (chemical or mechanical) into kinetic energy with the extremely rapid production or release of heat, gases, or mechanical pressure.

Explosion-proof apparatus Apparatus enclosed in a case that is capable of withstanding an explosion of a specified gas or vapor within it or preventing the ignition of a specific gas or vapor surrounding it by sparks, flashes, or explosion of fuels within and that operates at such an external temperature that a surrounding flammable atmosphere will not be ignited by it.

Explosive limits (Flammability limits) The lower and upper concentrations of an air/gas or air/vapor mixture in which combustion or deflagration will be supported.

F

Fahrenheit A temperature measurement system that designates the freezing point of water at 32° and the boiling point of water at 212°. Absolute zero is −459.69°.

Field model A computer fire model that uses *computational fluid dynamics (CFD)* to estimate the fire environment by dividing the compartment into uniform small cells, or control volumes. The computer program then attempts to predict conditions in every volume such as, for example, pressure, temperature, carbon monoxide, oxygen, and soot production.

Fire Rapid oxidation with the evolution of heat and light; uncontrolled combustion (*Kirk's* 7th ed.). A rapid oxidation process, which is a chemical reaction resulting in the evolution of light and heat in varying intensities (*NFPA 921,* 2011 ed., pt. 3.3.58).

Fire behavior The manner in which a fuel ignites, flame develops, and fire spreads. Unusual fire behavior may reveal the presence of added fuel or accelerants.

Fire load The total amount of fuel that might be involved in a fire, as measured by the amount of heat that would evolve from its combustion (expressed as units of heat).

Fire model A method using mathematical or computer calculations that describes a system or process related to fire development, including fire dynamics, fire spread, occupant exposure, and the effects of fire. The results produced by fire models can be compared with physical and eyewitness evidence to test working hypotheses.

Fire patterns Indicators of damage (or relative lack of damage) created by a combination of smoke, flames, and heat effects (*Kirk's* 7th ed.). The visible or measurable physical changes, or identifiable shapes, formed by a fire effect or group of fire effects. (*NFPA 921,* 2011 ed., pt. 3.3.64).

Fire plume   The buoyant convective column of hot gases generated by a fire; may include both flames and non-flaming products.

Fire point   The lowest temperature at which a liquid will ignite and achieve sustained burning when exposed to a test flame in accordance with defined test conditions, such as, for example, *ASTM D92: Standard Test Method for Flash and Fire Points by Cleveland Open Cup Tester (NFPA 30, 2012 ed.)*.

Fire-resistive   A structure or assembly of materials built to provide a predetermined degree of fire resistance as defined in building or fire prevention codes (calling for 1-, 2-, or 4-hour fire resistance).

Firestorm   Overwhelming progression of fire through structures or wildlands caused by a combination of convective and radiative processes.

Fire tetrahedron   A model that describes four components necessary for continued combustion: fuel, heat, oxygen, and an uninterrupted chemical chain reaction.

Fire wall   A solid wall of masonry or other noncombustible material capable of preventing passage of fire for a prescribed time (usually extends through the roof with parapets).

Flame front   The flaming leading edge of a propagating combustion reaction zone. (*NFPA 921*, 2011 ed., pt. 3.3.70).

Flame jets   Horizontal flame movements below ceilings or other restrictions resulting when convective heat transfer can no longer move upward and thus transfers laterally.

Flameover   The flaming ignition of the hot gas layer in a developing compartment fire (*Kirk's* 7th ed.). The condition where unburned fuel (pyrolysate) from the originating fire has accumulated in the ceiling layer to a sufficient concentration (i.e., at or above the lower flammable limit) that it ignites and burns. Can occur without ignition of, or prior to the ignition of, other fuels separate from the origin. Also known as *rollover* (*NFPA 921*, 2011 ed., pt. 3.3.71).

Flame plume   The buoyant column of incandescently hot gases from a fire visible to the unaided eye.

Flame point (Fire point)   The minimum temperature at which (under defined test conditions) a flame is sustained by evaporation or pyrolysis of a fuel after ignition.

Flame resistant   Material or surface that does not maintain or propagate a flame once an outside source of flame has been removed.

Flame spread   The rate at which flames extend across the surface of a material (usually under specific conditions).

Flammable   (same as *inflammable*) A combustible material that ignites easily, burns intensely, or has a rapid rate of flame spread.

Flaming combustion   The hot, light-emitting portion of a combustion process. The heat released is sufficient to raise the temperature of gases so that they emit energy waves in the visible light range.

Flammable limits   The minimum and maximum concentration of fuel vapor or gas in a fuel vapor or gas/gaseous oxidant mixture (usually expressed as percent by volume) defining the concentration range (flammable or explosive range) over which propagation of flame will occur on contact with an ignition source. The minimum concentration is known as the lower flammable limit (LFL) or the lower explosive limit (LEL). The maximum concentration is known as the upper flammable limit (UFL) or the upper explosive limit (UEL) (*NFPA 53*).

Flammable liquid   Any liquid that has a closed-cup flash point below 37.8°C (100°F), as determined by the test procedures and apparatus, and a Reid vapor pressure that does not exceed an absolute pressure of 276 kPa (40 psi) at 37.8°C (100°F), as determined by Sec. 4.4 of *ASTM D323: Standard Test Method for Vapor Pressure of Petroleum Products (Reid method)* (*NFPA 30*, 2012 ed.).

Flashback   The ignition of a gas or vapor from an ignition source back to a fuel source (often seen with flammable liquids).

Flash fire   A fire that spreads by means of a flame front moving rapidly through a diffuse fuel, such as dust, gas, or the vapors of an ignitable liquid, without the production of damaging pressure (*NFPA 921*, 2011 ed., pt. 3.3.76).

Flashover   The final stage of the process of fire growth. When all combustible fuels within a compartment are ignited, the room is said to have undergone flashover. (*Kirk's* 7th ed.). A transition phase in the development of a compartment fire in which surfaces exposed to thermal radiation reach ignition temperature more or less simultaneously and fire spreads rapidly throughout the space, resulting in full-room involvement or total involvement of the compartment or enclosed space (*NFPA 921*, 2011 ed., pt. 3.3.78).

Flash point   The minimum temperature at which (under defined test conditions) the vapor being produced by a material can be ignited to produce a brief flash (not sustained) of flame.

Fractional exposure concentration   A measurement that assesses the impact of toxic smoke and by-products of combustion on a subject. The FED depends on the concentration of particular toxins within the fire gases, and the duration of exposure.

Fragmentation   The fast-moving solid pieces created by an explosion. Primary fragmentation is that of the explosive container itself; secondary fragmentation is that of the target shattered by an explosion.

Fraud   Deception deliberately practiced in order to secure unfair or unlawful gain.

Fuel   A material that will maintain combustion under specified environmental conditions (*NFPA 921*, 2011 ed., pt. 3.3.80).

Fuel-controlled fire   A fire in which the heat release rate and growth rate are controlled by the characteristics of the fuel, such as quantity and geometry, and in which adequate air for combustion is available (*NFPA 921*, 2011 ed., pt. 3.3.83).

Fuel item   Any article that is capable of burning (*NFPA 921*, 2011 ed.).

Fuel load   All combustibles in a fire area, whether part of the structure, finish, or furnishings (*Kirk's* 7th ed.). The total quantity of the combustible contents of a building, space, or fire area, including interior finish and trim, expressed in heat units or the equivalent weight in wood (*NFPA 921*, 2011 ed., pt. 3.3.82).

Fuel package   A collection or array of fuel items in close proximity with one another such that flames can spread throughout the array of fuel items. Multiple fuel packages may be involved in a single compartment fire (*NFPA 921*, 2011 ed., pt. 5.6.2.1).

Full-room involvement   Condition in a compartment fire in which the entire volume is involved in fire (*NFPA 921*, 2011 ed., pt. 3.3.84).

Fully developed fire   Fuel and oxygen are readily available with sufficient heat to pyrolize the fuel. The fire reaches a steady burning state. Also called *free burning state* and *steady-state burning*.

Fully involved   The entire area of a fire building so involved with heat, smoke, and flame that entry is not possible until some measure of control has been obtained with hose stream attacks.

Furniture calorimeter   An instrumental system for measuring the amount of heat produced by the combustion of a moderate-sized fuel package, and the rate at which the heat is produced.

G

Ghost marks   Stained outlines of floor tiles produced on noncombustible floors by the dissolution and combustion of tile adhesive.

Ground   A conducting connection, whether intentional or accidental, between electrical circuit, equipment, and the earth, or to some conducting body that serves in place of the earth.

Ground fault   An interruption of the normal ground return path of electricity in a structure that leads to unintended current flows.

H

Haber's rule   The dosage of toxic gases assimilated by an individual is assumed to be equivalent to the concentration times the duration of exposure.

Heat   Thermal energy that flows between bodies of matter from a higher temperature to a lower temperature.

Heat flux   The rate at which heat energy is transferred to a surface per unit time per unit area, expressed in kW/m^2, $kJ/(m^2 \cdot s)$, or $BTU/(ft^2 \cdot s)$.

Heat horizon   The demarcation (usually horizontal) of fire damage revealed by the charring, burning, or discoloration of paint or wall coverings.

Heat of combustion   The quantity of heat released from a fuel during combustion measured in kilojoules per gram (kJ/g) or Btu per pound (Btu/lb) (*Kirk's* 7th ed.).

Heat release rate   The rate at which heat is generated by a source, usually measured in watts (W), joules/second (J/s), or Btu/second.

Heat transfer   The spread of thermal energy by convection, conduction, or radiation.

High explosive   Any material designed to function by, and capable of, detonation.

Hot set   Direct ignition of available fuels with an open flame (match or lighter).

Hydrocarbon   Chemical compound containing only hydrogen and carbon.

Hypothesis   A supposition or conjecture put forward to account for certain facts, and used as a basis for further investigation by which it may be proved or disproved (*ASTM E1138-89*, 1989).

Hypoxia   Condition relating to low concentrations of oxygen.

I

Ignitable liquid   Classification for liquid fuels including both flammable and combustible classes.

Ignition energy   The quantity of energy that must be transferred into a fuel/oxidizer combination to trigger a self-sustaining combustion.

Ignition temperature   (same as **autoignition temperature**) The minimum temperature to which (under specific test conditions) a substance must be heated in air to ignite independently of the heating source (i.e., in the absence of any other ignition source; nonpiloted ignition).

Incendiary fire   A deliberately set fire.

Incipient   Beginning stages of a fire.

Indicators   Observable (and usually measurable) changes in appearance caused by heat, flame, or smoke.

Inductive logic or reasoning   The process by which a person starts from a particular experience and proceeds to generalizations. The process by which hypotheses are developed based on observable or known facts and the training, experience, knowledge, and expertise of the observer (*NFPA 921*, 2011 ed., pt. 3.3.104).

Inorganic   Containing elements other than carbon, oxygen, and hydrogen.

Isochar   Line drawn on a scene diagram that connects points of similar char depth on wood surfaces.

J

Joule   The preferred SI unit of heat, energy, or work. A joule is the heat produced when one ampere is passed through a resistance of one ohm for one second, or it is the work required to move a mass of one meter against a force of one newton. There are 4.184 joules in a calorie, and 1055 joules in a British thermal unit (BTU). A watt is one joule/second (J/s) (*NFPA 921,* 2011 ed., pt.3.3.109).

K

Kelvin   A temperature measurement system that begins at absolute zero, the freezing point of water is at 273.16 K, and the boiling point of water is at 373.16 K.

L

Laser scanners   The use of technology that combines the accuracy of total station measurements with the completeness of panoramic digital photography. Three-dimensional (3D) laser scanners use LIDAR(LIght Detection And Ranging*)* technology to quickly capture millions of measurements at a scene.

Latent heat   The amount of heat required to change phase from a solid to a liquid (latent heat of fusion) or from a liquid to a vapor (latent heat of vaporization).

Latent heat of vaporization   The quantity of heat required to cause a mass unit of solid or liquid to convert to a vapor state. It is expressed in J/g or BTU/ lb. Also known as *heat of gasification.*

Liquid   Matter that possesses a definite volume but lacks definite shape; it assumes the shape of its containing vessel.

Lower explosive limit (LEL)   The lowest concentration of a fuel in air or oxygen in which a flame can be propagated Also known as lower flammability limit (LFL).

Low explosive   Any material designed to function (i.e., generate gases at high pressures) by deflagration.

M

Minimum ignition energy (MIE)   The minimum amount of energy released at a point in a combustible mixture that causes flame propagation away from the point, under specified test conditions (*NFPA 68*).

Moisture content   Amount of moisture intrinsic to an object expressed in percentage of total mass for that object.

Molotov cocktail   A breakable container filled with flammable liquid, usually thrown. It may be ignited by a flaming wick or by chemical means.

Motive   An inner drive or impulse that is the cause, reason, or incentive that induces or prompts a specific behavior (Rider 1980).

N

Neutral plane   In a developing fire, typically the location of the vertical plane or point at the door or window vent opening where the a zero pressure difference exists between the interface of the hot and cool gas layer.

Nonflammable   Material that will not burn under most conditions.

Normal hydrocarbons (*n*-Hydrocarbons)   Hydrocarbons having straight-chain structures with no side branching; aliphatics.

O

Ohm   Unit of measurement for electrical resistance.

Olefinic   Hydrocarbons containing double carbon–carbon bonds (denoted C — C); nonsaturated; alkenes.

Open wiring   Uninsulated conductors or insulated conductors without grounded metallic sheaths or shields, installed above ground but not inside enclosures or appliances.

Opinion   A belief or judgment based on facts and logic, but without absolute proof of its truth (*ASTM E1138-89,* 1989).

Organic   Compounds based on carbon.

Origin   The general location where a fire or explosion began (see **also point of origin** and **area of origin**) (*NFPA 921,* 2011 ed., pt. 3.3.119).

Outlet   A point on the wiring system at which current is taken to supply equipment or appliances by means of receptacles or direct connections.

Overcurrent   Current flow above the intended or design current.

Overhaul   The firefighting operation of eliminating hidden flames, glowing embers, or sparks that may rekindle the fire, usually accompanied by the removal of structural contents (*Kirk's* 7th ed.). A firefighting term involving the process of final extinguishment after the main body of the fire has been knocked down. All traces of fire must be extinguished at this time (*NFPA 921,* 2011 ed., pt. 3.3.121).

Overload   Operation of equipment in excess of normal, full-load rating or of a conductor in excess of rated ampacity that when it persists for a sufficient length of time would cause damage or dangerous overheating. An overload current is usually but might not always be confined to the normal intended conductive paths provided by conductors and other electrical components of an electrical circuit. Operation of the equipment or wiring under current flow conditions leading to temperatures in excess of the temperature rating of the equipment or wiring (*NFPA 921,* 2011 ed., pt. 3.3.122).

Oxidation   Combination of an element with oxygen; chemical conversion involving a loss of electrons.

P

Paraffinic   Hydrocarbon compounds involving no double or triple C — C bonds; alkanes; saturated aliphatic hydrocarbons.

Peer review   Peer review is a formal procedure generally employed in prepublication review of scientific or technical documents and screening of grant applications by research-sponsoring agencies. Peer review carries with it connotations of both independence and objectivity. Peer reviewers should not have any interest in the outcome of the review. The author does not select the reviewers, and reviews are often conducted anonymously. As such, the term *peer review* should not be applied to reviews of an investigator's work by coworkers, supervisors, or investigators from agencies conducting investigations of the same incident. Such reviews are more appropriately characterized as "technical reviews" (*NFPA 921,* 2011 ed., pt. 4.6.3).

Piloted ignition   The initiation of combustion by means of contact of gaseous material with an external high-energy source, such as a flame, spark, electrical arc, or glowing wire.

Point of origin   The specific location at which a fire was ignited (*Kirk's* 7th ed.). The exact physical location within the area of origin where a heat source and the fuel interact, resulting in a fire or explosion (*NFPA 921,* 2011 ed., pt. 3.3.127).

Premixed flame   A flame for which the fuel and oxidizer are mixed prior to combustion, as in a laboratory Bunsen burner or a gas cooking range. Propagation of the flame is governed by the interaction among the flow rate, transport processes, and chemical reaction (*NFPA 921,* 2011 ed., pt. 3.3.128).

Pyrolysis   The chemical decomposition of substances through the action of heat, in the absence of oxygen; oxidative pyrolysis occurs when oxygen comes in contact with the decomposing material.

Pyromania   Uncontrollable psychological impulse to start fires.

Pyrophoric   Capable of oxidizing on exposure to atmospheric oxygen at normal temperatures.

R

Raceway   Any channel for holding wires, cables, or busbars that is designed expressly for, and is used solely for, this purpose.

Radiation   Transfer of heat by electromagnetic waves.

Rankine   A temperature measurement system that begins at absolute zero. The freezing point of water is 491.67°R, and the boiling point of water is 671.67°R.

Receptacle   A contact device installed as the outlet for the connection of an appliance by means of a plug.

Rekindle   Reignition of a fire by latent heat, sparks, or embers after apparent but incomplete extinguishment.

Resistance   Opposition to the passage of an electrical current.

Rollover   See flameover.

S

Salvage   Procedures to reduce incidental losses from smoke, water, and weather following fires, generally by removing or covering contents.

Saturated   Hydrocarbons that have no double or triple C — C bonds.

Scene   The general physical location of a fire or explosion incident (geographic area, structure or portion of a structure, vehicle, boat, piece of equipment, etc.) designated as important to the investigation because it may contain physical damage or debris, evidence, victims, or incident-related hazards (*NFPA 921,* 2011 ed., pt. 3.3.143).

Scientific method   The systematic pursuit of knowledge involving the recognition and formulation of a problem, the collection of data through observation and experiment, and the formulation and testing of a hypothesis (*NFPA 921,* 2011 ed., pt. 3.3.139).

Seat of explosion   The area of most intense physical damage caused by high explosive pressures and shock waves in the vicinity of a solid or liquid explosive.

Seat of fire   Area where main body of fire is located, as determined by outward movement of heat, flames, and smoke.

Self-heating   An exothermic chemical or biological process that can generate enough heat to become an ignition source; spontaneous ignition.

Self-ignition   See autoignition.

Service   The conductors and equipment for delivering electricity from the supply system to the equipment of the premises served.

Service conductors   The supply conductors that extend from the street main or transformer to the equipment of the premises served.

Service drop   The overhead service conductors from the pole, transformer, or other aerial support to the service entrance equipment on the structure, including any splices.

Service equipment   The necessary hardware that constitutes the main control and means of cutoff of the electrical supply, usually consisting of circuit breakers, fuses, or switches located in a panel box near the point of entrance of supply conductors.

Set   Device or contrivance used to ignite an incendiary fire.

Short-circuit   Direct contact between a current-carrying conductor and another conductor.

SI   The system used for quantifying measurement that is accepted by the scientific community. Also known as *Système International.*

Smoke   The airborne solid and liquid particulates and gases evolved when a material undergoes pyrolysis or combustion, together with the quantity of air that is entrained or otherwise mixed into the mass (*NFPA 921*, 2011 ed., pt. 3.3.153).

Smoke horizon   Surface deposits that reveal the height at which smoke and soot stained the walls and windows of a room without thermal damage.

Smoldering combustion   The direct combination of a solid fuel with atmospheric oxygen to generate heat in the absence of gaseous flames; see **glowing combustion**.

Soot   The carbon-based solid residue created by incomplete combustion of carbon-based fuels.

Spall   Crumbling or fracturing of concrete or brick surface as a result of exposure to thermal or mechanical stress.

Spark   Superheated, incandescent particle.

Specific gravity (of a gas or vapor) (vapor density)   The ratio of the average molecular weight of a gas or vapor to the average molecular weight of air (*NFPA 921*, 2011 ed., pt. 3.3.160).

Specific gravity (of a liquid or solid)   The ratio of the mass of a given volume of a substance to the mass of an equal volume of water at a temperature of 4°C (*NFPA 921*, 2011 ed., pt. 3.3.161).

Specific heat   The ratio of the heat capacity of a substance to the heat capacity of water at the same temperature.

Spoliation   The destruction or material alteration of, or failure to save, evidence that could have been used by another in future litigation (*Kirk*'s 7th ed). Loss, destruction, or material alteration of an object or document that is evidence or potential evidence in a legal proceeding by one who has the responsibility for its preservation (*NFPA 921*, 2011 ed., pt. 3.3.162).

Spontaneous ignition   Chemical or biological process that generates sufficient heat to ignite the reacting materials. *See* self-heating.

Spontaneous ignition temperature (SIT)   See autoignition temperature.

Spot fires   Fires started by airborne embers some distance away from the main body of the fire.

Stoichiometry   Balance of chemical reactants and products.

Stoichiometric mixture   Chemical conditions where the proportion of reactants is such that there is no surplus of any reactant after the chemical reaction is completed (Babrauskas 2003).

Superposition   The combined effects of two or more fires or heat transfer effects, a phenomenon that may generate confusing fire damage indicators.

Suspicious   Fire cause has not been determined, but there are indications that the fire was deliberately set and all accidental fire causes have been eliminated.

T

Technical expert   A person educated, skilled, or experienced in the mechanical arts, applied sciences, or related crafts (*ASTM E1138-89*, 1989).

Temperature   Measurement of average kinetic heat energy of an object, which results from molecular motion.

Tenability   An assessment of the potential for harm that may be imposed from a fire based (1) on an analysis of the source, heat release rates, and rates of production of the combustion products; (2) when occupants might become exposed to these harmful conditions; and (3) the effects on the occupants to this exposure.

Thermal conductivity   The property of a material that relates to its ability to perform heat transfer by conduction. Materials with high thermal conductivity allow heat transfer at a higher rate than across materials of low thermal conductivity. Thermal conductivity is measured in watts per meter-kelvin $(W/(m \cdot K)$ or $W \cdot m^{-1} \cdot K^{-1})$.

Thermal inertia   The properties of a material that characterize its rate of surface temperature rise when exposed to heat. Thermal inertia is related to the product of the material's thermal conductivity (k), its density (ρ), and its heat capacity (c) (*NFPA 921*, 2011 ed., pt. 3.3.170).

Thermal protector   An inherent device against overheating that is responsive to temperature or current and will protect the equipment against overheating owing to overload or failure to start.

Thermally thick solid   A solid that while heated from one face, shows a negligible temperature rise at its opposite face. This characteristic is not simply a property of a material; it also depends on the time of exposure and the heat flux.

Thermally thin solid   A solid that while heated from one face, shows a back face temperature that is nearly identical to the temperature of the heated face. This characteristic is not simply a property of a material; it also depends on the time of exposure and the heat flux.

Thermoplastics   Organic materials that can be melted and resolidified without chemical degradation.

Thermosetting   Plastic or resin materials that once solidified will not melt but will chemically degrade as they are heated.

Torch   A professional fire setter.

Trailers   Long assemblages of combustible materials used to spread a fire throughout a structure.

U

Upper explosive limit (UEL)   The concentration of gas or vapor in air, above which a flame will not propagate upon exposure to an ignition source. Also known as the *upper flammable limit (UFL)*.

Upper layer   A buoyant layer of hot gases and smoke produced by a fire in a compartment (*NFPA 921*, 2011 ed.).

V

Vapor   The gaseous phase of a material that is a liquid or solid at normal temperatures and pressures.

Vapor density   The ratio of the molecular weight of a given volume of gas or vapor to that of an equal volume of air; or the ratio of the molecular weight of the gas or vapor to that of air (MW = 29).

Vent   An opening for the passage, or dissipation, of fluids, such as gases, fumes, smoke, and the like (*NFPA 921*, 2011 ed., pt. 3.3.179).

Ventilation   The supply of air or oxygen to support a fire.

Ventilation-controlled fire   A fire in which the heat release rate or growth is controlled by the amount of air available to the fire (*NFPA 921*, 2011 ed., pt. 3.3.181).

Verification and validation (V&V)   A formal process of establishing acceptable uses, suitability, and limitations of fire models. *Verification* determines that a model correctly represents the developer's conceptual description. Validation determines that a model is a suitable representation of the real world and is capable of reproducing phenomena of interest (Salley & Kassawar 2007g).

Volatile   A liquid having a low boiling point; one that is readily evaporated into the vapor state.

Volt   The basic unit of electromotive force.

Voltage, nominal   A value assigned to a circuit or system for the purpose of conveniently designating its voltage class (120/240, 480Y/277, 600, etc.).

W

Watt   Unit of power, or rate of work, equal to one joule per second, or the rate of work represented by a current of one ampere under the potential of one volt (*NFPA 921*, 2011 ed.).

Z

Zone model   A computer fire model based on the assumption that a fire in a room or enclosure can be described in terms of two unique zones known as the upper and the lower zone or layer, and the conditions within each zone can be predicted. Zone models use a set of differential equations for each zone solving for conditions such as, for example, pressure, temperature, carbon monoxide, oxygen, and soot production.

Sources

INDEX